D0930098

THE UPPER ATMOSPHERE

Meteorology and Physics

International Geophysics Series

Edited by

J. VAN MIEGHEM

Royal Belgian Meteorological Institute
Uccle, Belgium

Volume 1 BENO GUTENBERG. Physics of the Earth's Interior. 1959

Volume 2 JOSEPH W. CHAMBERLAIN. Physics of the Aurora and Airglow. 1961

Volume 3 S. K. RUNCORN (ed.). Continental Drift. 1962

Volume 4 C. E. JUNGE. Air Chemistry and Radioactivity. 1963

Volume 5 ROBERT C. FLEAGLE AND JOOST A. BUSINGER. An Introduction to Atmospheric Physics. 1963

Volume 6 L. DUFOUR AND R. DEFAY. Thermodynamics of Clouds. 1963

Volume 7 H. U. ROLL. Physics of the Marine Atmosphere. 1965

Volume 8 RICHARD A. CRAIG. The Upper Atmosphere: Meteorology and Physics. 1965

Volume 9 WILLIS L. WEBB. Structure of the Stratosphere and Mesosphere. 1966

Volume 10 MICHELE CAPUTO. The Gravity Field of the Earth from Classical and Modern Methods. 1967

Volume 11 S. MATSUSHITA AND WALLACE H. CAMPBELL (eds.). Physics of Geomagnetic Phenomena. (In two volumes.) 1967

Volume 12 K. YA. KONDRATYEV. Radiation in the Atmosphere. 1969.

In preparation

ERIK H. PALMEN AND CHESTER W. NEWTON. Atmosphere Circulation Systems: Their Structure and Physical Interpretation.

HENRY RISHBETH AND OWEN K. GARRIOTT. Introduction to Ionospheric Physics.

THE UPPER ATMOSPHERE

Meteorology and Physics

BY

RICHARD A. CRAIG

DEPARTMENT OF METEOROLOGY
THE FLORIDA STATE UNIVERSITY
TALLAHASSEE, FLORIDA

1965

ACADEMIC PRESS New York and London

ACADEMIC PRESS, INC.
111 Fifth Avenue, New York, New York 10003

United Kingdom Edition published by
ACADEMIC PRESS, INC. (LONDON) LTD.
Berkeley Square House, London W.1

LIBRARY OF CONGRESS CATALOG CARD NUMBER: 65-15768

Second Printing, 1969

PRINTED IN THE UNITED STATES OF AMERICA

Preface

During the past fifteen years, knowledge of the upper atmosphere of our planet has increased at a tremendous rate. The use of improved balloons, of rockets, and of satellites has resulted in a vast collection of observations and has greatly stimulated related research. New contributions appear in the scientific journals at the rate of many thousand pages per year. As a result, it is no longer practical to attempt a comprehensive book that will do justice to all or most of the specialized topics involved, as Professor Mitra was able to do in 1948. Instead a book must be written with a specific objective which will influence the choice of and relative emphasis on the topics treated. The objective of this book is to provide a suitable introduction to what I choose to call the meteorology of the upper atmosphere.

Neither "upper atmosphere" nor "meteorology" has an unambiguous meaning. I have found it convenient in this book to refer to the atmosphere above the tropopause as the "upper atmosphere," although the expression is often used elsewhere to refer to the atmosphere above some higher level. The word "meteorology" is quite broadly defined in dictionaries, but in practice is usually associated with studies of the atmosphere below some ill-defined limit. This limit has been moving upward in recent years and might now be taken as about 30 km. Meteorologists are, collectively, concerned with a broad range of difficult problems connected with this part of the atmosphere and have made relatively few contributions to studies of the upper atmosphere above, say, 30 km. The scientists who have contributed to such studies have been mostly physicists, chemists, and astronomers by training, and are often called "aeronomers."

However, I happen to feel that a distinction based on any altitude limit is a poor one, especially if it hinders understanding and communication. Meteorology embraces many different kinds of problems, from turbulence and diffusion to the general circulation, from cloud physics to numerical weather prediction. But most meteorologists would agree that our central problem is the understanding, prediction, and eventual control of the behavior of a fluid atmosphere subject to certain forces and boundary conditions on a rotating planet. This implies a basic interest in the structure, circulation, and interactions of the atmosphere as a whole, and especially in the time variations that occur, whether these are determined observationally or theoretically. This central underlying theme, however poorly I may have described it, distinguishes meteorology as I use the word. In this sense, there is no good reason why meteorology should be confined to the troposphere or even the troposphere and lower stratosphere.

I should like to predict that during the next ten to twenty years there will develop an active identifiable branch of atmospheric science that is concerned with the upper atmosphere from the point of view described above. Whether it will be called "meteorology of the upper atmosphere" I do not know, nor do I particularly care. Its development will require the cooperation of scientists who are presently labeled as meteorologists, aeronomers, atmospheric scientists, or perhaps something else. I hope that this book will to some extent stimulate and accelerate this development.

The difficulties, though, are formidable and the objective is correspondingly ambitious. Interdisciplinary studies are much easier to advocate in general terms than to carry out in practice. The meteorologist who attempts to extend his interests to the upper atmosphere must first of all learn that the astronomical, physical, and chemical background problems are quite different from those of the troposphere. Although he need not become an expert in all of these, he must understand and appreciate them and be able to communicate with people who are experts. Correspondingly, the aeronomer who wants to apply his observations to meteorology (or meteorology to an interpretation of his observations) must learn that a rotating, compressible, turbulent fluid is an extremely complex system not described satisfactorily by some of the simplifications of classical physics.

I have written this book with the meteorological reader primarily in mind. This means that I have emphasized those topics that I think might interest him the most. It also means that I have elaborated on related physical and chemical problems that I think might be the most unfamiliar to him, and at the same time, important for him to appreciate. On the other hand, I have assumed a mature reader with a good background in applied mathematics and in basic aspects of physics and chemistry, such as a graduate student or research worker in meteorology should have. Not all will agree with specific applications of these intentions, because no two readers will have the same interests and background. I have therefore thought it very important to include large numbers of references which will enable the reader to follow up subjects for which he finds my treatment deficient for his purposes. For various reasons the emphasis is on the atmosphere between the tropopause and 100 km, although a good deal of material is also included about the atmosphere between 100 and 300 km. Very little is said about the important problems at still higher levels, which are now the subject of active and vigorous research.

Despite a certain preoccupation with the difficulties of a meteorologist seeking to extend his view upward, I do hope that the book will be useful to all scientists interested in the upper atmosphere. The aeronomer who specializes in a certain branch of upper-atmospheric research will undoubtedly find shortcomings in the treatment of his specialty. But he should find it profitable to read about other aspects of the upper atmosphere from the point of view adopted here.

As always with a book of this sort, the author has been dependent on many other people for help and has incurred many debts in its preparation. First of all, I am very grateful to the following friends and colleagues who took the time to read various portions of the manuscript: R. M. Goody, B. Haurwitz, W. S. Hering, L. G. Jacchia, F. S. Johnson, J. London, and R. J. Reed. Their perceptive comments and helpful suggestions resulted in a considerable improvement of the manuscript over an earlier version. However, not all portions of the manuscript were reviewed, and I was unable for various reasons to take advantage of all suggestions about the portions that were reviewed. Therefore, responsibility for the faults that remain is clearly mine.

A majority of the figures in the text have been adapted from figures published elsewhere. The publishers involved have been most considerate in granting me the right to do this. Reference to sources is in all cases made in the captions.

Here at The Florida State University a number of people have contributed greatly to the preparation of the book. These include many graduate students who struggled through early course-note versions of the text. In particular, Dr. M. A. Lateef and Messrs. W. A. Bowman, S. Y. K. Li, and J. Bell have helped with text, figures, and references. Above all, I am indebted to Mrs. Janina Richards who has expertly typed the manuscript through several versions, perhaps a total of 3000 pages, and always without a word of complaint. In fact she somehow managed to make me feel guilty whenever progress was slow and there was no typing to be done. Without her help, the book would certainly have been delayed and perhaps never have been completed.

RICHARD A. CRAIG

December 1964

Contents

CHAPTER 1

Introduction

The term "upper atmosphere" is commonly used to designate the earth's atmosphere above some explicitly or implicitly defined altitude. However, the choice of altitude to separate "lower atmosphere" or "middle atmosphere" from "upper atmosphere" is not uniform. For example, some may regard air 30 km above the earth as being in the upper atmosphere while others may consider this air to be part of the lower atmosphere or middle atmosphere. Here the expression "upper atmosphere" refers to air above the tropopause. Thus, only air in the troposphere, which has the most direct thermal interaction with the earth's surface, is excluded. This is a rather more general use of the term than is customary, but there seems to be no better single expression to describe the part of the atmosphere with which we shall deal.

For purposes of reference, the upper atmosphere is often divided into subregions, each with a different name. The fact that several systems of nomenclature have been used has resulted in a good deal of unfortunate confusion and ambiguity. Most of these systems are based on the vertical temperature distribution, and it will be convenient to defer a detailed discussion of nomenclature (to Section 1.3) until after we have considered the broad outline of upper-atmosphere structure.

Since the end of World War II, several exciting developments have tended to focus broad scientific attention on the upper atmosphere of our planet. The increased ceiling and greater reliability of sounding balloons; the use of rockets to probe the upper atmosphere; cooperative international efforts during the International Geophysical Year; and especially the successes of Soviet and American scientists in launching instrumented satellites—all have broadened the horizons of upper-atmospheric research and uncovered significant new information. Nevertheless, it is a fact that the broad outlines of our knowledge of the upper atmosphere were already sketched before this era by a small group of dedicated men who were limited by the technology of their time to various indirect and ingenious methods of probing. As examples, one might mention the use of radio waves to study the ionized regions, the use of sound waves to deduce the vertical temperature distribution, the deduction of density distribution from meteor observations, measurements of solar ultraviolet radiation to learn about ozone, spectroscopic studies of aurora and airglow, and analyses of atmospheric tides and magnetic variations. These studies not only furnished a broad and generally accurate body of knowledge in the prerocket days, but also played a vital role in dictating the emphasis and direction of the recent research. Many of these techniques are still very useful.

Study of the upper atmosphere involves so many interrelated problems that it is difficult to discuss one without assuming some knowledge of the others. This introductory chapter deals in broad outline with the structure and composition of the upper atmosphere (Sections 1.1 and 1.2), covers nomenclature (Section 1.3), and, in Section 1.4, outlines the principal topics to be discussed in greater detail later. It is intended to define, to orient, and to facilitate cross-referencing in the more detailed chapters to follow.

1.1 Variables Describing Structure and Composition

The physical and chemical state of the atmosphere is partially described by the structure variables—temperature, pressure, and density. Much effort has been devoted to measuring, by various methods, how these parameters vary in space and time. Indeed, a significant portion of this book will be devoted to discussions of those methods and their results. Some methods of probing the upper atmosphere measure temperature directly, some measure pressure, and some measure density. The equation of state and the hydrostatic equation relate the three, and Subsection 1.1.1 considers how these relationships are usually written and used in the meteorological and aeronomical literature.

Two complications, not generally familiar to meteorologists, become important at high enough levels in the upper atmosphere. One is variable composition, which begins to affect the mean molecular weight of air above 80 km. This is discussed briefly in Subsection 1.1.2. The other is the variation with altitude of the acceleration of gravity, which must be considered in the upper atmosphere. This is discussed briefly in Subsection 1.1.3.

1.1.1 The Equation of State and the Hydrostatic Equation

The constituents of air obey the equation of state for an ideal gas very closely at the pressures existing in the atmosphere. Air itself therefore behaves very nearly as an ideal gas, provided that it is assigned a properly defined mean molecular weight. We can write

$$p = \rho RT/m \tag{1.1}$$

where p is the pressure, ρ the density, T the kinetic temperature, R the universal gas constant, and m the mean molecular weight. The value of R is 8.317×10^7 erg mole^{-1} deg^{-1}; the gram-molecular weight (for dry air) has the value 28.966 at sea level and is known to be essentially constant up to about 80 km. Above that level it undoubtedly decreases with altitude (see Subsection 1.1.2).

To a very good degree of approximation, air is in hydrostatic equilibrium, so that the downward-directed gravity force balances the upward-directed pressure-gradient force. Coriolis and acceleration terms are several orders of magnitude smaller, except perhaps occasionally, in the case of the latter, for small scales of motion. Therefore

$$\partial p/\partial z = -\rho g \tag{1.2}$$

where z is the vertical coordinate directed outward perpendicular to the earth's surface and g is the acceleration of gravity. The acceleration of gravity, by convention, combines the effect of Newtonian gravitation and centripetal acceleration due to the earth's rotation. It varies somewhat with both latitude and altitude (see Subsection 1.1.3).

Throughout the lower atmosphere and in some of the upper atmosphere, it is permissible to neglect variations of g and m with height. Then, according to (1.1) and (1.2), the pressure p at a height z above the bottom of an isothermal layer is related to the pressure p_0 at the bottom of the layer by

$$p = p_0 \exp(-gmz/RT) \tag{1.3a}$$

It is often convenient (although not so common with reference to the lower atmosphere as to the upper atmosphere) to introduce the *scale height* H, where $H = RT/mg$. In terms of H, (1.3a) may be written

$$p = p_0 \exp(-z/H) \tag{1.3b}$$

Also, for an isothermal layer,

$$\rho = \rho_0 \exp(-z/H) \tag{1.4}$$

Another case of considerable interest is a layer where the temperature varies linearly with z. In this case, for a temperature T_0 at the bottom of the layer and a lapse rate* Γ,

$$T = T_0 - \Gamma z \tag{1.5}$$

$$p = p_0(T/T_0)^{gm/R\Gamma} \tag{1.6}$$

$$\rho = \rho_0(T/T_0)^{(gm/R\Gamma)-1} \tag{1.7}$$

A convenient parameter, used frequently in the upper-atmosphere literature, is the *number density* n, the number of particles per unit volume. In a mixture of gases

$$n = \sum_i n_i \tag{1.8}$$

* In common meteorological usage, the lapse rate Γ is defined to be positive when temperature decreases with height; $\Gamma = -\partial T/\partial z$.

where n_i is the number density of the ith constituent. The equation of state is conveniently written for the mixture

$$p = nkT \tag{1.9}$$

or, for an individual constituent,

$$p_i = n_i kT \tag{1.10}$$

where k is Boltzmann's constant ($k = 1.380 \times 10^{-16}$ erg deg^{-1}) and p_i is the partial pressure of the ith constituent.

The density ρ_i of the ith constituent is given by $n_i\mu_i$, where μ_i is the molecular mass and is related to the gram-molecular weight m_i by $\mu_i = m_i/N$. Here N is Avogadro's constant, $N = 6.025 \times 10^{23}$ (g mole)$^{-1}$. Obviously, the universal gas constant R is related to k by $R = Nk$. The density of a mixture of gases may be written

$$\rho = \sum n_i\mu_i = n\mu \tag{1.11}$$

where μ, the *mean molecular mass*, is defined by $\mu = \Sigma n_i\mu_i / \Sigma n_i$.

The hydrostatic equation in this notation may be written

$$\partial p/\partial z = -n\mu g \tag{1.12}$$

and clearly (1.3a) and (1.6) may be used with k/μ substituted for R/m.

1.1.2 Variation of Mean Molecular Weight with Height; Scale Height, Molecular-Scale Temperature

At high enough elevations, the atmosphere is not mixed and μ (or m) decreases with height. In this circumstance, one may consider integrated forms of the hydrostatic equation for particular vertical variations of the scale height H rather than for particular vertical variations of T. For example, (1.3a) and (1.3b) are valid not for an isothermal layer but for a layer where the ratio T/m is constant (if vertical variations of g may be neglected). Corresponding to a constant lapse rate of temperature, one may consider a constant gradient* of scale height, such that $H = H_0 + \beta z$ and $\beta = \partial H/\partial z$.

The counterpart of (1.6) is

$$p = p_0 \left(\frac{H}{H_0}\right)^{-(1/\beta)} \tag{1.13}$$

and in such a layer

$$n\mu = n_0\mu_0 \left(\frac{H}{H_0}\right)^{-(1+\beta)/\beta} \tag{1.14}$$

* Note that by convention the variation of scale height in a linear layer is described by the vertical gradient β, which is positive when scale height increases upward.

A particularly interesting case, applicable at high enough levels, is that of diffusive equilibrium, where each gas individually obeys* an equation of the form (1.12):

$$\partial p_i/\partial z = -n_i\mu_i g \tag{1.15}$$

Then with the aid of (1.10) the vertical distribution of partial pressure p_i and number density n_i for each gas can be computed according to the vertical distribution of the scale height $H_i = kT/\mu_i g$ for that particular gas. The vertical distribution of the mean molecular mass can be determined by combination of the results for the individual gases.

Sometimes, especially in connection with standard atmospheres, the *molecular-scale temperature* T_m is used. This is defined by

$$T_m/m_0 = T/m \tag{1.16}$$

where m_0 has the sea-level value 28.966. The molecular-scale temperature is, of course, equal to the kinetic temperature when the mean molecular weight has its sea-level value. The equation of state can then be written

$$p = \rho R_0 T_m \tag{1.17}$$

where R_0 is the gas constant for (dry) air at sea level, $R_0 = R/m_0$.

1.1.3 Variations of g; Geopotential, Standard Geopotential

The acceleration of gravity g varies with both latitude and altitude. At sea level, it is about 0.5 per cent higher at the poles than at the equator. Its variation in the vertical is given approximately (but not exactly, because g includes not only Newtonian gravitation but also the effect of centripetal accelerations) by

$$g(z) = g(0)/[1 + (z/r_e)]^2 \tag{1.18}$$

where $g(0)$ is the value at sea level and r_e is the radius of the earth. For example, the value of g at 100 km is about 97 per cent of its value at sea level.

A vertical coordinate called *geopotential* is often introduced to absorb the variability of g. It is defined by

$$g_0\Phi = \int_0^z g\,dz \tag{1.19}$$

where Φ is the geopotential and g_0 is a constant chosen to make the numerical value of geopotential similar to the numerical value of the geometric altitude

* This is not quite correct but is sufficiently accurate for our purposes.

to which it corresponds. It has been customary to consider g_0 dimensionless so that Φ has the dimensions of energy per unit mass, not of distance, and represents the potential energy that would be gained by a unit mass lifted from the earth's surface to a height z against the local force of gravity. The numerical value of g_0 is assigned so that it is comparable in magnitude with the value of g in the units used.

In meteorology, for example, in the reduction of upper-air radiosonde observations, g_0 is usually assigned the value 9.8 if g is expressed in m sec^{-2} (or 980 if g is expressed in cm sec^{-2}, etc.); Φ is expressed in the units of z, prefixed by the word "geopotential"— for example, in geopotential meters if z is in meters. (Sometimes the value 10 is used, in which case "geodynamic" replaces "geopotential" as a prefix.)

In the ICAO (International Civil Aviation Organization) standard atmosphere, g_0 is given the value 9.80665. In this case Φ is expressed in the units of z, prefixed by the words "standard geopotential." Some values of Φ are given in Table 1.1.

TABLE 1.1

VALUES OF g (AT A LATITUDE WHERE g AT SEA LEVEL IS 9.80665 M SEC^{-2}) AND OF Φ IN STANDARD GEOPOTENTIAL KILOMETERS[a]

z (km)	Φ (sgkm)	g (m sec^{-2})
0	0	9.80665
50	49.610	9.65418
100	98.451	9.5052
150	146.542	9.359
200	193.899	9.217
250	240.540	9.078
300	286.480	8.942
350	331.735	8.809
400	376.320	8.679
450	420.250	8.552
500	463.540	8.428

[a] Taken from Minzner and Ripley (1957).

In Subsections 1.1.1 and 1.1.2, vertical variations of g were not explicitly considered in vertical integrations of the hydrostatic equation. This is strictly correct if the altitude z is understood to be expressed in geopotential units, that is, if $g\,dz$ is understood to be replaced by $g_0\,d\Phi$ in the hydrostatic equation. If z is geometric altitude (as determined, for example, in some upper-atmosphere experiments), then it is approximately correct to neglect the variation of g if the layer is not too thick and if g is represented by an appropriate mean value over the layer considered. The "appropriate mean value" must, of course, correspond to the mean altitude of the layer and not to the sea-level value (see Table 1.1).

1.2 The Composition of Air and Some Standard Atmospheres

Air is a mixture of gases. Near the surface of the earth, the relative proportions of most of the gases vary by no more than a few thousandths of 1 per cent. An important exception is water vapor, which may make up as much as 4 per cent of surface air by volume or may be almost totally absent from a given air sample. None of the other variable gases is ever present in sufficient amount to affect the mean molecular weight of air significantly.

Table 1.2 gives the composition of surface air, in the absence of water vapor, according to Gluekauf (1951). This composition is essentially the same as the one used to derive the standard-atmosphere value of 28.966 for the mean molecular weight of dry air. Variations due to the variable water-vapor content are accounted for in the lower atmosphere by introducing the virtual temperature; we shall not, however, consider that any such correction is necessary in the upper atmosphere.

TABLE 1.2

COMPOSITION OF DRY AIR AT SEA LEVEL[a]

Constituent	Proportion by volume (per cent)
Nitrogen (N_2)	78.084
Oxygen (O_2)	20.946
Argon	0.934
Carbon dioxide	0.033
Neon	0.00182
Helium	0.00052
Others	0.00066

[a] According to Gluekauf (1951).

The relative proportions of these permanent constituents are believed to remain essentially unchanged up to at least 80 km. This is not to say that the composition of the upper atmosphere below 80 km is of no concern. On the contrary, the abundances of certain trace constituents such as ozone and nitric oxide are of very great concern in connection with radiative and ionization phenomena and will be discussed later (especially in Chapter 5) in some detail.

However, in the present context of composition as it affects the molecular weight, these can be neglected.

Above 80 km, a significant change begins to take place. The production of atomic oxygen by photodissociation (which occurs to a very minor extent as low as 20 to 30 km) begins to become a significant factor in the gross aspects of composition. Atomic nitrogen is also certainly present in the upper atmosphere, but probably not in sufficient quantities to affect the mean molecular weight appreciably. Another factor, other than dissociation, that must eventually affect the composition of the upper atmosphere somewhere above 100 km is *diffusive separation*, a name given to the process by which heavier gases tend to distribute themselves at lower levels than the lighter ones. In the limit of diffusive equilibrium, as discussed in Subsection 1.1.2, each gas is separately distributed according to a hydrostatic equilibrium based on its own molecular weight. Such a condition, of course, is countered by mixing and air is believed to be well mixed below 100 km.

Unfortunately, as will be discussed at length in Chapter 6, the details of the decrease of mean molecular weight with altitude above 80 km are not well determined. This is doubly unfortunate. In the first place, the composition is a matter of great interest in itself. In the second place, a knowledge of the mean molecular weight is necessary to deduce values of kinetic temperature from measurements of pressure or density, as is easily seen from a consideration of the equations in Section 1.1. As a matter of fact, the temperature is never directly measured at such high elevations (although in some cases it can be inferred from spectroscopic data). The reader should therefore keep in mind, when considering results of standard atmospheres above 80 km, that the pressures, densities, and the ratios T/m are better determined than either T or m separately.

Temperature, density, and pressure at a given altitude in the atmosphere certainly vary from time to time and from place to place. Nevertheless, it is instructive and for many purposes very useful to consider the vertical distributions of the estimated average values of these variables, as represented in standard atmospheres. The official standard atmosphere for the United States and many other countries, called the ICAO standard atmosphere, was adopted in 1952 as a combination of two earlier versions that need not concern us. It extends only to an altitude of 20 sgkm (abbreviation for "standard geopotential kilometers"). Several proposals for extension to higher altitudes have been formulated. These include the "ARDC model atmosphere, 1956" (Minzner and Ripley, 1956), which was adopted up to 300 sgkm as a "U. S. extension to the ICAO standard atmosphere" (Minzner, Ripley, and Condron, 1958); the "ARDC model atmosphere, 1959" (Minzner, Champion, and Pond, 1959); and, more recently, the "U. S. standard atmosphere, 1962" (Sissenwine, Dubin, and Wexler, 1962). As newer and better observations (and interpretations thereof) have accumulated, there have been rather notable changes in the estimated properties of the upper atmosphere, especially above 100 km. No attempt is made at this point to discuss these changes or the reasons for them. Neither is it necessary here to give much detail. The purpose of this chapter is simply to introduce and orient; further discussions of upper-atmosphere structure and composition are contained in Chapters 2, 3, 5, and 6.

Figure 1.1 gives the main features of the U.S. standard atmosphere, 1962.

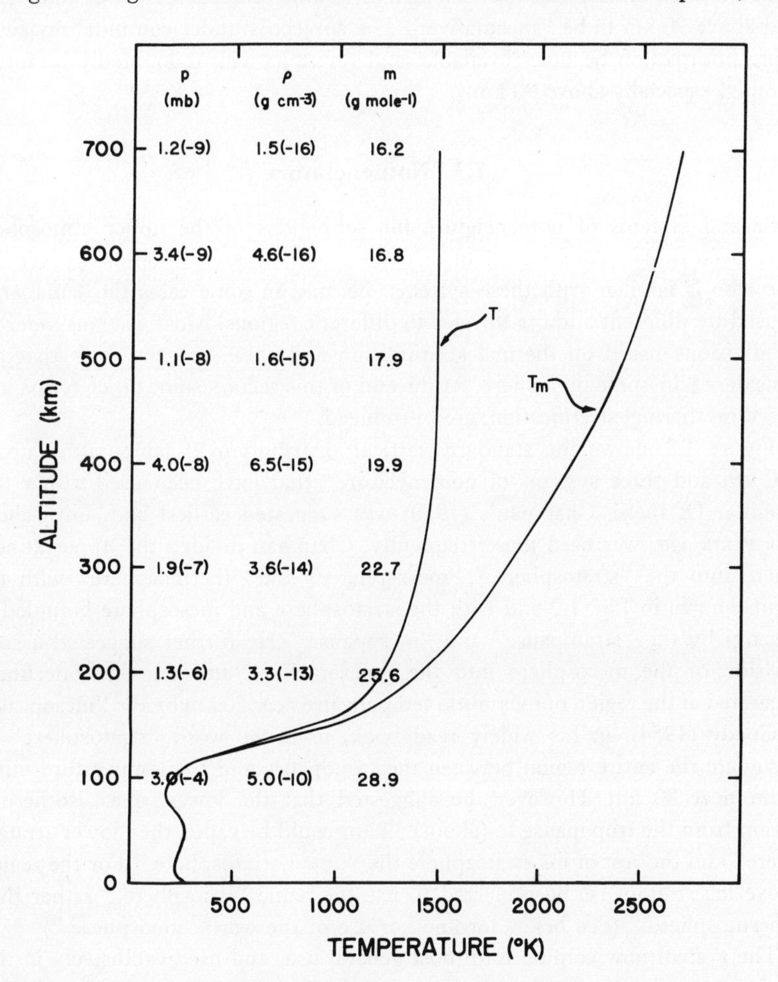

FIG. 1.1. The U. S. standard atmosphere, 1962. The curve labeled "T" gives kinetic temperature; the curve labeled "T_m" gives molecular-scale temperature. With respect to the tabulated values of pressure and density, a number in parentheses designates the power of 10 by which the preceding number must be multiplied; for example, 1.2(−9) means 1.2×10^{-9}.

This is intended to apply to midlatitude year-round conditions averaged for daylight hours and for the range of solar activity that occurs between sunspot minimum and sunspot maximum (the latter factor becomes important above 200 km; see Chapter 6). On the scale of Fig. 1.1, it is difficult to make out details of the vertical temperature distribution below 100 km. These are shown in Fig. 1.2, in connection with the discussion of nomenclature in Section 1.3. This standard atmosphere between 32 and 90 km is considered to be "tentative" and above 90 km to be "speculative." The subject is under continual review as more information becomes available, and revisions will undoubtedly be forthcoming, especially above 90 km.

1.3 Nomenclature

Several systems of nomenclature for subregions of the upper atmosphere have been in use. As a matter of fact, the situation is very confusing even for one who is familiar with these systems, because in some cases the same word is used by different authors to refer to different regions. Most systems refer to subdivisions based on thermal stratification and these are the ones that are considered in some detail here. At the end of this section some other terms, not based on thermal stratification, are introduced.

Figure 1.2 shows the standard vertical distribution of temperature up to 110 km and three systems of nomenclature* that have been used rather frequently. Of these, Chapman's (1950) was suggested earliest and, until about five years ago, was used most frequently. Chapman divided the upper atmosphere into the "stratosphere," "mesosphere," and "thermosphere" with the limits shown in Fig. 1.2 and with the stratosphere and mesosphere bounded at the top by the "stratopause" and "mesopause." He further suggested a subdivision of the mesosphere into the "mesoincline" and the "mesodecline," separated at the region of maximum temperature near 50 km by the "mesopeak."

Goody (1954), in his widely read book, used the word "stratosphere" to designate the entire region between the tropopause and the temperature minimum near 80 km. However, he suggested that the lower, quasi-isothermal region from the tropopause to (about) 32 km could be called the "lower stratosphere" and the rest of his stratosphere the "upper stratosphere." For the region above his stratopause he preferred to use the name "ionosphere," rather than "thermosphere." (See below for another use of the word "ionosphere.")

The system now coming into most general use, and used exclusively in this book, is usually associated with the name of Nicolet (see, for example, Nicolet, 1960). It was recommended by the International Union of Geodesy and Geo-

---

* For others, which have not been used so extensively, see Flohn and Penndorf (1950) and Gerson and Kaplan (1951).

physics at its Helsinki meeting in 1960 and is labeled "IUGG, 1960" in Fig. 1.2. In this system, the *stratosphere* extends from the tropopause to the temperature maximum near 50 km (*stratopause*) and the *mesosphere* lies between the stratopause and the temperature minimum near 80 km (*mesopause*). The *thermosphere* is above the mesopause. This system was also recommended for use

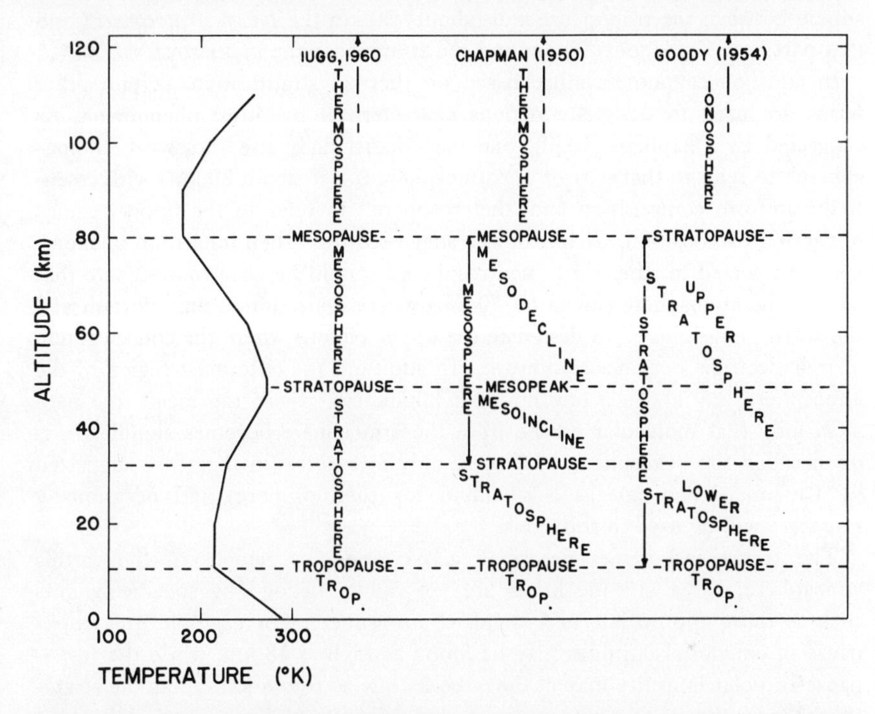

FIG. 1.2. Vertical distribution of temperature up to 110 km according to the U. S. standard atmosphere, 1962, and common systems of nomenclature. The latter are discussed in the text.

by the World Meteorological Organization's executive committee in 1962, as reported by Sawyer (1963). This author agrees with Sawyer that, "It is in the interests of all meteorologists that the new standard terminology should be widely known and adopted." Regardless of one's personal preference, a situation in which a word like "stratosphere" has three different and widely used meanings is ridiculous and intolerable. This book uses the IUGG, 1960 nomenclature henceforth.

In one respect, an extension of this system has been found to be useful and is adopted here. For many purposes of discussion, it is convenient to separate the stratosphere into two regions, bounded at approximately 30 km. For example, this level is the approximate ceiling of sounding balloons and routine meteorological data (see Chapter 2); it is an approximate boundary between an upper region where ozone tends to be in photochemical equilibrium and a lower region where it is not (see Chapter 5); and at least at some times and places it represents an approximate boundary between regions of different hydrostatic stability (see Chapter 3). For all these reasons we call that part of the stratosphere between the tropopause and (about) 30 km the *lower stratosphere** and that part between (about) 30 km and the stratopause the *upper stratosphere.**

In addition to nomenclature based on thermal stratification, certain other terms are used to designate regions characterized by other phenomena. As suggested by Chapman (1950), one may occasionally use the word "homosphere" to refer to that part of the atmosphere (up to about 80 km) with essentially uniform composition and "heterosphere" to refer to the upper regions of varying composition. Chapman also suggested that when ionization phenomena are referred to, the word "neutrosphere" should be used to designate that part of the atmosphere (up to 60–70 km) where ionization is unimportant and the word "ionosphere" to designate the upper regions where the concentration of free electrons becomes significant. In addition, the outermost region of the atmosphere, say above a few hundred kilometers, where the mean free path is so long that molecular escape from the atmosphere becomes significant, is often called the "exosphere." Of these only *ionosphere* (in the sense suggested by Chapman, and not as a synonym for thermosphere) and occasionally *exosphere* will be used in this book.

The terms and names described above designate regions of the earth's atmosphere whose altitude limits are not rigidly defined by specific heights such as those appropriate to a standard atmosphere. For example, the tropopause in equatorial latitudes may be found as high as 18 km, while the tropopause in polar latitudes may at times be as low as 6 to 8 km. Thus the stratosphere (and the upper atmosphere, by our definition) has a lower limit that depends on the thermal properties of the atmosphere at a given time and place and not on a definite defined altitude. Similar variability may be expected for the altitude limits of other higher regions, although the present state of knowledge leaves us in a poor position to define this variability.

1.4 Scope and Plan of this Book

The subject of the upper atmosphere encompasses so much material that a book of reasonable length necessarily must emphasize certain aspects and

* It should be noted that this nomenclature is not without its ambiguities, because some meteorologists prefer to divide the stratosphere into three regions: the lower stratosphere (tropopause to about 20 km), the middle stratosphere (about 20 km to about 30 km), and the upper stratosphere (about 30 km to the stratopause).

neglect others. The choice of topics depends on the author's interests and experience and also on his estimate of the interests and background of the intended audience (although these two factors cannot be completely separated). In this book, emphasis is placed on topics that appear to represent an extension to upper levels of problems that are of direct concern in the lower atmosphere. These include, but are not limited to, structure, composition, and circulation, especially from the point of view of their interrelationships and variability.

Just as in the lower atmosphere, these problems must be related to the physical and chemical processes that influence them, if a treatment is to be anything more than a mere compilation of observations. Many of these physical and chemical processes are quite different from those that are important in the lower atmosphere. For example, much more emphasis must be placed on detailed aspects of solar radiation and its absorption, which vitally affect the composition, structure, and circulation of the upper atmosphere. Certain chemical reactions, occurring as an aftermath of that absorption, must be considered in some detail. The heat-balance problem, involving both radiative and transport processes, has complications that are unfamiliar to the student of the lower atmosphere. The dynamics of the circulation, as yet barely studied, clearly differ in many ways from those in the lower atmosphere, for example, in the importance of tidal motions at high levels. All of these topics receive more or less detailed consideration in this book.

On the other hand, certain traditional problems of the upper atmosphere receive less emphasis than usual. These include the earth's magnetic field and its variations, problems of the aurora and airglow, and problems connected with the ionized regions. Much of this material has been covered recently in various review articles and in the comprehensive books of Ratcliffe (1960) and Chamberlain (1961). Furthermore, little is included here about the earth's outermost atmosphere, say above a few hundred kilometers, upon which much active investigation is now centered.

It is not easy, even after deciding general questions of objective, scope, and content, to organize material of this sort into a reasonable order of presentation. There are many problems, most of which are related to the general problem that composition and structure vitally affect the heat sources and sinks, which in turn vitally affect the composition and structure (as well as the circulation). There is the additional complication that the author has thought it desirable to discuss, at least superficially, certain physical problems not generally encountered in studies of the lower atmosphere. The purpose of the next few paragraphs is to outline the plan of this book and some of the reasons for it.

With regard to specific organization, one might say as a first approximation that the book starts at the tropopause and proceeds upward. Chapter 2 is concerned with the structure, composition, and circulation of the lower stratosphere and Chapter 3 with the structure and circulation of the upper stratosphere and mesosphere. These chapters refer primarily to methods and results of observations in the stratosphere and mesosphere and not to physical and chemical causes or effects.

One cannot go much further than this, in a discussion of the upper atmosphere, without detailed consideration of the sun, its ultraviolet radiation, the absorption of that radiation in the upper atmosphere, and the effects of that absorption. Chapter 4 takes up these topics and, in addition, includes discussions of certain aspects of spectroscopy, quantum mechanics, and radiative processes that are particularly relevant to the upper atmosphere. With this background, Chapters 5 and 6 continue the survey of composition and structure. Chapter 5 covers the composition of the upper stratosphere and mesosphere with particular reference to ozone. Chapter 6 takes up the composition and structure of the thermosphere, with emphasis on the effects of photodissociation, diffusion, and variability of solar radiation.

A fundamental problem of the upper atmosphere, just as of the lower, is the heat budget. One aspect of this is the gain of energy as the result of absorption of solar radiation, a subject surveyed in Chapter 4. More difficult aspects are the transfer of energy as a result of other radiative processes and various circulation processes. There are certain advantages to be gained by considering these with respect to the upper atmosphere as a whole rather than with respect to subregions separately. This is done in Chapter 7, and it is hoped that this approach serves to emphasize the similarities as well as the dissimilarities of the most important processes at various levels.

Little is known at present about the circulation of the thermosphere. A brief discussion of prevailing winds in the lower thermosphere is included in Chapter 3. But the wind field of the lower thermosphere is exceedingly complex. It involves not only a prevailing component but also large components related to atmospheric tides and to much smaller-scale disturbances (believed to represent a type of wave motion). In Chapter 8, these complications are discussed separately.

As mentioned earlier, this book places relatively little emphasis on certain topics that claim much attention in upper-atmospheric literature and in other books on the upper atmosphere. This is due not to a lack of appreciation for their importance, but to limitations of space, time, and the author's experience. Among other things, they are in some respects closely related to the topics of main concern in this book. Chapter 9 therefore attempts to introduce the reader to geomagnetism, the ionosphere, and the aurora and airglow. Some of this material is referred to in earlier chapters, with appropriate cross references. It must be emphasized that the objective and scope of this chapter are significantly different from those of the earlier chapters; and it is hoped that the interested reader will take advantage of other, more detailed treatments of these topics, references to which are included in Chapter 9.

At various places in this book, there will be discussions of atmospheric transport in connection with specific properties such as ozone and heat. The question of the atmospheric transport of properties has a great importance and a certain unity which tends to be obscured by these separate discussions. Therefore, a short Chapter 10 takes up the question of transport, partly by reference to the earlier discussions and partly by introduction of some new material.

REFERENCES

CHAMBERLAIN, J. W. (1961). "Physics of the Aurora and Airglow." Academic Press, New York.

CHAPMAN, S. (1950). Upper atmospheric nomenclature. *Bull. Amer. Meteor. Soc.* **31**, 288-290.

FLOHN, H., and PENNDORF, R. (1950). The stratification of the atmosphere. *Bull. Amer. Meteor. Soc.* **31**, 71-78, 126-130.

GERSON, N. C., and KAPLAN, J. (1951). Nomenclature of the upper atmosphere. *J. Atmos. Terr. Phys.* **1**, 200.

GLUEKAUF, E. (1951). The composition of atmospheric air. *In* "Compendium of Meteorology" (T. F. Malone, ed.), pp. 3–10. American Meteorological Society, Boston, Massachusetts.

GOODY, R. M. (1954). "The Physics of the Stratosphere." Cambridge Univ. Press, London and New York.

MINZNER, R. A., and RIPLEY, W. S. (1956). The ARDC model atmosphere, 1956. *AF Surveys in Geoph.* No. 86.

MINZNER, R. A., and RIPLEY, W. S. (1957). ARDC model atmosphere, 1956. *In* "Handbook of Geophysics for Air Force Designers," 1st ed., pp. 1–2 to 1–37. Air Force Cambridge Research Center, Bedford, Massachusetts.

MINZNER, R. A., RIPLEY, W. S., and CONDRON, T. P. (1958). "U. S. Extension to the ICAO Standard Atmosphere." U. S. Govt. Printing Office, Washington, D. C.

MINZNER, R. A., CHAMPION, K. S. W., and POND, H. L. (1959). The ARDC model atmosphere, 1959. *AF Surveys in Geoph.* No. 115.

NICOLET, M. (1960). The properties and constitution of the upper atmosphere. *In* "Physics of the Upper Atmosphere" (J. A. Ratcliffe, ed.), pp. 17–71. Academic Press, New York.

RATCLIFFE, J. A., ed. (1960). "Physics of the Upper Atmosphere." Academic Press, New York.

SAWYER, J. S. (1963). Note on terminology and conventions for the high atmosphere. *Quart. J. Roy. Meteor. Soc.* **89**, 156.

SISSENWINE, N., DUBIN, M., and WEXLER, H. (1962). The U. S. standard atmosphere, 1962. *J. Geoph. Res.* **67**, 3627–3630.

CHAPTER 2

Meteorological Conditions in the Lower Stratosphere

As recently as 1900, it was possible for a reputable scientist to conclude on the basis of the limited data of the time and his own deductions from radiation theory that the temperature must continue to decrease with height to the outer edge of the atmosphere. But even as such a conclusion was being reached, L. P. Teisserenc de Bort was in the process of carrying out a long series of observations that culminated with his announcement (1902) of the existence of an "isothermal layer" above about 11 km. He was even able with continued study to deduce the fact that the base of this layer, now called the tropopause, is lower over cyclones and higher over anticyclones.

Sixty years and many hundreds of thousands of observations later, considerable detail is available about the tropopause and the thermal and wind structure of the lower stratosphere. Of all the upper atmosphere, the lower stratosphere and the lowest few kilometers of the upper stratosphere are the only regions that can be reached by the usual meteorological measuring techniques. Although these techniques leave something to be desired as to accuracy, particularly near their ceilings, they have the tremendous advantage of being carried out routinely by the meteorological services of most nations of the world (although, in many cases, the top of the lower stratosphere is not reached).

This chapter summarizes the information about the tropopause and the lower stratosphere revealed by these observations and certain other specialized measurements. As an introduction, the first section contains a brief description of the standard observational equipment and data-reduction techniques, with particular emphasis on the types of inaccuracies that may be introduced. Because some of these inaccuracies become particularly serious at the altitudes we are considering, they require emphasis here. In Section 2.2, observed characteristics of the tropopause and various explanations of these are discussed. In Section 2.3, we consider specialized observations of water vapor and aerosols in the lower stratosphere. Sections 2.4 and 2.5 deal with the wind and temperature patterns of the lower stratosphere, first from a climatological point of view and then in terms of more detailed features, particularly those of the winter-time high-latitude regions.

2.1 Balloon Sounding Systems

The regular meteorological upper-air observing network provides the bulk of data pertaining to the meteorology of the lower stratosphere. Radiosonde reports of temperature and pressure, and wind data obtained by tracking the balloons by radar or radio direction-finding methods give in some regions of the world reasonably adequate information on a twice-daily basis. However, the spacing of the observing stations, viewed on a worldwide basis, is extremely irregular. Large oceanic areas and most of the Southern Hemisphere go relatively unobserved. With particular reference to stratospheric coverage, observations reaching to the top of the lower stratosphere are most numerous over North America. During the past few years, measurements reaching to these high levels have become increasingly frequent in much of western Europe, the Soviet Union, China, and Japan.

Sounding balloons are most commonly made of neoprene and inflated with helium or hydrogen. Being extensible, the balloon expands as it ascends, maintaining a relatively constant rate of rise. (This rate increases somewhat as the drag decreases at high elevations.) Eventually, of course, the balloon bursts. The altitude where bursting occurs, the "bursting height," places an upper limit on the usefulness of the system. During the past few years, the average bursting height of meteorological balloons has steadily increased as a result of (a) improved balloon materials and better methods of manufacture, (b) more attention to pretreating the balloons just before release, and (c) the use of larger balloons. There seems to be no reason why the average bursting height should not exceed 30 km within the next few years, and regular ascents to 40 km are possible but, for the time being, not economically feasible. It appears that 40–45 km represents a practical limiting ceiling for the use of sounding balloons.

2.1.1 Temperature and Pressure Measurements

Measuring and telemetering systems in use by various countries differ quite considerably in detail. Since we are primarily concerned with the upper atmosphere, we shall discuss only the United States rawinsonde, a system that has produced much of the data discussed in later sections of this chapter. The discussion is aimed to give an outline of the methods and especially the difficulties of making routine stratospheric observations.

A rawinsonde system includes a balloon-borne radiosonde, a ground receiver that serves also as a radio theodolite to track the balloon for wind information, and a recorder. With the GMD-1 equipment, the radiosonde transmits a frequency-modulated signal at a carrier frequency of 1680 megacycles. The temperature element is a thermistor, and humidity is measured by an electrolytic element, a lithium chloride strip. Neither this nor any other humidity element used in a routine meteorological sounding system is operable in the stratosphere. The pressure element is an aneroid, and serves as a baroswitch, periodically switching the temperature element, the humidity element, and certain reference resistors into the modulator as the pressure varies. The modulator is a blocking oscillator that interrupts the signal at an audio rate that depends on the resistance of the element currently switched in. The reference resistors are switched in at predetermined pressures.

The desired output from all radiosondes is a record of the altitude, temperature, and humidity as a function of pressure. However, no direct measurement of altitude is made. Instead, the altitudes of certain pressure levels are calculated from the hydrostatic equation and the observed relationship between temperature and pressure. For this purpose, (1.1) and (1.2) may be used to give

$$z_2 - z_1 = -\frac{R_0}{g}\int_{p_1}^{p_2} T\,d\ln p \tag{2.1}$$

where z_1 and z_2 are the altitudes* of the pressure levels p_1 and p_2.

At the ground, of course, the altitude (with reference to sea level) may be determined accurately once and for all and the pressure may be measured very accurately with a mercury barometer. The heights of successively more elevated pressure levels are then determined by evaluating the right side of Eq. (2.1) graphically from the sounding data. The pressure levels for which heights are determined and for which temperatures are read off the sounding and transmitted include (a) certain standard levels and (b) enough additional "significant" levels for each sounding to ensure that the reconstructed temperature curve will nowhere differ more than 1°C from the curve defined by all the original data. Standard levels in the lower stratosphere include 100 mb (about 16 km), 50 mb (about 21 km), and 30 mb† (about 24 km).

This procedure of using the hydrostatic relationship (a procedure common to all radiosonde systems) has interesting implications for the accuracy of derived data in terms of the accuracy of the basic measurements of temperature and pressure. The final usable information from the sounding system (if we neglect, for the time being, wind) consists of derived height and temperature values for each of several pressure levels. For illustrative purposes, it is convenient to write (2.1) in the form

$$z_2 = z_1 + \frac{R_0}{g}\bar{T}\ln\frac{p_1}{p_2} \tag{2.2}$$

* As explained in Subsection 1.1.3, it is customary to use a standard value of g in this calculation, so that altitude is expressed in geopotential units. However, this is not always shown explicitly in notation or discussion.

† The uppermost standard level in the United States was 25 mb until 1960. Consequently, most of the subsequent discussion refers to this level and not to 30 mb.

where $\bar{T}$ is an appropriate mean value of temperature between the two levels. Suppose that in Fig. 2.1 the level "1" is at the ground, where altitude and pressure are accurately known, and it is desired to find the altitude of the specified pressure level p_2. The observed data are used only to determine an appropriate value of $\bar{T}$ for solution of (2.2). An erroneous *pressure* measurement by the radio-

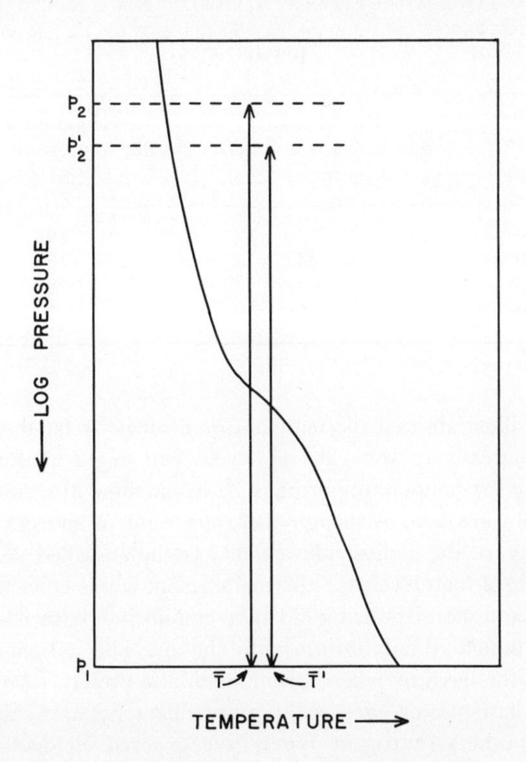

FIG. 2.1. Schematic representation of the effect of an erroneous pressure measurement on the determination of the mean temperature for a layer. The curve represents a measured sounding. The pressure element indicates that the balloon is at the pressure p_2' when it is really at p_2. The mean temperature between p_1 and p_2 is determined to be $\bar{T}'$ rather than $\bar{T}$.

sonde distorts the limits over which $\bar{T}$ should be determined; an erroneous *temperature* measurement, of course, affects $\bar{T}$ directly. The second error is by far the more serious.

Leviton (1954) has computed the cumulative effects of these errors, assuming (a) an error of 1°C in the measured temperature (either too high or too low at all levels), (b) an error in the pressure measurement of 3 mb up to the 100-mb level, and of $1\frac{1}{2}$ mb above the 100-mb level (either too high or too low at all levels), and (c) an average atmospheric sounding. Table 2.1 gives his results for some stratospheric levels.

TABLE 2.1

ERRORS IN THE HEIGHTS OF CONSTANT-PRESSURE SURFACES DUE TO (a) A TEMPERATURE ERROR AT ALL LEVELS OF 1°C AND TO (b) A PRESSURE ERROR OF 3 MB UP TO 100 MB AND OF $1\frac{1}{2}$ MB ABOVE 100 MB[a]

Pressure level (mb)	Height error due to temperature error (m)	Height error due to pressure error (m)
100	67	19
50	88	16
25	108	9
10	135	7

[a] After Leviton (1954).

These results illustrate that the temperature error is by far the more serious and becomes progressively worse at high levels. One might question the procedure of assuming the temperature error to be in the same direction at all levels, but this is amply justified. Although there are random sources of error, the greatest difficulty as the radiosonde ascends is the so-called radiation error. In the thinner air of high levels the thermal element is less efficiently ventilated and tends more and more toward a radiative equilibrium with its surroundings rather than a conductive equilibrium with the air. The radiation error is far more serious in the daytime when solar radiation is present than at night. To overcome the radiation error, some radiosondes (duct type) have shielded thermal elements, while others (outrigger type) have exposed elements with highly reflecting coated surfaces. The latter are more accurate and are now used more frequently. In addition standard radiation corrections are made to the observations. Nevertheless, the residual errors are still very serious at the high levels and meteorologists using the data (for example, Teweles and Finger, 1960) find it necessary to make additional empirical corrections based on the observed differences between night-time and day-time soundings.

The temperature value assigned to a given pressure level is, of course, affected directly by any temperature error. The effect of a pressure error on reported temperatures is to assign to a designated pressure level a temperature that was really measured at a different pressure level. The magnitude of the latter effect depends on the vertical variation of temperature with pressure (baric lapse rate) at the level in question, and is usually small. However, in the stratosphere the baric lapse rate can become large and sometimes a 2- to 3-mb pressure error results in very serious inaccuracies, especially near the ceiling of the balloons. The accuracy of pressure measurements at stratospheric levels can be greatly improved by use of a hypsometer element in place of the aneroid and a few United States observing stations are currently using this. This is because the percentage error of a hypsometer remains nearly constant so that the absolute error decreases as the pressure decreases.

Before turning to a discussion of wind-observing systems, let us consider briefly other radiosonde measurements. Most countries use bimetallic elements to measure temperature and hair elements to measure humidity. The bimetallic temperature element, unless carefully shielded, suffers very seriously from radiation error. The uncertainties in geopotential introduced by this type of error become painfully obvious when one analyzes the day-time observations at, say, 50 mb from several neighboring countries, each using a different type of instrument with a different amount of residual error. There is sometimes a systematic difference of 50–100 geopotential meters between observations on two sides of a national boundary.

2.1.2 Wind Measurements

The balloon moves horizontally with the prevailing wind as it rises through the atmosphere. It has small enough inertia so that its motion responds quickly to changes in wind speed and direction, and wind observations are obtained by tracking and observing its displacement. Various methods of tracking are in use.

In the United States GMD-1 system, the ground receiver serves as radio direction-finding equipment. The balloon's azimuth and elevation angles are observed and recorded throughout the ascent. To locate the balloon in space requires one additional bit of information. In this system, the height of the balloon above the surface, as determined from the radiosonde observation and the hydrostatic relationship, serves the purpose.

The errors inherent in this system are seriously magnified at high levels. Figure 2.2 illustrates the trigonometry of the calculation. Let the balloon be at the position A_1' at time t_1, and at the position A_2' at time t_2, the observer being at O. Let A_1 and A_2 be the projections of A_1' and A_2' on the earth's surface.* The average horizontal wind velocity between t_1 and t_2 is determined from the vector joining A_1 and A_2. The speed is given by the length, s, of the projected trajectory, divided by the time interval, and the direction by the angle ψ measured from some reference azimuth. With reference to the triangle OA_1A_2, the law of cosines gives

$$s^2 = d_1^2 + d_2^2 - 2d_1d_2\cos(\phi_2 - \phi_1) \tag{2.3}$$

* Here the earth is assumed to be flat. Corrections for the earth's curvature become important when the balloon is at a great distance.

and the law of sines gives

$$d_1 \sin(\psi - \phi_1) = d_2 \sin(\psi - \phi_2) \tag{2.4}$$

Thus d_1 and d_2 are needed to determine s from (2.3) and ψ from (2.4). One can determine d_1 and d_2 from the heights and the elevation angles, because $d_1 = h_1 \cot \epsilon_1$, and $d_2 = h_2 \cot \epsilon_2$.

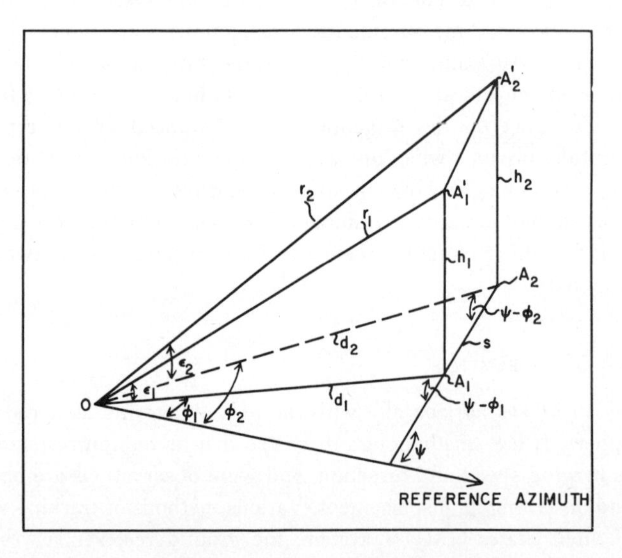

FIG. 2.2. Geometry of the calculation of wind when a balloon moves from point A_1' to point A_2' and is observed from location O. (Adapted from de Jong, 1958.)

There is not space to attempt a detailed discussion of the errors in s and ψ resulting from errors in the ϕ's, ϵ's and h's. De Jong (1958) has given a complete discussion of this question. However, it is important to note that the cotangent of the elevation angle enters into the calculation. When the elevation angles are small, as they usually are by the time the balloon is in the stratosphere, their cotangents are very large and changing rapidly. Thus, small observational errors of ϵ and h are greatly magnified and can introduce appreciable error into the derived value of wind velocity. Furthermore, when the elevation angle is small, it becomes increasingly difficult to measure, because the transmitted signal is increasingly subject to reflection by topographical irregularities. Winds are not reported when the elevation angles are less than 6°, even though the balloon may still be in range with radiosonde observations still being received.

In this connection, one should note that the value of h is considerably more uncertain than are the altitudes of specific constant-pressure surfaces. The latter, as we have seen, are worked up with the hydrostatic equation and are rather insensitive to errors in the pressure readings. However, the h that enters into the wind calculation requires knowing the height of the balloon at a specific time when specific elevation and azimuth angles were registered. For this purpose, pressure errors in the stratosphere are far more serious. According to Leviton, when the balloon is near 30 km, a 1-mb pressure error leads to an error of almost 600 m in height. This uncertainty gives another impetus for the use of hypsometers for high-level observations.

A notable improvement in wind observations is due within the next few years. During the 1960's, American equipment will gradually be converted to the GMD-2 system which will measure the *range* of the balloon (r_1 and r_2 in Fig. 2.2). With reference again to Fig. 2.2, it can be seen that $d_1 = r_1 \cos \epsilon_1$ and $d_2 = r_2 \cos \epsilon_2$. Even for small elevation angles, the *cosine* cannot exceed 1, so that this is inherently a better method than the present one. The range will be determined by a device in the radiosonde that upon interrogation will transmit a signal to the ground receiver.

The systems described above are not the only ones in use. The oldest method of tracking, still used in many countries, is visual, with the aid of a theodolite. These observations, called *pibals*, rarely reach the altitudes of interest here and are mentioned only to emphasize that some meteorological wind measurements are of little use for stratospheric studies. Alternatively some countries and some U. S. weather services use a system of tracking the balloons by radar. Although this method provides excellent data, its use to get stratospheric winds is limited by the range of the radar set. For balloons with the normal rate of rise (about 1000 ft min^{-1}) and for not unusual wind conditions, the target is sometimes lost below the tropopause and only infrequently can be followed to 30 km with some of the radar sets in use.

2.2 The Tropopause

The term *tropopause* is applied to the boundary between the troposphere and the stratosphere. The former is a region where the temperature, in the mean, decreases with height, and the latter is a region where the temperature, in the mean, is nearly independent of height. Therefore the tropopause separates layers of markedly different vertical temperature gradients and its height above the ground at a given time and place is, in principle, determined by that criterion.

In practice the precise definition and location of the tropopause is an arbitrary procedure. An individual vertical temperature profile may show not one, but often two and sometimes several levels where the lapse rate changes abruptly in the sense defined for the tropopause. Figure 2.3 gives several examples of soundings and illustrates that the selection of the tropopause level is not at all straightforward at times.

United States weather observers, encoding for teletype transmission the results of their rawinsonde observations, are instructed to select the tropopause level according to the following criteria (U. S. Weather Bureau, 1957):

At pressures of 500 mb or lower:

(a) The lowest level, with respect to altitude, at which the temperature lapse rate decreases to 2°C km^{-1} or less.

(b) The average lapse rate from this level to any point within the next higher two km does not exceed 2°C km^{-1}.

(c) The flight must extend at least 2 km above the tropopause level.

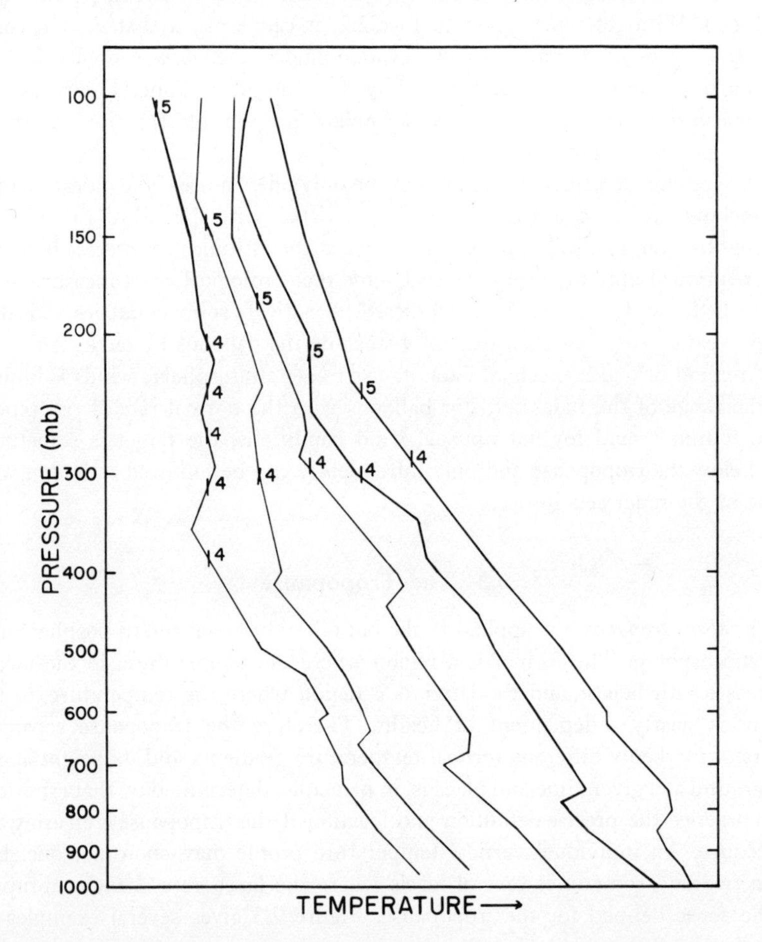

Fig. 2.3. Atmospheric soundings, some with tropopause ill defined. The vertical hatch-marks labeled "5" and "4" on each sounding represent the intersections with the −50° and −40°C isotherms, respectively.

These instructions are quoted to illustrate one simple operating procedure. Whether this or any other procedure is justified and adequate depends on the true physical nature of the tropopause. Is the transition from troposphere to stratosphere abrupt or gradual? Is the tropopause continuous in time and space?

2.2.1 Observed Characteristics of the Tropopause

Our concept of the tropopause has been steadily evolving during the past 30 years and it is fair to say that the more detailed and numerous the observations become, the more complex a phenomenon the tropopause appears to be. It is also fair to say that models of the tropopause depend to a great extent on the amount of averaging and smoothing of the observations on which the models are based. Let us review briefly some of these models and concepts and see for what scale of averaging they may still be appropriate.

In the years after Teisserenc de Bort's discovery of the stratosphere, the gradually accumulating observations pointed to a relatively simple model of the tropopause as a surface continuous in time and space. Bjerknes (1932) formalized this idea when he published a schematic meridional cross section, representing mean conditions, which showed the tropopause extending continuously from an altitude of about 17 km at the equator to an altitude of about 8 km at the pole, as in Fig. 2.4. However, even the observations available at that time showed that things were not quite so simple, if less averaging were done. Palmén in 1933 and Bjerknes and Palmén in 1937 pointed out that the tropopause in middle latitudes on individual days must have a more complicated structure. From their work came the concepts of the multiple tropopause and the "folded" tropopause.

Hess (1948) was apparently the first to present a multiple tropopause structure in a time-averaged meridional cross section and his model is illustrated in Fig. 2.4. Hess's data were for the meridian 80°W, averaged for the time period 1942–1945 inclusive. The high southern tropopause is called the tropical tropopause; the low northern tropopause is called the polar tropopause. The latitude where the two overlap is just the latitude where the westerly winds of the upper troposphere have their maximum speed, in the mean. The region between the two tropopauses is commonly referred to as a tropopause "break."

Among similar cross sections that have been constructed, for various meridians and time periods, Kochanski's (1955) results may also be mentioned. They are based on averages for the same meridian as Hess's but for a different period of time. Kochanski introduced still a third tropopause surface, a so-called tropopause "leaf" at high levels in middle latitudes. As shown in Fig. 2.4, his tropopause structure, except for the "leaf," is rather similar to Hess's even though based on a different period of time.

Defant and Taba (1957) in an analysis of the three-dimensional structure of the atmosphere for a short period of time found it desirable to represent the tropopause on individual days by three continuous surfaces. Figure 2.4 shows the tropopause structure in their latitudinally averaged north-south cross section for one day, 1 January 1956. Note the tropopause break at about 55°N between the polar tropopause and what they call the "middle" tropopause. Just as the

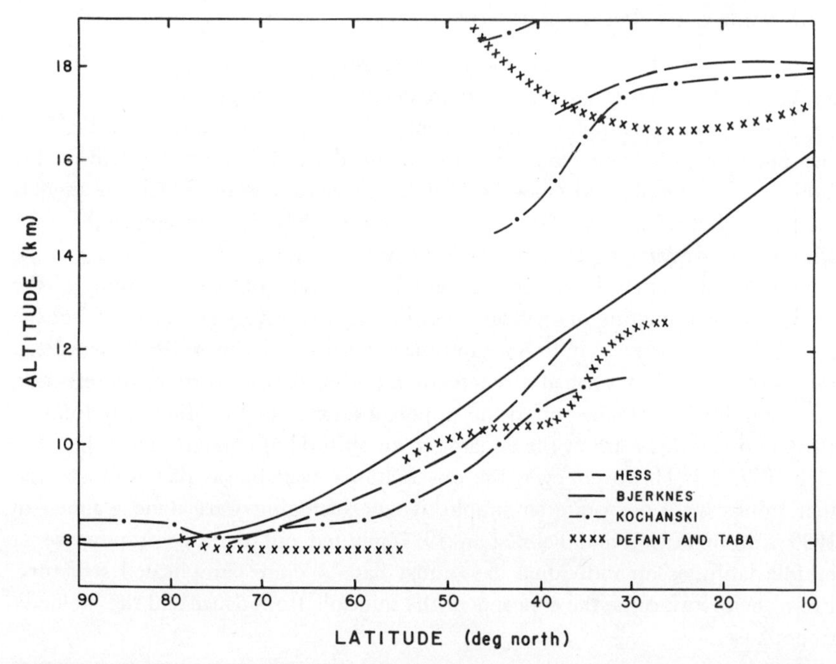

FIG. 2.4. North-south cross section showing average locations of the tropopause according to Bjerknes (1932) averaged for all seasons from meager early data, to Hess (1948) averaged for winter months at 80°W, to Kochanski (1955) averaged for January at 80°W, and to Defant and Taba (1957) averaged for all longitudes for 1 January 1956.

tropopause break on the time-averaged cross sections is at the latitude of maximum average speed in the westerlies, the two tropopause breaks on the Defant–Taba cross section correspond to two maxima of wind speed for the day studied, the southern one being the "subtropical" jet and the northern one the "polar-front" jet. The existence of two such maxima is the rule rather than the exception on daily maps; however, the polar-front jet, particularly, is found at varying latitudes from time to time, and does not show up in the long-period average. This, of course, is one illustration of the danger of drawing too general conclusions from time averages.

Defant and Taba's cross section represents conditions averaged around latitude circles. At one particular longitude on a particular day, the latitudes of the tropopause breaks may be quite different from the average shown in Fig. 2.4. Figure 2.5 shows the geographical locations of the breaklines on 1 January 1956. This distribution, furthermore, changes from time to time. The low

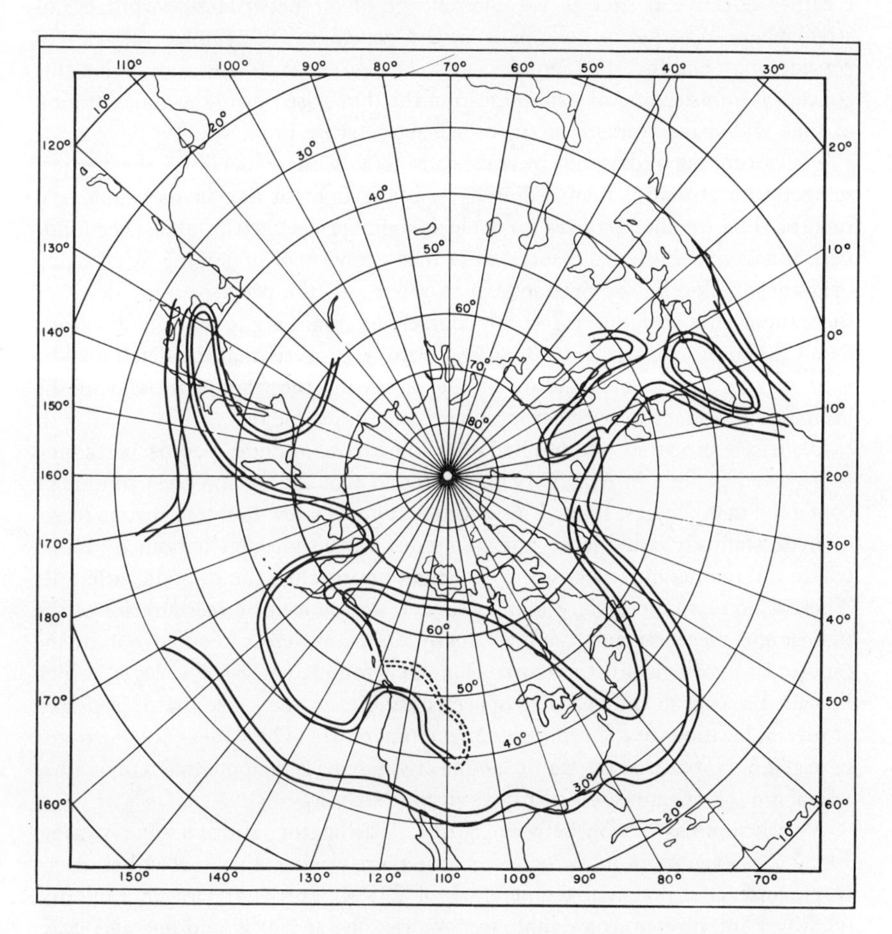

Fig. 2.5. Distribution of the tropopause breaklines on 1 January 1956. (After Defant and Taba, 1957.)

polar tropopause extends southward over regions of cyclonic circulation; the high tropical tropopause extends northward over regions of anticyclonic circulation. Teisserenc de Bort's early observation that the base of the "isothermal layer" was lower over cyclones than over anticyclones is amply justified by modern measurements.

Implicit in the above discussion is the concept that the tropopause at any given time may be thought of as consisting of a surface of (near) discontinuity in temperature lapse rate, continuous in space except at the one or two circumpolar tropopause breaks. It is a natural and frequent extension of this concept to think of the tropopause, except in the regions of the tropopause breaks, as a rather effective barrier to the interchange of air between troposphere and stratosphere. Whether or not this is a valid generalization requires very careful consideration and investigation. Recently, Danielsen (1959) has argued that this concept is a misleading oversimplification that has arisen from the consideration of time and space averages of smoothed temperature profiles.

The ascending radiosonde reveals considerably more detail of the vertical temperature structure than ordinarily sees the light of day in meteorological reports. The weather services, to avoid being overwhelmed by the sheer numbers of data in the original records, introduce a process of smoothing by choosing "significant" levels (see Section 2.1) to report. In the process, some detail of the temperature structure is lost. Of course, one might argue that the discarded detail is simply "noise." But Danielsen's study suggests that the detail may be vital to a proper understanding of the nature of the tropopause and of the exchange processes between troposphere and stratosphere.

Danielsen procured and studied the original radiosonde records pertaining to a period of time in March 1956. He found that the unsmoothed soundings contained many layers, limited in vertical extent to a few hundred meters or so, of hydrostatically stable lapse rates, bounded on the top and bottom by layers where the temperature decrease with height approached the dry-adiabatic rate. There is no reason to believe that these were introduced by inaccuracies of the radiosonde measurements, and yet many of them were smoothed out in the reported soundings. Furthermore, Danielsen found that a stable layer, which he called a "stable lamina," had space continuity in the sense that it appeared at several locations at the same potential temperature. These ideas are illustrated in Fig. 2.6. Notice that the stable lamina at whose base the potential temperature was about 320°K appeared clearly at all four stations.

A vertical cross section between Seattle, Washington, and Tucson, Arizona, Fig. 2.7, is even more revealing. At Seattle there was a rather low and clear-cut tropopause at a potential temperature of 291°K. Portland, Oregon exhibited what was interpreted as a double tropopause, one at 290°K and one at 315°K. This might correspond to the conventional tropopause break. The reported tropopauses at the other stations are difficult to imagine as forming a discontinuity surface separating troposphere from stratosphere. The stable lamina illustrated in Fig. 2.6, with base at 321°K, rose from a low altitude at Tucson to a maximum altitude near Oakland, where it merged with a 325°K lamina to form what was considered to be the tropopause at Medford. But it appears evident that there was no barrier whatsoever to flow along this isentropic lamina between the "troposphere" at Tucson and the "stratosphere" at Seattle. In fact close inspection of this figure will show that air flowing isentropically in the direction along this section could pass freely between the unambiguous stratosphere at Seattle and the troposphere of the more southerly stations. Danielsen traced several trajectories (not along this particular cross section) where such interchange clearly occurred.

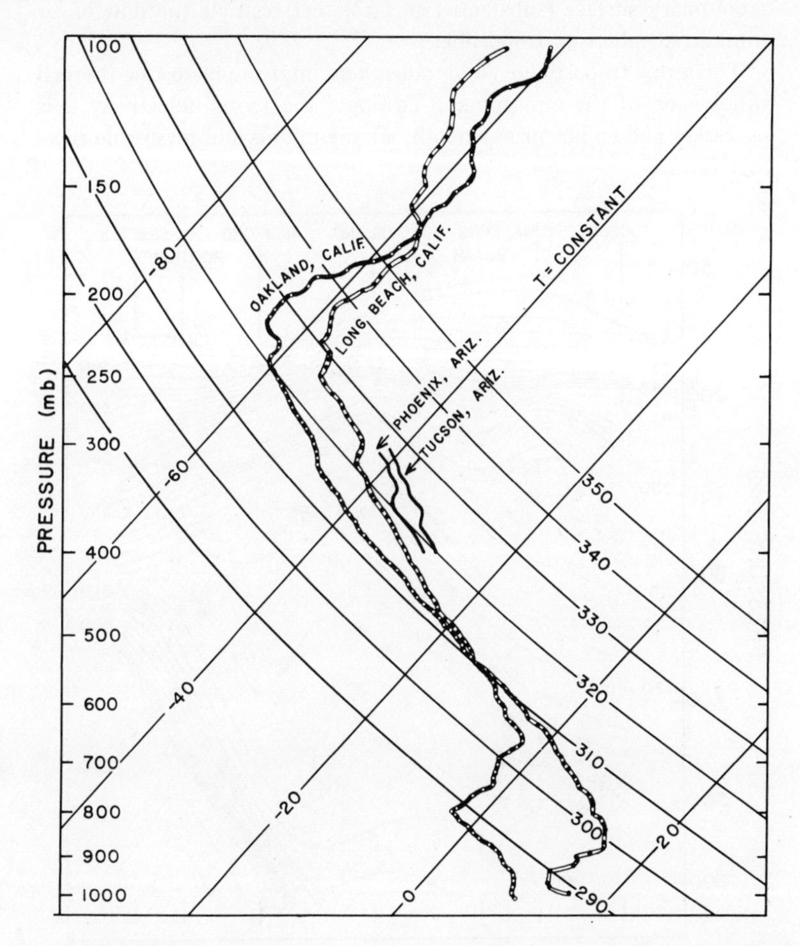

FIG. 2.6. Vertical soundings at the indicated locations on 27 March 1956. Isotherms are labeled in degrees C, isentropes (curves running from lower right to upper left) in degrees K. Each dot represents a temperature observation as computed by Danielsen from the original radiosonde records. Note the temperature inversion at each station based at a potential temperature of about 320°K. (After Danielsen, 1959.)

According to Danielsen:

> The thermal structure of the atmosphere is complicated by the existence of many stable laminae often bounded by layers of neutral

stability. The conventional concept of a tropopause requires that the base of one of these laminae be selected as having a special significance. Such a selection is necessarily arbitrary and in practice very subjective. The conventional concept is also misleading to the extent that it implies a boundary surface (substantial surface) that restricts the flow of air from stratosphere to troposphere.

The terms troposphere and stratosphere imply a macroscale thermal subdivision of the atmosphere. To apply the terms objectively it is necessary and sufficient to smooth out the micro- and mesoscale thermal features.

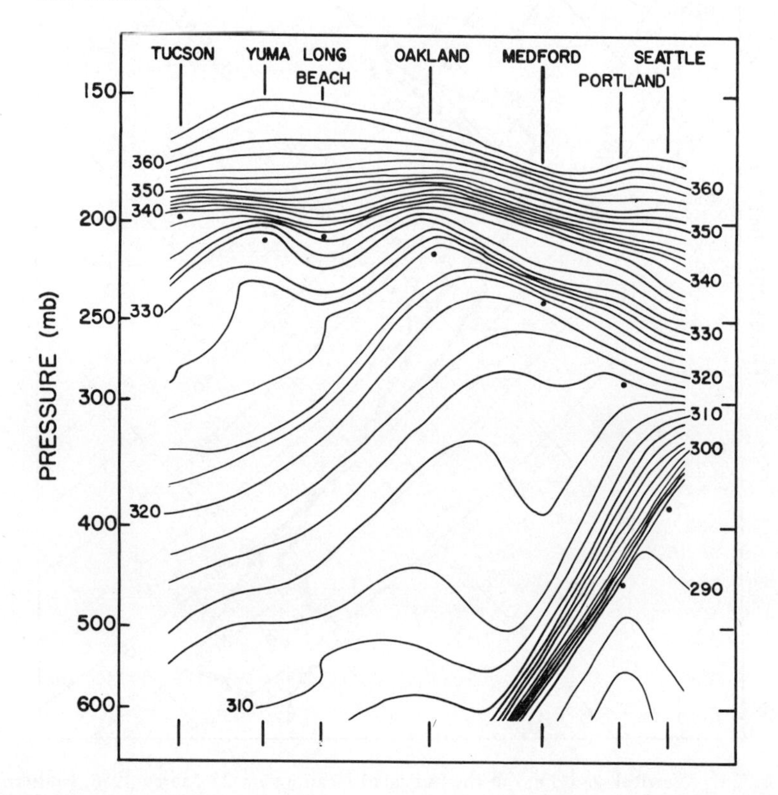

FIG. 2.7. Vertical cross section from Tucson, Arizona, to Seattle, Washington, on 27 March 1956. Lines of constant potential temperature are drawn and labeled in degrees K. The dots indicate the tropopauses coded by the radiosonde operators. (After Danielsen, 1959.)

Danielsen went on to suggest that, in the smoothed thermal structure, the tropopause is a barotropic surface and not a discontinuity surface that separates a region of small average stability and positive average baroclinity (the troposphere) from a region of large average stability and negative average baroclinity (the stratosphere).

This designation of the tropopause, in the mean, as a barotropic surface, appears to be too sweeping a generalization. Although it may be appropriate for middle latitudes, we shall see in Section 2.4 that even in the mean the west winds of high latitudes in winter continue to increase with height through what would appear to be a tropopause on the basis of average stability.

Danielsen's data, of course, cover only a limited period of time, at one season of the year, and in one latitude zone. However, the concept of stable laminae or layers within which air may flow unimpeded between stratosphere and troposphere is of the utmost importance. It has obvious application to the problem of interchange between stratosphere and troposphere. This is a problem of great theoretical interest and also of great practical concern in connection with the fallout of radioactive material deposited in the stratosphere by megaton nuclear explosions. Chapter 10 contains further discussion of the problem.

Whatever details further work may alter or verify, it is clear that the boundary between troposphere and stratosphere on the small scale is a very complex phenomenon. In the meantime, one may consider the time and space averages of Hess, Kochanski, and Defant and Taba to represent for various types of averaging some sort of large-scale division between troposphere and stratosphere.

2.2.2 Explanations of the Tropopause

There have been many attempts to "explain" the tropopause in terms of the physical processes responsible for it, and this question is still a controversial one. In view of the above discussion of the nature of the tropopause, it would appear wise to consider these explanations as they might apply to a progressively more complex description of the tropopause.

In the broadest sense, it is easy to understand why the atmosphere must have a temperature minimum of some description between the surface and the stratopause. Since atmospheric gases absorb directly very little of the visible radiation streaming in from the sun, most of the energy that is not reflected back to space goes to heat the earth's surface or to evaporate water from the surface. Thus the lower boundary of the atmosphere is a heat source. Emden (1913) long ago pointed out that if the lower atmosphere were in radiative equilibrium the resulting temperature lapse rate would be hydrostatically unstable and convection would set in. King (1952) has demonstrated the same thing more recently by more sophisticated methods. Therefore the temperature in the lower atmosphere must decrease with altitude. On the other hand, the upper stratosphere is a region, as we shall see in Chapter 4, where significant amounts of the sun's ultraviolet energy are absorbed by ozone and oxygen.

In this part of the atmosphere radiative considerations require that the temperature increase with height. Some temperature minimum in the intermediate region follows, then, simply from the radiative properties of the atmospheric gases.

These very general arguments do not at all explain the observed sharp inversion (or inversions) that often mark the transition from troposphere to stratosphere. Goody (1949, 1954) considers this to be the boundary between a lower layer where heat transfer is predominantly by convection and turbulence and an upper layer where the heat transfer is predominantly by radiation. (Appendix G discusses some early ideas on this subject.) The lowest possible level of this boundary is the level at which the radiation-equilibrium lapse rate in the upper layer is just hydrostatically stable. Goody showed that in this case a negative temperature discontinuity would have to exist at the boundary; even though the upper layer were just stable, there would be a shallow layer at the boundary where convection would occur. He showed that the tropopause must then occur at some particular higher level, with a *first-order* discontinuity in temperature, the stratosphere having the smaller lapse rate. Taking the troposphere as having everywhere a lapse rate of 6.5°C km^{-1}, using observed average values of surface temperature at different latitudes, and with some rather crude but reasonable approximations about the vertical distributions and radiating characteristics of atmospheric gases, he computed the height where this equilibrium would prevail. This calculation gives general agreement with the observed relationship between surface temperature and tropopause height, as indicated in Table 2.2.

TABLE 2.2

TROPOPAUSE HEIGHT AND TEMPERATURE AS A FUNCTION OF GROUND TEMPERATURE[a]

Ground temp. (°K)	Tropopause temp. (°K)		Tropopause height (km)	
	Calc.	Obs.	Calc.	Obs.
300	205	193	14.6	16.5
290	208	208	12.6	12.6
280	210	214	10.8	10.2
270	213	217	8.8	8.2
260	214	219	7.1	6.3

[a] After Goody (1949).

Staley (1957a,b) more recently has argued that Goody's radiation-convection hypothesis cannot explain the detailed characteristics of the tropopause. His principal argument is that the observed tropopause very often shows an extreme variation in lapse rate, sometimes approaching a temperature discontinuity. Infrared radiative processes act to smooth out such temperature minima, if there are no discontinuities in the distributions of the radiating gases. By sample calculations, using observed temperature and water-vapor profiles through the tropopause, Staley found substantial amounts of infrared heating at the tropopause level. Staley has proposed that the maintenance of a sharp curvature in the temperature profile at the tropopause is a balance between two processes: warming by radiation and cooling by divergence of vertical enthalpy flux. The latter effect requires that downward transfer of enthalpy from the stratosphere to the tropopause level be less than downward transfer of enthalpy from the tropopause level to the troposphere. The larger vertical gradient of potential temperature above the tropopause than below would ordinarily lead one to expect the opposite. However, the flux due to turbulent diffusion is known to depend not only on the vertical gradient of the quantity involved but also on hydrostatic stability. The greater hydrostatic stability above the tropopause than below may actually lead to the condition hypothesized by Staley.

One can observe in the increasing attention to details of these investigations a development somewhat parallel to the increasingly complex picture of the tropopause that has emerged from observations. Thus, Goody's model might be considered to be essentially correct in explaining the very broad features of tropopause distribution, but by its very nature as an equilibrium calculation, to be inapplicable to the multiple tropopauses and stable laminae that emerge in the mesoscale and microscale. Staley's hypothesis that vertical enthalpy flux will increase positive curvature once the latter is established by some other process is an interesting one. Although unproved by any observation, it has the merits of being reasonable and of attempting to explain the atmosphere's perverse and mystifying habit of establishing and maintaining near-discontinuities in its properties on a small scale while at the same time mixing itself up very efficiently on a large scale.

2.3 Composition of the Lower Stratosphere—Water Vapor and Dust

There is no reason to expect and no evidence from observations that the relative abundances of the permanent constituents of air in the lower stratosphere differ from those in the troposphere. There are, however, two gases that deserve special consideration, because their concentrations are variable in both time and space and also because they are extremely important to the radiative properties of air. These are ozone and water vapor.

The subject of ozone is so important and at the same time so complex that discussion of it is better postponed until Chapter 5. This discussion is more meaningful in the light of material introduced in Sections 2.4 and 2.5 and in

Chapters 3 and 4. Here, we simply point out that ozone is present in the lower stratosphere in amounts significant for many purposes; indeed ozone has its maximum density in the lower stratosphere.

Although relatively little condensation is observed in the lower stratosphere, the concentration of water vapor is nonetheless of great interest. However, measurement of this concentration is exceedingly difficult. We shall discuss the present unsatisfactory state of these observations and their interpretation in Subsection 2.3.1.

Another matter of considerable interest is the presence and distribution in the lower stratosphere of submicron dust particles—particularly those in the 0.1—1.0-μ range. These will be discussed in Subsection 2.3.2.

2.3.1 Water Vapor in the Lower Stratosphere

No measurements of water-vapor content are routinely available for the stratosphere. Humidity elements that are reasonably satisfactory for the lower troposphere are insensitive and slow to respond in the upper troposphere and lower stratosphere. The information on water-vapor content of the stratosphere is derived from a few series of observations with instruments especially designed for the purpose, which are, however, uneconomical or impractical for routine use.

This is exceedingly unfortunate. Water vapor plays an important role in the transfer of infrared radiation. Attempts to calculate the radiative balance of the stratosphere must be based on tenuous assumptions about the amount and distribution of water vapor. Usually, it is assumed to be absent altogether. Although the results under such an assumption appear to be not unreasonable, thus inferentially "justifying" the assumption, many more measurements of water vapor are needed to put the results on firmer ground.

By far the largest number of measurements have been made by British scientists with the Dobson-Brewer frost-point hygrometer. Unfortunately, this is a manually operated instrument and its use has been limited to the lowest part of the stratosphere by the ceilings of the aircraft in which it has been flown. Variants of the instrument, which is an ingenious development of the common dew-point hygrometer for use at very low temperature, have been described fully by Brewer, Cwilong, and Dobson (1948). The basic observation is the temperature of an aluminum thimble that can just maintain a deposit of hoar-frost in equilibrium with a narrow jet of air blown across its flat, blackened top.

Measurements to about 12 km made between 1945 and 1950 have been summarized by Bannon, Frith, and Shellard (1952). In 1954 and 1955, measurements to a maximum altitude of 15 or 16 km on jet aircraft were made on 81 different occasions. These have been summarized by Murgatroyd, Goldsmith, and Hollings (1955) and by Helliwell, Mackenzie, and Kerley (1957). Remarkably consistent results were obtained in these many measurements, so it is not out of order to give the results as an average vertical distribution of temperature and frost point for the 1954 and 1955 flights in Fig. 2.8. In addition to these measurements over England, the British workers have also used the same technique to gather data at other locations and latitudes (see summary by Gutnick, 1961).

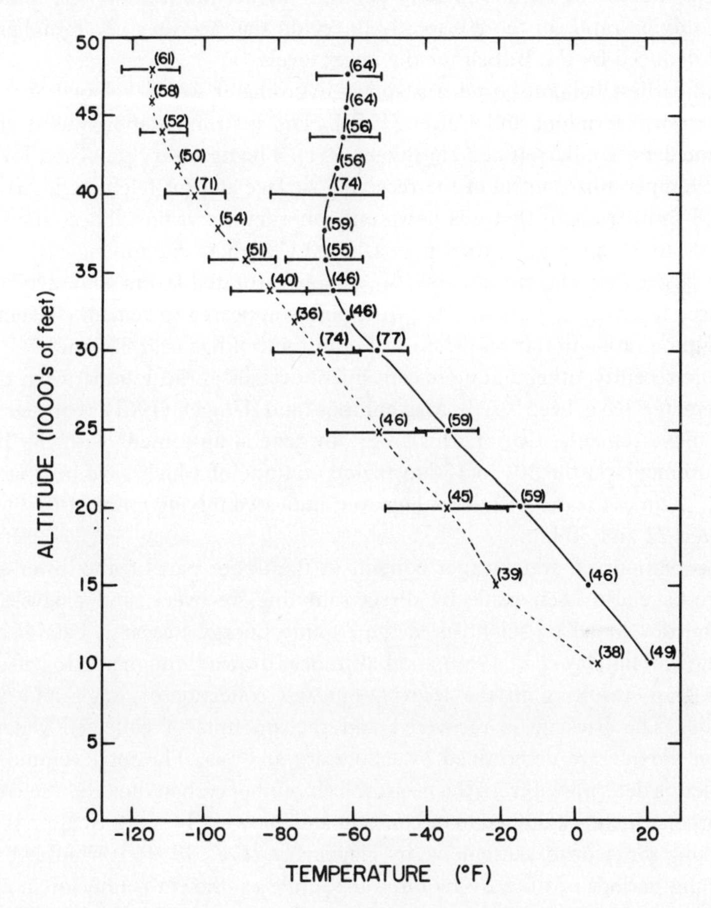

FIG. 2.8. Mean vertical distribution of temperature and frost point from 1954 and 1955 British measurements. Numbers of observations are shown in parentheses, standard deviations by horizontal lines. (After Helliwell *et al.*, 1957.)

The most surprising and consistent result of these measurements has been the remarkable dryness of the lowest few kilometers of the stratosphere. The frost point continues to decrease with height in the stratosphere, in many individual cases at a greater rate than in the troposphere, and eventually near the ceiling of the observations at about 15 km achieves a more or less steady value around $-80°$ to $-85°C$. This corresponds to a relative humidity on the order of 1 or 2 per cent and a mixing ratio of about 2×10^{-6}.

However, observations at these altitudes with balloon-borne frost-point hygrometers and by other methods have not always verified this extreme dryness. Furthermore, observations reaching higher altitudes almost uniformly show mixing ratios in the 20- to 30-km region that are very much higher than those deduced by the British for the lower levels.

The earliest balloon-borne frost-point hygrometer was developed and flown by Barrett, Herndon, and Carter (1950). The instrumentation was described in some detail by Barrett and Herndon (1951). The necessary provision for automatic temperature control of the receiving surface and for telemetering resulted in a 46-lb instrument that was flown only on a few occasions. These showed, in the 10- to 15-km region, frost points 10–20°C higher and mixing ratios 10–20 times larger than the aircraft results. Narrow saturated layers were detected on all three flights. Furthermore, the frost points appeared to remain constant and the mixing ratios to increase from 15 km to the ceilings near 30 km.

More recently, other and more convenient designs of the automatic frost-point hygrometer have been used. Mastenbrook and Dinger (1961) reported three such measurements, two of which were in general agreement with the British measurements in the 10- to 15-km region and one of which gave mixing ratios about 20 times greater. All three, however, indicated mixing ratios of 10^{-5} to 10^{-4} between 22 and 30 km.

Observations of water-vapor content in the upper part of the lower stratosphere have also been made by direct sampling, recovery, and analysis. In a method developed by scientists at the Atomic Energy Research Establishment in England (Barclay *et al.*, 1960), air at altitude is drawn through a nitrogen-cooled vapor trap, resulting in the freezing-out and collection of water and carbon dioxide. The package is recovered and the amounts of collected water and carbon dioxide are determined by laboratory analysis. The total volume of air sampled is determined from the measured amount of carbon dioxide, the concentration of which is assumed to be constant and known. (This information is based on a long series of measurements by Hagemann *et al.*, 1959). It should be noted that long periods of time are required to sample a sufficient volume of air at high altitudes, so that this instrument has been flown on constant-level balloons.

Seven determinations have been reported by Brown *et al.* (1961), all applying to levels between about 24 and 30 km. These gave mixing ratios averaging about 4×10^{-5}.

Steinberg and Rohrbough (1962) have described another sampling experiment, in which water vapor and carbon dioxide are collected by adsorption on synthetic zeolites. The volume of air sampled is measured by a flow meter. In the one experiment whose results have been published, the mixing ratio at about 22 km was found to be about 5×10^{-5}. The carbon dioxide determination was in satisfactory agreement with other measurements.

A third method of water-vapor study, other than frost-point determination or direct sampling, requires observations from high altitudes of solar radiation in spectral regions of strong water-vapor absorption. These spectra give only the integrated amount of water vapor between the instrument and the top of the atmosphere, and then only after very careful interpretation. The shape of the absorption lines in question is pressure dependent (see Chapter 7). According to Houghton and Seeley (1960, 1961) measurements on the 2.7-μ band from aircraft in England are consistent with the dry stratosphere, but do not definitively rule out the possibility of high mixing ratios at high levels.

Murcray *et al.* (1962) have given the results of measurements on the 6.3-μ band by balloon-borne spectrometer. On two occasions their results could have been produced by (vertically constant) mixing ratios of about 10^{-4} above about 28 km. On one of these occasions, a series of spectra obtained on the ascending balloon were interpreted to give a vertical distribution; this indicated mixing ratios of about 10^{-5} in the 10- to 15-km region.

Some of the results described above are illustrated in Fig. 2.9. There is at present no satisfactory explanation for the situation near 10 to 15 km, where several observers have on occasion found humidities significantly larger than any ever detected by the extensive and carefully executed British aircraft measurements. It has been suggested at various times that these higher humidities might result from contamination of the lower stratosphere by the balloon bearing the instrument, and that therefore the higher humidities are not representative. With regard to the upper levels of the lower stratosphere, there is apparently no observational evidence against the acceptance of mixing ratios of 10^{-5} to 10^{-4}. In addition, the occasional existence of clouds near this level is well known. The iridescent "mother-of-pearl" clouds occur at altitudes of 20 to 30 km in particular geographical regions. They have been extensively observed in Norway and also in Alaska. Presumably, they are clouds of ice crystals formed in the upward-moving air of an orographically induced standing wave.

The concept of a very dry region just above the tropopause and a relatively moist layer at higher levels, which seems at present to be emerging, gives rise to a very serious problem of interpretation. Unless there is a source of water vapor above 30 km, one is faced with the problem of explaining how water could be transported upward from the earth's surface to 25 km, while the lower region remains dry.

2.3.2 Dust in the Lower Stratosphere

Although dust (small particulate matter) in the troposphere has received considerable attention because of its role in the condensation process, there has

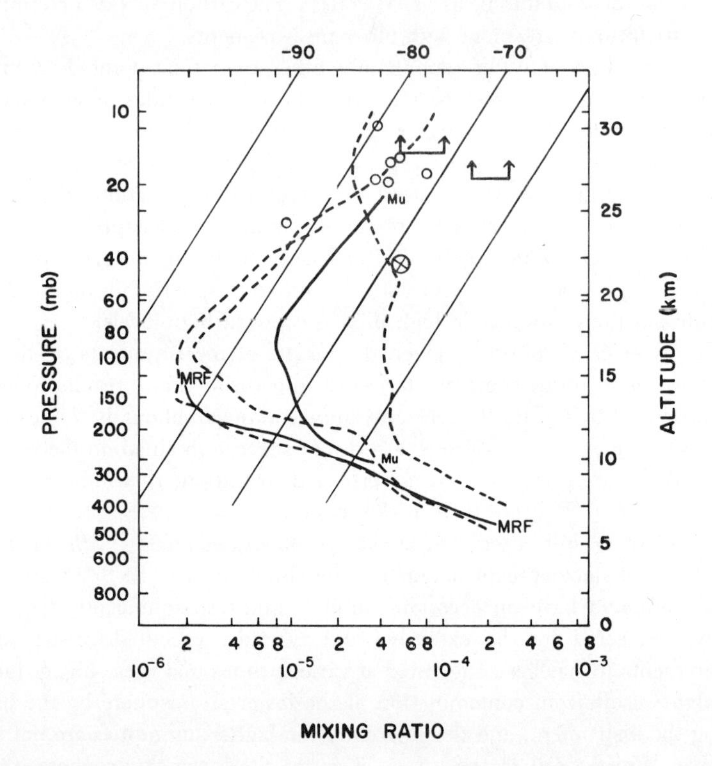

FIG. 2.9. Measurements of humidity in the lower stratosphere. Full curve labeled MRF is from Helliwell *et al.* (1957). Full curve labeled Mu is from Murcray *et al.* (1962). The three dashed curves are from Mastenbrook and Dinger (1961). The circles show measurements reported by Brown *et al.* (1961), and the circled cross shows a measurement of Steinberg and Rohrbough (1962). Horizontal lines with upward-pointing arrows indicate mixing ratios (assumed constant throughout a vertical column to the top of the atmosphere) consistent with spectrometer measurements of Murcray *et al.* (1962). Diagonal lines give the frost point (in degrees C) as a function of mixing ratio and pressure.

until recently been no quantitative information about dust in the upper atmosphere. A pioneering observational prgram by Junge and his collaborators has in the past few years made available a good deal of basic information about the lower stratosphere. Measurements have been obtained from both balloon and aircraft. The methods and results are most conveniently discussed in terms of sizes of the particles involved.

In one method of observation, particles were counted with an Aitken nuclei counter, which operates on the principle of an expansion cloud chamber with photography of the resulting cloud of water droplets (Junge, Chagnon, and Manson, 1961; Junge, 1961). Although this technique is not selective with regard to particle size, and gives no specific information on particle size, the results are assumed to apply to particles of radius less than 0.1 μ, because larger particles are present in insufficient number (as determined by independent measurements described below) to affect the results.

Figure 2.10 gives the average results of measurements made on seven different

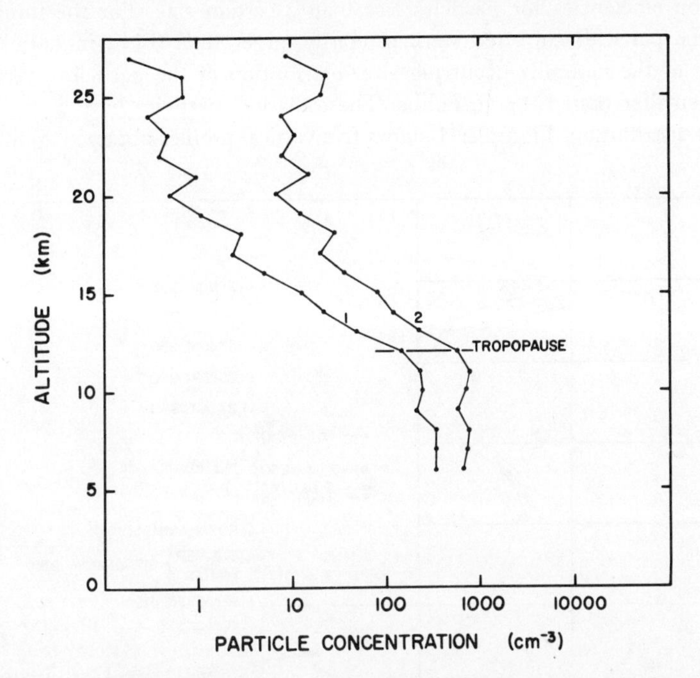

FIG. 2.10. Vertical distribution of small dust particles (radius less than 0.1 μ), according to Junge (1961). The curves give the average results of seven measurements, curve 1 in terms of particles per cubic centimeter of ambient air and curve 2 in terms of particles per cubic centimeter of air at standard temperature and pressure. Averages were obtained with the tropopause as datum level for each flight. (After Junge, 1961.)

occasions. The averaging was carried out with respect to the tropopause as a datum level. This average profile and also individual ascents show approximately constant concentrations a few kilometers below the tropopause and a rapid decrease of particle concentration with height above the tropopause. It is known from previous tropospheric studies that the concentration of these Aitken nuclei decreases rapidly with height in the lowest few kilometers of the atmosphere. It seems evident that the source of these small particles is at the earth's surface and those found in the lower stratosphere arrive from the troposphere.

Larger particles, with radius 0.1 to 1 μ, have been detected and measured by means of inertial impactors (Junge, Chagnon, and Manson, 1961; Chagnon and Junge, 1961). In these instruments, air is drawn through a narrow jet toward a specially coated gummed slide. As the air turns sharply to move toward an outlet, particles with sufficient inertia continue on and are imbedded in the gummed surface. Dimensions of the jet can be arranged to give vanishingly low collection efficiencies for particles less than a certain size. For the impactors used, the particles collected were primarily larger than 0.1 μ in radius, and because of the naturally occurring size distribution of the particles, were primarily smaller than 1.0 μ in radius. These "large" particles have a surprising vertical distribution. Figure 2.11 shows five vertical profiles obtained at different

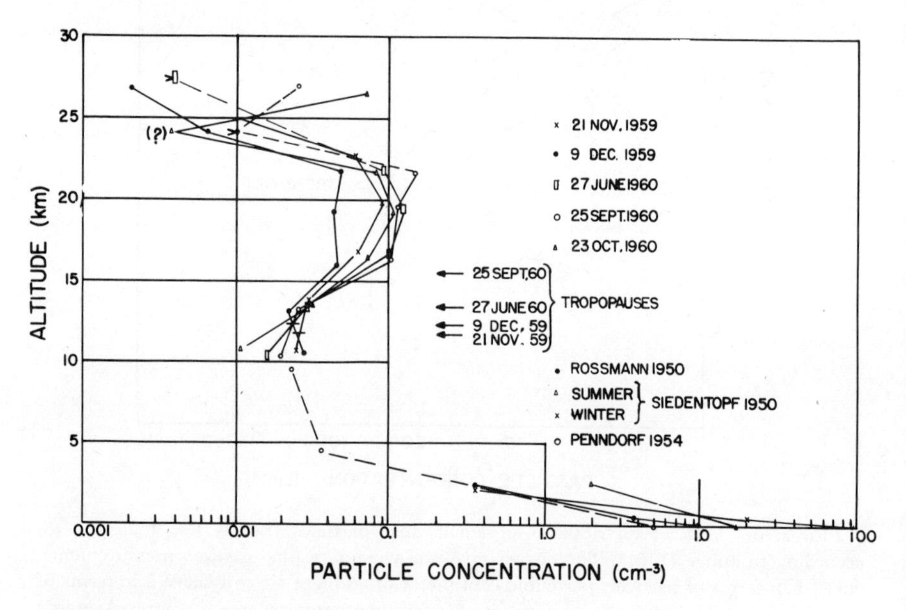

FIG. 2.11. Vertical distribution of "large" particles (radius greater than 0.1 μ) in the lower stratosphere, according to Chagnon and Junge (1961). Five profiles, obtained on the indicated dates, are shown. Data below 10 km show some earlier measurements of other investigators. (After Chagnon and Junge, 1961.)

times of the year; all show a maximum concentration near 20 km. Aircraft flights at 20 km from 63°S to 72°N (Junge and Manson, 1961) during March-November 1960 verified the presence near that altitude of an unexpectedly large number of particles at different times and altitudes.

Laboratory analyses of the particles show that they are composed mainly of sulfates. It appears from the observed vertical distributions that neither do they originate in the troposphere, nor are they of extraterrestrial origin. Junge and Manson (1961) suggested that they are formed in the lower stratosphere by oxidation of gaseous H_2S or SO_2. These gases are known to be present in the troposphere and are presumably transported into the lower stratosphere by mixing processes.

Volz and Goody (1962) have deduced from photometric observations at twilight the existence of a maximum of turbidity near 20 km, consistent with the observations discussed above. As a matter of fact, the first suggestion of a dust layer near 20 km was made by Gruner (1942) to explain optical effects of twilight (the "purple light").

2.4 Climatology of the Lower Stratosphere

Our information about temperature and wind distribution in the lower stratosphere is very much better than our information about water vapor. In spite of all the difficulties mentioned in Section 2.1, enough observations have accumulated so that certain large-scale features of the temperature and wind distributions are now clear. One may expect that the picture presented in this section will be changed, but not radically, as more observations become available.

Unfortunately, the observing techniques introduce a decided bias in two situations:

(a) When the stratosphere is particularly cold, the sounding balloons are more apt to burst at a low elevation. This situation has been improved in recent years by improvement of balloon materials.

(b) When the vertically averaged vector wind is large in magnitude, the balloon is blown to a great distance and the low elevation angle requires termination of the sounding at a lower level.

Analysis of the scarce data is greatly aided by the thermal-wind equation, relating vertical wind shear to horizontal temperature gradient, under the assumptions of geostrophic and hydrostatic equilibrium. In the usual tangent-plane, Cartesian system*

$$\frac{\partial u}{\partial z} = -\left(\frac{g}{fT}\right)\left(\frac{\partial T}{\partial y}\right) \tag{2.5a}$$

$$\frac{\partial v}{\partial z} = \left(\frac{g}{fT}\right)\left(\frac{\partial T}{\partial x}\right) \tag{2.5b}$$

where u is speed in the x direction (usually taken positive toward the east), v is speed in the y direction (usually taken positive toward the north), and f is the Coriolis parameter $2\omega \sin \phi$ where ω is the angular speed of rotation of

* Strictly speaking, the gradient of temperature in these equations should be taken along a constant-pressure surface. If it is measured on a surface of constant altitude, a term (usually small) involving the vertical gradient of temperature must be added to the right side of each of the equations (2.5a) and (2.5b).

the earth and ϕ is the latitude. For example, according to Eq. (2.5a) the west component of wind* increases with height when temperature decreases toward the north and decreases with height when temperature increases toward the north.

Meteorological conditions in the lower stratosphere vary with altitude, latitude, season, and longitude. To facilitate discussion of these many variables, Section 2.4 is divided into several parts. Subsection 2.4.1, as an introduction, is concerned with the gross features of temperature variation in the Northern Hemisphere and the discussion centers around vertical-latitudinal cross sections for summer and winter. Subsection 2.4.2 contains mean maps for one level (50 mb), which illustrate longitudinal features and more details of the seasonal variation. Subsection 2.4.3 goes into the question of the representativeness of monthly normals at stratospheric levels; particularly at high latitudes, conditions during a given winter month may be quite different from those portrayed on a "normal" map for that calendar month. These subsections are concerned with the Northern Hemisphere lower stratosphere, north of about 20°N. In Subsection 2.4.4, there is a discussion of the recently discovered 26-month oscillation of zonal wind direction in the tropical stratosphere. Finally, in Subsection 2.4.5, the scanty observations from the Southern Hemisphere are briefly considered.

2.4.1 Large-Scale Features of the Temperature Variation in the Lower Stratosphere

A simple and graphic way to portray the main meteorological features of the lower stratosphere is by means of a north-south vertical cross section, similar to that in Fig. 2.4. Such a cross section, of course, masks the longitudinal variations as well as the time variations, both of which must be discussed separately.

Several such temperature cross sections for longitudes 80°W or 90°W have been constructed. Unfortunately, none contains all the data from recent years that would be desirable. Among them, only those of Kochanski (1955) and Wege *et al.* (1958) were constructed with particular emphasis on the stratosphere. Here we show cross sections depicting average conditions during July for the years 1955–1958 and during January for the years 1956–1959. For pressure levels at and above 100 mb, these are based on the analyses of Muench and Borden (1962) for those years; some of the corresponding 50-mb maps for the

* In accordance with the universal meteorological convention, wind direction in this book is described by the direction from which the wind blows. Thus, a "west" wind blows from the west toward the east and may be spoken of as part of a "westerly" current. Meteorologists who may read of wind observations at high levels should be aware that the opposite convention is used by some aeronomers.

Northern Hemisphere are included in Subsection 2.4.2. For pressure levels below 100 mb, the necessary data were derived from "Monthly Climatic Data of the World" (U. S. Weather Bureau, 1955–1959).

Figure 2.12 contains the north-south cross section for July. There are three altitude intervals with different temperature characteristics.

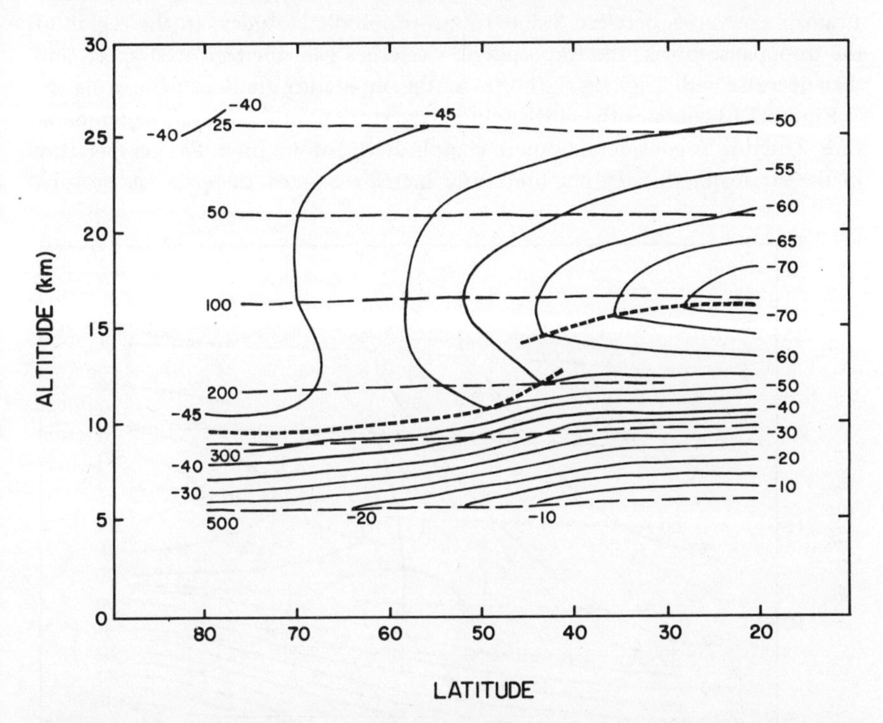

FIG. 2.12. North-south cross section for 90°W for July, average values for years 1955–1958. Full curves show the temperature (labeled in degrees C), dashed curves show certain constant-pressure surfaces (labeled in millibars), and thicker dashed curves define the approximate tropopause location.

(1) Below 9 km, all latitudes are in the troposphere. The temperature decreases with altitude and decreases poleward. The north-south temperature gradient is mainly concentrated in middle latitudes; in this region the normal west winds of the midlatitude troposphere increase in speed with increasing height.

(2) Between 16 and 30 km all latitudes are in the stratosphere. The temperature increases with height very markedly in the southern latitudes and increases very slightly or remains essentially constant in the northern latitudes. In contrast to the troposphere, highest temperatures are found near the pole; the poleward temperature gradient again has the greatest magnitude in middle

latitudes. The west winds of the troposphere decrease in speed with increasing altitude and above about 20 km are east at all latitudes.

(3) The 9- to 16-km layer is a transition zone between the two regions described above. It is part of the stratosphere at northerly latitudes, part of the troposphere at southerly latitudes. The transition between troposphere and stratosphere varies between 9 and 16 km in middle latitudes. In the region of the tropopause break, the tropospheric westerlies gain their greatest speed and then decrease with height as the north-south temperature gradient changes sign.

Figure 2.13 contains the north-south cross section for January. The temperature structure is considerably more complicated than in July. The temperature in the stratosphere does not uniformly increase toward the pole, as in July.

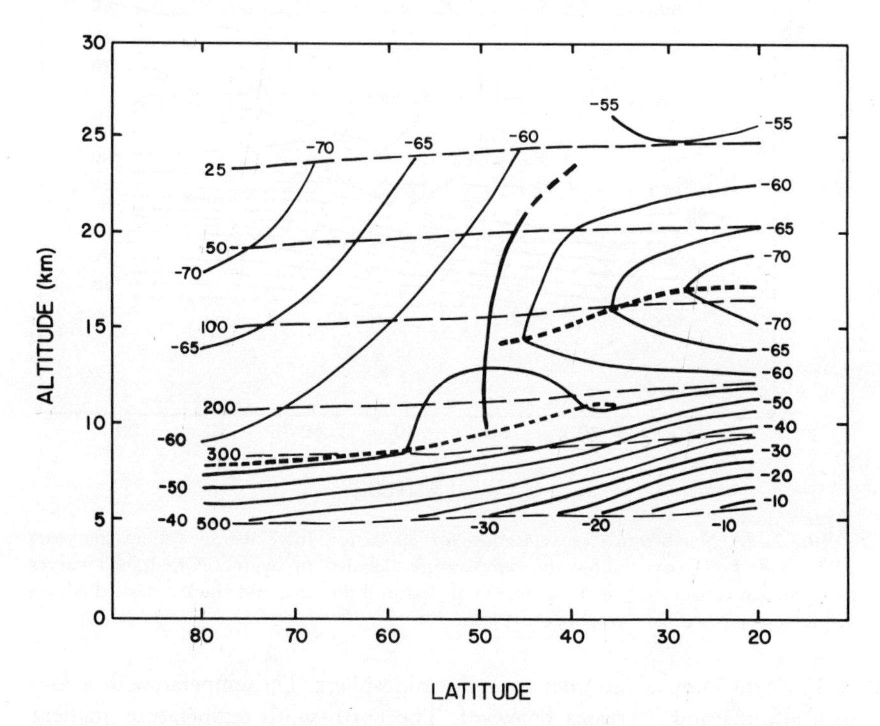

FIG. 2.13. Same as Fig. 2.12, except for January 1956–1959. The curved line extending upward from the tropopause near 50°N defines the boundary of the polar regime.

In northerly latitudes, the temperature decreases toward the pole; we shall follow the suggestion of Panofsky (1961) and refer to this situation as the "polar regime." The boundary of the polar regime, indicated on Fig. 2.13 by the heavy line, extends nearly vertically upward from the tropopause at about 50°N, and slopes toward the south and becomes indistinct at higher levels.

Except for the polar regime, the temperature distribution in the January cross section is qualitatively similar to that in the July cross section, although the tropopauses are at somewhat different elevations and the north-south gradients are somewhat greater in magnitude. In the troposphere the temperature decreases upward and poleward and west winds increase upward. In the lower stratosphere (outside the polar regime) the temperature increases upward and poleward and west winds decrease upward. The transition zone near the tropopause break shows a particularly striking reversal of poleward temperature gradient and the west-wind maximum of the upper troposphere, the "jet stream," damps out rapidly with height.

The polar regime has many of the characteristics of the midlatitude troposphere. Temperature decreases poleward from a maximum at the boundary

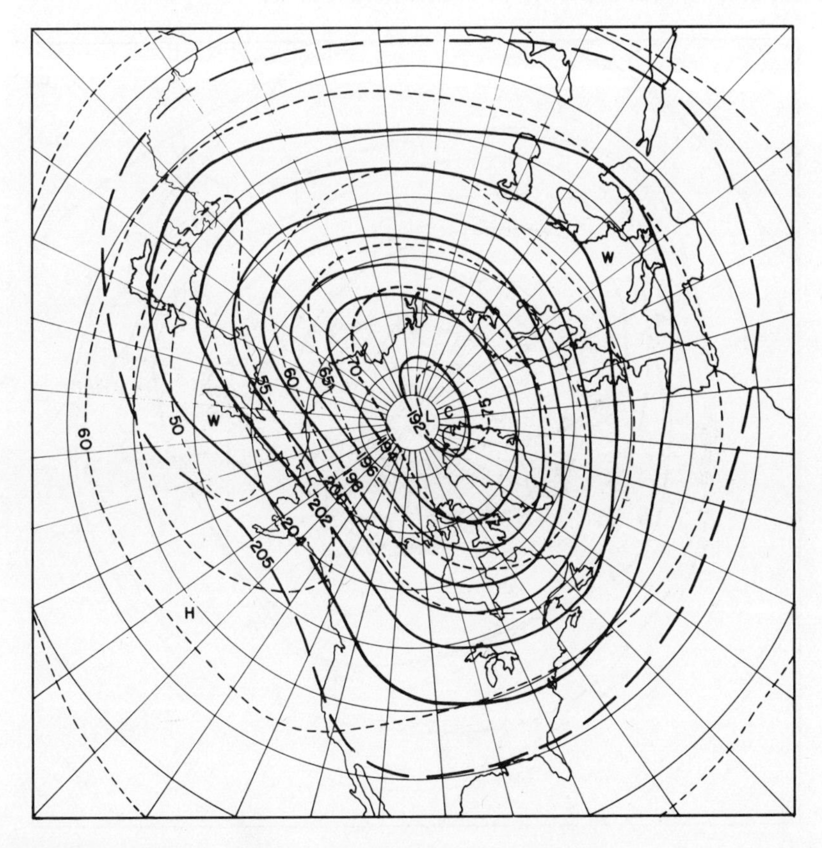

FIG. 2.14. Average distribution of temperature (dashed curves) and geopotential (full curves) at 50 mb for January 1956–1959. Contours are labeled in hundreds of meters, isotherms in degrees C with the minus sign omitted. (After Muench and Borden, 1962.)

of the polar regime. West winds increase with height, particularly between latitudes 60° and 70°, and often reach very high speeds above 20 km. Temperature decreases with height, although much less rapidly than in the troposphere.

2.4.2 Temperature and Wind at 50 mb in the Midseasonal Months

Figures 2.14, 2.15, 2.16, and 2.17 show 50-mb contours and isotherms over the Northern Hemisphere for, respectively, January, April, July, and October. These were prepared by Muench and Borden (1962) and are based on data for only four years (July 1955–June 1959). In view of the observed year-to-year fluctuations (except perhaps in July) this is too short a period for the establishment of quantitatively reliable normal maps; nevertheless the maps illustrate many features that appear to be qualitatively reliable.

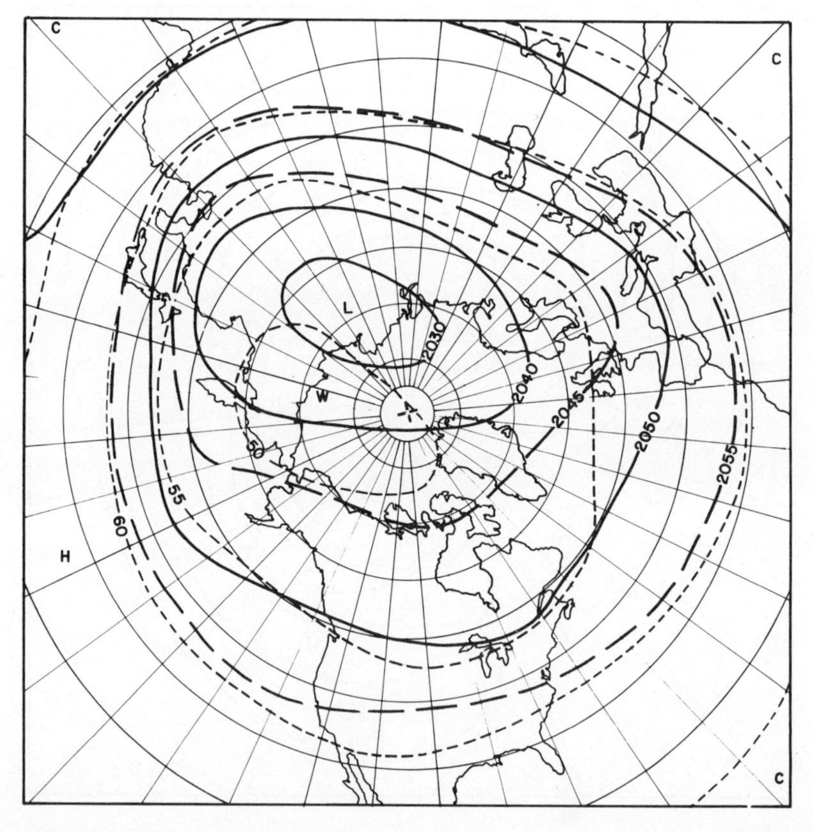

Fig. 2.15. Average distribution of temperature (dashed curves) and geopotential (full curves) at 50 mb for April 1956–1959. Contours are labeled in tens of meters, isotherms in degrees C with the minus sign omitted. (After Muench and Borden, 1962.)

In January, the circulation is marked by a cold low centered near the pole. There are west winds at all latitudes; within the polar regime, bounded by the north-south temperature maximum at latitude 40°–55° (depending on longitude), these west winds increase with height in accord with the poleward decrease of temperature. The longitudinal variations are of interest: note especially the high-latitude ridge and relative warmth in the Alaskan area. This appears to be a rather stable feature of the lower stratosphere in winter.

The July mean map (Fig. 2.16) shows strikingly different conditions. At this time of year at 50 mb there is a warm high centered near the pole with east winds at all latitudes. The poleward increase of temperature implies an easterly thermal wind and, indeed, the east winds at 25 mb (for example) have considerably greater speed than at 50 mb. The symmetry of the isotherm and contour pattern about the pole is rather striking and this is even more evident at 25 mb.

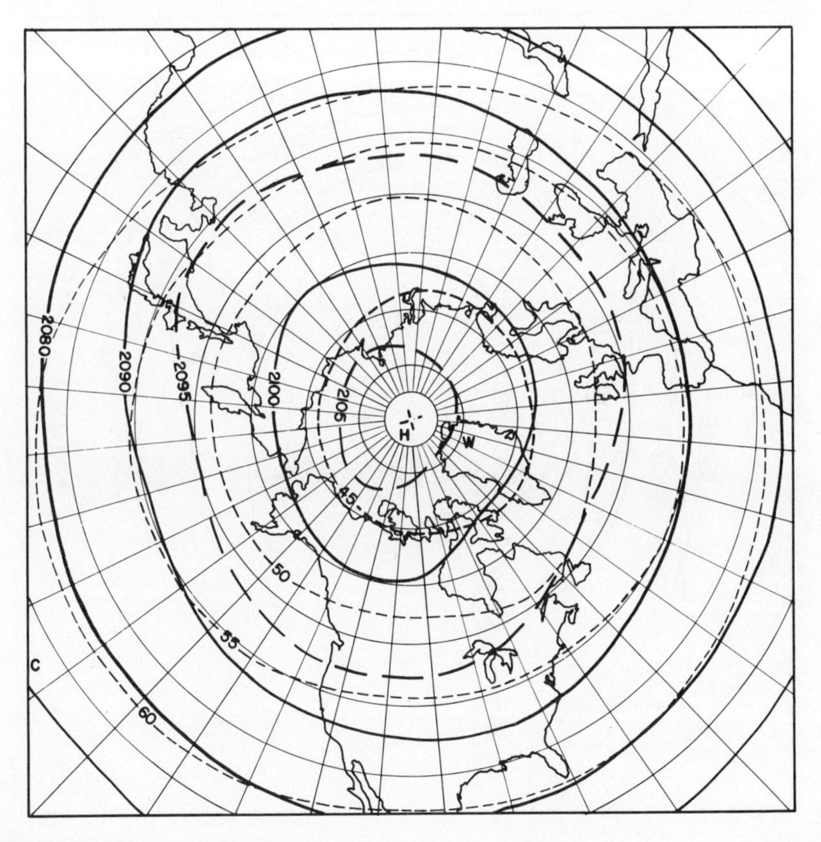

FIG. 2.16. Average distribution of temperature (dashed curves) and geopotential (full curves) at 50 mb for July 1955–1958. Contours are labeled in tens of meters, isotherms in degrees C with the minus sign omitted. (After Muench and Borden, 1962.)

April is a transitional month with a generally weak circulation, the details of which are variable from year to year. The four-year mean shown in Fig. 2.15 does not bear very much resemblance to the maps for the individual years on which it is based. According to Belmont (1962) the east winds of summer are not well established until some time in May, and April is characterized by cellular patterns and marked but light meridional flow.

On the other hand, in October, the reversal from summer easterlies to winter westerlies has already occurred. The four-year mean shown in Fig. 2.17 bears considerable resemblance to the means for the individual years on which it is based. Qualitatively, it is rather similar to the January pattern but with a weaker circulation. Belmont (1962) states that the fall reversal is quite rapid and, in the years studied, occurs during September.

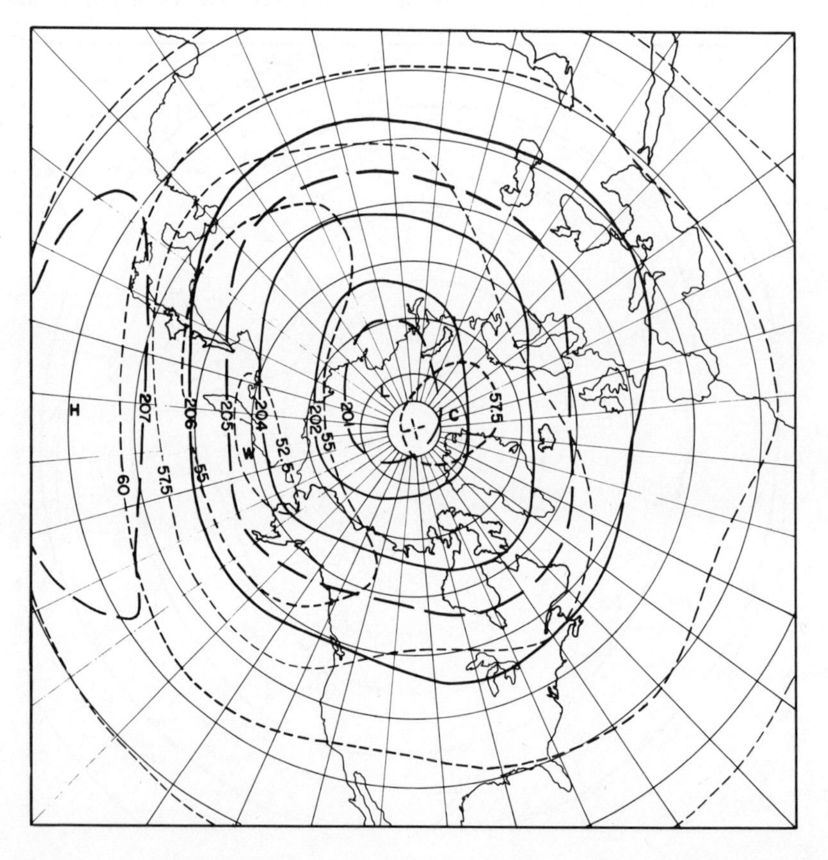

Fig. 2.17. Average distribution of temperature (dashed curves) and geopotential (full curves) at 50 mb for October 1955–1958. Contours are labeled in hundreds of meters, isotherms in degrees C with the minus sign omitted. (After Muench and Borden, 1962.)

2.4.3 Year-to-Year Variations

Even with the relatively few years of stratospheric data that are available, it is already apparent that there is much more variation from year to year during the winter and spring months than during the summer and autumn months.

The year-to-year variability of monthly mean maps for a given calendar month in winter and spring is partly associated with the phenomenon known variously as "stratospheric warming," "sudden warming," "explosive warming," or "impulsive warming." During most winters, at some time and at some longitudes, the polar regime disappears entirely and the high-latitude stratosphere reverts for a time to a condition more like that typical of July. The change from extreme polar-regime conditions to summer conditions begins with a rapid temperature increase near the boundary of the polar regime, usually at the highest levels observed, and the region of high temperature spreads within a week or two to lower levels and a large part of the polar region.

This phenomenon varies so much in detail from one year to the next that it is best to avoid too much generalization at this point and defer a fuller account to Section 2.5. In the context of the present essentially climatological discussion, it is sufficient to point out that the associated wind and temperature changes are

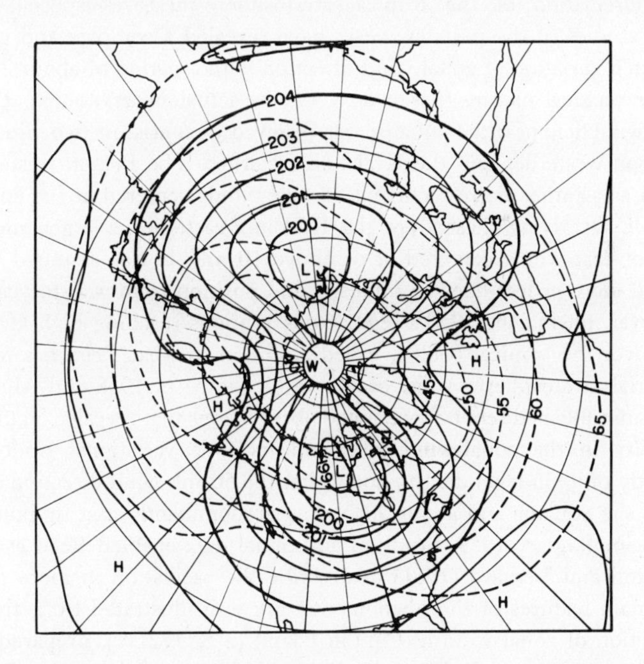

Fig. 2.18. Average distribution of temperature (dashed curves) and geopotential (full curves) at 50 mb for February 1957. Contours are labeled in hundreds of meters, isotherms in degrees C with the minus sign omitted. (After Muench and Borden, 1962.)

sometimes so widespread and persistent as to alter drastically the monthly mean patterns from the winter picture presented up to now in this section.

During the winters analyzed by Muench, February 1957, February 1958, and March 1959 were months when this occurred in pronounced fashion. As an example, the 50-mb map for February 1957 is shown in Fig. 2.18. Note the summerlike temperature distribution with warmest temperature near the pole, and also the weak circulation with east winds at some latitudes. Although appropriate monthly mean maps are not available, January 1953 and January 1955 appear on the basis of some daily charts and other data to have been months with a much less pronounced polar regime than indicated in Figs. 2.13 and 2.14.

On the other hand, it seems, on the basis of an admittedly small sample, that the months June through November have considerably less year-to-year variability of thermal and wind patterns. For those months climatological estimates appear to have considerable reliability.

2.4.4 Winds in the Tropical Stratosphere

The circulation of the tropical stratosphere deserves special attention, because studies of the past few years have revealed a complex and completely unexpected variation of zonal wind direction with a period of about 26 months.

In the classical picture (based on very few actual observations), the stratospheric wind field near the equator was believed to consist of two wind systems, the Berson westerlies and the Krakatoa easterlies. The Berson westerlies were pictured as a rather narrow current of west winds centered at the equator and at 50 mb. Overlying these were the Krakatoa easterlies, a larger and stronger system of east winds centered at or above 10 mb. It was assumed that these remained more or less fixed in intensity and position from year to year.

However, during the 1950's regular observations reaching high levels, particularly over the tropical Pacific, called attention to what Reed has termed the "equatorial stratospheric wind oscillation." Viezee in 1958 and McCreary in 1959, in unpublished reports, presented observations that revealed an unexpected complexity in the zonal-wind field and its time variations. Ebdon (1960), apparently quite independently, noted that the 50-mb wind direction at stations between the equator and about 20°N was predominantly east in January 1958 and west in January 1959. Study of additional data enabled Reed *et al.* (1961) and Ebdon and Veryard (1961) to describe the oscillation in more generality.

The main features of the phenomenon are well illustrated by a time-height cross section of zonal wind at Canton Island (3°S, 172°W), prepared by Reed *et al.* (1961) and shown in Fig. 2.19. Evidence from other locations (Reed *et al.*, 1961; Ebdon, 1960, 1961) indicates that similar variations occur with the same phase at other longitudes. The principal features of the oscillation are

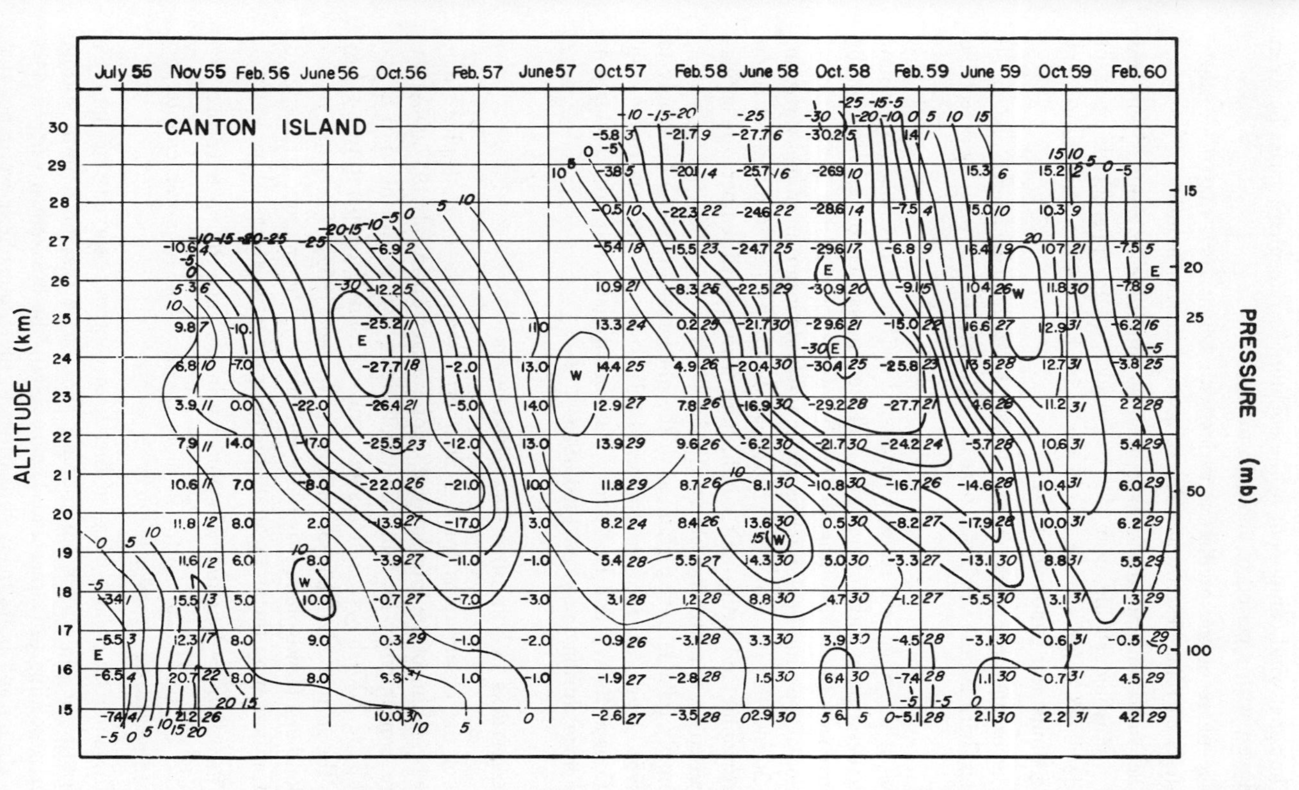

FIG. 2.19. Time-height cross section of mean monthly zonal wind at Canton Island, illustrating the equatorial stratospheric wind oscillation. Wind speeds are in meters per second and are plotted to the left of the verticals. The number of observations entering into each mean is shown to the right. (After Reed *et al.*, 1961.)

(a) The zonal wind component in the equatorial stratosphere varies between east and west with a period of about 26 months.

(b) The phase of the oscillation varies with height such that the easterly and westerly currents descend at a rate of about 1 km per month.

(c) The oscillation appears to have greatest amplitude at about 25 km and is nearly indistinguishable near the tropopause. (The apparent decrease of amplitude above 25 km may be due to a selective bias against the measurement of strong winds near balloon ceilings.)

(d) The oscillation occurs simultaneously in both hemispheres and at all longitudes, with largest amplitude near the equator. It is only barely detectable at 30°N.

Studies of the equatorial stratospheric wind oscillation have been up to now primarily descriptive and climatological. There is no satisfactory explanation for its occurrence. Reed and Rogers (1962) have constructed a mathematical description of the wind field in terms of a long-period mean, an annual variation, and the 26-month variation. Reed (1962) has shown that the 26-month variation is in geostrophic and hydrostatic equilibrium with a small temperature oscillation reported by Veryard and Ebdon (1961).

Reasonably reliable data pertaining to the phenomenon extend back only to about 1951. However, recently Ebdon (1963), from a study of scanty early observations at Batavia and East Africa, has shown that it was probably present in some form during the years 1908–1918.

2.4.5 Conditions in the Southern Hemisphere

Most of the observations for the Southern Hemisphere stratosphere date from the beginning of the International Geophysical Year in July 1957. Observations are concentrated in Antarctica and along two meridians, 170°E and 90°W. Figure 2.20 shows a north-south cross section of temperature for July 1957 at the former meridian, according to Wexler (1959). The polar regime in the Southern Hemisphere is marked and intense. The antarctic stratosphere, however, differs from the arctic stratosphere in winter in at least three ways:

(1) There is more longitudinal symmetry in the antarctic polar vortex;

(2) The antarctic polar vortex is more intense than its arctic counterpart (lower temperatures, stronger winds);

(3) The antarctic polar vortex, in the three winters observed thus far, is a stable feature of the winter months and does not break down until spring. However, according to Palmer and Taylor (1960), the breakdown when it does occur is abrupt and bears many similarities to the stratospheric warmings that occur during the winter in the arctic stratosphere.

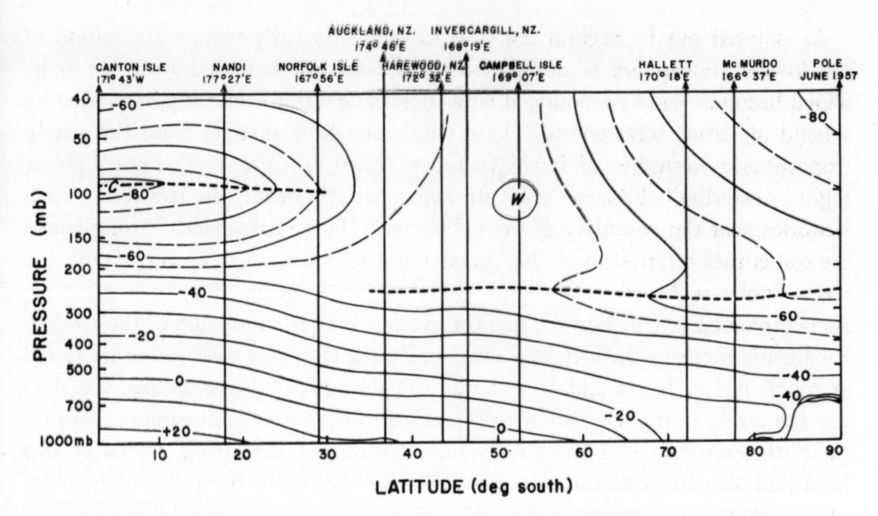

FIG. 2.20. North-south cross section of temperature in degrees C at about 170°E in the Southern Hemisphere averaged for June 1957. (After Wexler, 1959.)

2.5 The Disturbed Circulation of the Lower Stratosphere

Day-to-day and week-to-week changes of stratospheric circulation, in addition to seasonal and annual variations, are of great interest. Preparation and study of synoptic maps at least up to 25 mb during the past few years have given considerable information, although much undoubtedly remains to be revealed.

As might be expected from the vast difference between the mean patterns for summer and winter, there is also a vast difference between the disturbances of summer and winter. It is convenient, therefore, to divide this discussion into Subsection 2.5.1 referring to features of the winter circulation and Subsection 2.5.2 referring to features of the summer circulation.

2.5.1 Features of the Winter Circulation

It is a well-known fact that a tropospheric trough is generally accompanied by a relatively low and warm tropopause and a tropospheric ridge is generally accompanied by a relatively high and cold tropopause. The relatively warm air column overlying the trough and the relatively cold air column overlying the ridge lead, through the hydrostatic relationship, to a diminution of pressure contrasts as one proceeds upward into the lower stratosphere. The short-wave migratory troughs and ridges of the upper troposphere therefore damp out with height and are hardly discernible at and above about 50 mb. Instead, the normal winter flow is dominated by slowly moving troughs and ridges of long wave-length.

As pointed out in Section 2.4, the autumn and early-winter circulation of the lower stratosphere is dominated by a cold low centered near the pole, which becomes more pronounced with increasing altitude and is surrounded by a band of strong west winds. These winds are quite distinct from the strong tropospheric westerlies of lower latitudes. They are often called the "polar-night westerlies" because the strongest winds occur at relatively high latitudes near the boundary of the polar night. These particularly strong winds are sometimes referred to as the "polar-night jet stream." To some extent, the terms "polar-night westerlies" and "polar-night jet stream" may be misleading. As is discussed in Chapter 3, when the average winter circulation of the stratossphere and mesosphere is viewed as an entity, the strongest west winds are found at much higher levels and considerably farther south. Nevertheless, we shall use the terms in this chapter with reference to the lower stratosphere, because their use emphasizes the relatively high latitude of the strong winds at this level and also suggests the role that radiative cooling in the polar night must play in their development.

A detailed description of the polar-night westerlies, involving wind speeds, temperature gradients, wind shears, and the time variability of all these would be of very great practical and theoretical interest. Unfortunately, lack of data at high elevations, aggravated in this particular instance by the observational bias against low-temperature and strong-wind situations hampers such a description. Nevertheless, it is possible to construct vertical cross sections through these westerlies on occasions when wind data are barely sufficient to define the north-south wind maximum.

Figure 2.21 is such a cross section for 19 January 1957, when a particularly strong wind was measured at Goose Bay, Labrador (latitude 53°N). The cross section extends from Isachsen (79°N) SSE to a ship at 35°N, and is nearly normal to the WSW winds associated with the jet stream at this time. The strongest wind at about 26 km was observed to be 225 knots. Note the strong lateral and vertical wind shear. Note also the strong temperature gradient in the direction of the cross section at levels below the maximum wind. The fact that this gradient is still intense at the level of the highest reported wind indicates that the speed is still increasing upward to a stronger maximum at an unknown higher level. Other similar cross sections (Godson and Lee, 1958; Krishnamurti, 1959) show lower wind speeds some 10° to 20° farther north.

The reported maximum wind speeds are very apt to be underestimates because of observational bias. In the case represented by Fig. 2.21, only unusually light winds in the troposphere allowed the balloon to reach 26 km before its elevation angle became less than the critical value (6°) at which observations become hopelessly unreliable. In future years, with more representative observations, it will not be at all surprising to find that there are occasionally even higher wind speeds than this in the polar-night jet stream.

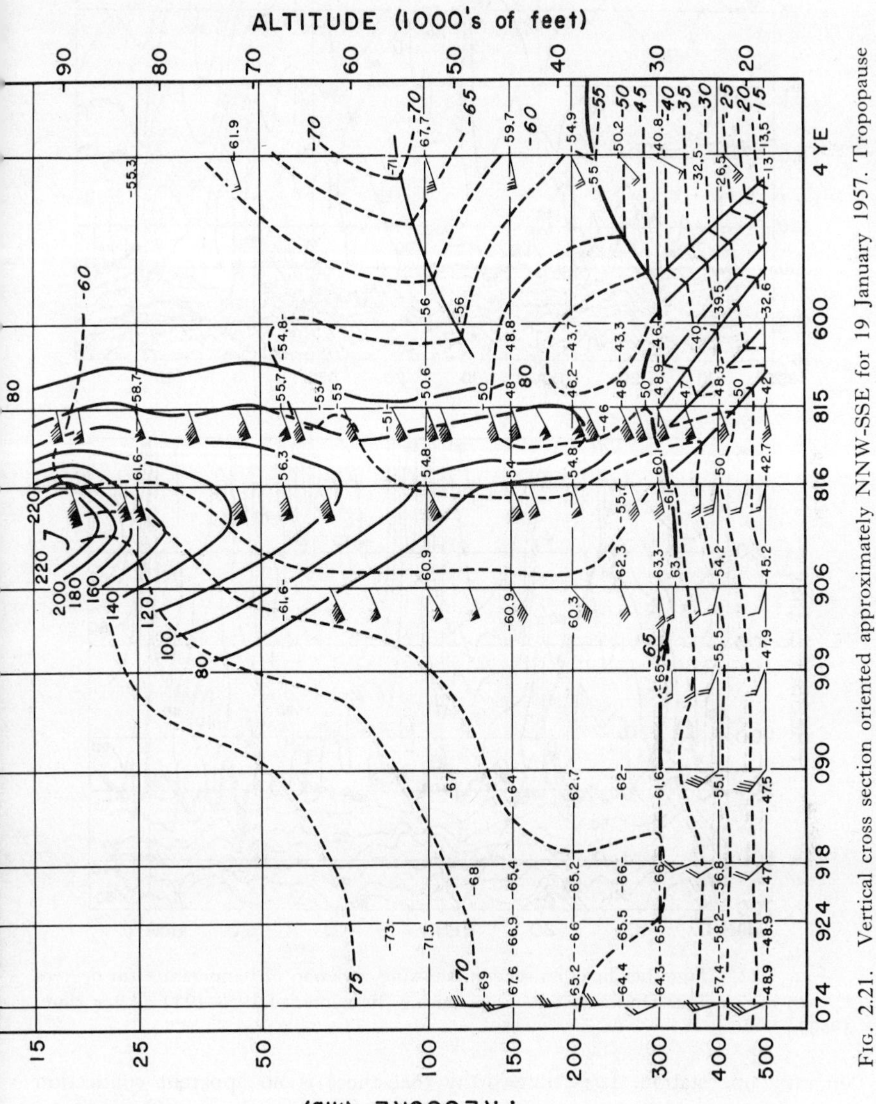

FIG. 2.21. Vertical cross section oriented approximately NNW-SSE for 19 January 1957. Tropopause is indicated by thick solid lines, isotherms (in degrees C) by broken lines, and isotachs of observed west-southwesterly winds (in knots) by solid lines. (After Godson and Lee, 1958.)

Studies of temperature variations in the zone of the polar-night westerlies reveal fluctuations of large amplitude and long period that appear to be quite distinct from disturbances in the troposphere. Austin and Krawitz (1956) were apparently the first to note these. Figure 2.22 shows a time-height cross section of temperature for Resolute during one autumn and winter and serves to illustrate certain features of these fluctuations in the context of their effect

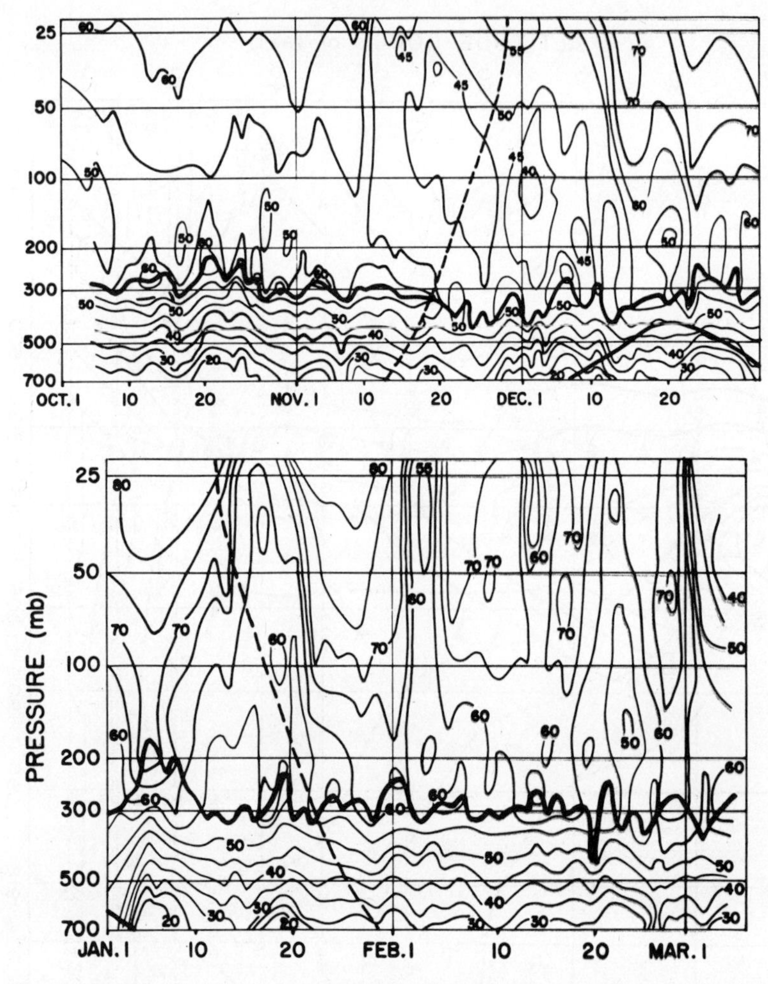

FIG. 2.22. Time-height cross section, showing variation of temperature (in degrees C with minus signs omitted) at Resolute during the winter of 1958–1959. (After Hare, 1960a.)

on only one station. It is noteworthy that there is no apparent connection between the temperature variations of the lower stratosphere and those of the troposphere. It is also noteworthy that the fluctuations are relatively small in the autumn and appear to increase in amplitude as the winter progresses. They tend to affect the entire lower stratosphere, with a tendency for the temperature rises and falls to occur earliest at the highest levels. The progression from higher to lower levels of the temperature changes is unambiguous when the full four-dimensional character of the most pronounced fluctuations is studied in detail.

The periods of rapid temperature rise, for example in mid-January, early February, and early March, are associated with the celebrated stratospheric warmings, first noted by Scherhag (1952) and extensively studied during the past 10 years. There is some confusion at the present time as to just how large and geographically widespread a temperature increase should be before it is called a "stratospheric warming." One might (and some do) refer by this name to all or most of the periods of temperature increase revealed by cross sections like the one in Fig. 2.22. However, it is clear from many studies of the phenomenon that some of the temperature increases so revealed reflect only localized and short-lived disturbances. At the other extreme, some may reflect events that affect large areas of the arctic stratosphere for long periods of time. The latter appear to occur at least once during each winter and to be accompanied by a dramatic weakening or disappearance of the polar-night westerlies over a substantial portion of the polar cap (see, for example, Fig. 2.18 and the discussion pertaining to it in Subsection 2.4.3). Sometimes these are referred to as "final warmings" because after their occurrence the polar regime, at least in some portions of the hemisphere, is only weakly re-established before the onset of summer conditions. Here we shall refer to them as "major stratospheric warmings," although this is not a commonly used term.

The timing, intensity, and character of major stratospheric warmings vary remarkably from year to year. It is to be emphasized that they are complex phenomena which can be described adequately only in terms of their three-dimensional evolution in time. However, the year-to-year variability of timing is usefully illustrated in Fig. 2.23, after Godson and Lee (1958), in terms of 10-day running means of the 100-mb temperature for stations at several latitudes in the Canadian Arctic. Several features are of interest in this presentation:

(a) During September and October, there is relatively little north-south temperature gradient as all stations show a gradual cooling.

(b) During November and December, the northernmost stations cool more rapidly than the lower-latitude stations and a strong baroclinic zone develops.

(c) During this period, long-period temperature waves are superimposed on the gradual cooling tendency.

(d) Finally, all stations participate in a major warming. Note the varied time of this major warming—early January in 1955, late March in 1956, and early February in 1957. Similar graphs would show it in early February of 1958 and mid-March of 1959.

Because these data represent only a single (and relatively low) level, a specific geographical area, and time-smoothed temperatures, they give only an approximate picture of what happens in the arctic stratosphere. Analysis of daily maps for several levels is necessary to establish the full picture. The 1957 warming has been studied in this way by Teweles (1958) and by Craig and Hering (1959). The 1957 warming actually began earlier than indicated in Fig. 2.23, at a different

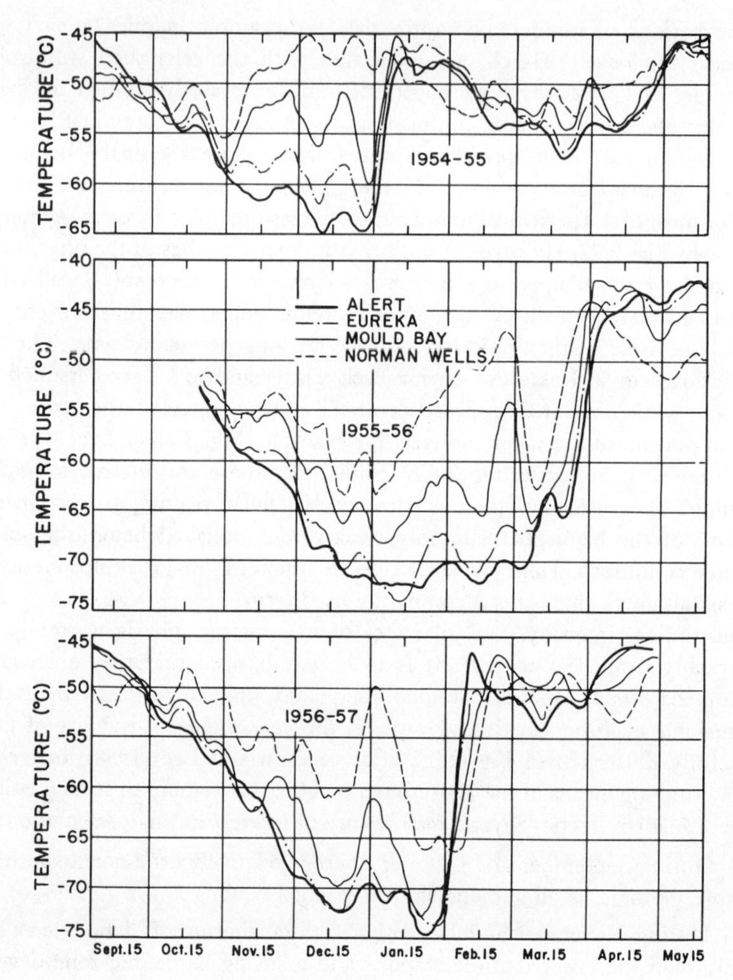

FIG. 2.23. Ten-day running means of 100-mb temperature for the stations and the winters indicated. (After Godson and Lee, 1958.)

location, and much higher than the 100-mb level. The highest level analyzed (25-mb level) showed a rapid warming just after 23 January in the vicinity of Labrador, Newfoundland, and southern Greenland. The temperature at Goose Bay, Labrador, rose from about −60°C to about −20°C between 23 January and 30 January. (There is evidence of an earlier temperature rise at 10 mb near Peoria, Illinois, but data at that time were insufficient to describe events so high up.) The most pronounced changes at 50 mb and 100 mb did not occur until early February. It must be emphasized that the 25-mb warming in its initial phases did not result from advection of warmer air along a constant-pressure surface. Rather, it appears to have resulted from downward motion and adiabatic compression during a long trajectory from the northwestern United States to the Labrador region. Craig and Lateef (1962) have computed the required vertical motions on the adiabatic assumption and found them to be quite reasonable.

The studies of this warming by Teweles, by Craig and Hering, and by Craig and Lateef emphasize the North American, North Atlantic region of relatively good stratospheric data. Nevertheless, a major stratospheric warming may affect the entire high-latitude Northern Hemisphere and data from all longitudes, when available, should be studied. In the case of the 1957 warming, despite scarcity of data from the Asian-Pacific sector of the hemisphere, Muench was able to construct hemispheric 50-mb charts, which have been published by Reed, Wolfe, and Nishimoto (1963). They reveal that there was a second center of warming approximately 180° of longitude from the one mentioned above.*

It is necessary to stress once again that no one major stratospheric warming, such as the one discussed here, is to be taken as representative of all. In particular, as emphasized by Wilson and Godson (1963), some seem to be characterized by a single main center of warming and result in a single-wave pattern with a warm anticyclone at some longitudes (usually in the western hemisphere) and a cold cyclone in the other. To appreciate the variations from year to year one must consider in detail case studies applying to different years. The reader is referred to Warnecke (1956), Teweles and Finger (1958), Scherhag (1960), and Hare (1960a). In the last reference, see figure 18 for what is apparently an example of the single-wave type of major warming.

There is no general agreement as to the "causes" of major stratospheric warmings. The temperature increases are ascribed by most investigators to downward motion and adiabatic compression; and the associated temperature decreases at other locations (which are not usually emphasized but which appear to be an essential part of the phenomenon) to upward motion and adiabatic expansion. However, this is only the starting point of an explanation because the associated vertical motions in turn require explanation. They presumably represent the growth of an unstable wave motion in the strong westerlies of the polar vortex. There has been discussion as to whether baroclinic instability or barotropic instability might be involved (Fleagle, 1958; Murray, 1960) or a combination of the two (Charney and Stern, 1962). None of these studies demonstrates a sufficient condition for instability that is clearly verified by observations. There is also the empirical fact that in the Southern Hemisphere (see Subsection 2.4.5) no breakdown of the vortex occurs until spring, despite the larger vertical and horizontal wind shears associated with the Southern Hemisphere vortex.

* See also Teweles (1958) for hemispheric 100-mb maps.

This suggests the importance, through some unexplained mechanism, of topography, which is the only difference between the two hemispheres that comes readily to mind.

Another point of considerable interest is the possible linkage between major warmings and tropospheric variations. In the absence of any theoretical demonstration of such linkage, and with the few cases available for empirical studies, one can say little about this. Julian (1961) has pointed out a rather similar tropospheric change associated with the warmings of 1956–1957, 1957–1958, and 1958–1959; namely, a rise of sea-level pressure in the polar regions. With so few cases, this must be considered to be simply suggestive. It is certainly reasonable to expect that such profound changes in stratospheric circulation would interact with the tropospheric circulation to some extent. How much and in what manner (indeed, whether always in the same manner) are unanswered questions.

Outside the polar regime, there are no changes comparable in magnitude to those discussed above. However, Riehl and Higgs (1960) have described the occurrence of a remarkable shear line above 30 km over the Caribbean during January 1960. This was similar in many aspects to phenomena observed in the troposphere.

2.5.2 Features of the Summer Circulation

In the summer stratosphere no such intense winds or violent changes occur. In the stratospheric easterlies above 20 km or so, the wind direction is surprisingly constant from day to day and latitude to latitude and seldom varies more than 10° or 20° from easterly. According to Flohn, Holzapfel, and Oeckel (1959), there is a maximum wind speed south of the large continents with speeds at least 20 m sec^{-1} and surprisingly high constancy.

Nevertheless, there is evidence of rather small-scale, small-amplitude fluctuations in the stratospheric easterlies. Flohn *et al.* (1959) have reported discovering oscillations with wavelength about 100 km from the trajectories of constant-level balloons and Mantis (1963), also from a study of constant-level balloon trajectories, reports that much of the variability of the wind near 30 km is associated with systems that are much smaller in scale than those found in the troposphere or indeed than can be revealed by conventional wind observations.

Hare (1960b) has discussed the summer circulation in the westerlies of high latitudes (below about 20 km and north of about 50°N). For this region he finds a temperature regime associated primarily with tropospheric disturbances and to some extent with closed systems near the pole and wholly within the region of westerly winds. These damp out rapidly with height and in the stratospheric easterlies above 15 to 20 km he finds the very steady conditions described above.

REFERENCES

Austin, J. M., and Krawitz, L. (1956). 50-millibar patterns and their relationship to tropospheric changes. *J. Meteor.* **13**, 152–159.

Bannon, J. K., Frith, R., and Shellard, H. C. (1952). Humidity of the upper troposphere and lower stratosphere over southern England. *Geoph. Mem.* **XI**, No. 3.

Barclay, F. R., Elliott, M. J. W., Goldsmith, P., and Jelley, J. V. (1960). A direct measurement of the humidity in the stratosphere using a cooled-vapour trap. *Quart. J. Roy. Meteor. Soc.* **86**, 259–264.

Barrett, E. W., and Herndon, L. R., Jr., (1951). An improved electronic dew-point hygrometer. *J. Meteor.* **8**, 40-51.

Barrett, E. W., Herndon, L. R., Jr., and Carter, H. J. (1950). Some measurements of the distribution of water vapor in the stratosphere. *Tellus* **2**, 302–311.

Belmont, A. D. (1962). The reversal of stratospheric winds over North America during 1957, 1958, and 1959. *Beitr. Phys. Atmos.* **35**, 126–140.

Bjerknes, J. (1932). Exploration de quelques perturbations atmosphériques à l'aide de sondages rapprochés dans le temps. *Geofys. Publ.* **9**, No. 9, 1-47.

Bjerknes, J. and Palmén, E. (1937). Investigations of selected European cyclones by means of serial ascents. *Geofys. Publ.* **12**, No. 2, 1-62.

Brewer, A. W., Cwilong, B., and Dobson, G. M. B. (1948). Measurement of absolute humidity in extremely dry air. *Proc. Phys. Soc.* **60**, 52–70.

Brown, F., Goldsmith, P., Green, H. F., Holt, A., and Parham, A. G. (1961). Measurements of the water vapour, tritium and carbon-14 content of the middle stratosphere over southern England. *Tellus* **13**, 407–416.

Chagnon, C. W., and Junge, C. E. (1961). The vertical distribution of sub-micron particles in the stratosphere. *J. Meteor.* **18**, 746–752.

Charney, J. G., and Stern, M. E. (1962). On the stability of internal baroclinic jets in a rotating atmosphere. *J. Atmos. Sci.* **19**, 159–172.

Craig, R. A., and Hering, W. S. (1959). The stratospheric warming of January–February 1957. *J. Meteor.* **16**, 91–107.

Craig, R. A., and Lateef, M. A. (1962). Vertical motion during the 1957 stratospheric warming. *J. Geoph. Res.* **67**, 1839–1854.

Danielsen, E. F. (1959). The laminar structure of the atmosphere and its relation to the concept of a tropopause. *Archiv Meteor., Geoph., Biokl.* **A11**, 293–332.

Defant, F., and Taba, H. (1957). The threefold structure of the atmosphere and the characteristics of the tropopause. *Tellus* **9**, 259–274.

de Jong, H. M. (1958). Errors in upper-level wind computations. *J. Meteor.* **15**, 131–137.

Ebdon, R. A. (1960). Notes on the wind flow at 50 mb in tropical and sub-tropical regions in January 1957 and January 1958. *Quart. J. Roy. Meteor. Soc.* **86**, 540–542.

Ebdon, R. A. (1961). Some notes on the stratospheric winds at Canton Island and Christmas Island. *Quart. J. Roy. Meteor. Soc.* **87**, 322-331.

Ebdon, R. A. (1963). Evidence of the permanency of the tropical stratospheric wind fluctuation. *Quart. J. Roy. Meteor. Soc.* **89**, 151–152.

Ebdon, R. A., and Veryard, R. G. (1961). Fluctuations in equatorial stratospheric winds. *Nature* **189**, 791–793.

Emden, R. (1913). Über Strahlungsgleichgewicht und atmosphärische Strahlung. *S. B. Akad. Wissenschaften, Munich* No. 1, 55–142.

Fleagle, R. G. (1958). Inferences concerning the dynamics of the mesosphere. *J. Geoph. Res.* **63**, 137–145.

Flohn, H., Holzapfel, R., and Oeckel, H. (1959). Untersuchungen über die stratosphärische Ostströmung auf der Sommerhalbkugel. *Beitr. Phys. Atmos.* **31**, 217–243.

Godson, W. L., and Lee, R. (1958). High-level fields of wind and temperature over the Canadian Arctic. *Beitr. Phys. Atmos.* **31**, 40–68.

Goody, R. M. (1949). The thermal equilibrium at the tropopause and the temperature of the lower stratosphere. *Proc. Roy. Soc.* **A197**, 487–505.

Goody, R. M. (1954). "The Physics of the Stratosphere." Cambridge Univ. Press, London and New York.

Gruner, P. (1942). Dämmerungserscheinungen. *In* "Handbuch der Geophysik," Vol. 8, pp. 432–526. Bornträger, Berlin.

Gutnick, M. (1961). How dry is the sky? *J. Geoph. Res.* **66**, 2867–2871.

Hagemann, F., Gray, J., Jr., Machta, L., and Turkevich, A. (1959). Stratospheric carbon-14, carbon dioxide, and tritium. *Science* **130**, 542–552.

Hare, F. K. (1960a). The disturbed circulation of the arctic stratosphere. *J. Meteor.* **17**, 36–51.

Hare, F. K. (1960b). The summer circulation of the arctic stratosphere below 30 km. *Quart. J. Roy. Meteor. Soc.* **86**, 127–143.

Helliwell, N. C., Mackenzie, J. K., and Kerley, M. J. (1957). Some further observations from aircraft of frost point and temperature up to 50,000 ft. *Quart. J. Roy. Meteor. Soc.* **83**, 257–262.

Hess, S. L. (1948). Some new mean meridional cross sections through the atmosphere. *J. Meteor.* **5**, 293–300.

Houghton, J. T., and Seeley, J. S. (1960). Spectroscopic observations of the water-vapour content of the stratosphere. *Quart. J. Roy. Meteor. Soc.* **86**, 358–370.

Houghton, J. T., and Seeley, J. S. (1961). Discussion of "Houghton and Seeley (1960)." *Quart. J. Roy. Meteor. Soc.* **87**, 251.

Julian, P. R. (1961). Remarks on "The disturbed circulation of the arctic stratosphere." *J. Meteor.* **18**, 119–121.

Junge, C. E.. (1961). Vertical profiles of condensation nuclei in the stratosphere. *J. Meteor.* **18**, 501–509.

Junge, C. E., and Manson, J. E. (1961). Stratospheric aerosol studies. *J. Geoph. Res.* **66**, 2163–2182.

Junge, C. E., Chagnon, C. W., and Manson, J. E. (1961). Stratospheric aerosols. *J. Meteor.* **18**, 81–108.

King, J. I. (1952). Line absorption and radiative equilibrium. *J. Meteor.* **9**, 311–321.

Kochanski, A. (1955). Cross sections of the mean zonal flow and temperature along 80°W. *J. Meteor.* **12**, 95–106.

Krishnamurti, T. N. (1959). A vertical cross section through the "polar-night" jet stream. *J. Geoph. Res.* **64**, 1835-1844.

Leviton, R. (1954). Height errors in a rawin system. *AF Surveys in Geoph.* No. 60.

McCreary, F. E. (1959). Unpublished report.

Mantis, H. T. (1963). Note on the structure of the stratospheric easterlies of midlatitude. *J. Appl. Meteor.* **2**, 427–429.

Mastenbrook, H. J., and Dinger, J. E. (1961). Distribution of water vapor in the stratosphere. *J. Geoph. Res.* **66**, 1437–1444.

Muench, H. S., and Borden, T. R. (1962). Atlas of monthly mean stratosphere charts, 1955–1959: Part I, January–June; Part II, July–December. *AF Surveys in Geoph.* No. 141.

Murcray, D. G., Murcray, F. H., and Williams, W. J. (1962). Distribution of water vapor in the stratosphere as determined from infrared absorption measurements. *J. Geoph. Res.* **67**, 759–766.

MURGATROYD, R. J., GOLDSMITH, P., and HOLLINGS, W. E. H. (1955). Some recent measurements of humidity from aircraft up to heights of about 50,000 ft over southern England. *Quart. J. Roy. Meteor. Soc.* **81**, 533–537.

MURRAY, F. W. (1960). Dynamic stability in the stratosphere. *J. Geoph. Res.* **65**, 3273–3305.

PALMÉN, E. (1933). Aerologische Untersuchungen der atmosphärischen Störungen mit besonderer Berücksichtigung der stratosphärischen Vorgänge. *Comm. Phys.-Math., Finska Veterskaps—Societeten* **7**, No. 6, 1-65.

PALMER, C. E., and TAYLOR, R. C. (1960). The vernal breakdown of the stratospheric cyclone over the south pole. *J. Geoph. Res.* **65**, 3319–3329.

PANOFSKY, H. A. (1961). Temperature and wind in the lower stratosphere. *Advances in Geoph.* **7**, 215–247.

REED, R. J. (1962). Evidence of geostrophic motion in the equatorial stratosphere. *Quart. J. Roy. Meteor. Soc.* **88**, 324–327.

REED, R. J., and ROGERS, D. G. (1962). The circulation of the tropical stratosphere in the years 1954–1960. *J. Atmos. Sci.* **19**, 127–135.

REED, R. J., CAMPBELL, W. J., RASMUSSEN, L. A., and ROGERS, D. G. (1961). Evidence of a downward-propagating, annual wind reversal in the equatorial stratosphere. *J. Geoph. Res.* **66**, 813–818.

REED, R. J., WOLFE, J. L., and NISHIMOTO, H. (1963). A spectral analysis of the energetics of the stratospheric sudden warming of early 1957. *J. Atmos. Sci.* **20**, 256–275.

RIEHL, H., and HIGGS, R. (1960). Unrest in the upper stratosphere over the Caribbean Sea during January 1960. *J. Meteor.* **17**, 555–561.

SCHERHAG, R. (1952). Die explosionsartigen Stratosphärenerwärmungen des Spätwinters 1951/1952. *Ber. Deut. Wetterd.* **6**, 51–63.

SCHERHAG, R. (1960). Stratospheric temperature changes and the associated changes in pressure distribution. *J. Meteor.* **17**, 575–582.

STALEY, D. O. (1957a). A study of tropopause formation. *Beitr. Phys. Atmos.* **29**, 290–316.

STALEY, D. O. (1957b). Some comments on physical processes at and near the tropopause. *Archiv Meteor., Geoph., Biokl.* **A10**, 1–19.

STEINBERG, S., and ROHRBOUGH, S. F. (1962). The collection and measurement of carbon dioxide and water vapor in the upper atmosphere. *J. Appl. Meteor.* **1**, 418–421.

TEISSERENC DE BORT, L. P. (1902). Variations de la température de l'air libre dans la zone comprise entre 8 km et 13 km d'altitude. *Comptes Rendus Acad. Sci. Paris* **134**, 987–989.

TEWELES, S. (1958). Anomalous warming of the stratosphere over North America in early 1957. *Mon. Wea. Rev.* **86**, 377–396.

TEWELES, S., and FINGER, F. G. (1958). An abrupt change in stratospheric circulation beginning in mid-January 1958. *Mon. Wea. Rev.* **86**, 23–28.

TEWELES, S., and FINGER, F. G. (1960). Reduction of diurnal variation in the reported temperatures and heights of stratospheric constant-pressure surfaces. *J. Meteor.* **17**, 177–194.

U. S. WEATHER BUREAU (1955–1959). "Monthly Climatic Data for the World," Vols. 8–12. U. S. Dept. of Commerce, Washington, D.C.

U. S. WEATHER BUREAU (1957). "Manual of Radiosonde Observations," Circular P, 7th ed. U. S. Dept. of Commerce, Washington, D.C.

VERYARD, R. G., and EBDON, R. A. (1961). Fluctuations in tropical stratospheric winds. *Meteor. Magaz.* **90**, 125–143.

VIEZEE, W. (1958). Unpublished report.

VOLZ, F. E., and GOODY, R. M. (1962). The intensity of the twilight and upper atmospheric dust. *J. Atmos. Sci.* **19**, 385–406.

WARNECKE, G. (1956). Ein Beitrag zur Aerologie der Arktischen Stratosphäre. *Meteor. Abhandlungen, Univ. Berlin* **III**, No. 3, 1-60.

WEGE, K., LEESE, H., GROENING, H. U., and HOFFMAN, G. (1958). Mean seasonal conditions of the atmosphere at altitudes of 20 to 30 km and cross sections along selected meridians in the Northern Hemisphere. *Meteor. Abhandlungen, Univ. Berlin* **VI**, No. 4, 1-28.

WEXLER, H. (1959). Seasonal and other temperature changes in the antarctic atmosphere. *Quart. J. Roy. Meteor. Soc.* **85**, 196–208.

WILSON, C. V., and GODSON, W. L. (1963). The structure of the arctic winter stratosphere over a 10-yr period. *Quart. J. Roy. Meteor. Soc.* **89**, 205–224.

CHAPTER 3

Structure and Circulation of the Upper Stratosphere and the Mesosphere

Above about 30 km, the quantity of conventional meteorological observations decreases very rapidly. Our knowledge of the temperature and wind structure of the upper stratosphere and the mesosphere is derived from diverse observational methods, only some of which are direct in the sense that instruments are carried up to the region. In this chapter, we undertake to review these methods and the information they have yielded. Discussion of the composition of the upper stratosphere and the mesosphere is deferred until Chapter 5.

From the historical point of view, one may distinguish three "periods" in the development of methods of observing this part of the atmosphere. From the first recognition by Lindemann and Dobson (1923) that temperature must increase with altitude somewhere above the tropopause, until the late 1940's, information was obtained exclusively from ground-based measurements and pertained primarily to temperature (or density). These observations, mostly of meteor trails and the anomalous propagation of sound, gave a picture of vertical temperature distribution that was qualitatively correct in that it showed a temperature maximum near 50 km (stratopause) and a temperature minimum near 80 km (mesopause). Quantitatively, however, the results were rather seriously in error with respect to the magnitude of the temperature maximum, which was believed to be about 350°K (Gutenberg, 1946; Warfield, 1947). During this period, almost the only wind information stemmed from scattered and rather unsystematic observations of the motions of noctilucent clouds (Chapter 5) and of meteor trains (long-enduring meteor trails). These pertained primarily to the upper mesosphere and low thermosphere.

The second period began in the late 1940's with the adaptation and use of rockets for upper-atmospheric research and ended at the beginning of the International Geophysical Year in 1957. The first few rocket measurements at White Sands, New Mexico (32°N), were sufficient to correct the temperature value at the stratopause (Rocket Panel, 1952). Careful and detailed acoustical studies were able to separate temperature and wind effects and shed some light on the circulation of the upper stratosphere. At the same time, the development of radar methods of observing ionized meteor trails began to give much more wind information than had been available from visual observations of meteor trains. Although this information applied primarily to the low thermosphere (80–100 km), it made possible, in conjunction with the sound data for lower elevations, some deductions about prevailing winds in the mesosphere. These and other scattered observations to be discussed below gave for the first time a reliable picture of the large-scale features of atmospheric structure and circulation between 30 km and 80 km. However, during this period observations were scanty, and furthermore, for various reasons, were concentrated in lower middle latitudes (30°–40°N).

During the third period, from 1957 until the present, some of the earlier rocket techniques have been used at other latitudes, particularly at Fort Churchill, Canada (59°N). Furthermore, this period has seen the development of relatively low-cost rockets and observing techniques and their frequent use at many more locations than was previously possible. As a result, it is now feasible, for the first time, to study day-to-day and week-to-week changes, so that the upper stratosphere and, to a lesser extent, the mesosphere are now being subjected to meteorological analyses that would have been impossible only a few years ago.

Section 3.1 contains a discussion of the large-scale features of temperature and wind variation—with altitude, season, and latitude. This is intended to serve as a background for and introduction to the more detailed material that follows. In Section 3.2, ground-based measuring techniques and their results are considered; this discussion is much shorter and less detailed than would have appeared appropriate only two or three years ago, because some of these methods seem now to have principally historical interest. In Section 3.3, the results of experiments designed for and used on large rockets are considered. Some of these experiments were brilliantly conceived and executed, and even though their frequent use is impractical for economic reasons they still have great potential for specialized studies. Section 3.4 describes the systems that are now capable of frequent application, principally the ones involving "meteorological rockets"; and the results of these measurements. Finally, in Section 3.5 we try to fill in some of the details of the large-scale picture presented in Section 3.1, and to summarize the present state of knowledge in this rapidly expanding area of study.

3.1 Large-Scale Features of the Wind and Temperature Variations

Murgatroyd (1957), after an exhaustive study of the data available at the time, prepared north-south vertical cross sections of zonal wind and temperature for both summer and winter, extending up to about 100 km. His method of analysis was such that the final cross sections of zonal wind and temperature were consistent not only with the wind and temperature data separately but also with the thermal-wind relationship [Eq. (2.5a)]. His results are shown in Figs. 3.1 and 3.2.

Although many more data are now available than were available in 1957, Murgatroyd's cross sections are nevertheless still very useful. They summarize results of observations prior to 1957; they serve as a convenient statement with which to compare later observations; and they reveal the large-scale features of wind and temperature variations (although some details have to be altered in the light of later observations).

The temperature field (Fig. 3.1) shows, in its vertical variation, the features

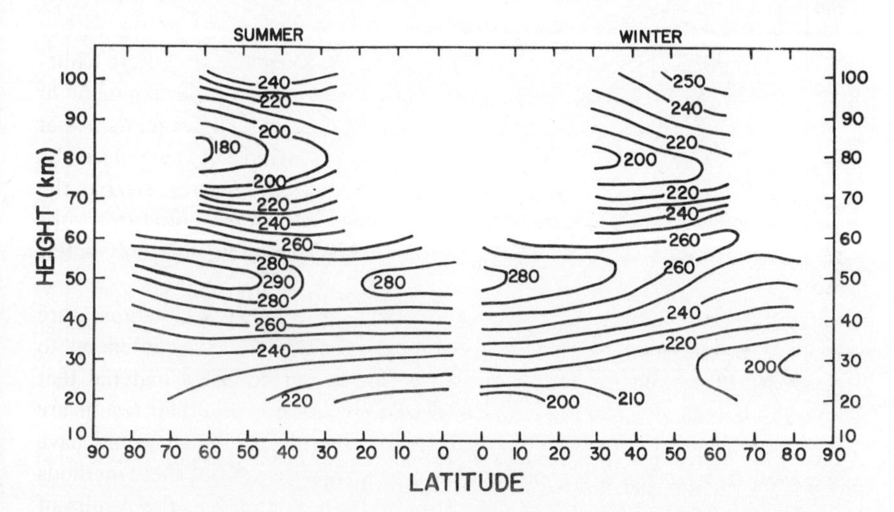

FIG. 3.1. Meridional cross sections of temperature (in degrees K) for summer and winter. (After Murgatroyd, 1957.)

described in Chapter 1: temperature increase from 30 km to the stratopause, temperature decrease from the stratopause to the mesopause, and temperature increase above the mesopause. Of interest are the depicted seasonal and latitudinal variations. The stratopause is shown to be warmest in high latitudes in summer and coldest in high latitudes in winter. The mesopause, on the other hand, is shown to be coldest in high latitudes in summer and warmest in high latitudes in winter. The suggested mesopause behavior (now evidently confirmed by data not available in 1957) must be considered to be quite surprising and to require considerable explanation (see Chapter 7). Murgatroyd had very little temperature data for the high-latitude mesopause region and his deduction was based primarily on thermal-wind considerations.

With respect to the zonal wind field (Fig. 3.2), of most interest are the systems of east winds in summer and west winds in winter with maximum speed near or just above the stratopause. Although the latitudes, altitudes, and magnitudes of the two wind maxima must be considered to be subject to revision, nevertheless, these wind systems are the central features of the circulation of the upper stratosphere and mesosphere. They are overlaid by westerlies in the summer and by easterlies in the winter.

More recently, Batten (1961) has prepared a cross section of zonal winds similar to Fig. 3.2. His results suggest that the summer wind maximum near the stratopause should be somewhat lower in altitude and farther south than shown

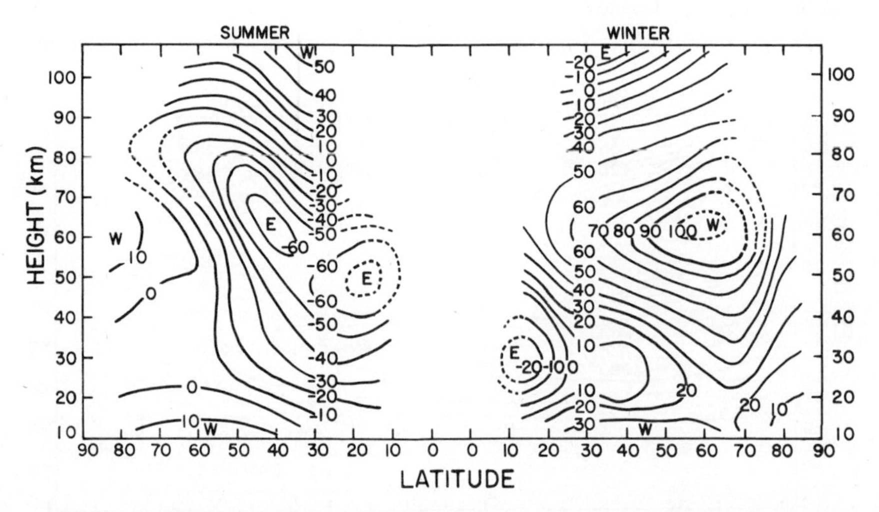

FIG. 3.2. Meridional cross sections of zonal wind (in meter $(\text{second})^{-1}$, west wind positive) for summer and winter. (After Murgatroyd, 1957.)

by Murgatroyd; and the winter wind maximum near the stratopause somewhat higher in altitude and considerably farther south than shown by Murgatroyd. They also suggest, with respect to higher altitudes, some changes in the altitude of zero zonal wind component, particularly in the summer months. However, Batten's winter wind field does not agree very well with the thermal-wind relationship and the observed temperatures and this discrepancy requires clarification.

Murgatroyd's cross sections extend a bit into the thermosphere, and we shall refer from time to time in this chapter to this 80- to 100-km region. However, these references will usually be confined to "prevailing" winds. Winds in the thermosphere (and, to some extent, the upper mesosphere) are considerably complicated by tidal components and by small-scale structure, matters that are left for discussion in Chapter 8.

3.2 Ground-Based Measurements

Until the availability of rockets after World War II, all of man's knowledge of the upper atmosphere above about 30 km had to be obtained from observations made at the ground, supplemented by theoretical considerations. In this circumstance a surprising amount of information was deduced by a relatively small group of scientists. Although in some other areas of upper-atmospheric research (ozone, airglow, aurora, ionosphere, geomagnetism) ground-based observations still retain considerable utility, in the present context they have been largely supplanted by direct measurements. This section, therefore, is mainly of historical interest and is curtailed accordingly.

During the first period of upper-atmospheric observations, two methods provided the bulk of data: observations of meteor trails to deduce density in the mesosphere and observations of the anomalous propagation of sound to deduce temperatures in the upper stratosphere. The latter method, refined to yield some information about the wind field, continued to be useful through the second period. Meteor observations (insofar as they pertain to density determinations) are discussed in Subsection 3.2.1 and sound observations in Subsection 3.2.2. During the second period, a few determinations of density in the upper stratosphere were made with what is called the searchlight experiment and these are considered in Subsection 3.2.3. Finally, visual observations of the motions of noctilucent clouds near 80 km and of enduring meteor trains have provided significant information about winds in the upper mesosphere; these data are considered in Subsection 3.2.4, along with some of the results of the powerful technique of meteor observation by radio.

Some of these techniques are related to more powerful and better controlled methods of observation using rockets. Thus, sound experiments involving ground-based explosions have given way to sound experiments involving grenades fired *in situ* from moving rockets; visual and photographic observations of naturally occurring phenomena such as noctilucent clouds and meteor trains have been supplemented by observations of artificial clouds released *in situ* from moving rockets. The latter experiments, however, are discussed in Section 3.3.

3.2.1 Meteors in the Mesosphere

A meteor trail, often popularly designated as a "shooting star," is produced by the interaction of the earth's upper atmosphere with a high-speed metallic particle of extraterrestrial origin. Meteors that produce visible trails are generally in the size range from a millimeter or so up to a few centimeters. Meteors more than 10 times smaller can be detected by radio probing of their ionized paths, and it is known that still smaller particles, called micrometeors, impinge on the atmosphere in very great numbers. On the other hand, relatively few are very much larger; although the typical meteor is vaporized in the upper atmosphere, these few survive to reach the earth's surface, in which case they are called *meteorites.*

The earliest studies of meteor trails were based on visual observations. Simultaneous observations by trained observers at two different locations can give the trajectory of the meteor and hence its height of appearance, height of disappearance, direction of motion, and (very roughly) its speed. It is evident that for a given meteor the altitude at which it appears (first gives off visible light) and the altitude at which it disappears (is completely vaporized) depend on the density of the atmosphere through which it travels. Lindemann and Dobson (1923) made use of a large number of visual observations of meteor trails to compute atmospheric densities at various heights of appearance and disappearance ranging from about 30 km up to about 150 km, but primarily in the mesosphere. The computed densities at altitudes above 60 km were clearly larger than the densities that would follow from the hydrostatic equation on the assumption of stratospheric temperatures near 220°K at all altitudes above the tropopause. Indeed, agreement between hydrostatically computed densities and meteor-computed densities was possible only if the mean temperature between the tropopause and 60 km was taken to be considerably higher than 220°K. Lindemann and Dobson surmised that the temperature might remain at 220°K up to 50 km and then increase to about 300°K at and above 60 km. That this was incorrect in detail does not detract from the significance of their work.

Lindemann and Dobson's remarkable paper revolutionized meteorological concepts of the upper atmosphere as thoroughly as Teisserenc de Bort's paper had done some 20 years earlier. Many details needed to be filled in, but the broad outlines were sketched. To illustrate further their insight, we quote the final paragraph of Lindemann and Dobson's paper, which suggested (correctly) that ozone might be responsible for the high temperature:

> The density above 60 km appears to be very much greater than corresponds to an isothermal atmosphere at 220° abs, and the temperature appears to be in the neighborhood of 300° abs. A tentative explanation is put forward to account for such a high temperature based on the radiative properties of ozone.

In the years following Lindemann and Dobson's work, the art of visual meteor observations was developed to a fine point, particularly by Öpik, and used extensively in the Harvard-Arizona meteor expedition (Shapley *et al.*, 1932). However, quantitative application of meteor observations to the computation of reasonably reliable upper-atmospheric densities had to await the development by Whipple of a reliable photographic technique for the recording of meteor trails. In this method of observation, the meteor trail is photographed simultaneously from two locations against the star background. A rotating shutter on each camera covers the lens at known short intervals and produces "breaks" in the photographed meteor trail. The combined information from the two photographs is sufficient to yield a rather accurate trajectory for the meteor, as well as values of meteor velocities and decelerations at various points on the trajectory (for further details, see F. L. Whipple and Hawkins, 1959).

In order to proceed from these data to the determination of atmospheric densities, Whipple built upon concepts of meteor-atmosphere interaction that had developed since the time of Lindemann and Dobson (F. L. Whipple, 1943, 1952a). Details of the theory are not reproduced here, but it should be pointed out that certain quantities entering into it were not known precisely and there was a residual uncertainty in the absolute values of the derived densities. F. L. Whipple in 1943 presented the results of several density determinations over Massachusetts and these were the primary basis for estimates of temperature in the mesosphere on which the NACA Tentative Standard Atmosphere (Warfield, 1947) was based. Jacchia (1948, 1949a,b) gave further results of the Massachusetts observations; these were summarized by F. L. Whipple (1952a).

Considering the imprecision of meteor-derived densities and the further difficulty that smaller meteors tend to fragment, in which case the simple theory originally used is not applicable (Jacchia, 1955), the tendency now in photographic meteor studies is to use values of atmospheric density based on other techniques of measurement to deduce meteor properties, such as mass and density (see, for example, Hawkins and Southworth, 1958; F. L. Whipple and Hawkins, 1959). Photographic meteor studies, however, still have the potential of contributing to our knowledge of variations in upper-atmospheric density, either seasonal or diurnal (for example, Jacchia, 1957).

3.2.2 The Anomalous Propagation of Sound

By "anomalous propagation of sound" is meant propagation of sound through the atmosphere in such a way that it is detected far beyond the range of the direct sound wave. Typically, for an intense surface explosion, the zone of audibility associated with direct travel of the disturbance is succeeded by a zone of silence, then by an anomalous zone of audibility, and sometimes by additional zones of silence and of audibility. The first anomalous zone of audibility is associated with sound waves that have traveled up to the upper stratosphere and been refracted back to the surface of the earth. Succeeding zones, if they occur, represent sound waves reflected from the earth's surface in the first anomalous zone and refracted down a second or third time in the same manner.

The speed of sound in an ideal gas with adiabatic changes of state is

$$C = (\gamma RT/m)^{\frac{1}{2}} \tag{3.1}$$

The symbol γ stands for the ratio of the gas's specific heat at constant pressure to its specific heat at constant volume, which ratio for dry air has the value 1.400. This expression gives the speed of sound relative to the medium through which it is traveling, in our case air. The speed of a sound wave relative to the ground is determined by adding to this speed the component of wind in the direction of propagation. Accordingly, a sound wave may be refracted back toward the ground in a region of the atmosphere where temperature increases with height and the component of wind in the direction of propagation increases with height. In the experiments with which we are concerned, this occurs in the upper stratosphere.*

Qualitative observations of this phenomenon date back to the early 1900's and were reported frequently in England during World War I when gunfire from France was sometimes audible. It was early recognized that refraction of this nature required the speed of sound to increase upward in some region of the atmosphere, but lack of quantitative data, and uncertainty about the effect of wind and of a possible upward decrease of mean molecular weight [see Eq. (3.1)] prevented a clear recognition of the significance of the phenomenon. However, after Lindemann and Dobson suggested an upward increase of temperature to explain meteor observations, F. J. W. Whipple† quickly recognized the connection between their hypothesis and the sound observations. Writing in *Nature*, F. J. W. Whipple (1923) commented:

> The work of Lindemann and Dobson on the theory of meteors, with the remarkable conclusion that the temperature of the atmosphere at heights such as 80 kilometers is about the same as that near the earth's surface, will be far-reaching in its influence. May I be allowed to point out that one of the phenomena for which an explanation will probably be provided is the occurrence of zones of audibility and zones of silence surrounding the scenes of great explosions.

Sound is a wavelike phenomenon. That is to say, the small pressure oscillation (or, if you will, the small longitudinal displacement of air molecules) associated with passage of the sound wave obeys the wave equation. It is customary and extremely useful to apply to the sound-propagation problem all of the concepts so familiar in geometrical optics—wave fronts, rays, Huygens' principle, and the resulting laws of reflection and refraction. Just as in geometrical optics, of course, these simplified techniques apply only when the wavelength is small compared with the characteristic distance associated with changes in the speed of the wave. The wavelengths of the sound waves we are concerned with are of the order of 10 m (this corresponds to a low-frequency sound wave,

---

* Downward refraction might also occur in the thermosphere. However, absorption of sound increases rapidly with decreasing density of the medium, and sound propagated to and refracted from such high levels would not ordinarily be detected; a few such instances have been reported (for example, Johnson and Hale, 1953).

† Not to be confused with F. L. Whipple, who was referred to extensively in Subsection 3.2.1.

which is used because higher frequencies are subject to more absorption during their atmospheric traverse). Most experiments are designed to determine average winds and temperatures in layers a few kilometers thick, so that diffraction effects on this scale are neglected.

Consider an idealized experiment in which the atmosphere is horizontally stratified and motionless, corresponding to conditions usually assumed in the early sound studies. The wave front from an explosion at the ground spreads upward and outward from the source and, a short time after the blast, proceeds to travel as a spherical wave front with the speed of sound, and effectively as a plane wave front when the radius of the sphere is large enough. Figure 3.3

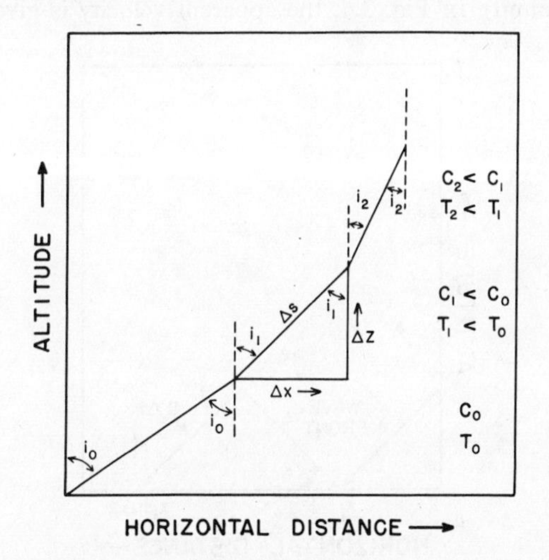

FIG. 3.3. A schematic representation of the path of a sound ray through the troposphere. The temperature is assumed to be constant in each of the three layers.

shows the path of one ray (normal to the wave front) through the troposphere. In this example, the temperature is assumed to have constant values T_0, T_1, T_2, ..., T_k in successively higher layers. By Snell's law,

$$\frac{C_0}{\sin i_0} = \frac{C_1}{\sin i_1} = \frac{C_2}{\sin i_2} = \cdots = \frac{C_k}{\sin i_k} \tag{3.2}$$

If $T_0 > T_1 > T_2 > \cdots > T_k$, then by Eq. (3.1) $C_0 > C_1 > C_2 > \cdots > C_k$ and the ray's path bends away from the earth's surface. If the atmosphere becomes isothermal at some higher level, then the ray follows a straight line; and if the temperature increases with height the ray bends back toward the earth's surface. In either of the latter cases Eq. (3.2) is still valid. Furthermore, its validity does not depend on the thickness of the layers in which we have taken T to be constant, so that in the limit of a continuous distribution we may conclude that for any level k

$$\frac{C_k}{\sin i_k} = \frac{C_0}{\sin i_0} \tag{3.3}$$

$$\frac{T_k}{\sin^2 i_k} = \frac{T_0}{\sin^2 i_0} \tag{3.4}$$

The ratio $C/\sin i$ is called the *apparent velocity* and represents the apparent velocity of the sound as it would be detected along a horizontal surface (which might be the ground). In Fig. 3.4, the apparent velocity is given by $\Delta x/\Delta t$. It

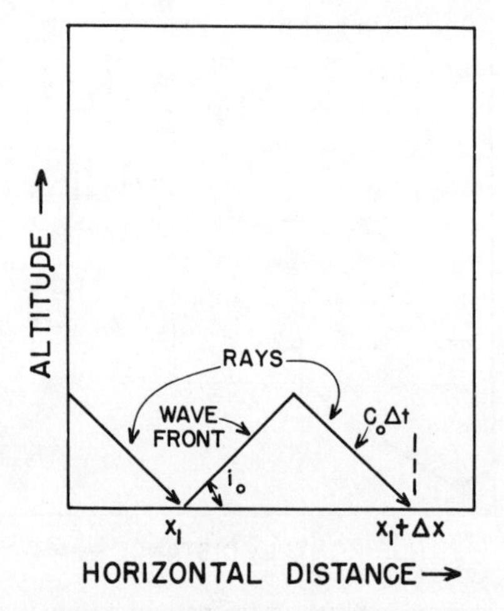

FIG. 3.4. A schematic representation of a portion of a wave front reaching the ground, illustrating the concept of apparent velocity. The wave front reaches the ground at distance x_1 at the time t_1; it reaches the ground at distance $x_1 + \Delta x$ at the time $t_1 + \Delta t$. The apparent velocity is given by $\Delta x/\Delta t$ and is equal to $C_0/\sin i_0$.

evidently remains the same for any portion of the wave front during its travel through the atmosphere.

The horizontal distance traveled by a ray while it is passing through one of the stratified layers in Fig. 3.3 (refer to diagram in the layer with subscript 1) is $\Delta x = \tan i \, \Delta z$ and the time for this travel is $\Delta t = \Delta s/C = \Delta z/C \cos i$. If a ray is at the point (x_1, z_1) at time t_1 and the point (x_2, z_2) at time t_2, we have

$$x_2 - x_1 = \int_{z_1}^{z_2} \tan i \, dz \tag{3.5}$$

$$t_2 - t_1 = \int_{z_1}^{z_2} \frac{dz}{C \cos i} \tag{3 6}$$

These integrals can be computed with the use of (3.1) and Snell's law, if the vertical temperature distribution and the value of i at one level are known.

If the ray is to be refracted back to earth, its path must be horizontal at the top of its trajectory. If we designate that level by the subscript h, $\sin i_h = 1$, and (3.3) and (3.4) give

$$C_h = \frac{C_0}{\sin i_0} \tag{3.7}$$

$$T_h = \frac{T_0}{\sin^2 i_0} \tag{3.8}$$

Equation (3.8) gives a simple but very useful result. It shows first of all that (in this idealized case) the temperature at the top of the trajectory must be at least as high as the temperature at the ground, if any ray is to return to earth. Furthermore, it can be used to compute T_h if the angle of incidence of the incoming ray can be measured. On the other hand, it does not tell the altitude z_h at which the temperature T_h is to be found.

The angle of incidence of an incoming ray at a given location can be determined by the use of an array of microphones with a suitable central recorder. At least three microphones are required for this and up to twelve have been used in some cases. From the differing times of arrival of the wave front at the various microphones, it is possible to determine i_0 and also the azimuth of the incoming ray [for a discussion of this geometrical problem, see Mitra (1952), pp. 68-69, or Murgatroyd (1956)]. This, together with the local temperature, gives T_h from Eq. (3.8).

F. J. W. Whipple (1935) described an approximate method of determining the altitude z_h corresponding to the inferred temperature T_h. In more recent experiments (see below), the situation is complicated by the consideration of wind effects, but the principle of the analysis and the nature of the approximations introduced are similar to Whipple's method for a motionless atmosphere.

Let the ray that is eventually detected at the observing site originate at the origin (see Fig. 3.5) with the angle of incidence i_0 at $t = 0$. This is the same angle, because of the horizontal stratification of the atmosphere, that is measured at the observing site at a distance x_m from the origin when the ray returns to the surface. Let the total time of travel from $x = 0$ to $x = x_m$ be t_m, which along with x_m is measured. Consider a datum level z_d, at and below which the temperature structure is known so that one can compute from Eqs. (3.5) and (3.6) the distance x_d and time t_d at which the ray reaches this level; thus

$$x_d = \int_0^{z_d} \tan i \, dz \tag{3.9}$$

$$t_d = \int_0^{z_d} \frac{dz}{C \cos i} \tag{3.10}$$

Let the ray reach its maximum altitude z_h at the time t_h and the distance x_h. Then since $x_h - x_d = x_m/2 - x_d$, and with (3.5),

$$\int_{z_d}^{z_h} \tan i \, dz = x_m/2 - x_d \tag{3.11}$$

Similarly

$$\int_{z_d}^{z_h} \frac{dz}{C \cos i} = t_m/2 - t_d \tag{3.12}$$

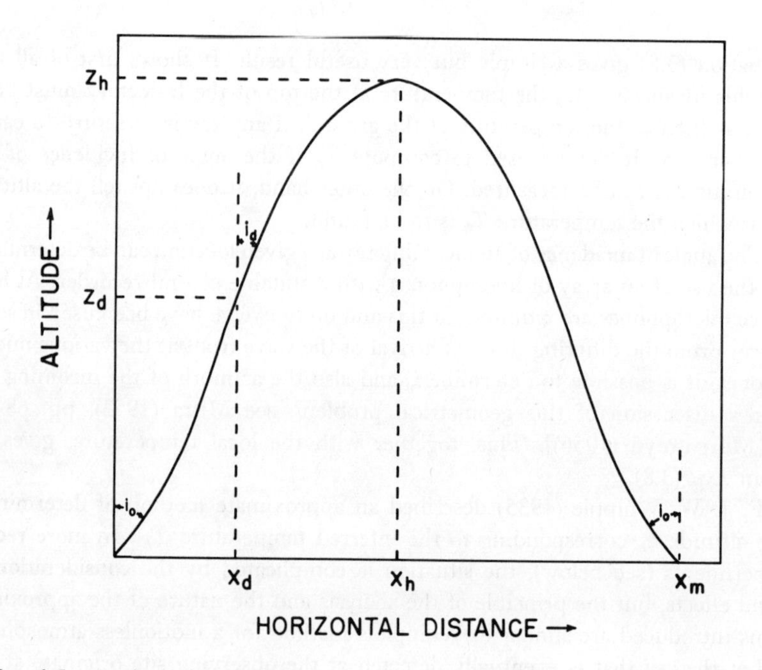

FIG. 3.5. A schematic representation of the path of a sound ray through a horizontally homogeneous atmosphere. The ray reaches maximum altitude z_h at the distance x_h and returns to the ground at the distance x_m. The path of the ray below the datum level z_d can be computed from meteorological data. Note that the horizontal scale is compressed relative to the vertical; in the actual atmosphere a ray that is returned to the ground has a larger initial angle of incidence i_0 than shown here.

The integrals in (3.11) and (3.12) depend on the unknown temperature distribution between z_d and z_h. Clearly there is no unique solution for this temperature distribution; but if it is assumed to have a simple form described by only two parameters, then these two parameters can be determined from (3.11) and (3.12). For example, Whipple assumed the temperature to be constant between z_d and an intermediate level h', and to vary linearly with lapse rate Γ between h' and z_h. The values of h' and Γ can be determined from (3.11) and (3.12); z_h is then that altitude at which the temperature T_h is reached.

This analysis neglects the effect of wind on the speed of the sound wave. If C_E is the speed of sound relative to the earth's surface in an atmosphere moving relative to the earth's surface, then

$$C_E = C + V_s \tag{3.13}$$

where C is the ordinary acoustic speed defined by (3.1) and V_s is the component of wind velocity in the direction of the ray. In this more general formulation, (3.3) and (3.7) become

$$\frac{C_{Ek}}{C_{E0}} = \frac{\sin i_k}{\sin i_0} \tag{3.14}$$

$$C_{Eh} = \frac{C_{E0}}{\sin i_0} \tag{3.15}$$

Equations (3.4) and (3.8) are not valid; wind and temperature effects cannot be separated with measurements at only one azimuth.

It would appear in retrospect that neglect of the wind effect on refraction was probably responsible in the early experiments for the deduction of stratopause temperatures that were too high. A simple calculation shows that wind may be very important under reasonable circumstances. Suppose that an anomalously propagated sound wave reaches the ground with an angle of incidence of 80°, where the ground temperature is 273°K and the wind is zero; and suppose further that at the level z_h for this particular ray the wind component in the direction of propagation is 30 m sec^{-1}. Then, according to Eq. (3.8), which neglects wind, T_h would be computed as 282°K; but according to Eq. (3.15) T_h is actually 235°K.

A number of sound experiments reported during the 1950's were designed to measure both temperature and wind (Crary, 1950, 1952, 1953; Crary and Bushnell, 1955; Richardson and Kennedy, 1952; Kennedy, Brogan, and Sible, 1955; and Murgatroyd, 1956). In all of these studies measurements were made at several azimuths from the explosion site. At least three azimuths are required to separate the three unknowns—temperature, wind speed, and wind direction. A method of analysis developed by Crary (1950) was used in most of the referenced experiments (Murgatroyd's reduction technique was somewhat different). Although the details of this method are not reproduced here, the method is similar in principle to the one described above, in that it involves assuming a particular form for the vertical variation of C_E [Eq. (3.13)] above a datum level. In combining the data from different azimuths, there is a certain amount of subjectivity involved.

Crary (1950) obtained wind and temperature data for Bermuda, the Canal Zone, and Alaska during the summer months and for Alaska during winter months. Crary (1953) reported data from Alaska between May 1950 and May 1951. From July 1950 to May 1951, Richardson and Kennedy (1952) conducted a series of experiments in eastern Colorado in the vicinity of 39°N. These were repeated (Kennedy *et al.*, 1955) during the period October 1951 to August 1952. Murgatroyd's (1956) experiments were conducted in England during the years 1944–1945.

During the second period mentioned in our introduction, these experiments provided a good deal of the information available at latitudes other than 30°N (where rocket observations were concentrated). Data from these experiments played an important role in Murgatroyd's construction of Figs. 3.1 and 3.2. They are therefore generally consistent with Murgatroyd's formulation and are not reproduced here.

It is possible in principle to obtain more information about the vertical distribution of temperature in the refracting layer and to avoid some of the approximations used in this method of analysis, by obtaining measurements at several different distances from the sound source along each azimuth. An elegant method borrowed from seismology was used in several of the earlier experiments (when, however, wind was not considered) by Gutenberg (1932, 1939), by Cox *et al.* (1949), and by Johnson and Hale (1953). This method, called the travel-time method, is not described here, but see, for example, Gutenberg (1951).

3.2.3 Measurements with a Searchlight

A beam of light directed into the atmosphere suffers a depletion of its intensity because of absorption and scattering. The latter is accomplished by particulate matter or by air molecules. In the upper atmosphere, only molecular scattering is believed to be significant. Furthermore, the intensity of light scattered in a direction which makes an angle ψ with the direction of the beam is predicted by the theory of molecular scattering, originally developed by Lord Rayleigh, to be proportional to $(1 + \cos^2 \psi)$ and to the air density. Therefore a measurement of the scattered light at a known angle ψ and a known altitude z, together with a suitable determination of the other factors in the relationship, gives a measure of the density at altitude z.

Various attemps have been made to utilize this principle but we shall discuss only the experiment of Elterman (1954a,b) since it was the only one to produce a significant number of temperature profiles. Figure 3.6 shows schematically the geometry of the experiment. A 60-inch searchlight located at Cedro Peak near Albuquerque, New Mexico, served as the source to direct a beam of light into the upper atmosphere at a fixed angle of 60° from the horizontal. The receiver at Sandia Crest, about 20 km away, was a photomultiplier tube mounted at the focus of a 60-inch parabolic mirror, identical to the searchlight mirror.

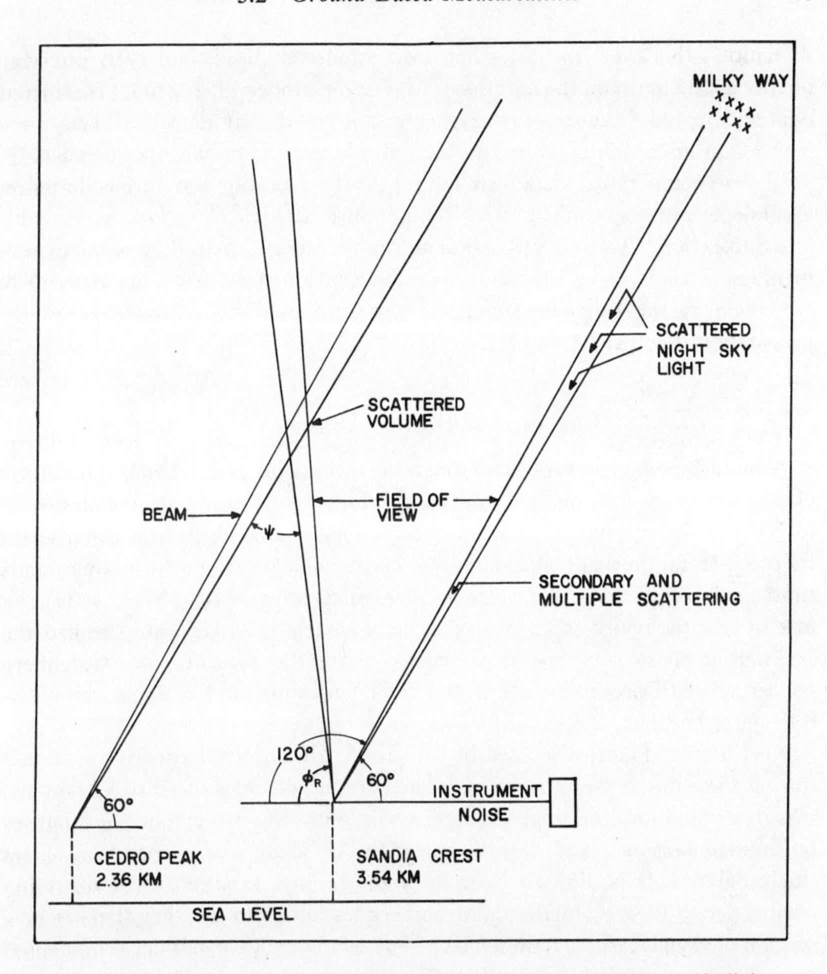

FIG. 3.6. Geometry of the searchlight experiment. (After Elterman, 1954b.)

This receiving apparatus could be moved in the vertical plane through the two sites to scan the beam at various altitudes z and corresponding angles ψ that were computed from the geometry of the situation.

The receiver, of course, detected not only light scattered directly from the beam, but also a background signal due to multiple scattering from the beam and to light from the stars and the night airglow. (Needless to say, the experiment was attempted only on moonless nights.) The greatest single experimental difficulty was to eliminate the background "noise" due to these sources. This was accomplished by suitable modulation and filtering and by calibration against the sky in a direction away from the beam.

Because of its restriction to moonless nights and some of the experimental difficulties, this experiment has not been extensively used. However, Elterman obtained data between the months of May and October 1952, primarily in June, September, and October. The experiment gives the air density directly, and conversion to a temperature profile introduces additional uncertainties. In particular, temperature data near the top of the sounding are unreliable unless an independent determination of temperature at high elevations is available (see Subsection 3.3.4 for a discussion of this problem). The ceiling of the experiment, as carried out by Elterman, was about 60 km. Data from the experiment are given in the original references and were used by Murgatroyd in the preparation of Figs. 3.1 and 3.2.

3.2.4 Noctilucent Clouds and Meteor Trails

Ground-based observations of naturally occurring phenomena, noctilucent clouds and meteor trails, provide wind information mainly at altitudes of 80 to 100 km. Interpretation of wind observations at this altitude is complicated by two factors: diurnal and semidiurnal oscillations associated with atmospheric tides have significant magnitude; and irregular components with a time scale of two or three hours and a vertical scale of several kilometers are also present. Neither of these components is treated here, the present discussion being concerned with prevailing winds and their latitudinal and seasonal variations. They are, however, discussed in Chapter 8.

Noctilucent clouds, discussed in Chapter 5, are observed selectively, namely during the summer months, at high latitudes, at altitudes of 80 to 85 km, and usually around local midnight. Therefore little information can be gained about latitudinal, seasonal, and vertical variations of wind; and recognition of the tidal components is difficult because of the limited variation of local time of occurrence. There is, furthermore, some question as to whether the observed motion of visual features represents the true wind or a wave motion. With respect to prevailing winds, Murgatroyd (1957) concluded from a study of the available observations:

> The measured winds vary considerably but in general they have the following common features.
>
> (a) The general drift is predominantly to the west indicating mainly easterly winds, although large north and south components are also found.
>
> (b) Measured speeds vary between 20 and over 100 m sec^{-1}, the mean being 40–50 m sec^{-1}.

Some relatively few meteors leave trails that are visible for a few minutes or even a few tens of minutes. These persistent trails are called meteor trains.

Observations of the displacements of the trains at various altitudes give a measure of the wind as a function of altitude. Optical (visual or photographic) observations of this type are naturally rather unsystematic and heterogeneous. Olivier (1942, 1947, 1957) has listed a large number of such observations. Murgatroyd (1957) studied those listed in the first two references (about 1600) and reached the following tentative conclusions:

(a) Winds at 80-100 km are predominantly westerly in summer, with easterly winds the least frequent.

(b) They are also predominantly westerly in winter, with northerly winds the least frequent.

Other results from visual studies of drifts are:

(i) Wind speeds between 50-100 m sec^{-1} are frequent between 70 and 100 km, and 30–50 m sec^{-1} between 30 and 70 km. Occasionally considerably higher wind speeds have been reported.

(ii) Some evidence has been advanced for significant changes at the 80-km level, Kahlke (1921) suggesting that wind directions are generally easterly below it and westerly above it. Trowbridge (1907), Hoffmeister (1946), and Fedynsky (1955) obtained results substantially in agreement.

(iii) There is evidence of large vertical wind shears with different wind directions at different levels and considerable turbulence at all levels from 40-100 km.

A technique for the study of wind between 80 and 100 km, based on radio-echo studies of the ionized trails left by meteors, has developed during the past 10 years. This technique is very useful because it is operable day and night and produces large numbers of data, so that components of the prevailing wind can be separated from components related to the diurnal variation. This is discussed in a good deal more detail in Chapter 8 but the results pertaining to the prevailing wind are summarized briefly here. Measurements have been obtained at Jodrell Bank, England (53°N), Adelaide, Australia (35°S), and Mawson, Antarctica (68°S).

Results at Jodrell Bank for the years 1953–1958 were given by Greenhow and Neufeld (1961). They refer to the 85–100-km region and may be summarized as follows:

(a) Winds are predominantly zonal, with west components of about 20 m sec^{-1} in both summer and winter, somewhat lighter east winds in the spring, and very weak east winds for a month in the autumn.

(b) The meridional component is small during most of the year but is consistently north during the summer months with magnitude 10–20 m sec^{-1}.

(c) Although the seasonal variation of wind is consistent from year to year, large irregular changes with periods of several days are sometimes superimposed. Greenhow and Neufeld (1960) have suggested that these correspond to large pressure systems near 90 km.

These results, insofar as they pertain to summer and winter and to the mean zonal component are in satisfactory agreement with Murgatroyd's zonal wind cross section.

Measurements at Adelaide for midseasonal months during 1953-1954, plus a few months at the end of 1952 and beginning of 1955, were summarized by Elford (1959). These data were stratified into three altitude regions centered at 80 km, 90 km, and 100 km. They show that

(a) prevailing winds are predominantly zonal and west;

(b) in (Southern Hemisphere) summer, winds are west and speed increases with altitude to about 75 m sec^{-1} at 100 km; extrapolation downward indicates zero zonal component at about 70 km;

(c) in winter, winds are west and speed decreases with altitude to zero at about 100 km;

(d) east winds occur only during the spring, with speeds of 10–25 m sec^{-1};

(e) meridional components are small and are consistently south in the summer and north in the winter.

Observations at Mawson, Antarctica (68°S), during December 1957, December 1958, and January 1959 (Elford and Murray, 1960) gave results different from the above in one important respect: Meridional components at this location during these months were much more prominent. Data were stratified to apply to two levels, one centered at 90 km, one at 100 km. During both Decembers, winds were approximately SE at the lower level with speeds 40–60 m sec^{-1} and nearly S at the upper level with speeds 25–40 m sec^{-1}. During the one January, winds were more nearly S at the lower level and SW at the upper level, with speeds comparable to the above. With respect to the zonal component, these results would imply a shift from east to west at 90 to 100 km at high latitudes in summer, a result not inconsistent with Fig. 3.2.

The behavior of the meridional component of wind at all three locations implies flow from the summer toward the winter pole, most pronounced at Mawson. Whether this represents real behavior of the atmosphere at the 80- to 100-km level or whether it results from the small data sample available is a most important question and awaits further elucidation.

3.3 Rocket Measurements

The use of rockets to carry instruments into the upper atmosphere has contributed in great measure to the advances of the past 15 years. Measurements of wind and temperature (or density or pressure) in the upper stratosphere and mesosphere represent only a fraction of the useful results. Other types of observation are discussed in other chapters. At this stage, however, it is appropriate to discuss briefly and in elementary fashion the characteristics of some of the rocket sounding systems that have been and are still being used. Those interested in the upper atmosphere, even if not directly concerned with rockets and rocket-instrumentation techniques, should at least be able to associate the names of various systems with the gross characteristics involved. Subsection 3.3.1 has the modest ambition of presenting enough information for this purpose. In Subsection 3.3.2, we consider methods and results of measuring pressure and density by means of pressure gages on the moving rocket. Subsection 3.3.3 takes up the falling-sphere experiment and Subsection 3.3.4 the rocket-grenade experiment. In Subsection 3.3.5 results (pertaining to wind) of the observation of artificial clouds formed by ejections from moving rockets are reviewed. Recent developments pertaining to "meteorological rockets" and their frequent employment at many missile ranges are deferred to Section 3.4.

3.3.1 Rockets for Atmospheric Research

A rocket is a jet-propelled vehicle that characteristically carries with it not only its fuel supply but also the means for burning it. The rocket, unlike the jet aircraft, does not depend on oxygen from the atmosphere for the latter function.

Operation of the rocket is based on the simple and fundamental principle that the total momentum of a physical system is changed only by the application of external forces. If we consider the physical system to include both the rocket body and the propellant fuel, then the ejection at high speed of a certain mass of fuel requires the rocket and the remainder of the fuel, in the absence of external forces and in order that the total momentum be conserved, to gain speed in the opposite direction from that traveled by the ejected fuel. Of course, the complete problem is much more complicated by the continuous rather than sudden ejection of mass, and by the presence of external forces such as gravity and drag of the surrounding atmosphere.

Problems of rocketry are beyond the scope of this book. Massey and Boyd (1959) give a clear, nonmathematical discussion of the characteristics of rockets. Newell (1959) gives a useful elementary technical discussion of rocket theory as well as considerable detail about the characteristics of specific sounding rockets. The latter publication contains ample references for the reader who desires to pursue the matter further. The objective here is simply to acquaint the reader with functional characteristics such as payload, peak altitude, and launching characteristics of some rockets that have been used for atmospheric studies. Tables 3.1a and 3.1b are designed for this purpose. No attempt is made to cover rocket systems of other countries than the United States, although certainly some fine systems have recently come into use. Neither is it attempted to cover rocket systems that are currently under development.

TABLE 3.1a

SOME SOUNDING ROCKETS AND THEIR CHARACTERISTICS[a]

Name	Payload (kg)	Peak altitude (km)	Launch	Remarks
V-2 (German A−4)	1000	160	Tower	First rocket used in upper-atmospheric research. Marginal stability when lightly loaded
Viking I	225	215	Tower	Too expensive for extensive use. Not currently used for sounding purposes
Viking II	550	250	Tower	
Aerobee	70	120	Tower	Used extensively during the 1950's
Aerobee-Hi	70	240	Tower	
Deacon Rockoon	Up to 25	60–120	Rocket launched from balloon at altitudes of 15 to 25 km	Plastic balloon and solid propellant rocket; inexpensive; balloon usually launched from shipboard; relatively easy to handle and assemble
Loki I Rockoon	3	60–100		
Loki II Rockoon (Hawk)	Up to 5	100–150		
Deacon-Nike	25	110	From ground or ship	Two solid-propellant motors

[a] The list is by no means comprehensive, but it includes most of the U. S. rockets mentioned in this book.

The trend in the development of sounding rockets over the years has been toward smaller, special-purpose devices. In part, this is due to increased ability to put the necessary instrumentation into smaller and smaller payloads; in part, it is due to the necessity of decreasing drastically the cost per rocket so that larger numbers of soundings can be made at a cost that is not prohibitive. The Viking rockets cost \$300,000–\$400,000 apiece, the Aerobees around \$30,000, and the Deacon rockoon system under \$2000. Another trend has been toward the use of solid propellants rather than liquid propellants, to achieve ease of handling and portability.

However, even with radically decreased cost and radically increased mobility, the sounding rocket as a tool for routine atmospheric measurement suffers from

TABLE 3.1b

A LIST OF SOUNDING ROCKETS USED BY THE AIR FORCE CAMBRIDGE RESEARCH LABORATORIES AS OF MARCH 1964[a]

Name	Payload (kg)	Peak altitude (km)	Launch	Remarks
Loki Dart	3.2	60	Rail	One solid-propellant motor with nonpowered dart
Arcas	4.5	66	Closed breech	Single solid-propellant motor
Black Brant II	145	144	Rail	Single solid-propellant motor
Nike Cajun	31.6	145	Rail	Two solid-propellant motors
Nike Apache	31.6	158	Rail	Two solid-propellant motors
Aerobee 150	68	250	Tower	Solid-propellant booster, liquid sustainer
Astrobee 200	68	282	Rail	Two solid-propellant motors
Exos	36.2	480	Rail	Three solid-propellant motors
Javelin	36.2	1120	Rail	Four solid-propellant motors

[a] Compiled with the help of Mr. Philip Gustafson and Mr. Robert Leviton of the Aerospace Instrumentation Laboratory, Air Force Cambridge Research Laboratories. Rockets used by other agencies are similar but may vary in detail.

a currently unsolved weakness: it cannot be fired in inhabited areas. Even at some of the fixed rocket-launching sites, certain rockets can be fired only when the wind direction is such as to preclude any possibility of their landing in nearby populated areas. Routine soundings of the upper stratosphere and mesosphere, similar in frequency to balloon soundings of the troposphere and lower stratosphere, are still years away.

3.3.2 PRESSURE AND DENSITY MEASUREMENTS FROM MOVING ROCKETS

Measurement of atmospheric structure (temperature, pressure, or density) from instrumentation on the moving rocket poses a difficult problem. Because of the small time periods available and the problems of aerodynamic heating, direct temperature measurements are never made. Various types of pressure gauges are available to measure rather accurately pressures at the surface of the rocket; but these observed quantities must be reduced to ambient-pressure, or ambient-density, values. Most of the early measurements were made in this manner, and it remains practically the only technique (other than deduction of atmospheric density from the drag on satellites) of sounding the atmosphere above about 100 km. Structure measurements in the upper stratosphere and mesosphere are now more commonly made by observing various objects ejected from rockets.

Measurements of atmospheric structure prior to 1952 were summarized and combined by the Rocket Panel (1952) into what has come to be called the "Panel Atmosphere" (see also F. L. Whipple, 1952b). The soundings on which this average was based were carried out at the White Sands Proving Ground (32°N) in New Mexico, with the exception of a single firing at the equator which showed no systematic deviation from the New Mexico results. All times of the year were represented with no evidence of a seasonal variation. The temperatures of the Panel Atmosphere were used by Murgatroyd. Hence, we shall not reproduce the results separately. It will be instructive, however, to discuss the measuring techniques that were used.

(a) Many of the data were obtained by the Naval Research Laboratory with V-2, Aerobee, and Viking rockets. These have been reported by Havens, Koll, and LaGow (1952). In all nine experiments, pressures were measured at a special location on the rocket (about 10 rocket diameters back from the tip, the exact distance depending on the rocket configuration) where wind-tunnel experiments have shown the pressure to be equal to the ambient pressure within a few per cent. From this information alone, temperature can be derived only from the *slope* of the curve of logarithm of pressure versus height,* according to Eqs. (1.1) and (1.2). This differentiation process, of course, involves taking differences of experimental quantities and may introduce errors considerably larger than the errors in the basic data. Havens, Koll, and LaGow did not themselves treat the data in this manner.

(b) During a half-dozen or so of the same flights stagnation pressures at the tip of the nose cone were also measured at some altitudes. Rayleigh's pitot-tube theory relates the pressure at this location to the ambient density and pressure and the speed of the rocket (Havens *et al.*, 1952). (At altitudes above 100 km, where the mean free path of the air molecules is large, a modification of this theory is necessary; see Subsection 6.4.1.) Thus, ambient density can be determined from this measurement together with the measurement of ambient pressure and the speed as determined by tracking or other data. (It is noteworthy that the ambient pressure need not be known precisely to give reasonable estimates of density by this procedure.) When both density and pressure are known, temperature can, of course, be computed with the aid of (1.1).

(c) Two Aerobee firings by the University of Michigan under an Air Force

* It is assumed throughout this chapter that at the altitudes being discussed the mean molecular weight of air has its sea-level value.

contract, measured pressures at the tip and at various locations on the conical nose, measurements which can be related by aerodynamic theory to the ambient density and pressure and hence to ambient temperature. These results and the method of reduction used have been discussed by Sicinski, Spencer, and Dow (1954).

(d) In one sounding by a University of Michigan group under a U. S. Army Signal Corps contract, the angle of the shock wave at the nose of the rocket was measured, and from it the Mach number deduced. The Mach number and the speed of the vehicle (from tracking data) yield the ambient temperature.

(e) Finally, the Panel Atmosphere was based on a half-dozen measurements by the U. S. Army Signal Corps Engineering Laboratories with the rocket-grenade method. In this method, the Aerobee rockets ejected a series of exploding grenades at regular altitude intervals throughout the upper stratosphere and the mesophere. The times and angles of arrival of the various sound waves at an array of ground-based microphones, together with a knowledge of the times and locations of the explosions, made it possible to deduce temperatures and winds in the upper stratosphere and the mesosphere. These results have been summarized by Ference *et al.* (1956) and by Weisner (1956). This useful technique, which has seen considerable application in more recent years, is discussed more fully in Subsection 3.3.4.

It is evident that methods (a) through (c) require the measurement of pressures over a wide range of magnitudes. Consequently several types of pressure gauges are often used on the same rocket. Details of the instrumentation are to be found in the original references. In all of the methods (a) to (d) the theory for reduction of the measurements assumes a zero angle of attack by the rocket. In practice, the longitudinal axis of the rocket usually precesses about the direction of the trajectory and corrections have to be made for this factor. Finally, it is necessary to know the trajectory and speed of the rocket rather accurately, and this is usually accomplished by visual or radar tracking.

During 1953 and 1954, the Naval Research Laboratory obtained measurements, this time at high latitudes (LaGow and Ainsworth, 1956). The vehicles were Deacon rockets fired from plastic balloons launched from shipboard. Two soundings in August 1953 gave ambient pressures, densities, and temperatures by method (b) up to 40–45 km. Two soundings in July 1954 obtained stagnation pressures only and ambient densities were determined approximately from the Rayleigh formula and estimates of the ambient pressure. None of these rockets was tracked, and all trajectories had to be estimated from telemetered values of launching altitude and fuel-burning time, together with estimates of the effect of atmospheric drag. The calculated trajectories were subject to the further constraint that ambient pressure must be equal at the same level during the upward and downward portion of the rocket's path.

During the IGY at Fort Churchill, Canada, personnel of the Naval Research Laboratory carried out two successful atmospheric-structure measurements with Aerobee-Hi rockets equipped with pitot-static tubes. Both stagnation pressures and ambient pressures were obtained. These experiments, and the theory and method of reducing such measurements to ambient structure parameters, have been discussed in detail by Ainsworth, Fox, and LaGow (1961). The results are given here in Fig. 3.7 in terms of ambient temperature, which

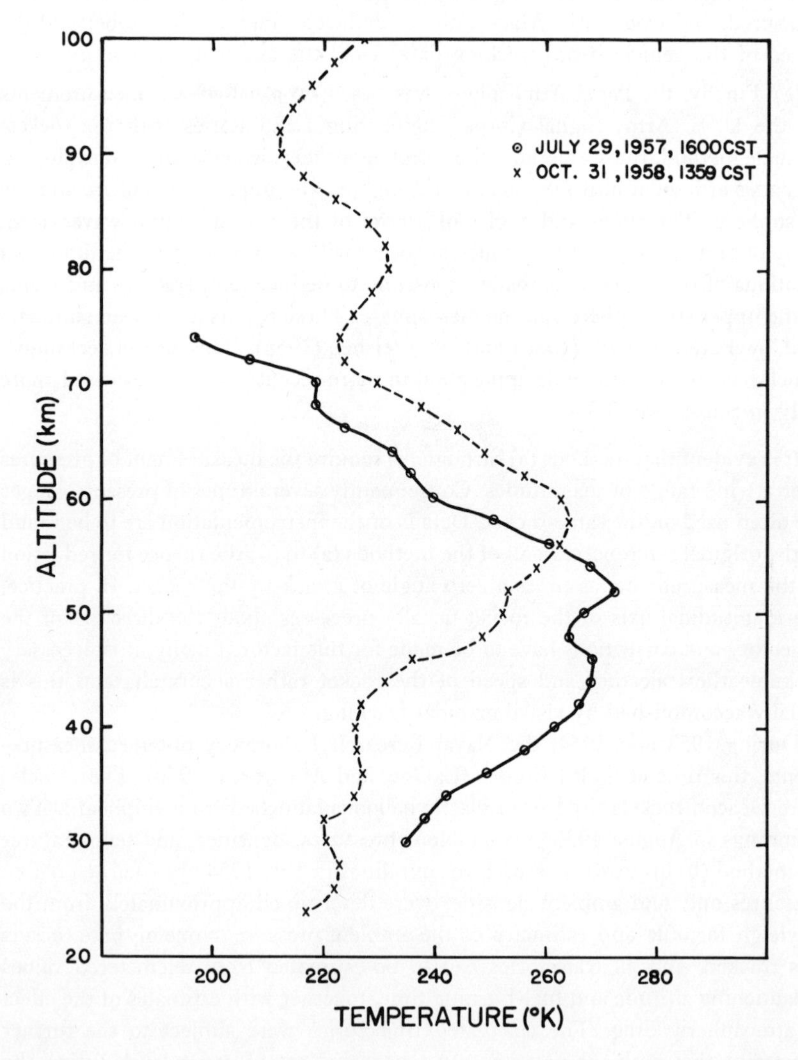

FIG. 3.7. Temperature profiles deduced from pitot-static tube measurements at Churchill on the dates indicated. (Prepared from data given by Ainsworth *et al.*, 1961.)

was not of course directly measured, but was deduced from the pressure and density determinations. Results from these same experiments that pertain to altitudes above 100 km are included in Chapter 6.

The result for the summer day shows a temperature at the stratopause that is some 20°C less than that indicated by Murgatroyd (Fig. 3.1). Temperatures in the vicinity of 30 and 70 km are, however, in satisfactory agreement with Fig. 3.1. The fall-day sounding shows a stratospause at a higher elevation and with a lower temperature than in summer, and agrees quite well with winter conditions shown in Fig. 3.1. Note the pronounced secondary temperature maximum at about 80 km. As we shall see in later parts of this section, secondary temperature maxima of this type appear to be the rule rather than the exception in the high-latitude winter mesosphere.

3.3.3 The Falling-Sphere Experiment

A method for the determination of density in the mesosphere and upper stratosphere is the "falling-sphere" method. Twelve successful measurements have been made by a group from the Department of Aeronautical and Astronautical Engineering of the University of Michigan. This technique uses the acceleration of a freely falling sphere ejected from a rocket near the top of its trajectory to determine the density and, from a knowledge of density as a function of altitude, the temperature. The effect of air motions being rather small, the principal forces on the sphere are a downward force due to gravity and an upward force due to drag of the surrounding air. The drag acceleration may be expressed as

$$a_D = \frac{\rho V^2 A c_D}{2M} \tag{3.16}$$

where V is the speed of the falling sphere, A is its cross-sectional area, M is its mass, ρ is the density of the surrounding air, and c_D is the drag coefficient. The latter, a function of V and of Reynolds number, is known from wind-tunnel studies. From a study of the sphere's trajectory and a knowledge of its size and mass, it is possible to determine ρ.

In the earliest version of the experiment, a 4-ft nylon sphere containing a miniature transponder was tracked by the DOVAP (Doppler velocity and position) radio-tracking system. Four measurements at White Sands at various seasons in 1952 and 1953 gave results generally consistent with the Panel Atmosphere (Bartman *et al.*, 1956). (The vertical temperature distribution at 30°N in Fig. 3.1 is rather similar to that in the Panel Atmosphere.) Later versions of the experiment utilized a 7-inch rigid sphere containing a specially designed accelerometer. The direct measurement and telemetering of accelerations obviated the need for tracking. The trajectory and speed of the sphere could be determined by integration of the acceleration records. The smaller size of the sphere allowed the firings to be carried out with the smaller Nike-Deacon or Nike-Cajun sounding rockets, four from shipboard in 1956 at various latitudes and four from Fort Churchill, Canada, during the International Geophysical Year.

Jones *et al.* (1959) gave the results of all of these measurements. Several late-autumn measurements at various latitudes showed that the density at this season becomes progressively less than the Panel Atmosphere density as latitude and altitude above 25 km increase. This implies lower temperature in the upper stratosphere than incorporated in the Panel Atmosphere.

A series of measurements at Churchill in early 1958 is particularly interesting. They occurred during the period of the major stratospheric warming of January 1958 and showed that, at least in this one case, large temperature changes extended to 50 or 60 km. Before discussing these, let us consider the method used to deduce temperatures from the basic density data.

Temperature in this technique is deduced from the density measurements with the use of the hydrostatic equation and the equation of state. Let the subscipt 1 refer to a level near the top of the sphere's trajectory and the subscript z refer to a level z for which temperature is to be determined. Integration of the hydrostatic equation gives

$$p_z - p_1 = \int_z^{z_1} g\rho \, dz \tag{3.17}$$

Substitution from the equation of state then yields

$$T_z = \frac{\int_z^{z_1} g\rho \, dz}{R_0 \rho_z} + \frac{\rho_1 T_1}{\rho_z} \tag{3.18}$$

Some arbitrary reasonable value has to be assigned for T_1, the temperature at the upper limit of the measurements (unless it is known from an independent measurement). However, any error in this initial "guess" damps out rapidly and has little effect on the temperatures derived for levels 15 km or more below the level, where the ratio ρ_1/ρ_z in the second term of (3.18) is small.

At first thought, one might consider it more reasonable to take the reference level z_1 near the bottom of the sounding and to use for T_1 an independent temperature measurement, say from radiosonde data. However, this procedure gives no better results near the top of the sounding. The ratio (ρ_1/ρ_z) in the second term of (3.18) then becomes increasingly large as altitude increases and the computed value of T_z becomes increasingly sensitive to a small uncertainty in the value of T_1. It should also be noted that in the development of Eq. (3.18) from Eq. (3.17) the mean molecular weight of air is assumed to have its sea-level value.

Some temperature distributions observed during and after the January 1958 warming are shown in Fig. 3.8. Note the large temperature rise at 40 to 50 km between 27 January and 29 January, and the return to "normal" temperature in early March. At Churchill, this temperature rise preceded large temperature rises in the lower stratosphere. However, large temperature rises had already

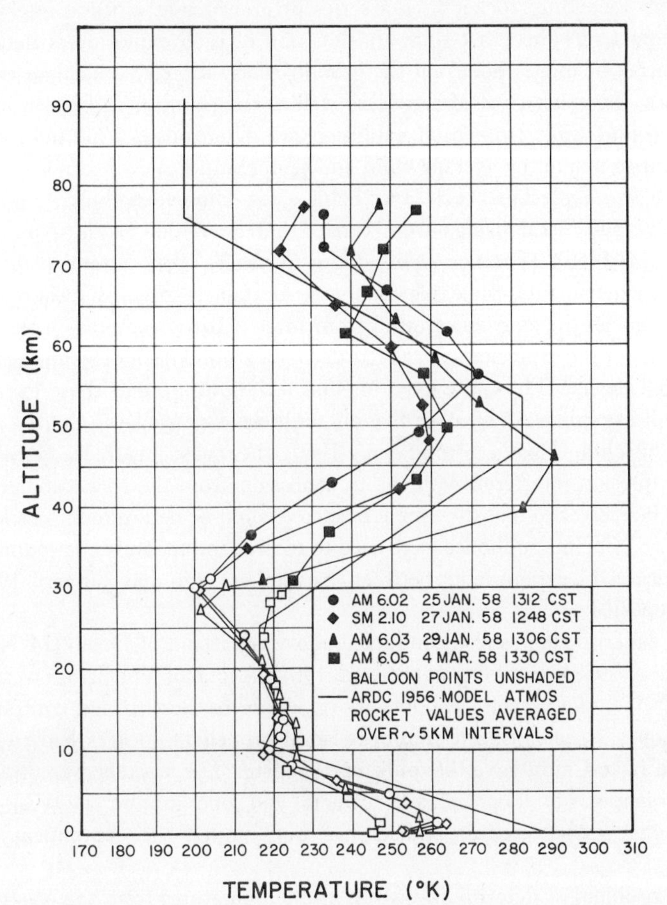

FIG. 3.8. Temperature as a function of altitude at Churchill on four days in early 1958. Temperatures above the balloon ceilings were deduced from density measurements by the falling-sphere method. Note the relatively high temperature at 40 to 50 km on 29 January 1958, presumably associated with anomalous warming in the lower stratosphere. (After Jones *et al.*, 1959.)

occurred in the lower stratosphere at other locations before 27 January; the first large temperature rises at 25 mb evidently took place in western Europe a week or so earlier (Scherhag, 1960). Note also that in all four soundings there are indications of secondary temperature maxima in the mesosphere.

3.3.4 The Rocket-Grenade Experiment

The rocket-grenade experiment is a very powerful method of determining winds and temperatures in the upper stratosphere and mesosphere. In this experiment, an ascending rocket fires a series of grenades at altitude intervals of a few kilometers. The exact time and location of each explosion is determined. At an array of microphones on the ground nearly directly under the exploding grenades, the exact time of arrival as well as the azimuth and elevation angles of the sound wave from each explosion are determined. This information is sufficient to define the average wind and temperature in each atmospheric layer between two explosions, each layer being a few kilometers thick.

Twelve successful firings were carried out at White Sands, New Mexico, during the 1950's (Ference *et al.*, 1956; Weisner, 1956; Stroud *et al.*, 1956). In these experiments, the explosions were located in space by means of photographs against the star background from three ballistic cameras on the ground. The times of the explosions were recorded by a ground flash detector, a recording photocell device. Thus, the experimental technique limited these experiments to cloudless nights. An array of five microphones was used.

At Churchill, Canada (Stroud *et al.*, 1960), the rocket was tracked by DOVAP. Each explosion interfered with the transmission from the rocket to the ground DOVAP receiver so that the time it occurred could be determined with sufficient accuracy. Nine microphones were used to receive and measure the sound signals. Ten successful experiments between November 1956 and January 1958 have been reported.

The experiment was also conducted above the island of Guam (14°N) during November 1958 (Nordberg and Stroud, 1961). A total of nine firings were carried out, of which only seven gave data above 50 km. In these firings, contrary to the earlier ones, which utilized Aerobees, solid-propellant rockets had to be used and the Nike-Cajun gave the bulk of the data. The rockets were tracked by a combination of one ballistic camera and one ground DOVAP station, which gave rather accurate trajectories but limited the experiment to clear nights.

The problem of determining winds and temperatures from the observables is somewhat complex in the most general case. The cited references and a paper by Groves (1956) discuss this problem in considerable detail.

A particularly simple case would occur if two grenades could be exploded directly above the microphone array and if there were no wind. Then the travel times of the two rays, t_2 and t_1, together with the heights of the two explosions, z_2 and z_1, would determine the average speed of sound in the intermediate region, by

$$C_{12} = \frac{z_2 - z_1}{t_2 - t_1} \tag{3.19}$$

In practice, it is impossible to achieve exact vertical alignment of the microphones and all the explosions. And, of course, wind is present and is indeed one of the variables to be measured.

Consider a grenade exploded at the altitude z_1, the sound being detected at the microphones a time t_1 after the explosion, and a second grenade exploded at a somewhat higher altitude z_2, the sound being detected at the microphones a time t_2 after the explosion. Assume that the temperature and wind structure between the surface and z_1 are known and it is desired to determine the mean wind and temperature between z_1 and z_2. Let

$\mathbf{V}_{01}$ = mean vector (horizontal) wind between the surface and z_1
$\mathbf{V}_{12}$ = mean vector (horizontal) wind between z_1 and z_2
$\mathbf{r}_1$ = displacement of the ray from z_1 due to the wind
$\mathbf{r}_2$ = displacement of the ray from z_2 due to the wind
$\mathbf{r}_2'$ = displacement of the ray from z_2 due to the wind between the ground and z_1
$\mathbf{r}_2''$ = displacement of the ray from z_2 due to the wind between z_1 and z_2
t_2' = time spent by the ray from z_2 between the ground and z_1
t_2'' = time spent by the ray from z_2 between z_1 and z_2

By definition,

$$t_2 = t_2' + t_2'' \tag{3.20}$$

$$\mathbf{r}_2 = \mathbf{r}_2' + \mathbf{r}_2'' = \mathbf{r}_2' + \mathbf{V}_{12}t_2'' \tag{3.21}$$

$$\mathbf{r}_1 = \mathbf{V}_{01}t_1\,; \qquad \mathbf{r}_2' = \mathbf{V}_{01}t_2' \tag{3.22}$$

By application of (3.6),

$$t_1\,\overline{\cos i_1} = t_2'\,\overline{\cos i_2} \tag{3.23}$$

where $\overline{\cos i_1}$ is an appropriate mean value of the angle of incidence of the ray from z_1 between z_1 and the ground and $\overline{\cos i_2}$ is an appropriate mean value of the ray from z_2 between z_1 and the ground. According to (3.22) and (3.23),

$$\mathbf{r}_1\,\overline{\cos i_1} = \mathbf{r}_2'\,\overline{\cos i_2} \tag{3.24}$$

The times t_1 and t_2 are measured and from the known locations of the explosions, the known temperature and wind field between the ground and z_1, and the known angles of incidence at the ground, it is possible to calculate $\mathbf{r}_1$, $\mathbf{r}_2$, $\overline{\cos i_1}$, $\overline{\cos i_2}$. With this information, $\mathbf{r}_2'$, t_2', t_2'', and $\mathbf{V}_{12}$ can be calculated from (3.20), (3.21), (3.23), and (3.24). Of these $\mathbf{V}_{12}$ is one of the desired variables and t_2'' is needed for the computation of the mean temperature of the layer.

The latter is obtained from Snell's law, Eq. (3.14). If V_{12} is the component of $\mathbf{V}_{12}$ in the direction from the location of the explosion at z_2 toward the microphones, and if i_{12} is the angle of incidence of the sound ray from z_2 within the layer z_1 to z_2, then the V_s of (3.13) is given by $V_s = V_{12} \sin i_{12}$ and (3.14) yields

$$\frac{C_{12}}{\sin i_{12}} + V_{12} = \frac{C_0}{\sin i_2} \tag{3.25}$$

where C_0 is the (acoustic) speed of sound at the ground, i_2 the measured angle of incidence at the ground (where the wind speed is assumed to be zero), and C_{12} is the (acoustic) speed of sound in the layer from z_1 to z_2. A further relation between C_{12} and i_{12} is obtained from

$$C_{12} \cos i_{12} = (z_2 - z_1)/t_2'' \tag{3.26}$$

From (3.25) and (3.26), C_{12} and hence T_{12} may be determined.

For discussion of probable error, see the references cited above. Generally the errors are believed to be less than $\pm 5°$K, ± 10 m sec^{-1}, and $\pm 18°$ in wind direction for altitudes less than 75 km.

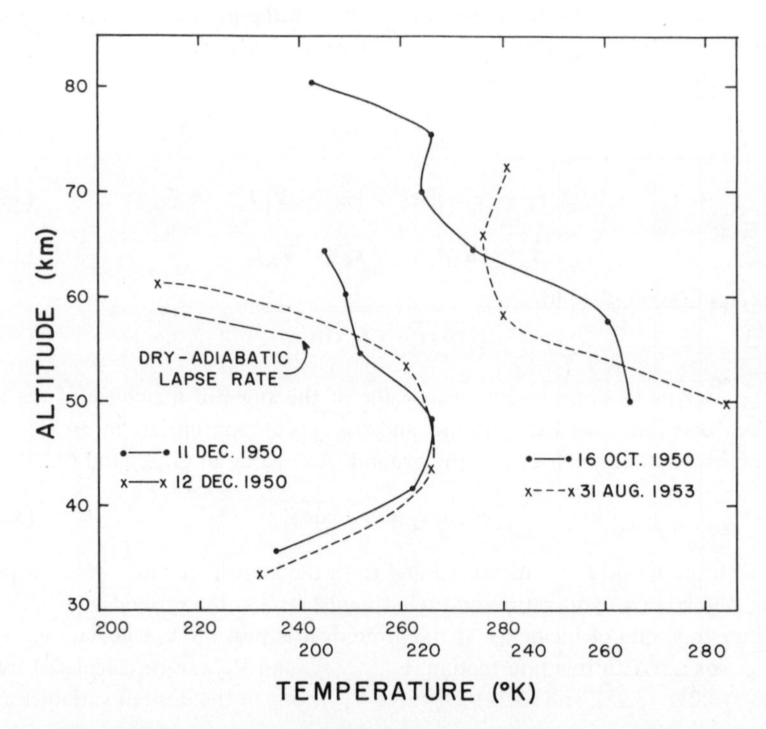

FIG. 3.9. Temperature as a function of altitude at White Sands on the dates indicated, determined from the rocket-grenade experiment. The lower abscissa scale applies to the two soundings at the right. The dry-adiabatic lapse rate is shown for comparison.

The White Sands data applicable to mean summer and winter conditions were included in the preparation of Figs. 3.1 and 3.2. The data showed maximum temperatures around 270°K at about 50 km with little seasonal variation at this latitude. Winds were strong and westerly in the months October to February, less strong and easterly during the months April to August. The lapse rate in the mesosphere was rather small by tropospheric standards, on the order of 2 to 3°C km^{-1}.

Individual soundings, however, showed temperature distributions that differed individually from one another or from the average picture depicted by standard atmospheres. By way of illustration, Fig. 3.9 contains four of the twelve profiles. Note the large temperature difference just above 60 km between two measurements one day apart. Note also the temperature inversions at 70 to 75 km on 16 October 1950 and at 65 to 70 km on 31 August 1953. These soundings also

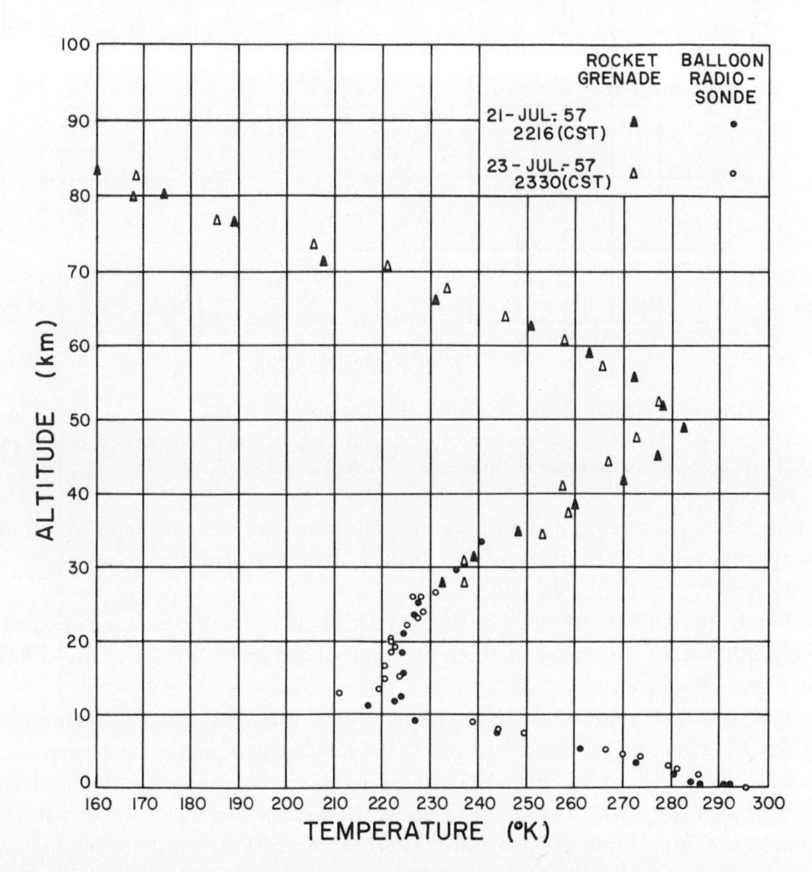

FIG. 3.10. Temperature as a function of altitude at Churchill, determined from the rocket-grenade experiment on the indicated days in July 1957. (After Stroud *et al.*, 1960.)

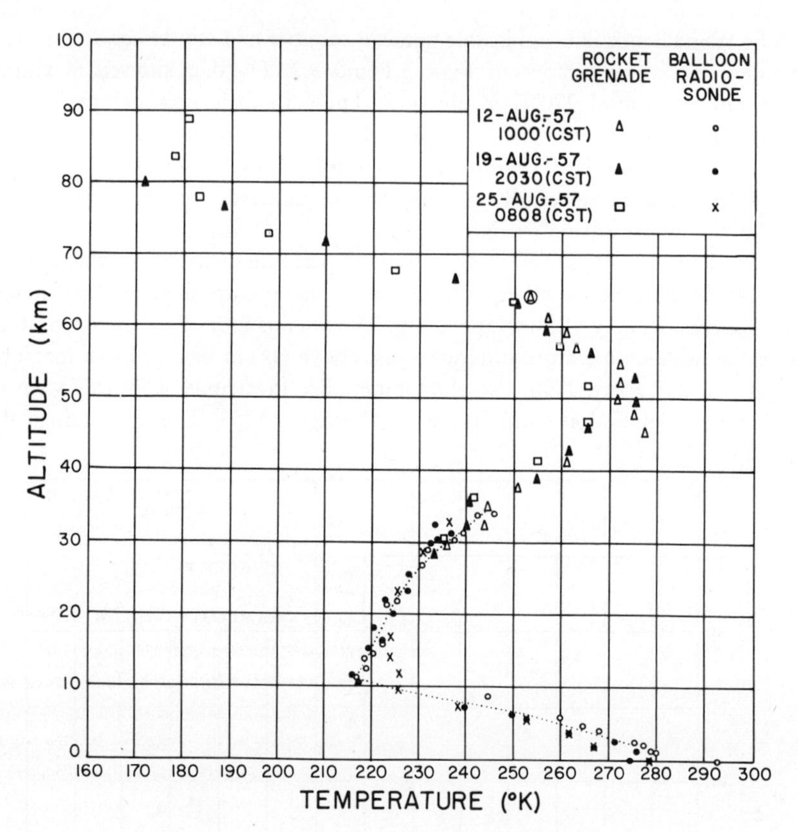

FIG. 3.11. Temperature as a function of altitude at Churchill, determined from the rocket-grenade experiment on the indicated days in August 1957. The circled point near 65 km represents an average over three layers. (After Stroud *et al.*, 1960.)

show occasions over limited altitude intervals when the dry-adiabatic lapse rate is approximated, even though the "mean" lapse rate in the mesosphere is considerably less than this value.

The Churchill data were not available to Murgatroyd and they are so exceedingly interesting that most of them are reproduced from Stroud *et al.* (1960) in Figs. 3.10–3.14.

The summer soundings show rather little variation among themselves at any elevation. Of particular interest is the low temperature at the mesopause, as low as 160°K on 21 July 1957. This is somewhat less than deduced by Murgatroyd (Fig. 3.1). The temperature at the stratopause is also somewhat less than in Fig. 3.1. Winds are generally easterly above 50 km, but there is a noticeable tendency for them to change to westerly above about 75 km, a considerably lower level than indicated in Fig. 3.2.

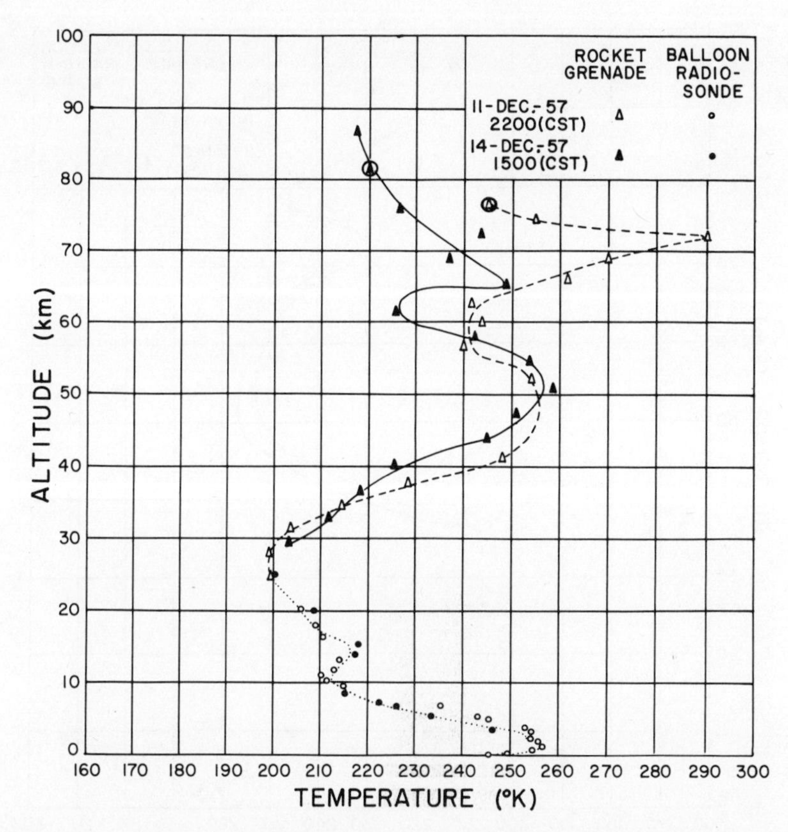

FIG. 3.12. Temperature as a function of altitude at Churchill, determined from the rocket-grenade experiment on the indicated days in December 1957. The lapse rate between 55 and 60 km on 14 December is very close to the dry-adiabatic rate. The circled points near 80 km represent averages over two layers each. (After Stroud *et al.*, 1960.)

The winter soundings (Figs. 3.12 and 3.13) show a well-marked change of lapse rate at about 30 km, stratopause at a somewhat lower elevation than shown in Fig. 3.1, and pronounced temperature inversions in the mesosphere, such as we have noted before. Note the large temperature change at about 70 km between 11 December 1957 and 14 December 1957. Wind speeds greater than 100 m sec^{-1} and occasionally exceeding 150 m sec^{-1} in the mesosphere are shown. The meridional wind components on 27 January are of considerable interest. North winds below 40 km and south winds above imply a marked east-west temperature gradient, with warmer air to the east of Churchill in the upper stratosphere and lower mesosphere. This inference may be combined with the observed temperature rise near the stratopause (Fig. 3.8) between 27 January and 29 January to indicate the progression of a warm center from

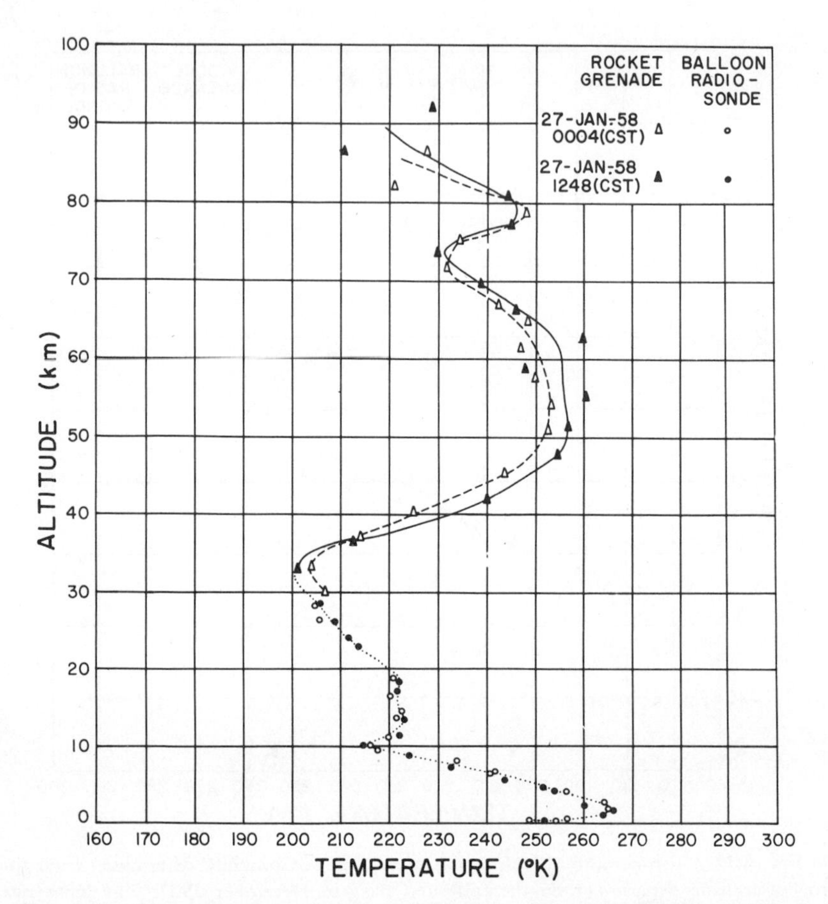

FIG. 3.13. Temperature as a function of altitude at Churchill, determined from the rocket-grenade experiment on the indicated days in January 1958. (After Stroud *et al.*, 1960.)

east to west, much as it progressed at lower elevations during the 1958 major warming. This point has been discussed by Teweles (1961).

The agreement between rocket-grenade and radiosonde measurements in the region of overlap is impressive, and on one occasion (27 January 1958) the rocket-grenade temperatures compared favorably with those measured at the same time by the falling-sphere method (see Fig. 3.8).

The Guam data for November 1958 are shown in Figs. 3.15a and 3.15b. The stratopause temperature of about 270°K at an altitude just below 50 km is a bit less than indicated on Fig. 3.1. Variation of temperature with altitude, in both upper stratosphere and mesosphere, is smoother than at high latitudes, and time changes appear to be less. The west winds above 30 km are in general

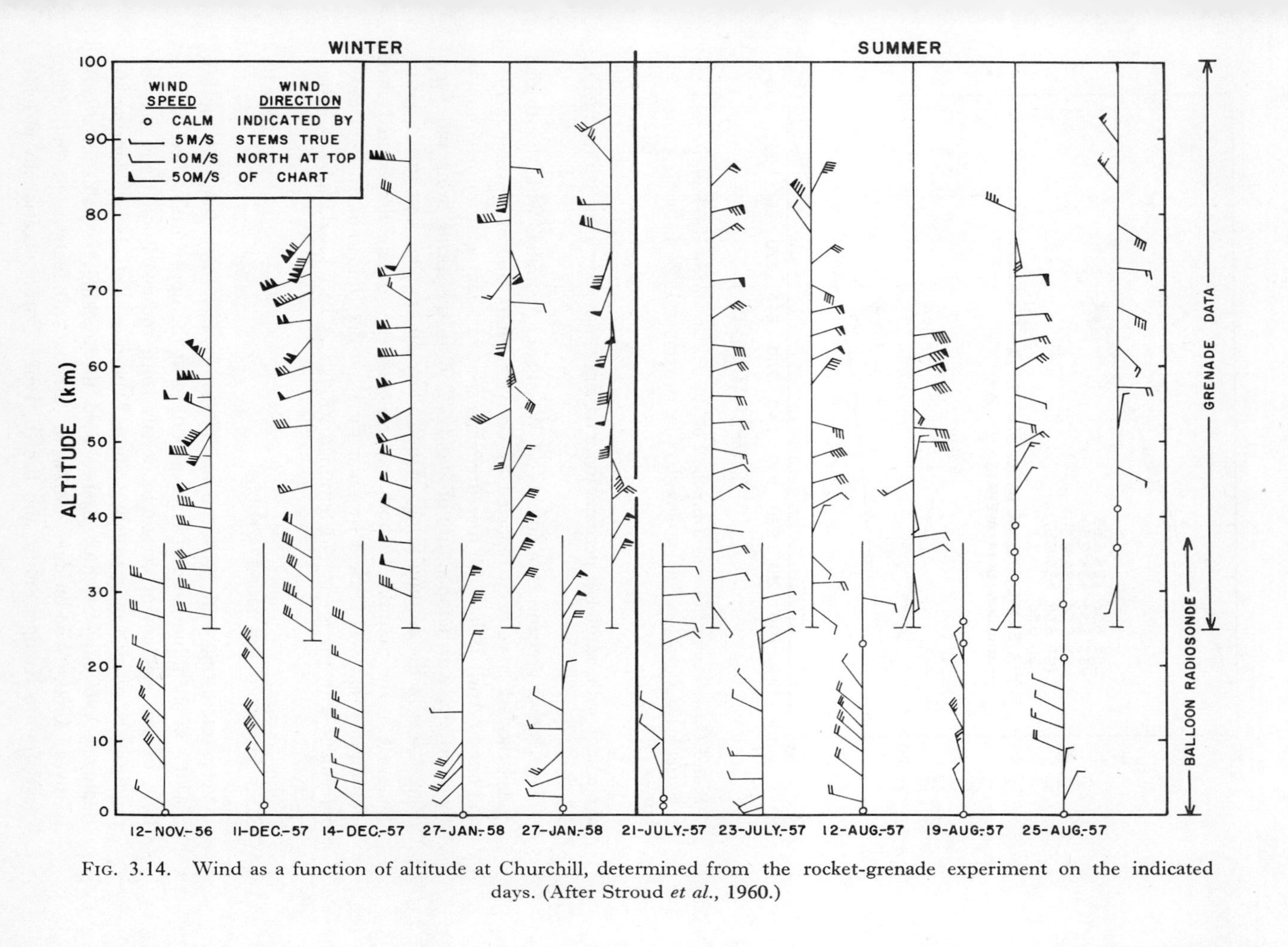

Fig. 3.14. Wind as a function of altitude at Churchill, determined from the rocket-grenade experiment on the indicated days. (After Stroud *et al.*, 1960.)

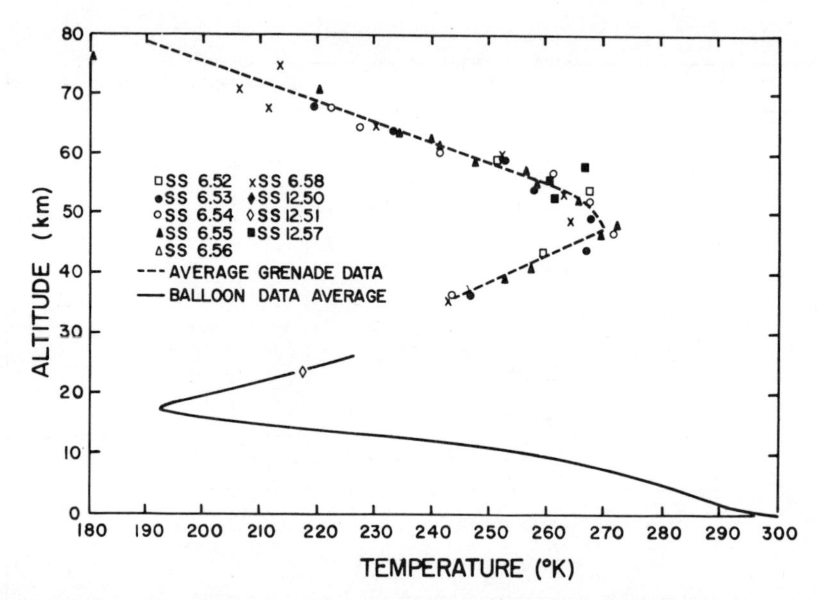

FIG. 3.15*a*. Temperature as a function of altitude at Guam, determined from the rocket-grenade experiment on several days in November 1958. (After Nordberg and Stroud, 1961.)

agreement with, although perhaps lighter than, those shown for winter in Fig. 3.2.

Two rocket-grenade experiments at Woomera, Australia, with the British rocket Skylark have been reported by Groves (1960), one in November 1957 and one in April 1958. Both measurements showed a temperature maximum at about 50 km, with temperature of about 272°K in the spring firing and a few degrees less in the autumn firing. Winds above 30 km were east in the spring and west in the autumn. In both experiments, the rockets also ejected chaff (aluminum) that was tracked by radar for wind determination. In both cases, agreement with the grenade-determined winds was very satisfactory.

3.3.5 Tracking Artificial Clouds

Very interesting and provocative information has come from the optical tracking of chemiluminescent trails produced in the upper atmosphere at twilight by the ejection of certain materials from rockets. Such trails may be produced by alkali vapor, usually sodium (Manring *et al.*, 1959; Manring and Bedinger, 1960), pyrotechnic flashes used for tracking by ballistic cameras (Groves, 1960; Woodbridge, 1962), or exhaust products from multistage rockets (aufm Kampe, Smith, and Brown, 1962). Nature occasionally provides

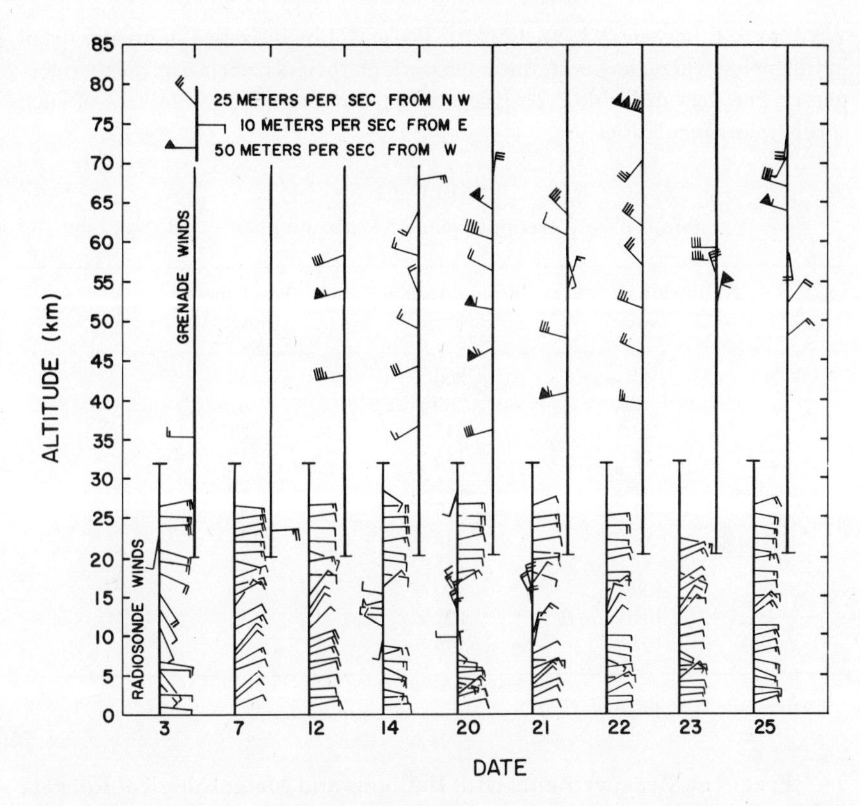

FIG. 3.15b. Wind as a function of altitude at Guam, determined from the rocket-grenade experiment on several days in November 1958. (After Nordberg and Stroud, 1961.)

such a chemiluminescent trail from a meteor, as discussed in Subsection 3.2.4.

Wind data obtained from photographic records of such trails are not numerous enough to contribute to our "climatological" knowledge of prevailing winds and their seasonal and latitudinal variations (which is the principal concern of the present chapter). Such data as are available, however, provide a unique picture of the instantaneous and unsmoothed vertical distribution of wind at a location, sometimes extending vertically over several tens of kilometers in the upper mesosphere and thermosphere.

Above all, these observations show very large vertical wind shears. As an illustration, Table 3.2 shows some data given by aufm Kampe *et al.* (1962), derived from ground triangulations on a rocket exhaust trail. Below 100 km, winds were observed to be generally west and above 100 km east. Notice, however, the large vertical wind shears on a relatively small scale—for example, the east component in a narrow layer near 75 km and the very rapid change from

WSW to SSE between 97 and 100 km. These will be discussed in greater detail in Chapter 8, where, however, the emphasis is on their occurrence in the thermosphere. The data in Table 3.2 suggest that the mesosphere is not free of such small-scale irregularities.

TABLE 3.2

WINDS OVER WALLOPS ISLAND, 27 FEBRUARY 1960[a]

Altitude (km)	Wind direction (nearest 5°)	Wind speed (m sec^{-1})
64	300	68
70	280	62
73.5	255	83
75	155	12
77	225	26
81	210	33
93	245	179
97	245	122
100	160	19
101	105	35
104	140	95

[a] After aufm Kampe *et al.* (1962).

3.4 Frequent Measurements with Balloons and Meteorological Rockets

Observations described in the previous section, primarily because of cost, have been carried out on an "occasional" basis. Our experience with the troposphere and lower stratosphere has taught us the value of regular, routine observations of the atmosphere from a large number of locations. For the upper stratosphere and mesosphere, this ideal is not yet possible. However, it has been possible to get fairly frequent data from a few locations by the use of special-purpose balloons or small "meteorological" rockets. The former have ceilings limited to 40 or 45 km. In Subsection 3.4.1 we consider a few of the balloon results available, and in 3.4.2 and 3.4.3 go into the tremendously interesting question of the development and use of small rockets for meteorological purposes.

3.4.1 Balloon Measurements

Most of the meteorological results of balloon soundings have been described in sufficient detail in the previous chapter. There are, however, a few examples of either special experimental programs with large balloons or special analyses of routine balloon soundings that lay emphasis on the upper stratosphere.

The highest altitude for which attempts have been made to construct regular synoptic charts from the usual radiosonde data corresponds to the 10-mb pressure surface (about 31 km). The U. S. Weather Bureau (1959) undertook the construction of thrice-monthly maps of this surface for the period of the International Geophysical Year. Daily 10-mb charts are now prepared routinely at the Free University of Berlin and published in their *Meteorologische Abhandlungen*, under the sponsorship of the U. S. Department of Army. Although data are scarce, the results undoubtedly outline reliably the large-scale features of the 10-mb circulation and temperature field. These have been discussed for the period July 1957 to June 1959 by Finger, Mason, and Corzine (1963).

At the 10-mb surface in summer, the temperature increases toward the pole, but less rapidly than at lower stratospheric levels. Winds are predominantly easterly at all latitudes with the strongest speeds at 20–25°N and no striking longitudinal variations. During September, the direction of the temperature gradient between, say, 30°N and the pole gradually reverses and the easterlies give way to westerlies that are well established throughout middle and high latitudes by October. The 10-mb surface evidently participates in stratospheric warmings, judging from the winters for which data are available. In short, qualitatively and generally speaking, the meteorological conditions at the 10-mb surface are similar to those in the lower stratosphere discussed in Sections 2.4 and 2.5.

A good deal of work has been done by the U. S. Army Signal Research and Development Laboratory to develop and apply balloon and radiosonde equipment for high-altitude measurements. For several years, scientists of this laboratory have carried out measurements principally at Belmar, New Jersey (40°N), daily during some periods and at midseasonal months during others. Measurements have been made at times at other latitudes. Some results have been reported in the scientific literature by Brasefield (1950), Conover and Wentzien (1955), Arnold and Lowenthal (1959), Panofsky (1961), and Conover (1961). High-altitude soundings at one, or a few, stations do not furnish a good basis for discussion of the synoptic situation. However, they may be examined for evidence of altitude and time variations.

Arnold and Lowenthal (1959) summarized the results of 16 months of daily soundings at Belmar from November 1952 to March 1954. The altitude and time variations of temperature (averaged by calendar month) are shown in Fig. 3.16. Two rather interesting features may be noted from this picture.

The annual temperature variation in the stratosphere according to these results exhibits a tendency to have an earlier maximum and minimum at the higher altitudes than at the lower. To study this quantitatively, Arnold and Lowenthal fitted the annual variation at each level with a sine curve and found that the dates of maximum temperature indicated by the phases of these curves progressed from mid-May at 2 mb to mid-July at 25 mb. The amplitudes of

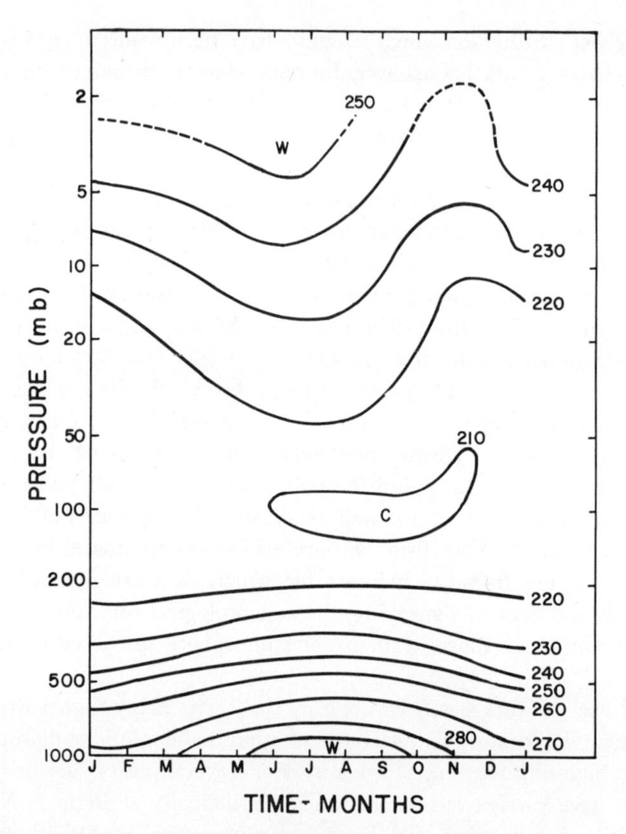

FIG. 3.16. Time cross section of monthly mean temperature (in degrees K) at Belmar, New Jersey. Means for April through October are based on one year's data, those for the other months on two years' data. (After Arnold and Lowenthal, 1959.)

the curves were greatest at the highest levels for which data were available. Standard deviations of temperature were greatest (up to 8–9°C) during the winter months at the highest levels probed.

Another point to notice is that the boundary between lower stratosphere and upper stratosphere, if it is to be defined as a level where the vertical temperature gradient changes more or less abruptly, disappears in the mean during the summer months. Conover and Wentzien (1955) discussing four years of observations, also at Belmar, noted the same thing. They pointed out that, from fall to spring, a "point of minimum temperature can often be found between 20 and 28 km." Inspection of some examples they gave of this point of minimum temperature indicates that it is generally not as pronounced as the relative minima associated with the tropopause or tropopauses of lower levels. In the mean, except for the summer months, they reported a slow rate of change of temperature between 10 and 26 km, with the temperature rising at the rate of approximately 2°C km^{-1} above 26 km.

With respect to day-to-day variations, a very interesting time-altitude section of temperature was published by Conover (1961) and is reproduced here in Fig. 3.17. The point of interest is the rapid warming and subsequent cooling

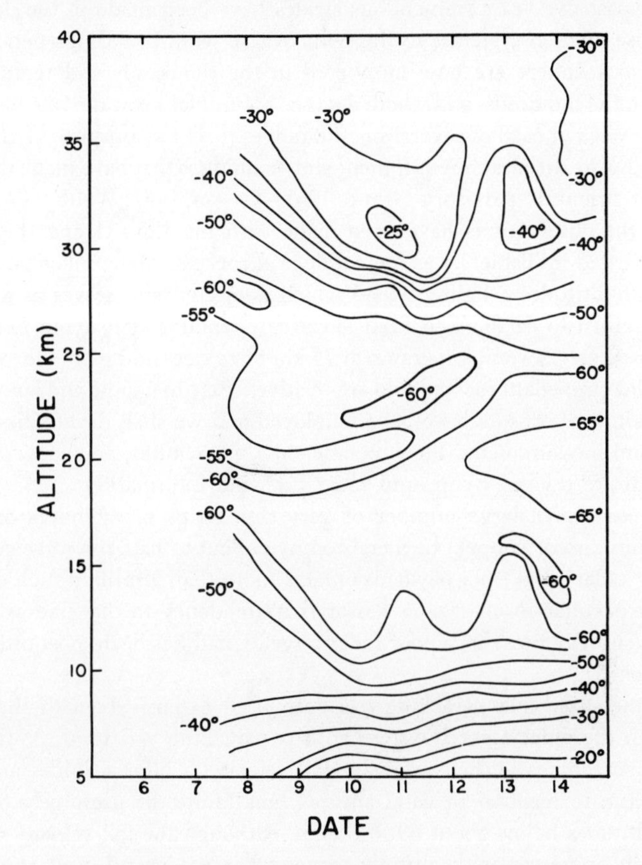

Fig. 3.17. Time-height cross section of temperature (in degrees C) at Belmar, New Jersey, for a period in January 1960. (After Conover, 1961.)

at levels near 30 km in the days centered around 11 January 1960. This is evidently a case of a limited stratospheric warming in the upper stratosphere, which in this case did not penetrate to lower levels (at least at this station). It is evident from this example, which we shall supplement with examples derived from rocket data, that the upper stratosphere is the seat of meteorological disturbances comparable to those that have been discovered in the lower stratosphere.

3.4.2 Meteorological Rockets—Equipment and Techniques

If the ceiling imposed by the bursting height of balloons is to be exceeded, rockets must be used. If observations are to be frequent in time and space, then low-cost systems with relatively simple instrumentation become a necessity. During the past five years tremendous strides have been made in the development and use of such systems; to the point where wind soundings penetrating the upper stratosphere are now numbered in the thousands and temperature soundings in the hundreds, and where data are accumulating at the rate of several soundings a week at each of several missile ranges. It is the purpose of this part of Section 3.4 to discuss the equipment and techniques that have made this possible. (Joint Scientific Advisory Group, 1961; Keegan, 1961; Webb *et al.*, 1962).

Most of the observations have been made with the Loki II and the Arcas, which were listed in Table 3.1 as meteorological rockets. Meteorological rockets carry to high altitudes a light payload which is ejected and serves as a sensor during its return to earth. The Loki II can carry about 1 kg to about 75 km and the Arcas can carry several kilograms to 75 km or so depending on the payload. Both are solid-propellant rockets and are relatively easy to handle and launch.

Various kinds of sensors have been employed and we shall discuss first those used for wind measurement. They include chaff, parachutes, and superpressure balloons. All are tracked by ground radar for wind information.

Chaff consists of a large number of very thin strips or cylinders of metal (usually aluminum or copper) or metalized nylon, cut to half the wavelength of the tracking radar. The Loki payload contains more than a million such dipoles. A difficulty peculiar to chaff as a sensor is its tendency to disperse as it falls so that it eventually presents a poor radar target. Chaff has been most often used with the Loki II.

A 15-ft metalized silk parachute has been used extensively with the Arcas rocket. With the older Loki I system an 8-ft parachute was used. A difficulty peculiar to parachutes is their excessive fall velocity at high altitudes and consequent failure to respond to wind shears. This limits the usefulness of parachutes to altitudes below about 60 to 65 km. Although the fall velocity of chaff also naturally increases with altitude, some types are useful well above this level. Parachutes have the advantages of presenting a discrete target and of being able to support instrumentation to sense and telemeter temperature (see below).

A third sensor that has been used is a 1-m superpressure mylar balloon, called the Robin. This is designed to provide density data (from observations of its rate of descent, a concept similar to that of the falling-sphere experiment) as well as wind data. The Robin is used with the Arcas rocket, but it has up to 1963 been used relatively infrequently for density determinations.

All wind sensors when expelled from the rocket near apogee have initially the forward speed of the rocket (several hundred feet per second). However, they lose this component quite rapidly and this is not believed to be a serious problem. All require fixed, elaborate tracking facilities at the ground.

Temperature measurements have up to now been made almost exclusively with the Gammasonde, whose temperature-sensing element is a 10-mil bead thermistor and whose telemetry is compatible with the GMD-1 ground equipment. It is used with the Arcas rocket and 15-ft parachute. Questions naturally arise about the accuracy of this element, in view of possible lag, radiative, and compressional-heating errors. According to Wagner (1961, 1962) who has analyzed this complex problem from a theoretical viewpoint, the measured temperatures after correction should be reliable within 1° or 2°C up to 50–55 km.

For various reasons, rocketsonde data have been gathered almost exclusively at the various missile ranges, particularly White Sands, New Mexico; Point Mugu, California; Cape Kennedy, Florida; Wallops Island, Virginia; Fort Churchill, Canada; Fort Greeley, Alaska; and Tonopah, Nevada. Among the reasons are the need for launching facilities and uninhabited impact areas, the need for accurate tracking facilities, and the opportunity at these locations to utilize the data to a degree compatible with their cost. Unfortunately most of these locations are in lower middle latitudes so that data are not well distributed with respect to latitude. Activities at the various locations have been coordinated to a rather remarkable degree through the Meteorological Working Group of the Inter-Range Instrumentation Group, and the collective of sites is often referred to as the Meteorological Rocket Network (MRN) (Webb *et al.*, 1961, 1962; Joint Scientific Advisory Group, 1961). The MRN at the present time is simply a cooperative venture among essentially independent groups, who exchange data and information and to the extent possible coordinate their firing schedules.

Although the emphasis in this discussion has been on the usual instrumentation of the MRN, Smith (1960, 1962) has utilized the Deacon-Arrow with chaff to get wind data to about 85 km and we shall mention his data shortly. Another interesting sounding system, developed by the Naval Ordnance Laboratory primarily for use aboard ships, is the HASP, which utilizes a modified Loki launched from standard 5-inch guns to deliver chaff which is then tracked by the ship's gun-director radar.

3.4.3 Meteorological Rockets—Some Results

Smith (1960) gave results from three soundings at Tonopah, Nevada (38°N), in May 1958 and from 20 soundings at Johnston Island (18°N) in July and August 1958. Ceiling of most of the soundings was between 75 and 90 km. The Nevada data showed winds predominantly from the east, but with some meridional components variable with altitude and time. The average speed of the zonal component, however, was somewhat less than indicated for summer at this latitude and altitude by Fig. 3.2. The Johnston Island data also revealed predominantly east winds up to about 80 km with indications of a shift to westerly at higher levels. According to Rapp (1960) these wind data have a standard vector error of about 10 m sec^{-1}.

The Rocket Network was activated in October of 1959 when firings were conducted at Fort Churchill and at the Pacific Missile Range, Point Mugu, California, which were joined later by the other stations mentioned above. The original plan was to conduct firings as often as possible during the midseasonal months, but in April of 1961 this was changed to a more regular routine of obtaining if possible three soundings per week throughout the year. The data have been made routinely available through the U. S. Army Electronics Research and Development Activity at White Sands and already many results and analyses have appeared in the scientific literature, of which we shall mention some.

Masterson, Hubert, and Carr (1961) have published all the Point Mugu data (110 soundings) from October 1959 through February 1961. These data show good qualitative agreement with the picture presented in Fig. 3.2, but with winter zonal wind speeds near the stratopause a bit lower. The temperature data show quite large day-to-day fluctuations in the upper stratosphere but agree well in the mean with the temperatures in Fig. 3.1 (below 50 km).

There have been various other studies of a climatological and descriptive nature based on the data from one or a few of the stations. Appleman (1962, 1963) has discussed the wind and wind variability between 45 and 60 km, and Miers (1963) has considered the spring and fall reversals of the zonal wind as a function of time and altitude. In both spring and fall, it appears that the reversal occurs first at the highest levels observed.

Keegan (1962) made use of wind data from the rocket soundings to illustrate the time variability of the zonal wind component during winter in the upper stratosphere. Figures 3.18 and 3.19 show time cross sections of zonal wind at Wallops Island and Point Mugu during the winter of 1960. Both show an anomalous period of east winds in late January and early February with maximum speed apparently at about 30 km. During the winter of 1961 at White Sands (Fig. 3.20) east-wind regimes occurred during January and again in February. It is clear from data like these that the prevailing westerlies of the winter upper stratosphere are subject to large-scale disturbances and variations of a most interesting nature. When data are plentiful enough to construct reliable and reasonably complete synoptic maps for these high levels, an exciting new branch of upper-atmospheric research will have been opened.

3.5 Summary and Discussion

In this section, we consider the data described and presented in this chapter. Subsection 3.5.1 is concerned with the mean structure and circulation of summer

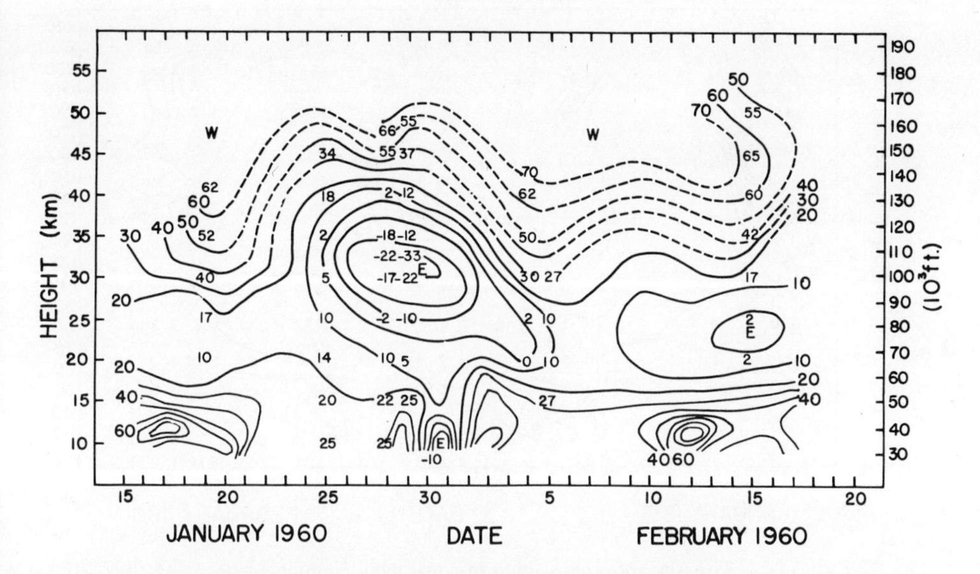

FIG. 3.18. Time-height cross section of zonal wind component (in meter (second)$^{-1}$, west wind positive) at Wallops Island in early 1960. (After Keegan, 1962.)

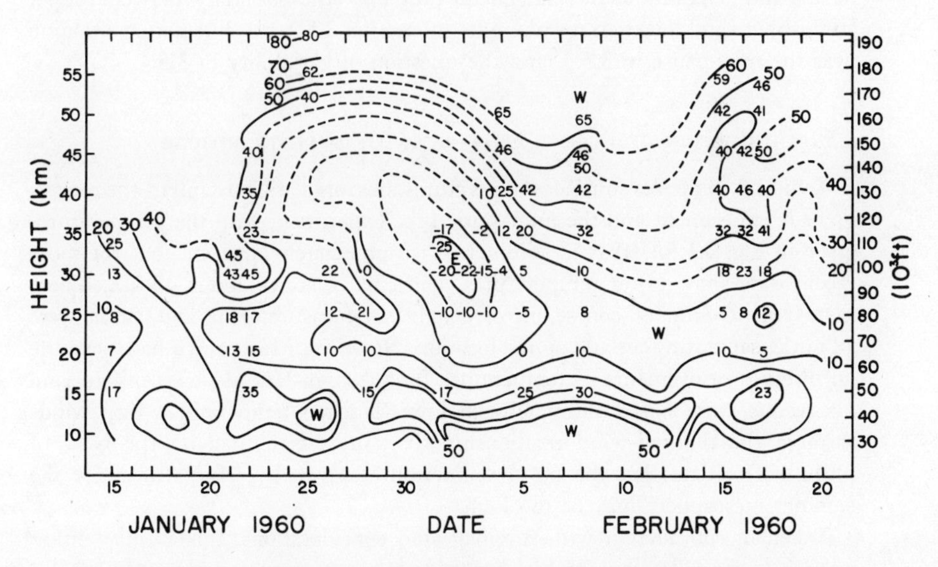

FIG. 3.19. Time-height cross section of zonal wind component (in meter (second)$^{-1}$, west wind positive) at Point Mugu in early 1960. (After Keegan, 1962.)

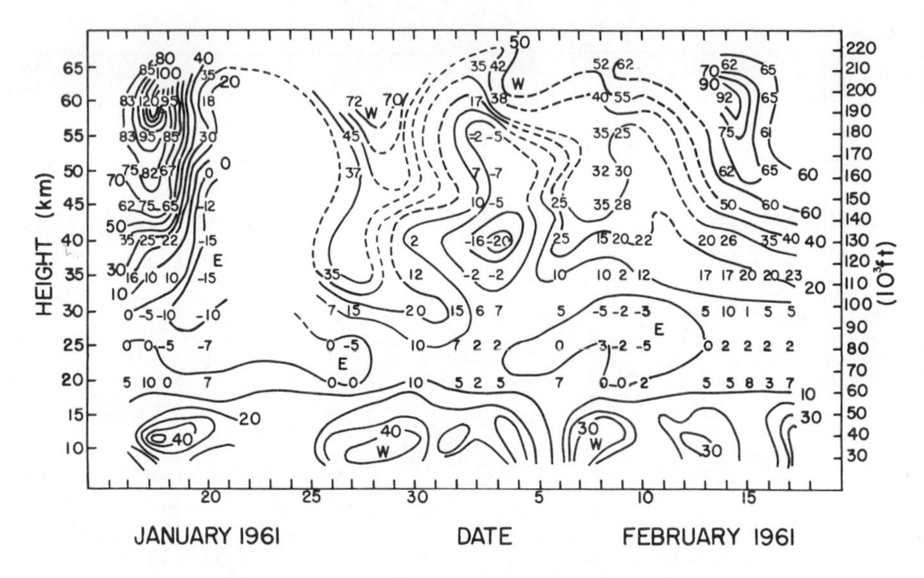

FIG. 3.20. Time-height cross section of zonal wind component (in meter (second)$^{-1}$, west wind positive) at White Sands in early 1961. (After Keegan, 1962.)

and winter in the context of Murgatroyd's statement (Figs. 3.1 and 3.2) and the results that have become available since then. Then we turn to some interesting details and elaborations of this general problem—the boundary between upper stratosphere and lower stratosphere in Subsection 3.5.2, the interesting situation near the mesopause in 3.5.3, and the question of variability in 3.5.4.

3.5.1 Comparison of Recent Data with Mean Distributions

Table 3.3 compares summer temperatures measured at Churchill in the rocket-grenade experiment and the pitot-static tube experiment with the temperatures given in Fig. 3.1 for 60°N in summer. It is immediately apparent that the more recent data show lower temperatures, particularly above about 45 km. These more recent data, of course, involve only six measurements and those were all in the same summer and at one location. However, Murgatroyd had available no direct measurements of temperature for the high-latitude mesosphere, and his results were based ultimately on lower-latitude temperatures and winds through the thermal-wind relationship. It would seem then on the basis of pitifully few data that the 60°N temperatures given by Murgatroyd for the summer mesosphere may be too high.

Taken in conjunction with thermal-wind considerations, this implies either or both of the following: Middle-latitude temperatures at and above 45 km are also lower than shown in Fig. 3.1; or the thermal wind above 45 km at, say,

TABLE 3.3

TEMPERATURES DERIVED FROM THE SUMMER SOUND-GRENADE EXPERIMENT AT FORT CHURCHILL AND IN THE PITOT-STATIC TUBE EXPERIMENT COMPARED WITH THE TEMPERATURES INDICATED BY FIG. 3.1 AT 60°N[a]

z (km)	Stroud[b] (°K)	Ainsworth[c] (°K)	Murgatroyd[d] (°K)
20	223	—	227
25	228	—	230
30	236	236	236
35	245	246	247
40	258	262	262
45	270	268	277
50	274	267	293
55	268	264	287
60	258	240	267
65	242	228	246
70	214	219	226
75	194	—	199
80	174	—	183

[a] Individual temperature values are subject to an uncertainty of a degree or two as a result of the author's interpolations from graphs and tables.

[b] Summer sound-grenade experiment at Fort Churchill (59°N). Temperatures are the averages of five soundings.

[c] Pitot-static tube experiment at Fort Churchill.

[d] From Fig. 3.1 (60°N).

50°N has a stronger westerly component than indicated in Fig. 3.2. There is some evidence that both factors may be involved.

(a) Regarding temperatures in middle latitudes, the summer temperatures at 40°N according to the balloon data of Arnold and Lowenthal (1959) are lower than Murgatroyd's between about 35 and 45 km. Furthermore, the stratopause temperatures indicated by the summer rocket-grenade experiments at White Sands are somewhat lower than those indicated by Fig. 3.1.

(b) Regarding wind at 50°N, Batten's (1961) analysis does show the maximum east wind at this latitude to occur at a lower elevation and with smaller magnitude than indicated in Fig. 3.2.

With regard to the winter conditions, the Churchill temperature data agree rather well with Murgatroyd's results for the upper stratosphere and the stratopause. On the other hand, both sphere data and grenade data show considerably higher temperatures between 70 and 80 km than does Fig. 3.1. In the mean for all four grenade soundings, the difference is more than 20°K. These higher temperatures, if typical, would imply a rather strong east thermal wind between

70 and 80 km in middle latitudes so that the change from west to east winds would occur at a lower altitude than the 90–100 km indicated in Fig. 3.2. There seems to be no good evidence that this does indeed occur. Batten, as mentioned briefly in Section 3.1, was unable to reconcile his more recent wind analysis with these high temperatures.

3.5.2 Boundary between Lower and Upper Stratosphere

In view of the nomenclature problem, it is interesting to consider whether and under what circumstances there is a pronounced change of lapse rate in the region of 25 to 30 km, which separates our lower stratosphere and upper stratosphere (the stratosphere from the mesosphere under the older Chapman system).

There are good reasons to believe that there is no such change in low latitudes. Above the level that we call the "tropical tropopause," the rate of increase of temperature with altitude may remain essentially constant up to the stratopause. The "mean tropical sounding" presented by Palmer *et al.* (1955) shows a remarkably constant lapse rate between 80 mb (about 18 km) and 10 mb (about 31 km) with a value of $-2.84°K\ km^{-1}$. Extrapolation of the temperature to 50 km with this same lapse rate would give a temperature of about 290°K, a rather reasonable value, without assuming a larger rate of temperature increase above 10 mb. Furthermore, summer data in lower middle latitudes, in the presence of the "tropical tropopause," show no clear-cut change in the lapse rate up to the stratopause (for example, balloon data at Belmar and searchlight data at White Sands). With regard to the former data, Conover and Wentzien (1955) specifically commented:

> Although the tropical tropopause rarely disappears from over this station at any time throughout the year, the arctic tropopause appears in the fall and disappears in the spring. During the existence of the arctic tropopause, a third point of minimum temperature can often be found between 20 and 28 km. The presence of several points of minimum temperature such as these, produces, on the average, a slow rate of change of temperature between 10 and 26 km. Above 26 km, the temperature rises at approximately 2°C per kilometer. During the summer months, this almost isothermal layer is rarely observed, the temperature profile is characterized by a pronounced minimum around 16 km. Above this point, the temperature increases at an average rate of about 1.6°C per kilometer.

On the other hand there appears to be an obvious lapse-rate change in high latitudes, particularly in winter, in the vicinity of 25 to 30 km. The Churchill rocket-grenade data (Figs. 3.10–3.14) and the Churchill falling-sphere data (Fig. 3.8) illustrate this.

This rather abrupt change in lapse rate, which is most apparent at high latitudes, especially in winter, and exists also in middle latitudes for a good part of the year, is a reasonable boundary between the lower stratosphere and upper stratosphere. One might even call it the "lower stratopause," although this writer has refrained from using a new term in view of the already deplorably confused situation with respect to nomenclature. At low latitudes and at middle latitudes in summer, where no clear-cut lapse-rate change appears to exist, the distinction between "lower stratosphere" and "upper stratosphere" is still convenient. Here the boundary might be taken arbitrarily to be 30 km, since this level represents the approximate ceiling of routine meteorological observations.

3.5.3 The High-Latitude Mesopause

Perhaps the most interesting phenomenon revealed by observations of the past few years is the temperature stucture of the high-latitude mesosphere. At Churchill a peculiar and unexpected second maximum of temperature shows up somewhere between 65 and 80 km on all four winter grenade soundings, on the autumn pitot-static tube sounding, and is suggested by all the winter falling-sphere soundings. On the other hand, the summer observations show a steady temperature decrease through the mesosphere to an unexpectedly low temperature at the mesopause.

The fact that winter temperatures are higher than summer temperatures near 80 km is contrary to expectations based on radiative considerations and some attention has already been given to this problem, as is discussed in Chapter 7. If it should turn out from additional data that an actual secondary maximum of temperature is typical of the high-latitude winter mesosphere, then the problem will be considerably more complicated.

3.5.4 Temperature and Wind Variability

We have had frequent occasion to remark on day-to-day and week-to-week variability of wind and temperature in the upper stratosphere and mesosphere. This has been illustrated, for example, in Figs. 3.8–3.14, and 3.17–3.20.

Table 3.4 gives a measure of quantitative expression to this. This table contains standard deviations of temperature by season (of daily values about the seasonal mean) at Belmar, New Jersey (Arnold and Lowenthal, 1959). They increase upward and are especially high in winter. Appleman (1963) found rather large standard vector deviations of wind at 45 to 60 km, especially in winter. Additional statistical studies of this variability must await the gathering of more data.

TABLE 3.4

STANDARD DEVIATIONS OF TEMPERATURE, IN DEGREES K, AT BELMAR, NEW JERSEY[a]

Pressure (mb)	Dec. '52 to Feb. '53	Mar. '53 to May '53	Jun. '53 to Aug. '53	Sept. '53 to Nov. '53	Dec. '53 to Feb. '54
3	—	—	—	7.6 (24)	—
4	—	—	4.6 (29)	6.2 (37)	8.7 (23)
5	—	5.9 (33)	4.2 (47)	5.8 (58)	7.4 (50)
6	6.4 (28)	5.8 (43)	4.2 (54)	5.7 (67)	6.4 (60)
7	5.7 (39)	5.5 (45)	3.8 (57)	5.4 (67)	6.9 (63)
8	5.5 (55)	4.8 (50)	3.4 (59)	5.2 (68)	7.1 (66)
9	5.5 (58)	4.2 (55)	3.3 (60)	5.0 (69)	7.0 (69)
10	5.7 (66)	4.5 (68)	3.1 (66)	4.9 (70)	7.0 (73)
15	3.8 (76)	3.2 (78)	2.4 (73)	4.5 (77)	4.8 (79)
20	3.3 (80)	2.9 (87)	2.5 (78)	4.5 (79)	3.8 (85)
25	2.9 (79)	2.7 (91)	2.3 (80)	4.5 (80)	3.8 (85)
40	2.9 (84)	2.0 (90)	2.1 (87)	3.8 (82)	3.4 (85)
50	2.7 (86)	2.1 (90)	2.2 (90)	3.2 (82)	3.6 (85)
75	3.6 (87)	3.2 (91)	3.0 (88)	2.5 (85)	3.8 (86)
100	3.6 (87)	3.6 (92)	3.4 (87)	3.5 (86)	4.5 (86)

[a] The numbers of observations are shown in parentheses. After Arnold and Lowenthal (1959).

In view of the many interesting problems that are already being suggested by scanty observations, one can look forward with anticipation to the exciting developments that will follow upon adequate observation of the upper stratosphere and mesosphere.

REFERENCES

AINSWORTH, J. E., FOX, D. F., and LAGOW, H. E. (1961). Upper-atmosphere structure measurement made with the pitot-static tube. *J. Geoph. Res.* **66**, 3191–3212.

APPLEMAN, H. S. (1962). A comparison of climatological and persistence wind forecasts at 45 kilometers. *J. Geoph. Res.* **67**, 767–771.

APPLEMAN, H. S. (1963). The climatological wind and wind variability between 45 and 60 kilometers. *J. Geoph. Res.* **68**, 3611–3617.

ARNOLD, A., and LOWENTHAL, M. J. (1959). A sixteen-month series of mid-stratospheric temperature measurements. *J. Meteor.* **16**, 626–629.

AUFM KAMPE, H. J., SMITH, M. E., and BROWN, R. M. (1962). Winds between 60 and 110 kilometers. *J. Geoph. Res.* **67**, 4243–4257.

BARTMAN, F. L., CHANEY, L. W., JONES, L. M., and LIU, V. C. (1956). Upper-air density and temperature by the falling-sphere method. *J. Appl. Phys.* **27**, 706–712.

BATTEN, E. S. (1961). Wind systems in the mesosphere and lower ionosphere. *J. Meteor.* **18**, 283–291.

BRASEFIELD, C. J. (1950). Winds and temperatures in the lower stratosphere. *J. Meteor.* **7**, 66–69.

CONOVER, W. C. (1961). An instance of a stratospheric "explosive" warming. *J. Meteor.* **18**, 410–413.

CONOVER, W. C., and WENTZIEN, C. J. (1955). Winds and temperatures to forty kilometers. *J. Meteor.* **12**, 160–164.

COX, E. F., ATANASOFF, J. V., SNAVELY, B. L., BEECHER, D. W., and BROWN, J. (1949). Upper-atmosphere temperatures from Helgoland Big Bang. *J. Meteor.* **6**, 300–311.

CRARY, A. P. (1950). Stratosphere winds and temperatures from acoustical propagation studies. *J. Meteor.* **7**, 233–242.

CRARY, A. P. (1952). Stratosphere winds and temperatures in low latitudes from acoustical propagation studies. *J. Meteor.* **9**, 93–109.

CRARY, A. P. (1953). Annual variations of upper air winds and temperatures in Alaska from acoustical measurements. *J. Meteor.* **10**, 380–389.

CRARY, A. P., and BUSHNELL, V. C. (1955). Determination of high-altitude winds and temperatures in the Rocky Mountain area by acoustic soundings, October 1951. *J. Meteor.* **12**, 463–471.

ELFORD, W. G. (1959). A study of winds between 80 and 100 km in medium latitudes. *Plan. Space Sci.* **1**, 94–101.

ELFORD, W. G., and MURRAY, E. L. (1960). Upper atmosphere wind measurements in the Antarctic. *In* "Space Research I" (H. Kallmann Bijl, ed.), pp. 158–163. North-Holland Publ. Co., Amsterdam.

ELTERMAN, L. (1954a). Seasonal trends of temperature, density, and pressure to 67.6 km obtained with the searchlight probing technique. *J. Geoph. Res.* **59**, 351–358.

ELTERMAN, L. (1954b). Seasonal trends of temperature, density, and pressure in the stratosphere obtained with the searchlight-probing technique. *Geoph. Res. Papers* No. 29, 1-70.

FEDYNSKY, V. V. (1955). Meteor studies in the Soviet Union. *In* "Meteors" (T. R. Kaiser, ed.), pp. 188–192. Pergamon Press, New York.

FERENCE, M., STROUD, J. R., WALSH, J. R., and WEISNER, A. G. (1956). Measurement of temperatures at elevations of 30 to 80 km by the rocket-grenade experiment. *J. Meteor.* **13**, 5–12.

FINGER, F. G., MASON, R. B., and CORZINE, H. A. (1963). Some features of the circulation at the 10-mb surface, July 195° through June 1959. *Mon. Wea. Rev.* **91**, 235–249.

GREENHOW, J. S., and NEUFELD, E. L. (1960). Large scale irregularities in high altitude winds. *Proc. Phys. Soc.* **75**, 228–234.

GREENHOW, J. S., and NEUFELD, E. L. (1961). Winds in the upper atmosphere. *Quart. J. Roy. Meteor. Soc.* **87**, 472–489.

GROVES, G. V. (1956). Introductory theory for upper atmosphere wind and sonic velocity determination by sound propagation. *J. Atmos. Terr. Phys.* **8**, 24–38.

GROVES, G. V. (1960). Wind and temperature results obtained in Skylark experiments. *In* "Space Research I" (H. Kallmann Bijl, ed.), pp. 144–153. North-Holland Publ. Co., Amsterdam.

GUTENBERG, B. (1932). Die Schallausbreitung in der Atmosphäre. *In* "Handbuch der Geophysik," Vol. 9, pp. 89–145. Bornträger, Berlin.

GUTENBERG, B. (1939). The velocity of sound waves and the temperature in the stratosphere in southern California. *Bull. Amer. Meteor. Soc.* **20**, 192–201.

GUTENBERG, B. (1946). Physical properties of the atmosphere up to 100 km. *J. Meteor.* **3**, 27–30.

GUTENBERG, B. (1951). Sound propagation in the atmosphere. *In* "Compendium of Meteorology" (T. F. Malone, ed.), pp. 366–375. American Meteorological Society, Boston, Massachusetts.

HAVENS, R. J., KOLL, R. T., and LAGOW, H. E. (1952). The pressure, density, and temperature of the earth's atmosphere to 160 kilometers. *J. Geoph. Res.* **57**, 59–72.

HAWKINS, G. S., and SOUTHWORTH, R. B. (1958). The statistics of meteors in the earth's atmosphere. *Smiths. Contr. Ap.* **2**, No. 11, 349–364.

HOFFMEISTER, C. (1946). Die Strömungen der Atmosphäre in 120 km Höhe. *Z. Meteor.* **1**, 33–41.

JACCHIA, L. G. (1948). Ballistics of the upper atmosphere. *Harvard Coll. Obs. Reprint Series* **II**, No. 26, 1–30.

JACCHIA, L. G. (1949a). Photographic meteor phenomena and theory. *Harvard Coll. Obs. Reprint Series* **II**, No. 31, 1–36.

JACCHIA, L. G. (1949b). Atmospheric density profile and gradients from early parts of photographic meteor trails. *Harvard Coll. Obs. Reprint Series* **II**, No. 32, 1–12.

JACCHIA, L. G. (1955). The physical theory of meteors. VIII, Fragmentation as cause of the faint-meteor anomaly. *Ap. J.* **121**, 521–527.

JACCHIA, L. G. (1957). A preliminary analysis of atmospheric densities from meteor decelerations for solar, lunar and yearly oscillations. *J. Meteor.* **14**, 34–37.

JOHNSON, C. T., and HALE, F. E. (1953). Abnormal sound propagation over the southwestern United States. *J. Acoust. Soc. Amer.* **25**, 642–650.

JOINT SCIENTIFIC ADVISORY GROUP (1961). The meteorological rocket network—an analysis of the first year in operation. *J. Geoph. Res.* **66**, 2821–2842.

JONES, L. M., PETERSON, J. W., SCHAEFER, E. J., and SCHULTE, H. F. (1959). Upper-air density and temperature: some variations and an abrupt warming in the mesospere. *J. Geoph. Res.* **64**, 2331–2340.

KAHLKE, S. (1921). Meteorschweife und hochatmosphärische Windströmungen. *Ann. Hydrog. Mar. Meteor.* **49**, 294–299.

KEEGAN, T. J. (1961). Meteorological rocketsonde equipment and techniques. *Bull. Amer. Meteor. Soc.* **42**, 715–721.

KEEGAN, T. J. (1962). Large-scale disturbances of atmospheric circulation between 30 and 70 kilometers in winter. *J. Geoph. Res.* **67**, 1831–1838.

KENNEDY, W. B., BROGAN, L., and SIBLE, N. J. (1955). Further acoustical studies of atmospheric winds and temperatures at elevations of 30 to 60 kilometers. *J. Meteor.* **12**, 519–532.

LAGOW, H. E., and AINSWORTH, J. (1956). Arctic upper-atmosphere pressure and density measurements with rockets. *J. Geoph. Res.* **61**, 77–92.

LINDEMANN, F. A., and DOBSON, G. M. B. (1923). A theory of meteors, and the density and temperature of the outer atmosphere to which it leads. *Proc. Roy. Soc.* **A102**, 411–436.

MANRING, E., and BEDINGER, J. (1960). Winds in the atmosphere from 80 to 230 km. *In* "Space Research I" (H. Kallmann Bijl, ed.), pp. 154–157. North-Holland Publ. Co., Amsterdam.

MANRING, E., BEDINGER, J. F., PETTIT, H. B., and MOORE, C. B. (1959). Some wind determinations in the upper atmosphere using artificially generated sodium clouds. *J. Geoph. Res.* **64**, 587–591.

MASSEY, H. S. W., and BOYD, R. L. F. (1959). "The Upper Atmosphere." Philosophical Library, New York.

MASTERSON, J. E., HUBERT, W. E., and CARR, T. R. (1961). Wind and temperature measurements in the mesosphere by meteorological rockets. *J. Geoph. Res.* **66**, 2141–2151.

MIERS, B. T. (1963). Zonal wind reversal between 30 and 80 km over the southwestern United States. *J. Atmos. Sci.* **20**, 87–93.

Mitra, S. K. (1952). "The Upper Atmosphere," 2nd ed. Asiatic Society, Calcutta.

Murgatroyd, R. J. (1956). Wind and temperature to 50 km. over England. *Geoph. Mem.* No. 95, 1-30.

Murgatroyd, R. J. (1957). Winds and temperatures between 20 km and 100 km—a review. *Quart. J. Roy. Meteor. Soc.* **83**, 417–458.

Newell, H. E. (1959). "Sounding Rockets." McGraw-Hill, New York.

Nordberg, W., and Stroud, W. G. (1961). Results of IGY rocket-grenade experiments to measure temperatures and winds above the island of Guam. *J. Geoph. Res.* **66**, 455–464.

Olivier, C. P. (1942). Long enduring meteor trains. *Proc. Amer. Phil. Soc.* **85**, 93–135.

Olivier, C. P. (1947). Long enduring meteor trains: second paper. *Proc. Amer. Phil. Soc.* **91**, 315–327.

Olivier, C. P. (1957). Long enduring meteor trains: third paper. *Proc. Amer. Phil. Soc.* **101**, 296–315.

Palmer, C. E., Wise, C. W., Stempson, L. J., and Duncan, G. H. (1955). The practical aspect of tropical meteorology. *AF Surveys in Geoph.* No. 76, 1-95.

Panofsky, H. A. (1961). Temperature and wind in the lower stratosphere. *Advances in Geoph.* **7**, 215–247.

Rapp, R. R. (1960). The accuracy of winds derived by the radar tracking of chaff at high altitudes. *J. Meteor.* **17**, 507–514.

Richardson, J. M., and Kennedy, W. B. (1952). Atmospheric winds and temperatures to 50-kilometers altitude as determined by acoustical propagation studies. *J. Acoust. Soc. Amer.* **24**, 731–741.

Rocket Panel (1952). Pressures, densities, and temperatures in the upper atmosphere. *Phys. Rev.* **88**, 1027–1032.

Scherhag, R. (1960). Stratospheric temperature changes and the associated changes in pressure distribution. *J. Meteor.* **17**, 575–582.

Shapley, H., Öpik, E., and Boothroyd, S. (1932). The Arizona expedition for the study of meteors. *Proc. Nat. Acad. Sci.* **10**, 16–23.

Sicinski, H. S., Spencer, N. W., and Dow, W. G. (1954). Rocket measurements of upper atmosphere ambient temperature and pressure in the 30-to 75-kilometer region. *J. Appl. Phys.* **25**, 161–168.

Smith, L. B. (1960). The measurement of winds between 100,000 and 300,000 ft by use of chaff rockets. *J. Meteor.* **17**, 296–310.

Smith, L. B. (1962). Monthly wind measurements in the mesodecline over a one-year period. *J. Geoph. Res.* **67**, 4653–4672.

Stroud, W. G., Nordberg, W., and Walsh, J. R. (1956). Atmospheric temperatures and winds between 30 and 80 km. *J. Geoph. Res.* **61**, 45–56.

Stroud, W. G., Nordberg, W., Bandeen, W. R., Bartman, F. L., and Titus, R. (1960). Rocket-grenade measurements of temperatures and winds in the mesosphere over Churchill, Canada. *J. Geoph. Res.* **65**, 2307–2323.

Teweles, S. (1961). Time section and hodograph analysis of Churchill rocket and radiosonde winds and temperatures. *Mon. Wea. Rev.* **89**, 125–136.

Trowbridge, C. C. (1907). On atmospheric currents at very great altitudes. *Mon. Wea. Rev.* **35**, 390–397.

U. S. Weather Bureau (1959). "10-Millibar Synoptic Weather Maps." U. S. Dept. of Commerce, Washington, D. C.

Wagner, N. K. (1961). Theoretical time constant and radiation error of a rocketsonde thermistor. *J. Meteor.* **18**, 606–614.

Wagner, N. K. (1962). Unpublished technical reports.

WARFIELD, C. N. (1947). Tentative tables for the properties of the upper atmosphere. *Nat. Adv. Comm. Aeron. Tech. Notes* No. 1200.

WEBB, W. L., HUBERT, W. E., MILLER, R. L., and SPURLING, J. F. (1961). The first meteorological rocket network. *Bull. Amer. Meteor. Soc.* **42**, 482–494.

WEBB, W. L., CHRISTENSEN, W. I., VARNER, E. P., and SPURLING, J. F. (1962). Inter-range instrumentation group participation in the meteorological rocket network. *Bull. Amer. Meteor. Soc.* **43**, 640–649.

WEISNER, A. G. (1956). Measurement of winds at elevations of 30 to 80 km by the rocket-grenade experiment. *J. Meteor.* **13**, 30–39.

WHIPPLE, F. J. W. (1923). The high temperature of the upper atmosphere as an explanation of zones of audibility. *Nature* **111**, 187.

WHIPPLE, F. J. W. (1935). The propagation of sound to great distances. *Quart. J. Roy. Meteor. Soc.* **61**, 285–308.

WHIPPLE, F. L. (1943). Meteors and the earth's upper atmosphere. *Rev. Mod. Phys.* **15**, 246–264.

WHIPPLE, F. L. (1952a). Exploration of the upper atmosphere by meteoritic techniques. *Advances in Geoph.* **1**, 119–154.

WHIPPLE, F. L. (1952b). Results of rocket and meteor research. *Bull. Amer. Meteor. Soc.* **33**, 13–25.

WHIPPLE, F. L., and HAWKINS, G. S. (1959). Meteors. *In* "Handbuch der Physik" (S. Flügge, ed.), Vol. 52, pp. 519–564. Springer, Berlin.

WOODBRIDGE, D. D. (1962). Ionospheric winds. *J. Geoph. Res.* **67**, 4221–4231.

CHAPTER 4

The Sun's Radiation and the Upper Atmosphere

Our atmosphere is largely transparent to radiation in the visible part of the solar spectrum. In striking contrast, it is completely opaque in the ultraviolet for wavelengths less than about 3000 A. Furthermore, in most of the ultraviolet parts of the spectrum, absorption by atmospheric gases is so intense that the available supply of solar energy is essentially all absorbed in the upper atmosphere—at various altitudes for various wavelengths. The upper atmosphere, therefore, in contrast to the troposphere, derives most of its energy directly by absorption and not indirectly after reradiation or convection from the earth's surface. Although the energy so absorbed represents only a few per cent of the total solar energy reaching the earth, its effects are extremely important. In the first place, the heat capacity of the upper atmosphere is small, because of the low density; in the second place, the radiation involved is at high enough frequencies to dissociate or ionize some of the absorbing molecules.

Many important attributes of the upper atmosphere cannot be understood or discussed except in terms of interaction with solar radiation. Composition, structure, and ionization are directly related to the characteristics of the impinging radiation, the absorption properties of the various atmospheric gases, and the effects upon them of that absorption. There are, furthermore, secondary processes, such as emission of radiation and chemical reactions, which depend ultimately on the sun for their energy.

In considering all of these topics, some familiarity with certain aspects of spectroscopy, quantum mechanics, photochemistry, and radiative transfer is essential. Some of these are outlined in Sections 4.1–4.3 of this chapter. Further detail is contained in Appendix B. Depending on their backgrounds, some readers may need to consider all of this material, others only some of it, and still others none of it. Other authors have recognized the need for a review of these topics in connection with upper-atmospheric studies and the reader may consult helpful discussions by Chapman (1951) and Mitra (1952). For those who desire still more detail, particularly useful references are by Herzberg (1944, 1950).

In Section 4.4 is a discussion of the vertical distribution of absorbed energy under certain idealized conditions. Section 4.5 considers some characteristics of the sun and its ultraviolet radiation and Section 4.6 takes up the absorption spectra of some important constituents of the upper atmosphere and contains a discussion of the resulting distribution of absorbed energy in the upper atmosphere.

It should be noted that this chapter is concerned only with electromagnetic radiation and not with the effects of particles emitted by the sun.

4.1 Some Applicable Aspects of Spectroscopy and Quantum Mechanics

Some substances of meteorological importance emit continuous radiation and even approximate black bodies in some parts of the spectrum (for example, the ground, most clouds, and in some spectral regions the sun). However, the gaseous constituents of the atmosphere have *characteristic* spectra. Far from having radiative properties that are continuous with wavelength, they absorb and emit in discrete, narrow wavelength intervals called *lines*. Molecular, as opposed to atomic, spectra exhibit groups of lines within larger wavelength intervals, called *bands*. Under low dispersion, bands may even appear to represent continuous emission or absorption. For both atoms and molecules, however, there are spectral regions called *continua* in which the rate of absorption or emission varies slowly with wavelength. A particular kind of atom (or molecule) possesses a unique distribution of spectral lines (or bands) and continua. The interpretation of observed spectra in terms of atomic and molecular structure and behavior, largely accomplished during the first half of the twentieth century, accompanied and stimulated the development of quantum mechanics.

In this section we consider some fundamental concepts and definitions (Subsection 4.1.1); some further details with respect to atoms, especially hydrogen, helium, nitrogen, and oxygen (Subsection 4.1.2); and some further details with respect to diatomic molecules, especially nitrogen and oxygen (Subsection 4.1.3). It should be noted that this section is concerned primarily with radiative effects; that is, with changes in the internal energy of atoms or molecules that are accompanied by the absorption or emission of radiation. Such changes may also take place in connection with collisions, but these are discussed more specifically in Section 4.2.

4.1.1 Some Fundamental Concepts and Definitions

The study of radiation places emphasis on the internal energy associated with a particular *energy level* of the atom or molecule (not including kinetic energy associated with translational motion of the particle* as a whole). For the

* Hereafter, in this section the word "particle" is occasionally used for brevity to mean either atom or molecule, neutral or ionized. However, "particle" in other contexts may have a somewhat more general meaning; for example, it may include free electrons.

atom, this energy consists primarily of the kinetic energy of the electrons and the potential energy associated with the forces binding them to the nucleus. For molecules, there are additional energies associated with vibrations of the individual atoms relative to their equilibrium positions, and with rotation of the molecule as a whole. Contrary to expectations based on classical mechanics, and in order to interpret observed spectroscopic (and other) phenomena, the quantum theory postulates:

(a) An atom or molecule cannot exist with any arbitrary value of energy, but only with certain discrete values characteristic of the particle. Energies intermediate between these discrete values are not observed.

(b) Absorption or emission of electromagnetic radiation of frequency ν is associated with a change of energy of the atom or molecule from one discrete level to another, the frequency of the radiation being given by

$$E' - E'' = h\nu \tag{4.1}$$

Here E' is the energy of the higher level, E'' of the lower level ($E' > E''$), and h is Planck's constant, $h = 6.625 \times 10^{-27}$ erg sec. The quantity $h\nu$ is often called a *quantum* of energy. A particle absorbs electromagnetic radiation and jumps to a higher level only when the frequency of the radiation corresponds to the difference between the initial energy and that of some permitted higher energy level. Conversely, a particle with energy larger than its lowest permitted value tends to revert spontaneously to some permitted lower energy level with the emission of electromagnetic radiation of frequency proportional to the difference between the initial energy and that of the lower level.

(c) According to wave mechanics, there is a definite probability associated with each possible energy change, or *transition*. Those transitions with the greatest probability are associated with a change in the electric dipole moment of the atom or molecule and are identified by the *selection rules*. The probabilities of other transitions, which are called *forbidden transitions*, are small, although not usually exactly zero. In fact some forbidden transitions are important in the upper atmosphere.

In these circumstances, it is clear that the spectrum of an atom or molecule can be described more concisely by specifying its energy levels and the selection rules than by attempting to list transitions corresponding to all the possible energy differences. The energy levels are often represented graphically on an *energy-level diagram* with energy as ordinate and horizontal lines representing the discrete possible energy levels. The vertical distance between two energy levels, according to Eq. (4.1), is then proportional to the frequency of the radiation emitted or absorbed in the transition. Pertinent examples of energy-level diagrams are given later in this section.

Both atoms and molecules contain groups of closely spaced energy levels, the energy differences between the groups being characteristically greater than the energy differences between the levels within each group. In the case of molecules, such a group is referred to as a molecular *state*; a state contains *vibrational levels*, each of which contains *rotational levels*. In the case of atoms, such a group is referred to as a *term*, or sometimes state*; a term contains only a few *levels* (sometimes only one). Often only the energy associated with (the lowest level of) a state or a term is shown on energy-level diagrams.

The *ground term* (for an atom) or *ground state* (for a molecule) is the one of lowest energy. Other terms or states may be referred to as *excited*. An excited term or state from which radiative transitions to lower levels are forbidden by the selection rules is called *metastable*.

For any particle a characteristic amount of energy is required to remove an electron and this amount (with respect to the ground level) is called the *ionization energy* or *ionization potential* of that particle. For atoms or molecules with more than one electron, ionization potential usually refers to the most loosely bound electron, the one that requires the least energy for removal. But there may be second, third, etc., ionization potentials corresponding to removal of the second, third, etc., most easily removed electrons. A singly ionized atom or molecule is often designated by a superscript $+$, a doubly ionized particle by a superscript $++$, etc. Alternatively, a neutral particle is sometimes designated by a following Roman one, a singly ionized particle by a following Roman two, etc. Thus, for example, H (or H I), He^+ (or He II), and Li^{++} (or Li III) each have one electron. On the other hand, certain gases (in the atmosphere, O or O_2) may form negative ions, designated as O^- or O_2^-; the energy necessary to detach the excess electron from a negative ion is called *attachment energy* or *electron affinity*.

A radiative transition from one level to another is marked in the absorption or emission spectrum by a line whose frequency is related to the energy difference by Eq. (4.1). However, in the ionization process the atom or molecule may absorb more than the minimum energy required to remove the electron. This additional energy may be thought of as supplying kinetic energy to the freed electron and it is not quantized. Hence, on the high-frequency side of a frequency corresponding to ionization, there is an *ionization continuum* in which absorption is truly continuous. The ionization continuum in an emission spectrum would, in an analogous manner, correspond to recombination of an electron and an ionized particle.

In the particular case of a molecule, a certain amount of energy is required to *dissociate* or separate the atoms and this (with respect to the ground level)

* This usage of "state" is avoided in this book because the word is often used in another sense with respect to atoms. (Condon and Shortley, 1951, Preface).

is called the *dissociation energy*, or *dissociation potential*. In general a molecule has more than one dissociation potential, the lowest corresponding to the case where the separated atoms are both in their ground terms, and the others to cases where the separated atoms are in various excited terms. Just as in the case of ionization, the molecule may absorb or emit somewhat more energy than the minimum required for dissociation, so that molecular spectra may exhibit a *dissociation continuum* on the high-frequency side of the frequency corresponding to a particular dissociation energy.

Certain levels leading to one ionization potential may lie as far above the ground level as the continuum associated with another (lower) ionization potential. An atom or molecule in one of these levels may in some cases undergo ionization without further absorption of energy. This process is called *preionization*. A similar process leading to dissociation of molecules is called *predissociation*.

With each energy level is associated a set of *quantum numbers*, integral or half-integral, and a symbol that specifies these numbers and other characteristics of the energy level. Although a full explanation of these is well beyond the scope of the present treatment, some familiarity with the nomenclature and symbolism is necessary, because selection rules, energy levels, and transitions that are important in the upper atmosphere are often described in terms of this symbolism. The situation is naturally more complicated for molecules than for atoms. This subject is discussed further in Subsections 4.1.2 and 4.1.3.

The energy of an atom or molecule is specified only relative to some arbitrary zero point. In the case of an atom, the zero is often taken to correspond to the lowest ionization potential; thus on this scale discrete levels may have "negative" energy, the ground level being the most negative. Nevertheless, energy-level diagrams for atoms often have the ordinate labeled, for convenience, with the ground level as zero. The zero for molecules is chosen to correspond to the level of lowest energy.

In spectroscopy, energy is usually specified in units of *electron volts* (ev) or in units of *wave number*. An electron volt is the energy acquired by an electron accelerated through a potential difference of one volt. Since the electron has a charge of 4.803×10^{-10} electrostatic units and the volt corresponds to $1/299.8$ electrostatic units, one electron volt is equivalent to 1.602×10^{-12} erg. The wave number $\tilde{\nu}$ is simply the reciprocal of wavelength *in vacuo* and is usually expressed in (centimeter)$^{-1}$. Obviously it is related to frequency by $\tilde{\nu} = \nu/c$, where c is the speed of light, 2.998×10^{10} cm sec^{-1}; 1 cm^{-1} is equivalent to 1.986×10^{-16} erg or to 1.240×10^{-4} ev. On the other hand, the positions of spectral lines are often given in angstrom units (A). It is convenient to remember that the wavelength in angstrom units of a spectral line is 10^8 divided by the energy difference in (centimeter)$^{-1}$ or 12,397 divided by the energy difference in electron volts.

4.1.2 Atomic Energy Levels; Hydrogen, Helium, Nitrogen, Oxygen

The hydrogen atom, with one electron and a nucleus, is the simplest of all atoms and molecules. For this reason, it is convenient to discuss it first. Furthermore, the hydrogen atom plays an important role in the solar emission in the ultraviolet.

To a good approximation, the energy levels for this atom are given by

$$E_n = -R_H hc/n^2; \qquad n = 1, 2, 3, ... \tag{4.2}$$

where R_H is the *Rydberg constant* for hydrogen, with the value 1.097×10^5 cm^{-1}. There are energy levels for all positive integers n. The integer n is called the *principal* quantum number. The lowest (most negative) energy level corresponds to $n = 1$ and zero energy (ionization) is approached as n becomes large.

It follows from Eqs. (4.1) and (4.2) (and since $\tilde{\nu} = \nu/c$) that the wave numbers of the spectral lines of the hydrogen atom are given by

$$\tilde{\nu} = R_H \left[\frac{1}{(n'')^2} - \frac{1}{(n')^2}\right] \tag{4.3}$$

where n' is the principal quantum number of the upper level and n'' the principal quantum number of the lower level, $n' > n''$. This simple equation quite accurately represents the wave numbers of all lines of the hydrogen atom. As a matter of fact, it was first determined empirically, and also followed from the Bohr model of the hydrogen atom. In terms of wave mechanics, the energy levels represented by (4.2) are the eigenvalues of the Schrödinger wave equation for an electron moving in a pure Coulomb field (see Appendix B.1).

Figure 4.1 shows the energy-level diagram for the hydrogen atom, according to Eq. (4.2). Note that the energy levels corresponding to different values of n lie closer and closer together as n gets larger. In the limit as n gets very large, the energy approaches zero on our scale. The ground level, with $n = 1$, corresponds to an energy of -13.60 ev.

Transitions occur between all values of n. Those transitions for which the quantum number of the lower level, n'', has a fixed value produce a series of spectral lines. For example, those transitions with $n'' = 1$ produce the Lyman series, those with $n'' = 2$ produce the Balmer series, those with $n'' = 3$ produce the Paschen series. Lines with $n' = n'' + 1$, $n'' + 2$, $n'' + 3$, etc., are referred to as the α, β, γ, etc., lines of the series. Figure 4.2 shows the position in the spectrum of some lines in and the continua corresponding to the Lyman and Balmer series. The other series occur in the infrared and are of no interest here. Note that the lines in each series converge as the continuum is approached. Note also that the Lyman lines, on a linear wavelength scale, are closer together than the Balmer lines. Series of lines based on transitions between

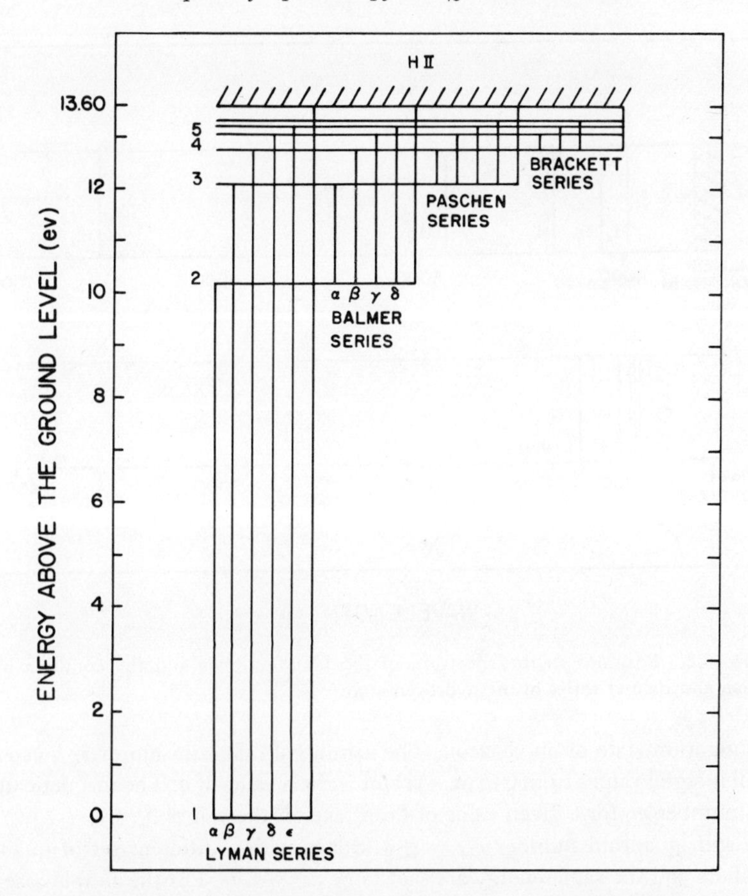

FIG. 4.1. Energy-level diagram of the hydrogen atom, computed from Eq. (4.2). The principal quantum numbers are shown beside the energy levels.

energy levels with different principal quantum number, even in more complicated atoms and molecules, are called *Rydberg series* and have the same general characteristics with regard to line spacing.

Other single-electron atoms such as singly ionized helium and doubly ionized lithium have spectra similar to that of the hydrogen atom but with two variations: the Rydberg constant is slightly different and the right side of Eq. (4.2) must by multiplied by Z^2, where Z is the number of elemental charges (of magnitude e) on the nucleus. The latter difference corresponds to a strong displacement of corresponding lines toward shorter wavelengths. For example, the Lyman α line of H is found at about 1216 A and the corresponding line of He^+ is near 304 A; both are very important lines in the sun's emission spectrum.

In general, other quantum numbers in addition to n are required to specify

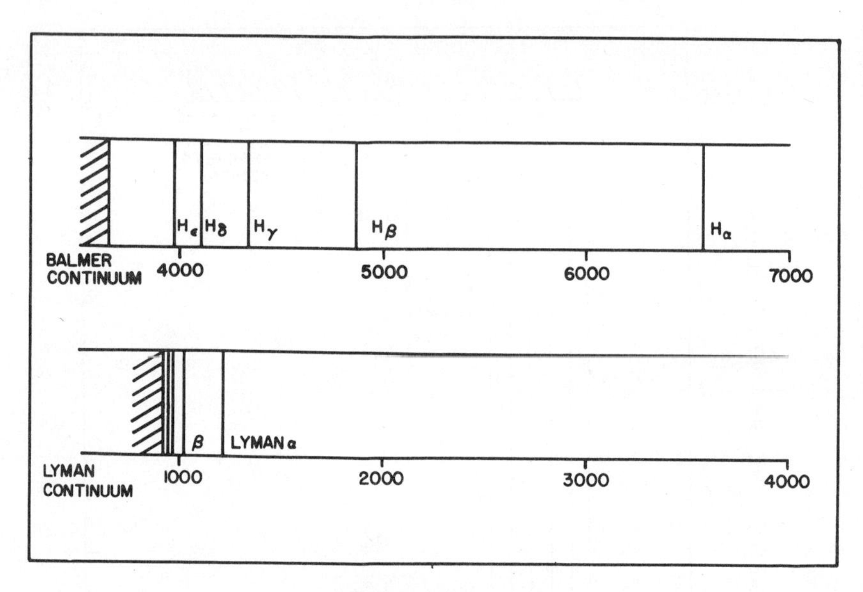

FIG. 4.2. Positions in the spectrum of the first few lines and the continua of the Lyman and Balmer series of the hydrogen atom.

the quantum state of an electron. The azimuthal quantum number, l, can take on all integral values from 0 to $(n - 1)$ for a given value of n. The magnetic quantum number m_l for a given value of l can take on the values 0, ± 1, ± 2, ..., $\pm l$. The spin quantum number $m_s = \pm \frac{1}{2}$. States with a different set of quantum numbers and the same energy are said to be *degenerate*. For the simple case discussed above, a single electron moving in a pure Coulomb field, the energy depends only on n. However, even in this case, the degeneracy with respect to l is removed if a relativistic correction is made, but this is of no practical importance for our purposes. For alkali atoms in which a single electron moves in the field of the nucleus and the innermost electrons, levels with the same n and different l can have quite different energies. In the case of m_l, the degeneracy is removed in the presence of an external field, electric or magnetic, however slight. Appendix B.2 contains a further discussion of these electronic quantum numbers.

An electron is designated according to the values of its quantum numbers n and l. For the quantum number l, a letter designator is used: s for $l = 0$, p for $l = 1$, d for $l = 2$, f for $l = 3$, g for $l = 4$. This is preceded by a number designating n. Thus a 1s electron has $n = 1$, $l = 0$; a 2s electron has $n = 2$, $l = 0$; a 2p electron has $n = 2$, $l = 1$; etc. Statement in this manner of the n and l quantum numbers for all the electrons in an atom is called an *electron configuration*; pertinent examples are given below. Electrons with the same value of n and the same value of l are spoken of as *equivalent electrons.*

According to the *Pauli exclusion principle*, no two electrons in the same atom can have all four quantum numbers, n, l, m_l, and m_s, the same. In the ground configuration of the helium atom, both electrons have $n = 1$ and must have $l = 0$ and $m_l = 0$ according to the way the possible values of these last two are limited by the value of n. Therefore, one electron must have $m_s = +\frac{1}{2}$ and the other $m_s = -\frac{1}{2}$. Note that no atom can have more than two electrons with $n = 1$. It may, however, have eight electrons with $n = 2$, and 18 electrons with $n = 3$. Electrons with $n = 1$ are said to be in the K shell, with $n = 2$ in the L shell, with $n = 3$ in the M shell, etc. The K shell is closed with helium and atoms with more electrons must have some of them in other shells.

The electron configuration of an atom is related to the energy level of the atom, but does not in general uniquely describe it. In addition, atomic quantum numbers, which are derived in a certain way* from the quantum numbers of the individual electrons, are required. The quantum number L is related to the resultant orbital angular momentum, the quantum number S to the resultant spin angular momentum, and the quantum number J to the resultant total angular momentum. The term symbol consists of a letter† designating L (S for $L = 0$, P for $L = 1$, D for $L = 2$, F for $L = 3$) and a preceding superscript which is equal to $2S + 1$, which combination preceded by the electron configuration specifies a *term.* Within each term there may be different energy *levels* associated with different values of J, and the value of J may be written as a succeeding subscript to the term symbol to specify a level. Thus $^3S_1$ refers to an atom with $L = 0$, $S = 1$, $J = 1$ and is read "triplet S one."

The atomic quantum numbers are important not only because they distinguish energy levels occurring with the same electron configuration but also because the selection rules are stated in this symbolism. These are:

(a) $\Delta S = 0$;

(b) $\Delta L = 0, \pm 1$;

(c) $\Delta J = 0, \pm 1$ (except transitions between $J' = J'' = 0$ are excluded);

(d) if the sum of the l values of all the electrons in one term is odd, it must be even in the other term of the transition. Odd terms are sometimes distinguished from even by a following superscript o; sometimes the subscripts g (*gerade* = even) and u (*ungerade* = odd) are used in place of the J values.

Let us now consider the helium atom in the light of the above discussion. In the ground configuration, both electrons have $n = 1$, $l = 0$ and the electron

* See Appendix B.3 for some elaboration.

† The letter S which designates $L = 0$ should not be confused with the spin quantum number S. The same symbol is used with the two different meanings.

configuration is written $1s^2$, where the following superscript specifies that there are two equivalent $1s$ electrons. In this case, the only level possible is $^1S_0$. For the consideration of excited terms and levels, it is sufficient to consider the usual case where one electron remains in the inner shell with $n = 1$, $l = 0$, $m_l = 0$, and $m_s = \pm\frac{1}{2}$, and the other electron (called the *emission electron*) is in an outer shell. Table 4.1 gives some possible values of n, l, and m_s for the emission electron, and the resulting values of L, S, J for the atom, as well as the term symbol that goes with each. Figure 4.3 shows the energy-level diagram for the helium atom. A brief discussion of the helium spectrum with reference to this diagram and the selection rules is given in Appendix B.4.

The nitrogen atom with seven electrons normally has two in the K shell, two in the L shell with $l = 0$ (sometimes called the L_1 shell), and three in the

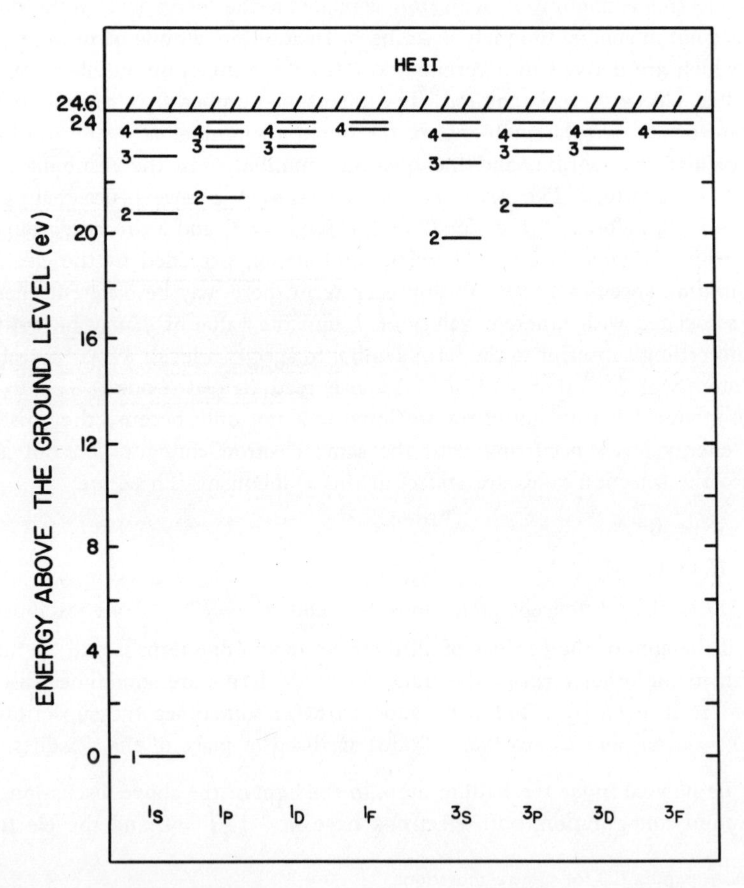

Fig. 4.3. Energy-level diagram of the helium atom. Principal quantum number of the emission electron is shown beside each term.

L shell with $l = 1$ (the L_2 shell). The electron configuration is written $1s^2 2s^2 2p^3$; sometimes the innermost electrons are not specifically mentioned, that is, the electron configuration for the normal nitrogen atom might be written simply $2s^2 2p^3$. Of the two $2s$ electrons, both must have $m_l = 0$ and one must have $m_s = +\frac{1}{2}$ and the other $m_s = -\frac{1}{2}$, according to the Pauli exclusion principle, and the L_1 shell can hold no more electrons. The L_2 shell, however, can hold altogether six electrons ($m_l = 0, \pm 1$ with $m_s = \pm\frac{1}{2}$ for each value of m_l),

TABLE 4.1

POSSIBLE VALUES OF l AND m_s FOR AN ELECTRON IN THE EXCITED HELIUM ATOM WITH $n = 2, 3$, AND 4[a]

n	l	m_s	L	S	J	Term
2	0	−	0	0	0	$2s\ ^1S$
2	0	+	0	1	1	$2s\ ^3S$
2	1	−	1	0	1	$2p\ ^1P^o$
2	1	+	1	1	0, 1, 2	$2p\ ^3P^o$
3	0	−	0	0	0	$3s\ ^1S$
3	0	+	0	1	1	$3s\ ^3S$
3	1	−	1	0	1	$3p\ ^1P^o$
3	1	+	1	1	0, 1, 2	$3p\ ^3P^o$
3	2	−	2	0	2	$3d\ ^1D$
3	2	+	2	1	1, 2, 3	$3d\ ^3D$
4	0	−	0	0	0	$4s\ ^1S$
4	0	+	0	1	1	$4s\ ^3S$
4	1	−	1	0	1	$4p\ ^1P^o$
4	1	+	1	1	0, 1, 2	$4p\ ^3P^o$
4	2	−	2	0	2	$4d\ ^1D$
4	2	+	2	1	1, 2, 3	$4d\ ^3D$
4	3	−	3	0	3	$4f\ ^1F^o$
4	3	+	3	1	2, 3, 4	$4f\ ^3F^o$

[a] When the other electron is in the K shell with $l = 0$ and $m_s = +\frac{1}{2}$, the indicated values of L, S, and J (and the corresponding term notations) are possible. In the column for m_s, "−" or "+" designates "$-\frac{1}{2}$" or "$+\frac{1}{2}$."

only three of which are present in the normal nitrogen atom. The two $1s$ electrons are equivalent, the two $2s$ electrons are equivalent, and the three $2p$ electrons are equivalent in the nitrogen atom. The contribution of equivalent electrons in a *closed* shell to the L and S values of the atomic term is zero; that is, the atomic quantum numbers can always be determined from the quantum numbers of the electrons outside the closed shells, in this case the three $2p$ electrons. There are three terms corresponding to this electron configuration, with the term symbols $^4S^o$, $^2D^o$, and $^2P^o$. Of these, $^4S^o$ has the least energy and represents the ground term of the nitrogen atom. It includes only one level, with $J = \frac{3}{2}$ (see Appendix B.3). The $^2P^o$ and $^2D^o$ terms are both metastable terms, transitions to the ground term being forbidden by selection rules (a) and (d) and, in the latter case, also by (b). Each of these terms includes two levels, the former with $J = \frac{3}{2}$ and $\frac{1}{2}$, the latter with $J = \frac{5}{2}$ and $\frac{3}{2}$.

Other excited levels of the N atom arise when one of the three $2p$ electrons goes to a shell with $n \geqslant 3$. To find the new energy level, one must first consider which of the three electrons is involved. This distinction is made by consideration of the energy associated with the remaining electrons. To put it another

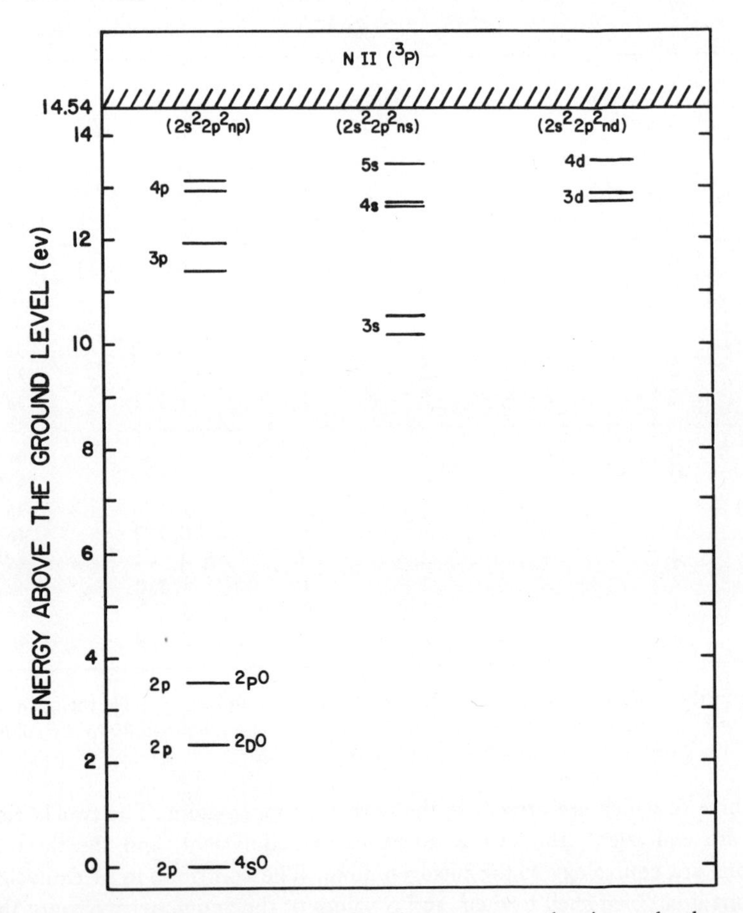

FIG. 4.4. Energy-level diagram of the nitrogen atom, showing only those terms leading to NII ^3P. Each column corresponds to a different value of l for the emission electron. For each configuration np, various energy levels lie within the limits shown; some are doublets, some quartets.

way, if the emission electron were completely removed, the remaining two $2p$ electrons in the ion could give rise to three possible terms ^1S, ^1D, and ^3P of which the last would be the ground term. Evidently, removing the electron that leaves a ^3P ion requires the least energy and corresponds to the lowest ionization potential. In the case of nitrogen, this is about 14.54 volts, corresponding to about 852 A. Therefore, we expect no continua in the absorption spectrum of N at a wavelength longer than 852 A. Different series of terms with different ionization potential result for the other possibilities.

Figure 4.4 shows a partial energy-level diagram for N, in which the emission electron is assumed to be the one for which the resulting ion is a $2p^2\,{}^3P$. A large

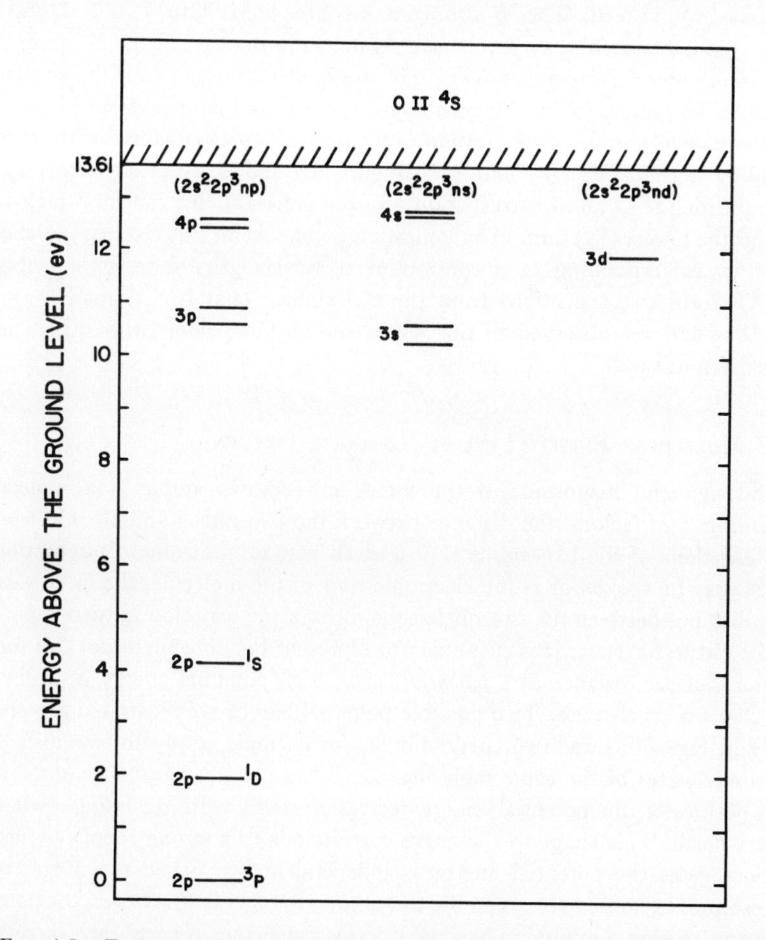

FIG. 4.5. Energy-level diagram of the oxygen atom, showing only those terms leading to OII ^4S. Each column corresponds to a different value of l for the emission electron. For each configuration np, the higher term is a triplet, the lower a quintet.

number of energy levels results according to various values of n, l, and m_s for the emission electron. Evidently the nitrogen atom is far more complicated than the helium atom which in turn is far more complicated than the hydrogen atom.

We should mention in passing that the possibilities of excited levels are not exhausted by the consideration of one emission electron. It is possible but much less likely for two electrons to leave the L_2 shell; terms arising from this condition are called anomalous terms. It is also possible to have only one electron in the L_1 shell and four in the L_2 shell, but this again is rare.

The oxygen atom with eight electrons has normally the configuration $1s^2 2s^2 2p^4$. The four equivalent $2p$ electrons can give rise to the terms ^3P, ^1D, ^1S of which ^3P is the ground term. The ^3P term contains three levels with $J = 2$, 1, and 0; the others contain only one level each. Again, as in the case of the N atom, the ionization potential depends on which of the $2p$ electrons serves as the emission electron. The singly ionized oxygen atom in its ground configuration obviously has the same electronic configuration as the neutral nitrogen atom, with corresponding terms $^4S^o$, $^2D^o$, and $^2P^o$. Figure 4.5 shows a partial energy-level diagram for the O atom corresponding to the emission electron for which the ion has the lowest (^4S) term. The ionization potential (for this electron) is about 13.61 ev corresponding to a continuum at wavelengths shorter than about 911 A. Forbidden transitions from the metastable ^1D and ^1S terms ($^1D \leftarrow {}^1S$ and $^3P \leftarrow {}^1D$) are observed in the aurora and airglow, along with other, permitted, transitions.

4.1.3 Molecular Energy Levels; Nitrogen, Oxygen

The sign and magnitude of the force between two atoms in a molecule depend on two factors: the distance between the two nuclei and the electronic configurations of the two atoms. For a given pair of electronic configurations, which may be spoken of as the electronic *state* of the molecule, the force varies with distance between the two nuclei; the form of the variation in turn depends on the electronic state. It is customary to represent the dependence of this force on internuclear distance as a *potential curve*, with potential energy as ordinate and distance as abscissa. Two possible potential curves are illustrated schematically in Fig. 4.6; these two curves might, for example, apply to two different electronic states of the same molecule.

In both cases, the potential energy decreases rapidly with increasing r when r is very small. This shape of the curve corresponds to a strong repulsive force. In both cases the potential energy is independent of r when r is large; this corresponds to no force between the two atoms. In one case, however, the potential curve reaches a minimum at some intermediate value of r and then increases with increasing r corresponding to an attractive force. At the distance corresponding to the minimum, the atoms exert no net force on one another and a

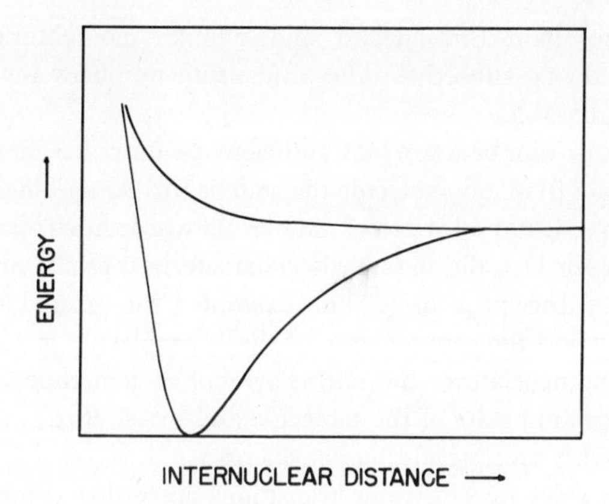

FIG. 4.6. Schematic representation of potential curves for a diatomic molecule. Typically, the vertical scale would encompass a few electron volts, the horizontal scale a few angstroms.

stable molecule can be formed. When no minimum occurs, the atoms repel one another at all distances and no stable molecule is possible. It is customary to define the *electronic energy* of a molecular state as the potential energy at the minimum of the potential curve. The ground state is taken as the zero point of the molecular energy scale, a definition somewhat different from the case of atoms.

A molecule, in contrast to an atom, can have two additional kinds of energy: (a) the individual atoms can vibrate about their equilibrium positions relative to one another, and (b) the molecule as a whole can rotate about an axis through the center of mass and perpendicular to the internuclear axis. The total energy of a molecule can be written as the sum of the electronic, vibrational, and rotational energies,

$$E = E_e + E_v + E_r \tag{4.4}$$

although these are not completely independent of one another. In the space available, it is impossible to discuss these energies and their dependence on the appropriate quantum numbers in any detail. We shall confine ourselves to a brief discussion of notation and of the qualitative spectral results, and, furthermore, to diatomic molecules.

As with atoms, the electronic states of molecules are classified by means of a quantum number that is associated with the orbital angular momentum of the electrons. By analogy with the atomic notation, this quantum number is called Λ, and $\Lambda = 0$ is represented by Σ, $\Lambda = 1$ by Π, $\Lambda = 2$ by Δ, $\Lambda = 3$ by Φ. The spin quantum number S is related to the resultant angular momentum of electron spin, and as with atoms, $2S + 1$ is called the multiplicity, and is written as a preceding superscript to the symbol for Λ. A third quantum number Ω, related to the resultant (orbital and spin) angular momentum, is sometimes written as a following subscript. These quantum numbers are discussed a bit more in Appendix B.5.

These quantum numbers are not sufficient to describe an electronic state. For Σ states ($\Lambda = 0$) of any molecule the state is further specified by a following superscript + or −. For all states of a molecule whose nuclei have equal charge (for example, N_2 or O_2), the state is also characterized as *odd* or *even*, indicated by a following subscript *u* or *g*. For example, the ground state of the O_2 molecule is $^3\Sigma_g^-$ and the ground state of the N_2 molecule is $^1\Sigma_g^+$. In addition to this formal nomenclature, the entire symbol is sometimes preceded by *X* to indicate the ground state of the molecule and by *A*, *B*, *C*, ... or *a*, *b*, *c*, ... to indicate states with successively higher energies.

The selection rules for electronic transitions are rather complicated and can be most concisely illustrated by listing the transitions that are generally allowed, as in Table 4.2. The transitions listed, as with atoms, are likely to occur only when the two states have the same multiplicity.

TABLE 4.2

ALLOWED ELECTRONIC TRANSITIONS FOR DIATOMIC MOLECULES WHEN BOTH STATES HAVE THE SAME MULTIPLICITY[a]

Equal nuclear charge		Unequal nuclear charge	
$\Sigma_g^+ \leftrightarrow \Sigma_u^+$,	$\Sigma_g^- \leftrightarrow \Sigma_u^-$	$\Sigma^+ \leftrightarrow \Sigma^+$,	$\Sigma^- \leftrightarrow \Sigma^-$
$\Pi_g \leftrightarrow \Sigma_u^+$,	$\Pi_g \leftrightarrow \Sigma_u^-$	$\Pi \leftrightarrow \Sigma^+$,	$\Pi \leftrightarrow \Sigma^-$
$\Pi_u \leftrightarrow \Sigma_g^+$,	$\Pi_u \leftrightarrow \Sigma_g^-$		
$\Pi_g \leftrightarrow \Pi_u$		$\Pi \leftrightarrow \Pi$	
$\Pi_g \leftrightarrow \Delta_u$,	$\Pi_u \leftrightarrow \Delta_g$	$\Pi \leftrightarrow \Delta$	
$\Delta_g \leftrightarrow \Delta_u$		$\Delta \leftrightarrow \Delta$	

[a] Transitions between states with different multiplicity are forbidden.

The vibrational energy of a diatomic molecule is connected with the vibratory motions of the two atoms about their equilibrium positions. The equilibrium separation of the two, r_E, corresponds to the minimum in the potential curve (and, of course, varies with the electronic state). If the actual separation r is greater than r_E, then a force acts to decrease r; if the actual separation r is less than r_E, then a force acts to increase r. According to quantum mechanics, not all energies of vibration are possible.* The vibrational energy levels are given to a fair approximation by

$$E_v = h\nu_e(v + \tfrac{1}{2}); \qquad v = 0, 1, 2, \ldots \tag{4.5}$$

* See Appendix B.1.

where ν_e is the vibrational frequency, which depends on the electronic state of the molecule, and v, the vibrational quantum number, is zero or a positive integer. Note that the vibrational energy levels, in contrast to the energy levels in a Rydberg series, are equally spaced (in this approximation). Furthermore, only a limited number of vibrational levels is possible; if the vibrational energy is large enough (corresponding to large v), the atoms exceed the distance at which a restoring force acts and the molecule dissociates. Since the energy with which the atoms may separate is not quantized, a dissociation continuum may occur in addition to the vibrational bands. Such dissociation continua are sometimes observed.

Transitions from one vibrational level to another are specified by the v values of the two levels, that for the upper level (higher energy) being given first. The quantum number of the upper level is distinguished by one prime, that of the lower level by two primes. When vibrational transitions occur without an electronic transition, the selection rule for v in the approximation discussed above is $\Delta v = v' - v'' = 1$; a more realistic model shows that other changes in v are possible but less likely. At normal atmospheric temperatures, most molecules are in the ground electronic state with $v = 0$; their vibrational absorption spectrum therefore consists primarily of a single vibrational transition, the (1–0) transition with $v' = 1$ and $v'' = 0$, and much weaker (2–0), (3–0), etc. transitions. The bands corresponding to these transitions generally lie in the near infrared, the (1–0) transition at the longest wavelength, while electronic transitions generally give rise to bands in the visible and ultraviolet. The energy associated with vibrational transitions is much less than that associated with the typical electronic transition.

We say "bands" and not "lines" because vibrational transitions are generally accompanied by transitions between various rotational levels, as is discussed below. These bands are called *vibrational-rotational bands.* Molecules made up of two similar atoms, such as N_2 and O_2, have no pure vibrational-rotational transitions and no significant absorption in the infrared. This is because, both classically and quantum-mechanically, absorption and emission of radiation accompany a changing dipole moment and the dipole moment of such a molecule is the same in different rotational and vibrational states if the electronic state remains the same. This fact is of far-reaching significance for the atmosphere's heat budget and leads to the rather strange situation that infrared transfer depends on minor atmospheric constituents such as H_2O, CO_2, and O_3.

Vibrational transitions may also occur in conjunction with electronic transitions. The totality of vibrational bands accompanying one electronic transition is called a *band system*; according to what was said above, a band system may be accompanied at the short-wavelength end by a dissociation continuum. A band system may contain bands in which $v' < v''$; in this case the vibrational energy in the upper state is less than in the lower, although the total energy, including electronic energy, is greater. When vibrational transitions accompany electronic transitions, there is no easily stated selection rule for the vibrational transitions. The most intense vibrational bands are found near some particular value of Δv, but whether Δv is small, intermediate, or large depends, according to the Franck–Condon principle, on the potential curves of the two electronic states.

An additional kind of quantized molecular energy is that of rotation. Rotational energy is characterized by the rotational quantum number J. To a first approximation, the energy associated with rotation of the molecule is given by (see Appendix B.1)

$$E_r = hcB_v J(J+1); \qquad J = 0, 1, 2, \ldots \tag{4.6}$$

The rotational "constant," B_v varies slowly with the vibrational quantum number, and Eq. (4.6) must be modified somewhat when the electronic motions are considered. Such details need not concern us here. For our purposes it is sufficient to note that energy differences associated with different values of J are small compared with the differences between vibrational levels and especially electronic states. Consequently, the pure rotation spectrum of a molecule consists of a series of lines in the far infrared. Rotational transitions accompanying vibrational transitions give rise to a series of lines that make up the various bands. The selection rule for J is $\Delta J = 0, \pm 1$ (except when both states have $\Lambda = 0$, in which case $\Delta J = 0$ is forbidden). According to whether $\Delta J\,(= J' - J'')$ has the value -1, 0, or $+1$ the lines corresponding to rotational transitions are said to belong to the P, Q, or R branches of the band.

Figure 4.7 gives the energy-level diagram for the N_2 molecule. Also shown are the dissociation potentials together with the terms of the resulting atoms. Note that there are several such possibilities, the lowest of which corresponds to two atoms in the N ground term; in the other cases, one of the atoms is excited.

It is to be especially noted that no dissociation continua have been observed for the nitrogen molecule. In all cases the highest vibrational energy observed is insufficient for dissociation. The N_2 molecule can be dissociated, however, as a result of predissociation. Predissociation of the N_2 molecule has been observed in the ${}^1\Pi_g$ and the ${}^3\Pi_g$ states at the 9.76-ev limit; and in the ${}^3\Pi_u$ state at the 12.14-ev limit.

On the other hand, ionization continua are observed. The lowest ionization potential at 15.58 ev (about 796 A for transitions from the ground state) leaves N_2^+ in its ground state ${}^2\Sigma_g^+$, while the second at 16.94 ev leaves N_2^+ in the excited ${}^2\Pi_g$ state. Both join a Rydberg series of absorption bands. These correspond to energy levels in which one of the electrons of the nitrogen molecule has higher and higher principal quantum number, just as in the case of ionization of an atom. Note that the lower energy states are compounded of atoms with terms ${}^4S$, ${}^2D$, and ${}^2P$, all of which correspond to the lowest possible electron configuration for the nitrogen atom, namely $1s^2 2s^2 2p^3$.

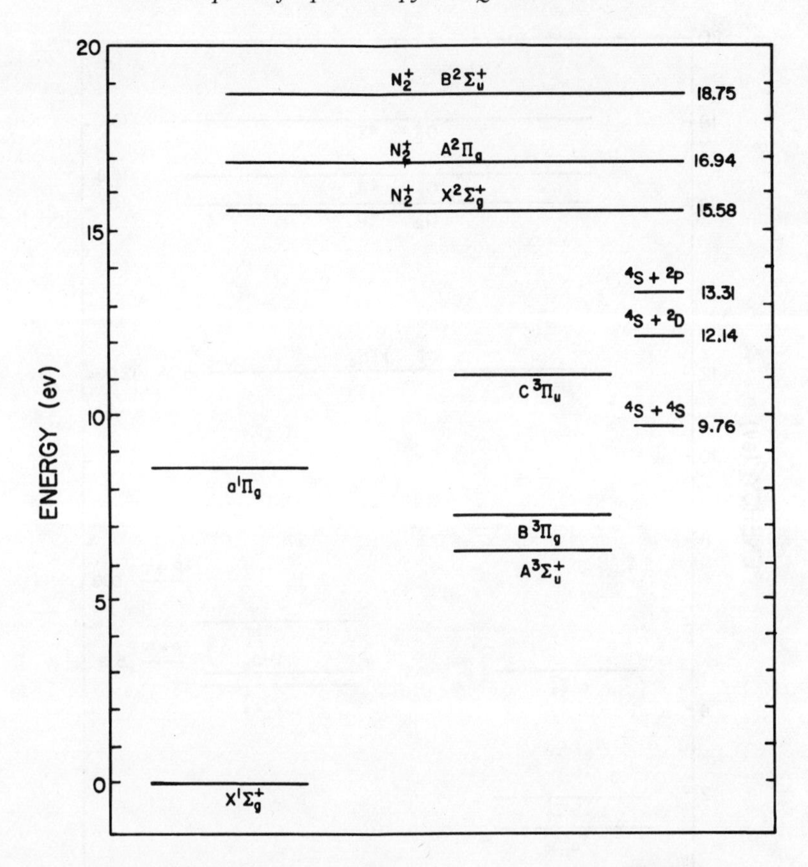

FIG. 4.7. Energy-level diagram of the nitrogen molecule, showing some of the observed states and the lowest dissociation and ionization potentials.

Figure 4.8 shows the energy-level diagram for the O_2 molecule and the O_2^+ ion. In contrast to N_2, two dissociation continua have been observed. The first occurs in the ${}^3\Sigma_u{}^+$ state and leaves two ${}^3P$ atoms. Transitions from the O_2 ground state to this dissociation continuum are forbidden but nevertheless a weak series of bands and a very weak continuum between 2000 and 2423 A have been observed. The second dissociation occurs at about 1750 A (for transitions from the ground state) and leaves one normal ${}^3P$ atom and one excited ${}^1D$ atom. Absorption leading to this dissociation (${}^3\Sigma_u{}^- \leftarrow {}^3\Sigma_g{}^-$) is very strong. Both of these continua are very important with respect to the upper atmosphere.

Four ionization continua join Rydberg series of bands at about 1026, 771, 736, and 682 A. The first leaves O_2^+ in its ground state ${}^2\Pi_g$ and is observed only very weakly. The others leave O_2^+ in various excited states and give strong absorption in the extreme ultraviolet.

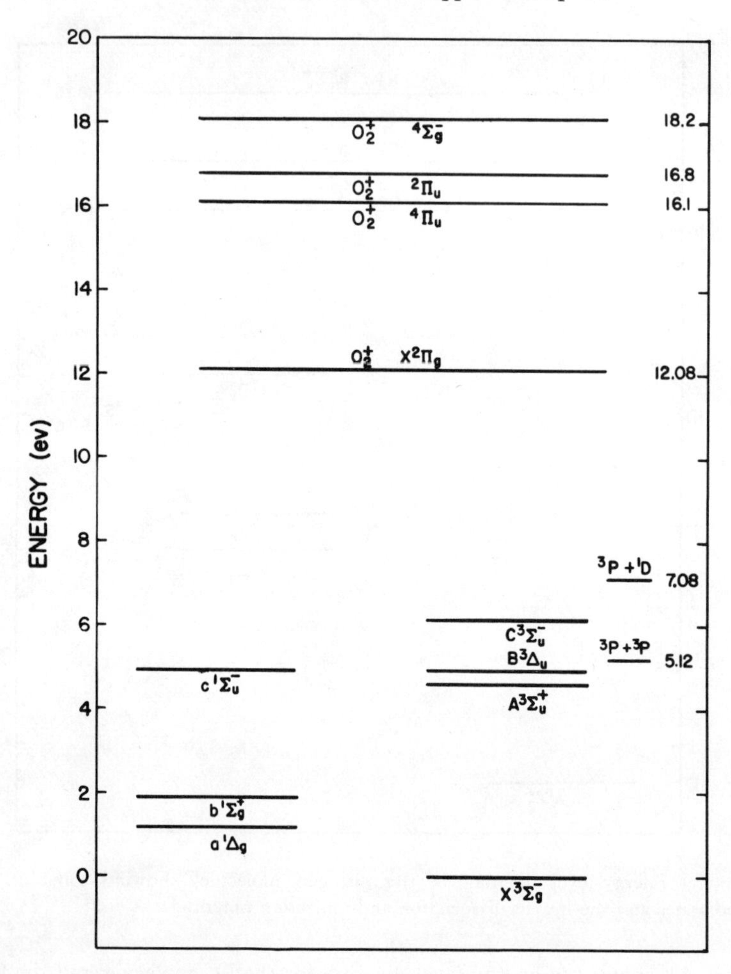

FIG. 4.8. Energy-level diagram of the oxygen molecule, showing some of the observed states and the lowest dissociation and ionization potentials.

We shall refer back later to Figs. 4.4, 4.5, 4.7, and 4.8 in further discussion of the emission and absorption spectra of N, O, N_2, and O_2.

4.2 Energy Exchange by Collision

In the previous section we spoke primarily of energy changes accompanied by the absorption or emission of electromagnetic energy. In addition to these radiative processes, there are a large number of reactions that can take place in conjunction with collisions between particles, and some of these reactions are of very great importance in the upper atmosphere. In a radiative process, the change of internal energy of the atom or molecule is just balanced by the energy associated with the absorbed or emitted quantum. In a collision process, on the other hand, the kinetic energy of translation of the particles involved can serve as a source or sink for transitional energy. A reaction that results in an increase of kinetic energy is called *exothermic*, one that results in a decrease of kinetic energy is called *endothermic*.

Thus, an atom or molecule in an excited level can lose its excess energy, in entirety or in part, by a radiative transition or by collision with another particle. The probability that a radiative transition will occur in the time interval (t) to ($t + dt$) is given by dt/T, where T is called the *half-life* or *lifetime* and is constant for a given excited level of a given particle. On the other hand, the probability that a collision reaction will take place depends in part on the frequency of collisions and therefore on the number densities of the particles involved. A radiative transition with a given lifetime T may be quite unlikely under conditions where the collision interval is much less than T, but on the other hand may be quite common at high enough levels in the atmosphere where the density is low and the collision interval is large.

In any collision, not only energy but also momentum must be conserved. That is, the total translational momentum of the particles involved must be the same after the reaction as before. This requirement makes certain reactions, which we shall discuss below, quite unlikely to take place even when collisions occur frequently.

In the following discussion, we shall consider certain important kinds of reactions, giving, where appropriate, examples that occur in the earth's upper atmosphere. In the general discussion, we shall let the symbols A, B, C refer to atoms, appropriate combinations of A, B, C refer to molecules, and the symbols L, M refer to particles that may be either atomic or molecular. The symbol e will refer to an electron. A prime will indicate an excited atom or molecule.

In excitation by collision, the required excitation energy comes from the translational kinetic energy of the particles. For example,

$$\mathrm{L} + \mathrm{M} \rightarrow \mathrm{L}' + \mathrm{M} \tag{4.7a}$$

$$\mathrm{AB} + \mathrm{M} \rightarrow \mathrm{A} + \mathrm{B} + \mathrm{M} \tag{4.7b}$$

$$\mathrm{L} + \mathrm{M} \rightarrow \mathrm{L}^+ + \mathrm{e} + \mathrm{M} \tag{4.7c}$$

At the temperatures obtaining in the upper atmosphere only a minute fraction of the atmospheric particles has sufficient kinetic energy to effect dissociation or ionization or even to produce electronic transitions. However, excitation to low-lying vibrational, rotational, or J levels is important at times (see Chapter 7). Particles (including electrons) impinging from outside the atmosphere, or from the very high atmosphere, may have sufficient energy to excite higher levels. Excitation by impact of these particles is widely accepted as being responsible for much of the energy that eventually appears as auroral emission.

We have already mentioned briefly above the possibility of de-excitation by collision. Combination (of an atom with an atom or an electron with an ion) is a particularly important case of this and may take place in a two-body collision with emission of a photon (*radiative combination*)

$$A + B \rightarrow AB + h\nu \tag{4.8a}$$
$$L^+ + e \rightarrow L + h\nu \tag{4.8b}$$

or in a three-body collision (*three-body combination*)

$$A + B + M \rightarrow AB + M \tag{4.9a}$$
$$A + BC + M \rightarrow ABC + M \tag{4.9b}$$
$$L^+ + e + M \rightarrow L + M \tag{4.9c}$$

In the case of radiative combination one particle is formed from two and its momentum (and therefore its kinetic energy) is uniquely determined by the initial momenta of the particles (the momentum of the photon is negligible). Conservation of energy requires the emission of exactly the right energy to account for the energy of combination and the change of kinetic energy of translation. The probability that this can occur during the very short period of contact is usually rather small.

On the other hand, in three-body combination the conservation of momentum and energy may be achieved more easily. Of course, the likelihood of a three-body collision occurring at all is less than that of a two-body collision, and this likelihood decreases quite rapidly as the density decreases. In the upper stratosphere and mesosphere, three-body combinations play an important role; for example,

$$O + O_2 + M \rightarrow O_3 + M \tag{4.10}$$
$$O + O + M \rightarrow O_2 + M \tag{4.11}$$

Howerer, at high enough levels, because of the decreasing density, radiative combination must become more important.

In *transfer* reactions, internal energy of one kind or another is transferred from one colliding partner to the other. In *chemical* transfer, this involves the breaking of a molecular bond and the forming of another, for example,

$$AB + C \rightarrow AC + B \tag{4.12}$$

where either or both of the resulting particles may be in an excited state. At upper-atmosphere temperatures, this type of reaction can be important only when it is exothermic, that is, when the energy required to break the initial bond is less than that released when the new one is formed. Furthermore, in this type of reaction, it is found that the available kinetic energy must exceed a certain threshold called the *activation energy* (even though the process is exothermic). One important example of a chemical transfer reaction is

$$O + O_3 \rightarrow O_2 + O_2 \tag{4.13}$$

which plays an important role in the upper stratosphere and mesosphere.

There may also be transfer reactions in which energy of excitation is transferred

$$L' + M \rightarrow L + M' \tag{4.14}$$

For example,

$$O\,(^1D) + O_2\,(^3\Sigma_g^-, v'' = 0) \rightarrow O\,(^3P_2) + O_2\,(^1\Sigma_g^+, v' \leqslant 2) \tag{4.15}$$

may be an important process in the lower thermosphere.

A reaction that is of great importance in the thermosphere is *dissociative recombination*,

$$AB^+ + e \rightarrow A + B \tag{4.16}$$

where some of the energy released by the molecule-electron combination is used to dissociate the molecule into atoms (which may themselves be excited). This is an efficient process for the recombination of electrons and ions but, of course, is applicable only when the ion is molecular. However, molecular ions may be formed from atomic ions by *charge transfer*

$$A^+ + BC \rightarrow A + BC^+ \tag{4.17}$$

or by *ion-atom interchange*

$$A^+ + BC \rightarrow B + AC^+ \tag{4.18}$$

For example, the free electron and the ion formed in the photoionization of an oxygen atom may eventually recombine through the sequence

$$O^+ + O_2 \rightarrow O + O_2^+ \tag{4.19a}$$

$$O_2^+ + e \rightarrow O + O \tag{4.19b}$$

or the sequence

$$O^+ + N_2 \rightarrow N + NO^+ \tag{4.19c}$$

$$NO^+ + e \rightarrow N + O \tag{4.19d}$$

4.3 Black-Body Radiation and Some Simple Aspects of Radiative Transfer

Apart from or in addition to considerations of the details of molecular energy changes, it is often desirable to discuss the interchange of radiative energy from a macroscopic point of view in terms of experimentally determined rates of energy emission, absorption, or flux. This section introduces some of the concepts involved in the application of such a macroscopic point of view. The names and symbols introduced here and used throughout the book represent the author's choices from among a variety of usages. Appendix C is devoted to some of the problems and ambiguities of nomenclature in radiation studies.

First, let us consider the energy radiated from an extended surface—for example, the ground, a cloud top, or a layer of the atmosphere. We may be interested in the energy radiated in a particular direction, or the energy radiated into a hemisphere whose equatorial plane is along the radiating surface. Consider an elemental area dA of such a surface. Let $(dQ)_1$ be the energy radiated from this area into the hemisphere in the time dt and in the frequency interval ν to $\nu + d\nu$. Define the *emittance* M_ν as the energy radiated into the hemisphere per unit area, per unit time, per unit frequency. Then

$$(dQ)_1 = M_\nu \, dA \, dt \, d\nu \tag{4.20}$$

Now let $(dQ)_2$ be that part of $(dQ)_1$ contained within the solid angle $d\omega$ in a direction that makes an angle θ with the normal to the surface and an angle ϕ with some arbitrary reference direction in the plane of dA (see Fig. 4.9). Define

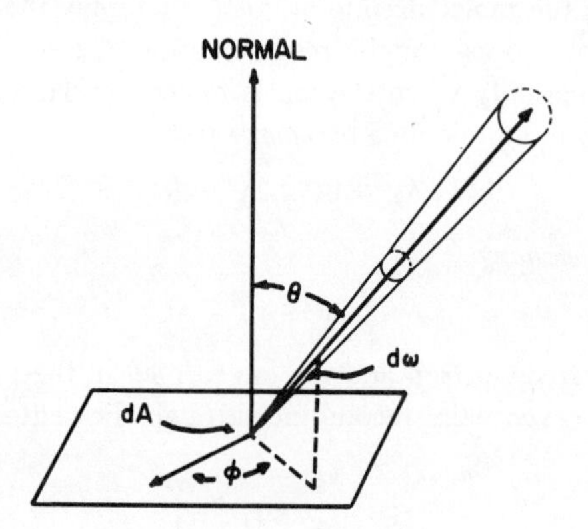

FIG. 4.9. Diagram illustrating definition of radiance.

the *radiance* L_ν as the energy radiated per unit area normal to the direction specified, per unit time, per unit frequency, per unit solid angle. Then

$$(dQ)_2 = L_\nu(dA \cos \theta) \, dt \, d\nu \, d\omega \tag{4.21}$$

We can find a relationship between L_ν and M_ν by noting that

$$(dQ)_1 = \int (dQ)_2$$

where the integration is carried out over the hemisphere. Therefore,

$$M_\nu = \int L_\nu \cos \theta \, d\omega = \int_0^{2\pi} \int_0^{\pi/2} L_\nu \sin \theta \cos \theta \, d\theta \, d\phi \tag{4.22}$$

since $d\omega = \sin\theta \, d\theta \, d\phi$. For the case of an *isotropic* radiator, L_ν is independent of direction and the integration gives

$$M_\nu = \pi L_\nu \tag{4.23}$$

We assume this to be true for the radiators that we shall consider.

According to Planck's law, the radiance of a black body, $L_{\nu B}$, is a function of temperature and frequency only. It is given by

$$L_{\nu B} = \frac{2h}{c^2} \frac{\nu^3}{e^{h\nu/kT} - 1} \tag{4.24}$$

where h is Planck's constant, c is the speed of light, and k is Boltzmann's constant.

The integrated *black-body radiance* contained within a finite frequency interval ν_1 to ν_2 is, of course, given by

$$L_B = \int_{\nu_1}^{\nu_2} L_{\nu B} \, d\nu$$

When ν_1 and ν_2 encompass the entire spectrum, then we have the familiar Stefan-Boltzmann law,

$$L_B = \sigma T^4/\pi; \qquad \text{or} \qquad M_B = \sigma T^4 \tag{4.25}$$

where σ is the Stefan–Boltzmann constant, 5.669×10^{-5} erg cm^{-2} sec^{-1} deg^{-4}. The wavelength of maximum emission for a black body at temperature T is given by the Wien displacement law

$$\lambda_{\max} = C/T \tag{4.26}$$

with $C = 0.2898$ cm deg.

The ratio of the real radiance of a radiating object to the black-body radiance is called the *emissivity* ϵ_ν. Thus

$$L_\nu = \epsilon_\nu L_{\nu B} \tag{4.27}$$

where the emissivity is in general frequency dependent. A black body has $\epsilon_\nu = 1$ at all frequencies and a *gray body* has ϵ_ν constant but less than 1 at all frequencies. According to *Kirchhoff's law*, the emissivity of an object is equal to its *absorptivity*, which is the ratio of absorbed to incident radiation.

In atmospheric radiation problems, it is customary to treat the ground and cloud surfaces as black bodies, and to calculate the radiance of a given mass of air with the use of Kirchhoff's law. However, Kirchhoff's law is applicable only under conditions of local thermodynamic equilibrium and cannot be applied in the very high atmosphere. This question is discussed further in Sections 7.1 and 7.2.

Let us now consider some fundamental definitions and concepts with respect to the transfer of radiation through a medium which absorbs, scatters, and emits radiation. The standard reference for this general problem is the work of Chandrasekhar (1950).

The radiation field is described in terms of the *specific intensity*, I_ν. This is defined, in a manner analogous to the definition of radiance, such that the energy transported across an element of area dA, within the frequency interval ν to $\nu + d\nu$, within the solid angle $d\omega$ in a direction making an angle θ with the normal to the area, and in the time dt, is given by

$$I_\nu \, dA \cos\theta \, d\nu \, d\omega \, dt$$

We may speak of the energy so defined as a *pencil* of radiation.

We may also define the *flux*, F_ν, as the energy transported across a surface per unit area, per unit frequency interval, per unit time from all directions in the hemisphere, so that, just as in (4.22),

$$F_\nu = \int_0^{2\pi} \int_0^{\pi/2} I_\nu \sin\theta \cos\theta \, d\theta \, d\phi \tag{4.28}$$

A pencil of radiation traversing an atmosphere is weakened by its interaction with matter. If the specific intensity changes from I_ν to $I_\nu + dI_\nu$ while the pencil traverses a distance ds in its direction of propagation, then

$$dI_\nu = -\kappa_\nu \rho I_\nu \, ds \tag{4.29}$$

where ρ is the density of the medium and κ_ν is the *extinction coefficient*, which is defined by this relation. In general, some of the radiation so lost may reappear (at the same frequency) in other directions; this is the case of *scattering*. On the other hand, some of it may be lost to the radiation field by transformation into other forms of energy, or radiation at other frequencies; this is the case of *absorption*.

Radiation in the pencil may also be strengthened by emission. This emission may in general include radiation of the same frequency scattered from all other directions into the pencil under consideration. Define the *emission coefficient* j_ν such that the change of specific intensity due to emission is

$$dI_\nu = j_\nu \rho \, ds \tag{4.30}$$

The *equation of transfer* is obtained by combining the contributions represented by (4.29) and (4.30), so that

$$\frac{dI_\nu}{ds} = -\kappa_\nu \rho I_\nu + j_\nu \rho \tag{4.31}$$

This may be written in terms of the *source function* J_ν, defined by

$$J_\nu \equiv j_\nu/\kappa_\nu \tag{4.32}$$

so that the equation of transfer becomes

$$-\frac{dI_\nu}{\kappa_\nu \rho \, ds} = I_\nu - J_\nu \tag{4.33}$$

In the general case of an absorbing, scattering, and emitting atmosphere, problems of radiative transfer are extremely complicated. We shall usually be concerned with relatively simple problems of transfer, in which scattering is neglected and the source function takes a simple form. Nevertheless, for some upper-atmospheric problems, such as photometry of the aurora and airglow, and for many astrophysical problems, scattering must be considered.

A particularly simple case of radiative transfer occurs when both scattering and emission can be neglected. This is often the case when one is considering the passage of ultraviolet solar radiation through the upper atmosphere. Then the source function is zero and

$$dI_\nu = -I_\nu k_\nu \rho \, ds \tag{4.34}$$

where k_ν is the *absorption coefficient.* Integration gives

$$I_{\nu s} = I_{\nu 0} \exp\left(-\int_0^s k_\nu \rho \, ds\right) \tag{4.35}$$

where $I_{\nu 0}$ is the value of the specific intensity at some point where s is taken to be zero.

Following the practice of astrophysicists and aeronomers, we shall refer to the quantity $\int_0^s k_\nu \rho \, ds$ as the *optical thickness* or *optical depth.* Often in atmospheric radiation problems (for example, Elsasser, 1960, or Huschke, 1959) the name optical thickness is used to designate the quantity $\int_0^s \rho \, ds$, which, however, we shall call the *integrated mass* (of a unit-area column) or sometimes the *path length.* When k_ν is independent of distance s, we shall refer to Eq. (4.35) as *Beer's law.*

A quantity which plays an important role in the theory of infrared transfer in the atmosphere is the *transmissivity* τ_ν, which is the ratio of transmitted to incident radiation, emission being neglected. The transmissivity of the absorbing column which is visualized in the derivation of Eq. (4.35) is simply

$$\tau_\nu = \frac{I_{\nu s}}{I_{\nu 0}} = \exp\left[-\int_0^s k_\nu \rho \, ds\right] \tag{4.36}$$

The transmissivity is obviously a function of the optical thickness. A functional relationship such as Eq. (4.36) that defines this dependence is called a *transmission function.*

We shall return in Chapter 7 to the question of infrared transfer. Here we simply note that in much of the earth's atmosphere the equation of transfer for infrared radiation, although more complicated than Eq. (4.34), is still much simpler than the general case. This is because scattering can be neglected and the emission can be obtained from Kirchhoff's law. The radiance of the mass $\rho\, ds$ is $k_\nu \rho\, ds\, L_{\nu B}$ and the equation of transfer is

$$-\frac{dI_\nu}{k_\nu \rho\, ds} = I_\nu - L_{\nu B} \tag{4.37}$$

It is very important to note that all the equations in this section have been formulated for monochromatic radiation. If, however, variations with frequency are slow enough, one may apply these equations to small finite spectral intervals, using average values of k_ν and I_ν. This is sometimes done in problems involving absorption of solar radiation by the upper atmosphere, but is not permissible in infrared-transfer problems. Furthermore, all of the transfer equations in this section have been formulated for radiation confined to an infinitesimally small solid angle. To determine flux one must integrate the specific intensity over direction according to (4.28), as has to be done in infrared-transfer problems.

However, when the energy is confined to a small solid angle and the flux, as defined by (4.28), may be considered constant (from a geometrical point of view) over distances of interest, we may speak of *parallel radiation.* For example, in the case of solar radiation it may be shown from Eq. (4.28) that the flux at the outer edge of the atmosphere through an area normal to the solar beam is given by

$$F_\nu = M_\nu (r_s/d)^2 \tag{4.38}$$

where M_ν is the solar emittance, r_s is the radius of the sun, and d is the distance to the sun. Solar radiation passing through the earth's atmosphere is considered to be parallel radiation because the value of d in (4.38) is essentially the same at the outer edge of the atmosphere as at the earth's surface. Nevertheless, the solar flux at the distance of, say, Jupiter is quite different from that at the distance of the earth and in this sense solar radiation is not parallel. For parallel radiation, one may use (4.35) to describe the variation of flux along the beam, because formally one needs only to multiply both sides by (effectively) the same small solid angle $\Delta\omega$ to convert the specific intensities to fluxes. As a matter of fact, the flux of a parallel beam of radiation (through a surface perpendicular to the beam) is sometimes referred to as the "intensity" of the beam and given the symbol I. To avoid confusion with specific intensity as defined above (which is itself sometimes called "intensity"), it is preferred here to refer to the flux of a parallel beam by the symbol F or $I\,\Delta\omega$.

4.4 Absorption of Solar Radiation by an Exponential Atmosphere

The ultraviolet solar energy incident on the upper atmosphere is absorbed at various altitudes and by various gaseous constituents before it reaches the troposphere. As discussed in Section 4.3, Beer's law is often applicable and it is instructive to consider the vertical distribution of absorbed radiation for a representative vertical variation of density. Strictly speaking the discussion is applicable only to monochromatic radiation, but for simplicity the subscripts ν will be dispensed with.

Chapman (1931a) first treated this problem. Although Chapman's results were based on a number of simplifications, and although his treatment has since been extended and generalized in various ways, he captured the essence of the problem and his results have often been used (in particular) in interpretations of ionospheric data.*

The idealized problem treated by Chapman involves the absorption of a beam of parallel, monochromatic radiation impinging on an atmosphere of uniform composition in which the density varies exponentially with height. Let us consider first the absorption as a function of height and of the zenith angle of the impinging radiation. Further, let us consider the earth to be flat; the solution for a spherical earth, which must be used when the zenith angle is large, is discussed in Appendix D (after Chapman, 1931b).

Let the flux of radiation be F, and the zenith angle Z. Let the air density vary with height z according to

$$\rho = \rho_0 \exp(-z/H) \tag{4.39}$$

where H is the scale height and where ρ_0 is the value of ρ at some arbitrary level at which z is taken to be zero. The change of flux while the radiation travels the (slant) path between z and $z + dz$ is, according to (4.34),

$$dF = kF\rho_0 \exp(-z/H) \sec Z \, dz \tag{4.40}$$

Note that the minus sign in (4.34) is here omitted because the coordinate system is such that flux decreases in the negative z direction. The absorption coefficient k refers to unit mass of air, since ρ_0 and H refer to the total air density.

Integration of (4.40) gives, with $F = F_\infty$ outside the atmosphere,

$$F = F_\infty \exp[-k\rho_0 H \sec Z \exp(-z/H)] \tag{4.41}$$

The absorbed energy per unit time and volume is $dF/(\sec Z \, dz)$ and we call this q. From (4.40) and (4.41)

$$q = k\rho_0 F_\infty \exp[-(z/H) - k\rho_0 H \sec Z \exp(-z/H)] \tag{4.42}$$

* Chapman considered the amount of absorbed energy per unit volume to be proportional to the number of ionizations (or dissociations) in the volume, and, further, he considered the recombination problem. In this discussion we refer only to absorption.

At high levels, q must decrease upward because of the decreasing air density. At low enough levels (if the radiation is essentially all absorbed during its passage through the atmosphere), q must be small and must increase upward because the flux of the beam is small and increasing upward. At some intermediate level, z_m, there is maximum absorption, q_m. These may be obtained from (4.42) by standard methods and are related to the other variables by

$$\exp(z_m/H) = k\rho_0 H \sec Z \tag{4.43}$$

$$q_m = \frac{F_\infty \cos Z}{H \exp(1)} \tag{4.44}$$

It is interesting to note that z_m is independent of the initial flux of the beam, and q_m is independent of the absorption coefficient. With regard to zenith angle, z_m increases and q_m decreases as Z increases. The relationship between z_m and Z is shown in Fig. 4.10.

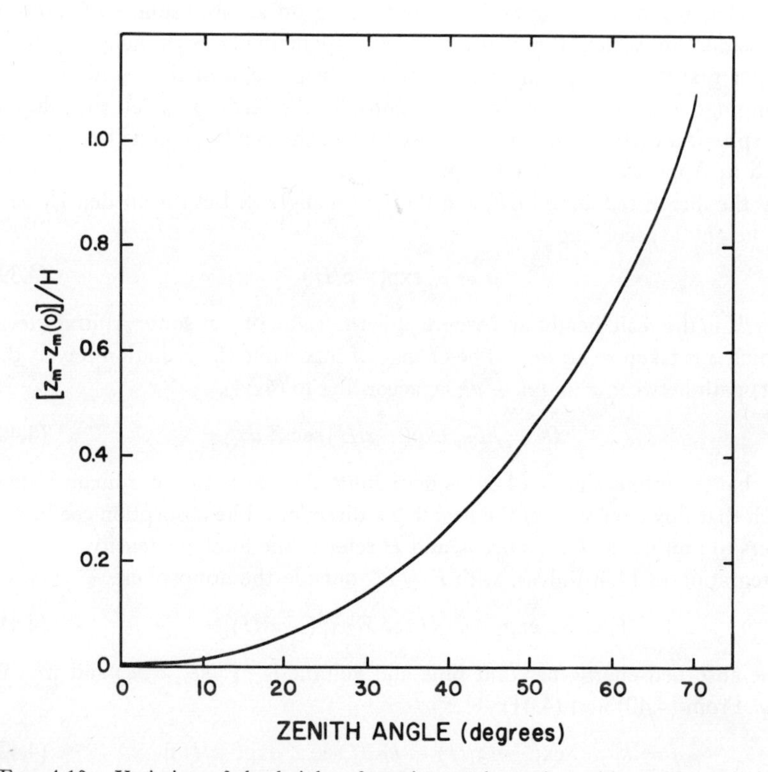

FIG. 4.10. Variation of the height of maximum absorption with zenith angle in an exponential atmosphere. Height of maximum absorption z_m is measured, in units of the scale height, from the height where absorption is maximum when the zenith angle is zero.

It is convenient to express q as a function of the distance from the level of maximum absorption, in terms of H as a unit of height; that is, as a function of a variable z_1, where

$$z_1 = (z - z_m)/H \tag{4.45}$$

With this variable, (4.42) becomes

$$(q/q_m) = \exp[1 - z_1 - \exp(-z_1)] \tag{4.46}$$

This function is plotted in Fig. 4.11, which shows that most of the absorption

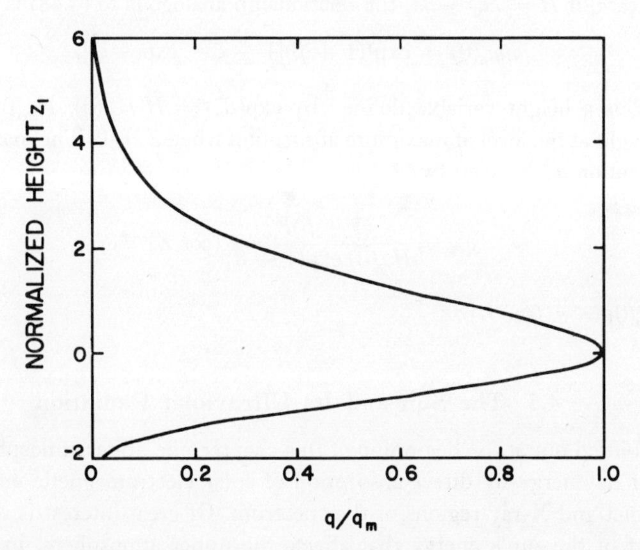

FIG. 4.11. Variation of absorption q with height in an exponential atmosphere. See text for definitions of symbols.

takes place within four scale heights above and two scale heights below the level of maximum absorption. The shape of the layer is independent of zenith angle, although, of course, q_m and therefore q vary as cos Z according to (4.44).

Another convenient way to represent the vertical variation of q (used in Chapman's original paper) is to show its variation as a function of the distance from the level where maximum absorption occurs when $Z = 0$, also in terms of H as a unit. Let $z_m(0)$ be the value of z_m when $Z = 0$ and let

$$z_2 = [z - z_m(0)]/H \tag{4.47}$$

Then

$$q/q_m(0) = \exp[1 - z_2 - \sec Z \exp(-z_2)] \tag{4.48}$$

where $q_m(0)$ represents the maximum rate of absorption when $Z = 0$.

For large zenith angles, say $Z > 75°$, Chapman (1931b) has shown that "sec Z" in (4.41) and (4.42) must be replaced by a more complicated expression, which is derived in Appendix D.

Chapman (1939) extended the treatment to an absorption band, with a particular wavelength dependence of absorption coefficient symmetrical about a wavelength of maximum absorption. He found that the layer, particularly its lower boundary, would in this case be less sharply defined.

Several authors (Nicolet and Bossy, 1949; Gledhill and Szendrei, 1950; Nicolet, 1951) have looked into the effect of a scale-height gradient on the results. In this case, if $H = H_0 + \beta z$, the relationship analogous to (4.48) is

$$q/q_m(0) = \exp\{(1 + \beta)[1 - \zeta - \exp(-\zeta)]\} \tag{4.49}$$

where ζ is a height variable defined by $\exp(\beta\zeta) = H/H_m(0)$, $H_m(0)$ being the scale height at the level of maximum absorption when $Z = 0$. The maximum rate of absorption q_m is given by

$$q_m = \frac{(1 + \beta)F_\infty}{H_m(0)\exp(1 + \beta)}(\cos Z)^{1+\beta} \tag{4.50}$$

and $q_m(0) = q_m(\sec Z)^{1+\beta}$.

4.5 The Sun and Its Ultraviolet Radiation

As pointed out at the beginning of this chapter, the upper atmosphere derives most of its energy by direct absorption of solar electromagnetic energy in the ultraviolet and X-ray regions of the spectrum. Of great interest is the fact that the part of the sun's energy that affects the upper atmosphere, in contrast to visible and infrared radiations that affect the troposphere, is variable in some spectral regions. Many cyclic and aperiodic changes in the upper atmosphere are directly related to this variability.

In this discussion, we consider first some physical characteristics of the sun, particularly those associated directly or indirectly with energy variation (for a more complete discussion of this subject, see Athay and Warwick, 1961), and then turn to a discussion of the energy itself.

4.5.1 Some Physical and Optical Features of the Sun

From an astronomical point of view, the sun is rather a run-of-the-mill star, not particularly large or particularly small, not unduly bright or unduly dim. Its mass is approximately 2×10^{33} g, the diameter of its visible disk about 1.4×10^6 km. In the interior of the sun, where the temperature is perhaps as high as 2×10^7°K, energy is produced as the net result of a chain of nuclear reactions which convert H to He. This energy is radiated toward the surface, but is absorbed and re-emitted countless times by the gradually cooler gases away from the interior so that the energy finally leaving the sun is characteristic of the cooler outer portions and not of the hot interior portions.

Most of the energy reaching the earth originates in a comparatively thin layer of gas several hundred kilometers thick called the *photosphere*. Although the sun is gaseous throughout, we may speak of the photosphere as the "surface" of the sun, meaning the visible surface in white light. An important part of the sun's mass, as far as the earth is concerned, lies well above the photosphere, but is largely transparent to photospheric radiation. Observations of the photosphere show that it is not at all uniform in brightness. Apart from larger and more concentrated features, such as the sunspots and faculae which are discussed below, the photosphere is marked by relatively bright *granules*. These have diameters on the order of 1000 km and are separated by narrower dark regions. The granules are distributed more or less uniformly over the solar disk and are presumably associated with a zone of intense convection in and below the photosphere.

Radiation from the photosphere is essentially continuous (however, see below) and in the visible part of the spectrum resembles radiation from a black body at a temperature in the vicinity of 6000°K, according to (4.24). The fit to such a distribution is, however, not at all perfect; particularly at the short-wave end of the visible and in the near ultraviolet, above the short-wave cutoff near 3000 A, the energy is less than would be expected from the above statement. We shall see that rocket measurements of the ultraviolet region below 3000 A have shown much less energy from the photosphere than would be expected from a black body at 6000°K.

Superimposed on the continuum radiation from the photosphere in the visible part of the spectrum are many relatively dark lines, the *Fraunhofer lines*. These are produced by selective absorption and re-emission of light by gases in the upper photosphere and overlying chromosphere (which is discussed below). According to a simple interpretation, the energy at the centers of the Fraunhofer lines, being less intense than in neighboring parts of the spectrum, would be expected to be radiated from levels with lower temperatures than those characteristic of the photosphere. Such a layer of cooler gas lying above the photosphere was once identified as the "reversing layer." It is now known, however, that many of the lines originate at levels of higher temperature, in the chromosphere, where the absence of thermodynamic equilibrium results in radiating temperatures that are much lower than the prevailing kinetic temperatures.

The region above the photosphere is called the *chromosphere*, extending approximately 5000 km above the photosphere. In the chromosphere the temperature increases with altitude, perhaps less markedly near the lower boundary, and eventually at the upper level reaches a value of about 10^6°K. Energy originates in the chromosphere as line emission, mostly from H, He, and Ca, and is quite weak in the visible compared with the radiation from the photosphere. When the photosphere is eclipsed by the moon or instrumentally, the chromosphere becomes visible and has a characteristic reddish appearance which is due to the large amount of energy emitted in the Hα line at 6563 A. Another way to view the chromosphere is to isolate light in one of the emission lines, such as Hα or one of the Ca II lines, with a spectrograph, blocking out the more intense light from the photosphere in adjoining regions of the spectrum. The light thus seen originates above the photosphere.

Merging with the top of the chromosphere and extending outward for many solar diameters is the *corona*. The corona is visible as a faint white halo during total eclipses. This white light originates in the photosphere and is scattered by free electrons in the corona. In addition, spectral examination of light from the corona shows a series of emission lines originating in the corona itself. These have been shown to originate from the emissions of highly ionized metals, such as nickel, iron, and calcium. For example, the most intense coronal line, at 5303 A (sometimes called the "green coronal line"), is emitted by Fe XIV, the iron atom minus 13 of its usual 26 electrons. This and other ions are produced by collisions. Since very large energies are required to strip these atoms of so many electrons (355 ev for Fe XIV, even more for some other ions whose presence is known), the kinetic temperature must be very high. It is generally estimated to be about 10^6 °K and may be considerably higher in localized regions at times. The *coronagraph* is an instrument especially designed to view the corona in the absence of a natural eclipse, by artificially eclipsing the rest of the solar disk. With the aid of the coronagraph the intensities of some of the coronal emission lines are measured routinely.

Several observable features of the sun are of particular interest in that their transient occurrence may be indicators, to some degree, of variability in the emission of solar energy that affects the upper atmosphere. When these features are present with more than usual frequency or intensity, the sun is said to be "active" or "disturbed"; when they are absent, the sun is said to be "quiet."

The best known and longest observed of these variable features are the *sunspots*. A sunspot is a relatively dark region of the photosphere, ranging in size from barely visible to areas some 50,000 km in diameter. Sunspots generally occur in groups, a sunspot group consisting of a "leader"spot (leading in the direction of the sun's rotation), which generally appears first, and one or several smaller follower spots. Sunspots generally occur between 5° and 35° from the solar equator in either hemisphere. As their dark appearance indicates, sunspots are regions of cooler photospheric gas. Thus, they are not, in themselves, sources of enhanced emission, but, as we shall discuss shortly, they are often accompanied by other transient features that are indicative of enhanced emission.

The localized areas characterized by these phenomena are usually referred to as *active regions*. Sunspot groups, once formed, may last only a day or less or may persist for several solar rotations, each rotation as viewed from the earth and at the latitude where the sunspots predominate lasting about 27 days.

The number of sunspots that appear on the solar disk, averaged over, say, a month or a year, is highly variable in a quasi-periodic fashion. Although there are various possible ways of counting the spots, weighting them according to their association in groups or their size in one way or another, all show that there are periods of time called *sunspot maxima* when the spots are relatively numerous and periods of time a few years later called *sunspot minima* when spots occur hardly at all. The average length of time between sunspot maxima has been observed to be a little more than eleven years, but it must be emphasized that upon occasion during the 200 years of observations this period has been as much as 4 to 5 years less or more than the 11-year average. Moreover, the number of spots associated with different sunspot maxima varies by a factor of 2 or more. At the beginning of a new cycle, just after sunspot minimum, new spots begin to appear at relatively high latitudes (for spots, that is) and as the cycle progresses they are more apt to appear nearer and nearer the equator.

Sunspots are the seats of intense localized magnetic fields with polarities varying typically (but not always) as indicated schematically in Fig. 4.12. Note that leader spots and follower spots have opposite polarities. Leader spots in the same hemisphere have the same polarity during one solar cycle but have opposite polarities from one solar cycle to the next. This characteristic variation from cycle to cycle has led to the designation of the "22-year" or "double" sunspot cycle, which thus refers also to magnetic characteristics and not only to sunspot number.

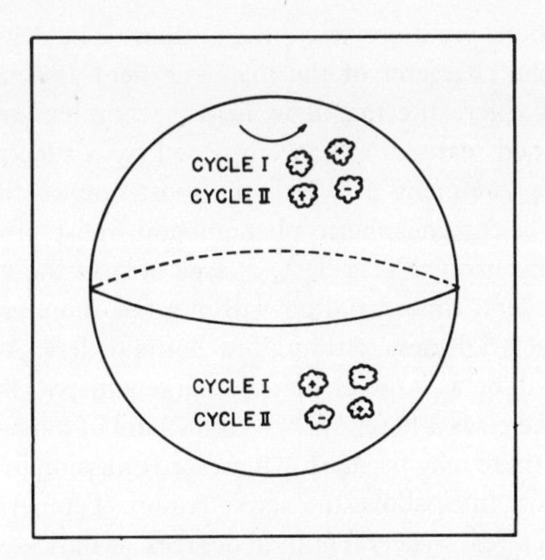

FIG. 4.12. Diagram illustrating polarities of the magnetic fields of leader and follower spots in the two hemispheres in successive cycles of sunspot number.

Associated with and surrounding sunspot groups are bright regions of the photosphere called *faculae*. A facula sometimes appears before and persists after the visible appearance of the sunspots. Faculae are also observed at high solar latitudes in the absence of sunspots.

Sunspots and faculae are only the easily visible evidences of a complex class of phenomena that characterize an active region. Others are chromospheric *plages*, characteristic types of *prominences*, enhanced coronal emission lines, and solar *flares*.

As mentioned above, it is possible to observe the chromosphere by isolating the light from certain emission lines such as Hα. At these particular wavelengths, the light reaching the earth originates in the chromosphere. Examination of the sun in this manner reveals large bright regions, called plages, associated with sunspot groups. Plages are characteristically slow to change their appearance, size, or brightness as contrasted to the flares to be discussed below.

When the bright photosphere is eclipsed by the moon or instrumentally by means of the coronagraph, the top of the chromosphere is seen to be in continual eruption, even in quiet regions. Thousands of small jetlike protuberances, the *spicules*, extend upward, the lifetime of each being only a few minutes. Very much larger and more stable "clouds" of bright gas are sometimes seen extending outward from the chromosphere or suspended above it. These are called prominences. Prominences in the vicinity of active regions are typically smaller and behave quite differently from ordinary prominences, changing much more rapidly in appearance, brightness, and extent. If the light from the coronal emission lines is observed, enhanced emission is observed in the vicinity of active regions, particularly in the green coronal line and the red line at 6374 A.

Some active regions are more active than others. The degree of activity seems to be related to the character of the magnetic field associated with the active region. In regions where the magnetic field is complex and rapidly changing, the above-mentioned features are accompanied by a succession of short-lived, intense, eruptive phenomena of which the most noteworthy is the solar flare. The solar flare is a chromospheric phenomenon, most prominent in the light of Hα. With the occurrence of a flare, a large area in the vicinity of the active region becomes several times brighter within a few minutes and then gradually subsides to normal brightness within a few hours or less. According to the area covered, which may be as large as that of a large sunspot, flares are classified as class 1, 2, or 3. The class 3 flares are the largest and of longest duration. Accompanying the flare, there may be sharply increased emission in the already brighter than normal coronal lines above the active region. Temperatures in the corona above the flare often reach several million degrees, as indicated by the appearance of emission lines of such high-energy ions as Ca XV. The flare is sometimes accompanied also by a particularly active prominence of the surge type, in which material is apparently ejected outward from the chromosphere at tremendous speeds.

Solar emission at the radio frequencies, in the wavelength range 1 cm to 1 m, also shows variability. There is a variation in net integrated emission over the solar cycle and observations show individual long-lived emission centers occurring in the vicinity of plages. There are also shorter periods of enhanced emission, sometimes referred to as radio bursts. These are often associated with chromospheric and coronal events, such as solar flares, and because of the relative ease with which they can be observed serve as a valuable indicator of solar activity.

4.5.2 Electromagnetic Energy from the Sun

Of the electromagnetic energy received from the sun, according to Johnson (1954), about 50 per cent occurs in wavelengths longer than the visible region, about 40 per cent in the visible region (4000–7000 A), and about 10 per cent in wavelengths shorter than the visible. With some minor exceptions, such as weak ozone absorption in the visible and between 3000 and 3500 A, the upper atmosphere is affected directly only by the approximately 1 per cent of the total energy below 3000 A. Because the upper atmosphere absorbs all this energy, the spectral distribution and indeed the integrated value of energy below 3000 A cannot be observed from the earth's surface and were a matter of conjecture until rockets became available to carry instruments above the absorbing gases.

After about 15 years of painstaking and difficult experimentation, there is now available a good deal of quantitative information on this 1 per cent of the sun's electromagnetic energy. It is of particular interest from a geophysical point of view to know accurately the absolute value (and the variability thereof) of solar flux at the outer edge of the atmosphere at all ultraviolet and X-ray wavelengths. Although this ideal is far from being reached, great strides have been made during the past few years. A comprehensive and authoritative review of this progress (excluding studies of the X-ray region) has been given by Tousey (1963), one of the men principally responsible for it.

In considering data of this type, one must keep in mind the several sources of uncertainty involved. Some of these are the presence even at high levels of a certain mass of overlying atmosphere which may be significant in spectral regions of strong absorption; observations (at certain times) of radiation that may vary with the sunspot cycle or with the occurrence of active regions; and the difficulty of establishing an absolute scale for quantitative measurement of flux.

A detailed treatment of instrumentation is beyond the scope of this discussion, but certain fundamental aspects need to be mentioned. Aerobee-Hi rockets have usually been employed, along with a pointing control to keep the spectrograph trained on the sun. The first successful pointing control was developed at the University of Colorado (Stacey *et al.*, 1954) and that basic design is still in use. There have been several different designs of spectrographs used, but all have employed a reflection-type grating as the basic dispersing element. Spectrographic designs have gradually been refined to improve resolution and extend the lowest detectable limit far into the extreme ultraviolet. For example, Tousey (1962) has recently reported a spectrograph showing excellent resolution down to 170 A. Normal-incidence spectrographs have been used at the longer wavelengths but grazing-incidence spectrographs have usually been employed for wavelengths less than about 1000 A. The majority of experiments have made use of photographic film as a detector, a method which necessitates instrument recovery and photographic photometry. Hinteregger and his collaborators (see, for example, Hinteregger, 1961a) have used a photoelectric detector with direct telemetry, a method which provides more reliable absolute determinations but less resolution. In the X-ray region it has up to now been necessary to use photon counters which accept energy in spectral intervals that are quite wide compared with the resolution attainable at longer wavelengths (see, for example, Friedman, 1961a). For much more detailed discussions of instrumentation, see Friedman (1960, 1961a), Tousey (1961, 1963), Rense (1961), and Hinteregger (1961a).

Down to about 2085 A the solar spectrum is qualitatively similar to that in the visible, a continuum with numerous strong Fraunhofer lines. Below this wavelength, Fraunhofer lines become faint and disappear altogether at about 1550 A. Emission lines begin to appear at about 1900 A and become relatively more important compared with the continuum as the wavelength decreases. The identifications and relative intensities of these lines are of the utmost importance to the astrophysicist and have been discussed at length. See, for example, Wilson *et al.* (1954), Malitson *et al.* (1960), Detwiler *et al.* (1961), Violett and Rense (1959), and Pagel (1963). Here, we shall consider only the values of flux integrated over rather wide spectral intervals.

Johnson (1954) and Detwiler *et al.* (1961) have summarized the work of the Naval Research Laboratory on solar flux and its wavelength variability down to 850 A. The former gave earlier data down to about 2200 A; the latter extended them down to 850 A. Table 4.3 gives the results for spectral intervals 50 A in width. As pointed out above, such "smoothing" obscures much detail in the spectrum. It should also be pointed out that there is still some uncertainty about the absolute scale of the measurements, especially below 2000 A and increasingly so as the wavelength gets shorter.

With reference to Table 4.3, and in terms of equivalent black-body temperature, the reported fluxes lie somewhat above 5000°K in the interval 2600–2100 A and fall gradually to a minimum level of about 4700°K at about 1400 A. From there toward shorter wavelengths, the equivalent black-body temperature of the continuum increases more or less steadily to a value of about 6600°K in the Lyman continuum short of 912 A. Superimposed is the large amount of energy associated with Lyman α at 1216 A. This is shown graphically in Fig. 4.13.

Below 1300 A the dominant feature of the solar spectrum is the Lyman α line of hydrogen at 1216 A. Lyman α was first photographed in 1952 (Rense, 1953). Since then, it has been studied extensively by several methods. The

TABLE 4.3

FLUX OF SOLAR ENERGY AT THE OUTER EDGE OF THE ATMOSPHERE[a]

λ (A)	F [erg cm^{-2} sec^{-1} (50 A)$^{-1}$]	λ (A)	F [erg cm^{-2} sec^{-1} (50 A)$^{-1}$]
3000	3050	1900	41
2950	3150	1850	28
2900	2600	1800	19
2850	1700	1750	12
2800	1200	1700	8.2
2750	1100	1650	5.0
2700	1250	1600	3.2
2650	1000	1550	1.7
2600	700	1500	0.95
2550	560	1450	0.50
2500	380	1400	0.26
2450	390	1350	0.26
2400	340	1300	0.18
2350	320	1250	0.15
2300	360	1200	5.7
2250	350	1150	0.08
2200	310	1100	0.06
2150	240	1050	0.10
2100	145	1000	0.18
2050	90	950	0.15
2000	70	900	0.25
1950	55	850	0.11

[a] Given as integrated values (including both lines and continuum) over 50-A intervals centered at the wavelengths specified. After Johnson (1954) for wavelengths above 2600 A; after Detwiler *et al.* (1961) for wavelengths below 2600 A.

Naval Research Laboratory (Purcell and Tousey, 1960) has obtained a high-resolution spectrogram of the line showing a width of about 1 A and a sharp absorption core presumably caused by hydrogen in the outer reaches of the terrestrial atmosphere. The sun's disk has been photographed in the light of Lyman α (Purcell *et al.*, 1960) and shows enhanced emission in the vicinity of plages.

The energy flux in the Lyman α line is very much greater than in surrounding regions of the spectrum and is responsible for the large value listed in the

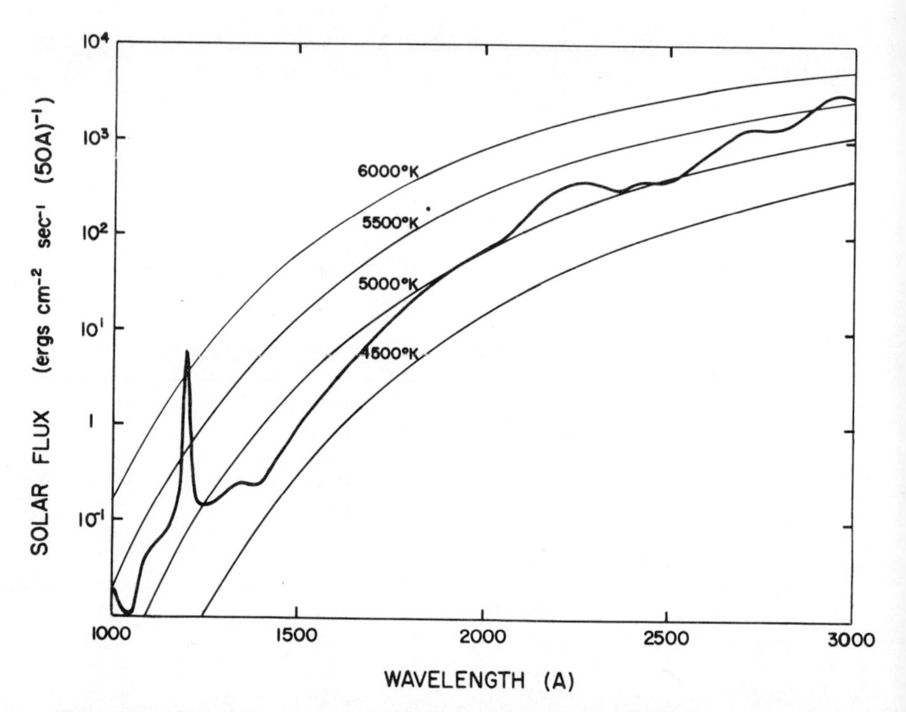

FIG. 4.13. Solar flux outside the earth's atmosphere derived from rocket measurements (based on data reported by Johnson, 1954, and by Detwiler *et al.*, 1961). The measurements are smoothed over 50-A spectral intervals. Black-body curves are shown for comparison.

1175–1224 A region in Table 4.3. Friedman (1960, 1961b) has discussed these measurements. Ion chamber measurements since 1955 have given rather consistent values between 5 and 6 erg cm^{-2} sec^{-1}. Earlier measurements by other techniques between 1950 and 1955 gave lower values, one as low as 0.1 erg cm^{-2} sec^{-1}. According to Friedman, the later measurements are more reliable and the earlier data may in whole or in part represent experimental uncertainties rather than a difference associated with the sunspot cycle. The variability (or lack of it) of Lyman α emission during solar flares is discussed below.

Table 4.4, adapted from Watanabe and Hinteregger (1962), gives some preliminary values for solar flux in the extreme ultraviolet. See also Hinteregger (1961b). Especially below 850 A, these are based on rather few data and are subject to correction as time goes on.

Measurements with photon counters or ion chambers in various spectral regions within the X-ray region have been made on a number of occasions (Byram *et al.*, 1956; Friedman, 1960; Kreplin, 1961). The instruments used

TABLE 4.4

FLUX OF SOLAR ENERGY AT THE OUTER EDGE OF THE ATMOSPHERE[a]

λ or $\lambda_1 - \lambda_2$ (A)	F (10^{-2} erg cm^{-2} sec^{-1})	Remarks
1025.7	5	Lyman β line of H
1000–1027	3	Ionization of O_2 below 1027 A
989.8	1	N III line
977.0	6	C III line
972.5	2	Lyman γ line of H
949.7	1	Lyman δ line of H
911–1000	8	Ionization of O and Lyman continuum below 911 A
850–911	22	Ionization of N below 850 A
796–850	6	Ionization of N_2 below 796 A
700–796	13	—
600–700	15	—
584.3	10	Resonance line of He I
500–600	11	—
400–500	11	—
303.8	28	Resonance line (Lyman α) of He II
300-400	17	—
230–300	24	—
170–230	33	—
110–170	7	—

[a] Given as values for certain lines centered at λ or as integrated values over interval $\lambda_1 - \lambda_2$. The latter values include continua and all lines not separately listed. After Watanabe and Hinteregger (1962).

have had responses within various spectral regions such as 0–8 A, 8–20 A, 44–60 A, and 44–100 A. Even within these rather limited bands, however, they have quantum yields that are strongly wavelength dependent, so that to convert these measurements to flux in ergs cm^{-2} sec^{-1} one must also know or assume the wavelength dependence of the incident radiation. The data are usually reduced under the assumption that the spectral distribution of solar energy in the X-ray region corresponds to that of a gray body at temperatures ranging from 0.5×10^6 to 2×10^6°K.

The data from several rocket measurements are consistent with the following picture of the X-ray spectrum. The "normal" spectrum resembles in shape a black-body curve corresponding to temperatures in the range $0.5–1.0 \times 10^6$°K, but with very low emissivity*. The estimated total energy associated with this emission may vary between 0.1 and 1 erg cm^{-2} sec^{-1}. For this type of distribution, there would be very little energy below 20 A, and indeed on some occasions very little has been observed. The region below 20 A, however, has turned out to be highly variable. Kreplin (1961) has concluded from the data that from 1953 to 1959 there was an increase by a factor of about 650 in the energy below 8 A, of about 60 in the 8–20-A band, and of about 7 in the 44–60-A band. Thus the variation from sunspot minimum to sunspot maximum is not only in terms of total energy but also in terms of a "hardening" of the spectrum.

Furthermore, the Naval Research Laboratory group has been active in the investigation of X-ray emission during solar flares (Friedman, 1959; Kreplin, 1961) and has found unmistakable evidence of an increase in X-ray emission at these times. Rocket observations indicated particularly large increases in the spectral range below 8 A, and this region was further monitored by means of an X-ray photometer mounted on the NRL SR-1 satellite in 1960 (Kreplin *et al.*, 1962). Their conclusions with regard to X-ray flux in this region follow:

> From the SR-1 experiment we can draw the following general conclusions. First, we can say that the sun does not normally emit an X-ray flux in the spectral range below 8 A which falls above our limit of measurement, namely, 0.6×10^{-3} erg cm^{-2} sec^{-1} of 2×10^{6}°K radiation. Second, most of the time during which solar 0–8 A flux exceeds 0.6×10^{-3} erg cm^{-2} sec^{-1} one or more distinct solar events are visible on the sun, although on one or two days a measurable flux was observed that appeared to be associated only with a generally high level of activity. Third, when the 0–8 A flux exceeds 2×10^{-3} erg cm^{-2} sec^{-1}, short-wave fadeout and other ionospheric effects are noticeable on earth. Fourth, significant changes in X-ray emission can take place in a time scale of the order of 1 minute.

In another interesting X-ray experiment, the Naval Research Laboratory personnel obtained an X-ray picture of the sun on April 19, 1960 (Blake *et al.*, 1963). They found that at least 75 per cent of the X-radiation originated in the lower corona in regions above calcium plages. These regions correlated well with regions of decimeter radio emission.

In the same experiments (rocket and satellite) that monitored X-ray flux

---

* Since this was written, scientists at the Naval Research Laboratory have obtained spectra down to 14 A. These show that the solar X-ray spectrum is characterized by line emission with little contribution from a continuum. Some preliminary results about the spectrum down to 33 A are discussed in *Sky and Telescope*, **27**, April 1964, 1-4. The author is indebted to Dr. R. W. Kreplin of the Naval Research Laboratory for calling his attention to these results.

during periods of solar activity, provision was made to detect variability in the Lyman α line. Somewhat unexpectedly, no significant variation has been found. Kreplin *et al.* (1962) conclude:

> Fifth, the total change in Lyman-α radiation accompanying distinct solar events is a small fraction of the average solar Lyman-α emission from the whole solar disk; hence, rapid changes in solar Lyman α are not geophysically significant.

4.6 Absorption by the Upper Atmosphere

Ultraviolet energy absorbed by the upper atmosphere produces electronic transitions, and sometimes dissociation or ionization. The transition probabilities for the various possible changes determine whether the absorption is "strong" or "weak," that is, whether the energy is depleted after passing through a relatively thin layer of the upper atmosphere or only after penetrating to lower levels. Although it is possible in some special cases to compute these probabilities from the quantum theory, the strength of the absorption is usually expressed by the empirically determined absorption coefficient.

The expression for optical thickness, $\int k\rho \, ds$, is, of course, dimensionless. However, the integrated mass of absorbing material $\int \rho \, ds$ and therefore the absorption coefficient k are expressed in varying units in the upper-atmosphere literature. The two most common units are (a) number of molecules of absorbing gas and (b) centimeters NTP.*

(a) The number of molecules per unit volume, n, of the absorbing gas is related to the density of the absorbing gas by (see Subsection 1.1.1)

$$n = \frac{N\rho}{m}$$

where N is Avogadro's number and m is the gram-molecular weight. When the amount of absorbing gas is expressed in terms of n, it is customary to write the optical thickness as $\int \sigma n \, ds$, where σ is called the *absorption cross section* and has the dimensions of area; usually it is expressed in (centimeter)^2, n in (centimeter)$^{-3}$, and s in centimeters.

This is sometimes written, especially with reference to laboratory data, as simply (σns), since n is often independent of s in the laboratory. In the upper atmosphere this is generally not true, and even σ may occasionally vary with s

* The unit "cm NTP" refers to the height in centimeters at normal temperature and pressure of a unit-area column containing mass equivalent to $\int \rho \, ds$. Normal temperature and pressure are taken to be 0°C and 1013 mb. The unit is sometimes written "cm STP," sometimes "atm-cm" or "cm-atm," or sometimes simply "cm."

(for example, if it is pressure dependent) so that $\int \sigma n \, ds$ is a more precise formulation.

(b) When the integrated mass of absorbing material is expressed in centimeters NTP, the symbol u is often used for $\int \rho \, ds$. Then the optical thickness is ku (if k is independent of s), where k is the absorption coefficient in (centimeters NTP)$^{-1}$, u is the integrated mass in centimeters NTP. The absorption cross section is related to k in this unit by $k = \sigma n_0$, where n_0 is Loschmidt's number (2.687×10^{19} particles cm^{-3} at NTP), and u is $(1/n_0) \int n \, ds$.

We shall use either set of units as convenient.

4.6.1 Absorption by Atomic Nitrogen and Atomic Oxygen

Although atomic nitrogen is probably not abundant enough to be an important absorber in the upper atmosphere, there is no doubt that atomic oxygen plays a significant role in the absorption of solar radiation in the thermosphere. Unfortunately, there is practically no experimental information about the absorption coefficients of either.

However, it is clear that no absorption continuum is to be expected for O above 911 A, corresponding to the first ionization limit of 13.61 ev, and none for N above 852 A corresponding to its first ionization limit of 14.54 ev. Absorption coefficients have been computed for both atoms at shorter wavelengths by Bates and Seaton (1949) and by Dalgarno and Parkinson (1960). Results of the latter are shown in Fig. 4.14.

For the absorption lines at longer wavelengths, there is no such information. However, the permitted transition $2p^4\,{}^3P \rightarrow 2p^3\,3d\,{}^3D^o$ of oxygen (see Fig. 4.5) corresponds very closely to the Lyman β line in the solar spectrum and may play an important role (Kato, 1954).

4.6.2 Absorption by Molecular Nitrogen

No dissociation continua have been observed for the nitrogen molecule and no ionization continua exist above 796 A, corresponding to the first ionization potential of 15.58 ev. No absorption whatsoever has been observed above 1450 A, the wavelength of the (0–0) band of the forbidden transition ${}^1\Pi_g \leftarrow X\,{}^1\Sigma_g^+$ (see Fig. 4.7).

This band system, containing the Lyman–Birge–Hopfield bands, is the only observed feature of the absorption spectrum down to about 1120 A and consists of 14 (Tanaka, 1955) rather sharp bands (less than 5 A, according to Watanabe *et al.*, 1953) requiring high resolution to get much detail. According to the authors just quoted, the maximum absorption cross section (which occurs in the 2–0 band) is 4×10^{-21} cm^2 (coefficient of 0.11 cm NTP^{-1}), although Watanabe (1958) quotes a value of 7×10^{-21} cm^2 from more recent measure-

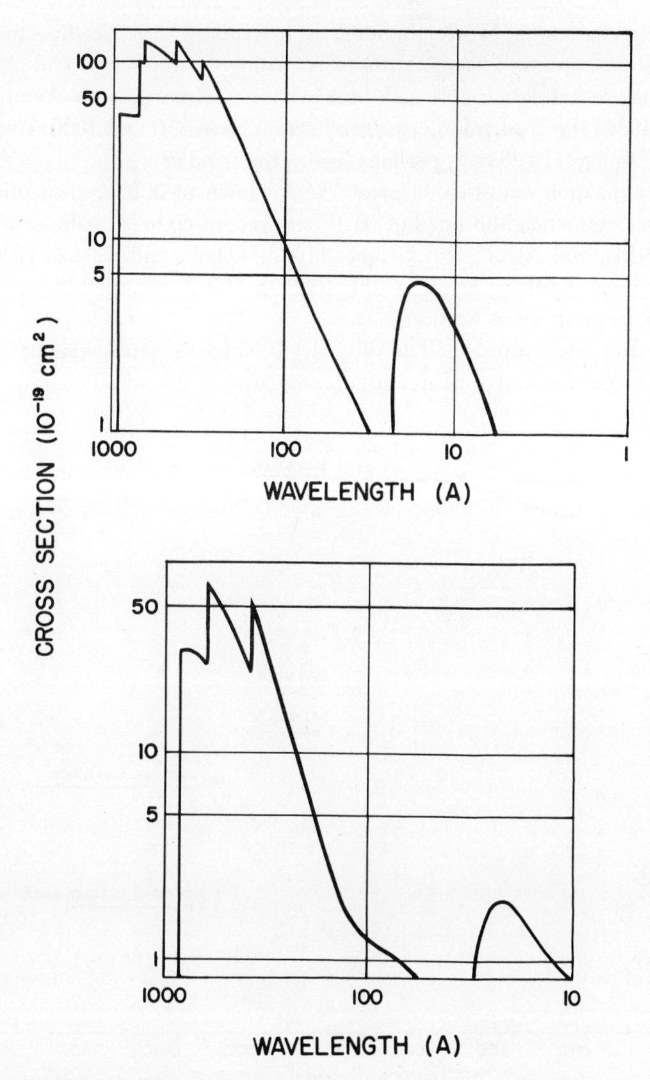

FIG. 4.14. Photoionization cross sections of O (above) and N (below), as calculated by Dalgarno and Parkinson (1960).

ments. Between the bands, according to Watanabe *et al.* (1953), the maximum value of the cross section is 3×10^{-22} cm^2 (coefficient of about 0.01 cm NTP^{-1}). The upper limit at Lyman α, according to Ditchburn *et al.* (1954), is even lower, 6×10^{-23} cm^2.

Between the shortest-wavelength Lyman–Birge–Hopfield band and the ionization continuum, the nitrogen spectrum consists of various other sharp bands (Tanaka bands, Worley bands) and a Rydberg series leading to the first ionization continuum. The spectrum is exceedingly complicated and absorption coefficients are highly variable and poorly known. According to Watanabe and Marmo (1956) the absorption cross sections between 850 and 1000 A vary from 3×10^{-17} to 3×10^{-20} cm^2, perhaps less in the windows.

In the ionization continua short of 796 A, down to 200 A, absorption cross sections are in the neighborhood of 10^{-17} cm^2, according to Weissler *et al.* (1952), Curtis (1954), and Astoin and Granier (1957). Ogawa and Tanaka (1962) have studied several Rydberg series in the 1000 to 600-A region, but without obtaining absorption cross sections.

These data are summarized in Fig. 4.15, which is semischematic and not intended to be the source of quantitative data.

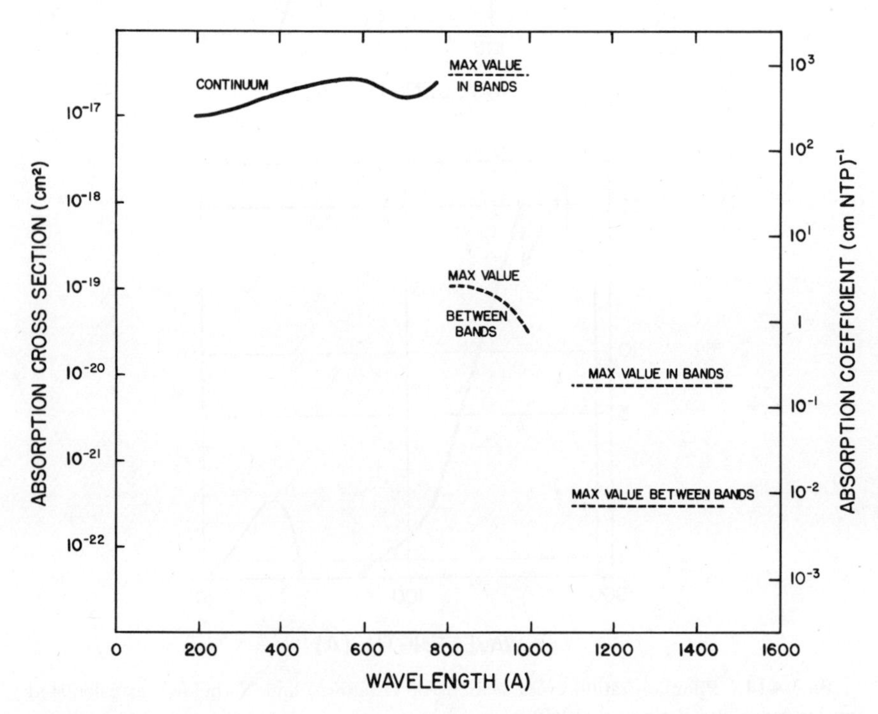

FIG. 4.15. Semischematic representation of N_2 absorption. See text and references for details.

4.6.3 Absorption by Molecular Oxygen

The ultraviolet absorption spectrum of O_2 has been extensively studied in the laboratory and is reasonably well established. Figure 4.16 shows on a coarse scale the variation of absorption cross section with wavelength. (On the scale

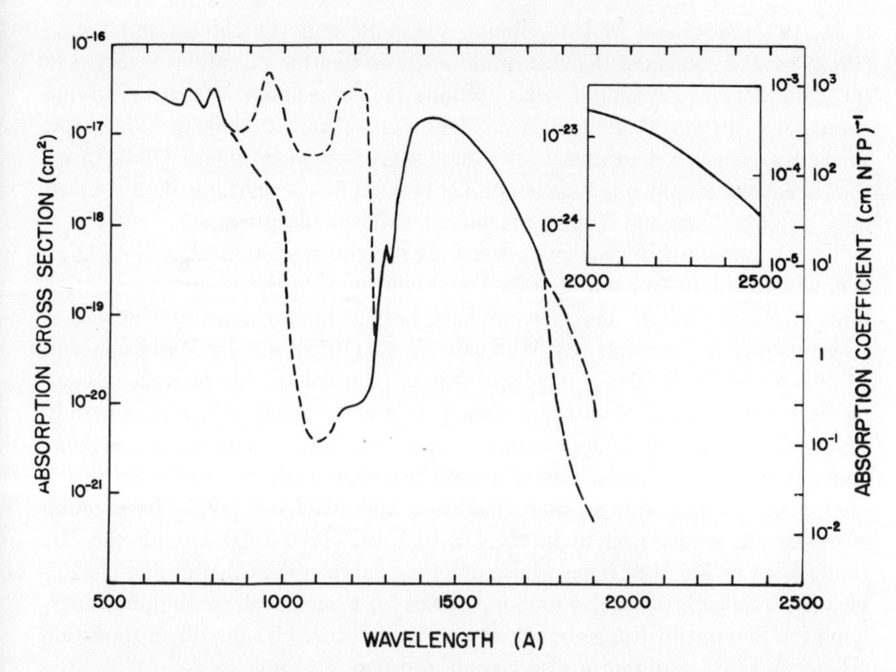

FIG. 4.16. Semischematic representation of O_2 absorption. Full curves represent regions of continuous absorption, dashed portions represent approximate maxima and minima in spectral regions where bands are important. In the Herzberg continuum (inset), where absorption cross section is pressure dependent, the curve refers to atmospheric pressure.

of Fig. 4.16, it is impossible to show detail in the regions of band absorption and, as for Fig. 4.15, it is intended to illustrate and orient, not to serve as a source of quantitative data.)

The very weak absorption above about 2000 A is due primarily to the forbidden transition from the ground state ${}^3\Sigma_g^-$ to ${}^3\Sigma_u^+$ (refer to Fig. 4.8). The first dissociation limit of O_2 leading to two ${}^3P$ atoms lies at 5.115 electron volts or about 2423 A, according to Brix and Herzberg (1954). The vibrational bands leading to this limit have been discussed extensively by Herzberg (see, for example, Herzberg, 1952) but their intensity is so weak that no quantitative values of absorption coefficient are available. Oxygen absorption in this region of the spectrum may be neglected compared with absorption by ozone (see below). Herzberg (1953) has described the band systems leading to the same dissociation limit but associated with transitions from the ground state to ${}^3\Delta_u$ and ${}^1\Sigma_u^-$.

In the Herzberg continuum short of 2423 A, absorption is also very weak. Certain measurements have been made through long paths of air (Buisson *et al.*, 1933; Götz and Maier-Leibnitz, 1933). Recently Ditchburn and Young (1962) have investigated the absorption cross section for this spectral region in the laboratory. They found cross sections (at atmospheric pressure) varying from 1.4×10^{-24} cm^2 at 2500 A to 2.64×10^{-23} cm^2 at 2000 A. They also verified a pressure dependence of σ first suggested by Heilpern (1941) from measurements at only one wavelength (2144 A). The curve in Fig. 4.16 is based only on Ditchburn and Young's results at atmospheric pressure.

The Schumann-Runge band system, due to the transition ${}^3\Sigma_u^- \leftarrow X\,{}^3\Sigma_g^-$, begins with the (0–0) band at 2026 A and converges to the second dissociation limit at about 1750 A. These bands have been studied extensively by several investigators, for example, by Watanabe *et al.* (1953) and by Wilkinson and Mulliken (1957). Rather good dispersion is required in this interval because of the rapid variation of absorption coefficient with wavelength. Figure 4.16 in this region simply gives approximate upper and lower limits for the maxima and minima of the bands. Carroll (1959) has suggested predissociation in the (4–0) band of this system, and Wilkinson and Mulliken (1957) have found evidence for predissociation in the (12–0) band (1790–1800 A) with the ${}^3\Pi_u$ (not shown in Fig. 4.8) state, whose potential curve crosses that of ${}^3\Sigma_u^-$ in this vicinity. The latter have also found evidence for a very weak continuum underlying the Schumann-Runge bands and probably caused by the direct transition ${}^3\Pi_u \leftarrow X\,{}^3\Sigma_g^-$, resulting in dissociation into two ${}^3P$ atoms.

The Schumann-Runge continuum from about 1250 to 1750 A involves dissociation into one ${}^3P$ and one ${}^1D$ atom. Measurements by Ladenburg and Van Voorhis (1933) and by Watanabe *et al.* (1953) differ by about 25 per cent at some wavelengths with a peak value in the neighborhood of 1450 A of about 10^{-17} cm^2.

Absorption at shorter wavelengths has been discussed by Watanabe and Marmo (1956), Lee (1955), and Aboud *et al.* (1955). There is some discrepancy among the results, and the values in Fig. 4.16 were taken from the summary by Watanabe (1958). Down to 1000 A the absorption coefficient varies greatly with wavelength, various bands (most of them poorly understood in terms of molecular energy states) being superimposed on various weak dissociation continua. The first ionization potential of O_2 is at about 1026 A, but this gives rise to a relatively weak continuum which according to Lee (1955) has a maximum absorption cross section of 4×10^{-18} cm^2 at 920 A. Overlying this continuum down to about 800 A are various strong bands, mostly Rydberg bands converging to other ionization continua at about 771, 736, and 682 A. The absorption bands between 900 and 1000 A lead to preionization, according to Watanabe and Marmo (1956). At wavelengths shorter than 682 A, the principal ionization continuum gives rise to very strong absorption, although there is some disagreement as to the exact values of the absorption coefficient. Below 400 A, the absorption coefficient decreases markedly down to 100 A (Aboud *et al.*, 1955).

Of very great interest is the value of the absorption coefficient at 1216 A, the wavelength of Lyman α, and accordingly this spectral region has been extensively studied. It turns out that this wavelength corresponds almost exactly to a "window" in the absorption spectrum. Figure 4.17, after Watanabe (1958), illustrates this. Watanabe's results show good agreement with earlier measurements by Preston (1940). This small absorption at precisely the wavelength of large solar emission means that molecular oxygen presents no barrier to the penetration of Lyman α radiation down into the mesosphere. We saw earlier that N, O, and N_2 are similarly transparent at this important wavelength.

There is some evidence that, in certain parts of the spectrum, the absorption coefficient is pressure dependent so that Beer's law is not obeyed. Herzberg (1952) has pointed this out for the Herzberg bands, we have already mentioned it for the Herzberg continuum, and Watanabe *et al.* (1953) have pointed out an apparent pressure effect for the Schumann-Runge bands. On the other hand, according to Schneider (1937) and Watanabe *et al.* (1953), there is no appreciable pressure effect in the Schumann–Runge continuum. The pressure effects are in most cases rather poorly defined by measurements, particularly at pressures obtaining in the upper atmosphere, and give rise to a considerable source of uncertainty in some applications of laboratory results to atmospheric conditions.

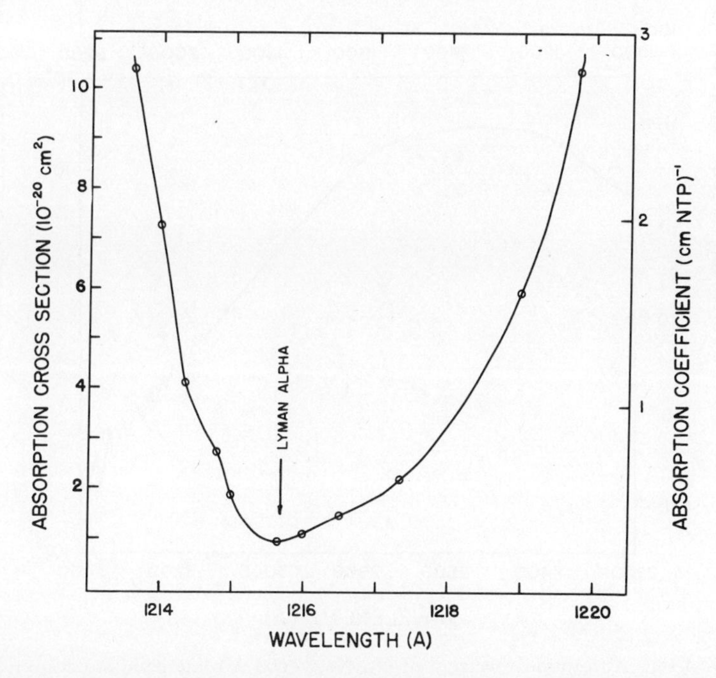

FIG. 4.17. Absorption cross section of O_2 near Lyman α. (After Watanabe, 1958.)

4.6.4 Absorption by Ozone

Of the principal constituents of the upper atmosphere discussed above, only molecular oxygen (and that very weakly) absorbs between 2000 and 3000 A. This part of the solar spectrum is mainly absorbed by minute quantities of ozone in the upper stratosphere and mesosphere. Postponing until Chapter 5 consideration of the formation and vertical distribution of ozone, we discuss briefly here ozone's absorption spectrum.

The ozone absorption spectrum in the ultraviolet has been studied many times in the laboratory. Earlier work was summarized by Craig (1950). Since then, more reliable and more extensive data have become available. In particular, the measurements of Vigroux (1953) between 2300 and 3400 A, and of Inn and

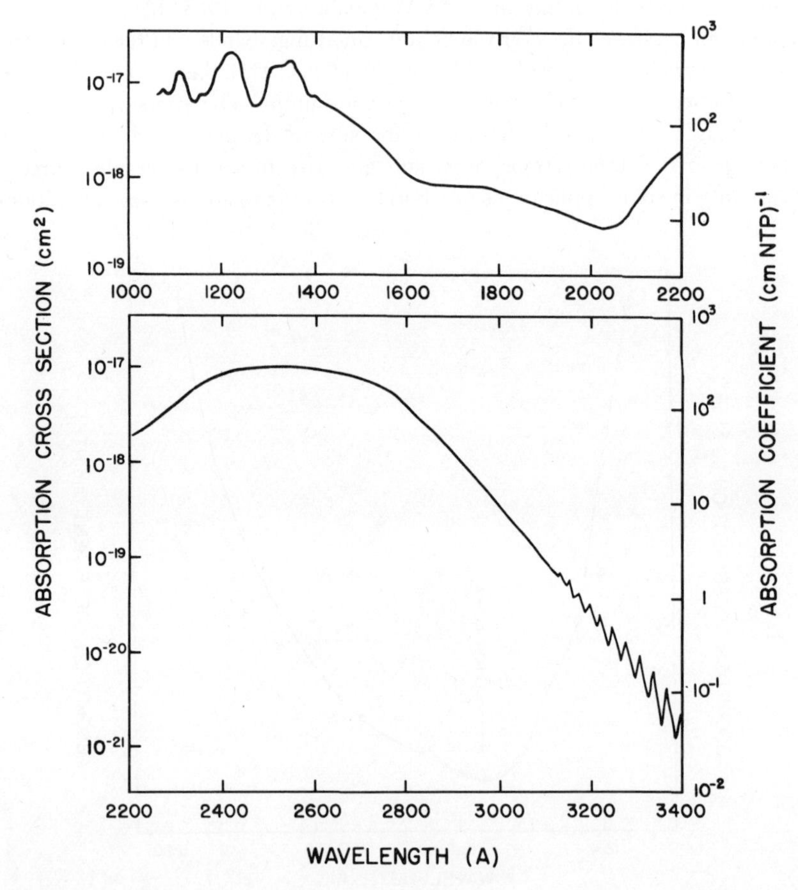

Fig. 4.18. Absorption spectrum of O_3. Near 2600 A some band structure has been smoothed out.

Tanaka (1953) between 2000 and 3500 A, give absorption coefficients consistently smaller than those that were formerly accepted. In addition, Tanaka *et al.* (1953) have studied the 1050–2200 A region, about most of which no information was formerly available.

Figure 4.18 illustrates these data. The Hartley region from about 2000 to 3000 A is the most important in the heat budget of the upper atmosphere. Here less than 10^{-1} cm NTP of ozone is sufficient to absorb the solar radiation and none reaches the earth's surface. In the Huggins bands above 3000 A the atmosphere is semitransparent. (Of course, this somewhat arbitrary division between "opaque" and "semitransparent" depends on the solar zenith angle and the amount of ozone.) Short of 2000 A, the unexpectedly large absorption coefficients determined by Tanaka *et al.* (1953) are not as significant as might at first appear, because the available solar energy is absorbed by O_2 at levels above those where ozone occurs in significant amounts. A possibly important exception to this statement is the wavelength of Lyman α, where ozone may play an important role in the absorption of solar energy, if it occurs in sufficient amounts at sufficiently high levels.

4.6.5 Absorption by Other Constituents

The absorption cross sections of a number of other minor constituents of the upper atmosphere have been discussed by Watanabe (1958). Included among these are NO, N_2O, H_2O, and CO_2. Generally speaking, either because they occur in too small quantities or because they are dissociated at high levels, they absorb relatively little energy. Nevertheless, some of these may play important roles in the upper atmosphere in spite of their low concentrations. For example, NO has a low ionization potential and may contribute significantly to the formation of the *D* region, since it absorbs rather strongly at the wavelength of Lyman α.

4.6.6 Altitudes and Amounts of Absorbed Energy

The chief source of radiative energy for the stratosphere is ozone absorption, in the Hartley region for the upper stratosphere and in the Huggins bands for the lower stratosphere. If, as an order-of-magnitude approximation, all the solar energy between 2000 and 3000 A is absorbed by ozone between 30 and 50 km, then according to the data in Table 4.3 about 1.8×10^4 ergs are absorbed each second by each (centimeter)^2 column in the direction of the solar beam between these altitude limits. This number is worth remembering for comparison with the very much smaller amounts of energy available to higher regions of the atmosphere at shorter wavelengths.

It is possible to make much more refined computations of the absorbed energy as a function of solar zenith angle and vertical ozone distribution with the use of Fig. 4.19. This diagram, first introduced by Craig (1951), has as its ordinate the integrated mass u of ozone above level z at which the absorption is to be computed. (u is the vertical path length times the secant of the solar zenith angle, unless the latter approaches 90°.) The abscissa is q/n, with q the energy absorbed per unit volume and unit time and n the density of ozone (expressed in this case in centimeters NTP per centimeter depth of a vertical column of unit cross section). The curve in Fig. 4.19 is based on some unpublished calculations of Mr. P. R. Sticksel of the Florida State University, who made use of the absorption coefficients of Vigroux (1953) and of Tanaka *et al.* (1953) (see Subsection 4.6.4 and Fig. 4.18) and of Johnson's (1954) solar data. It therefore differs somewhat (but never more than about 15 per cent in the values of q/n) from the curve originally given by Craig (1951). It is worth

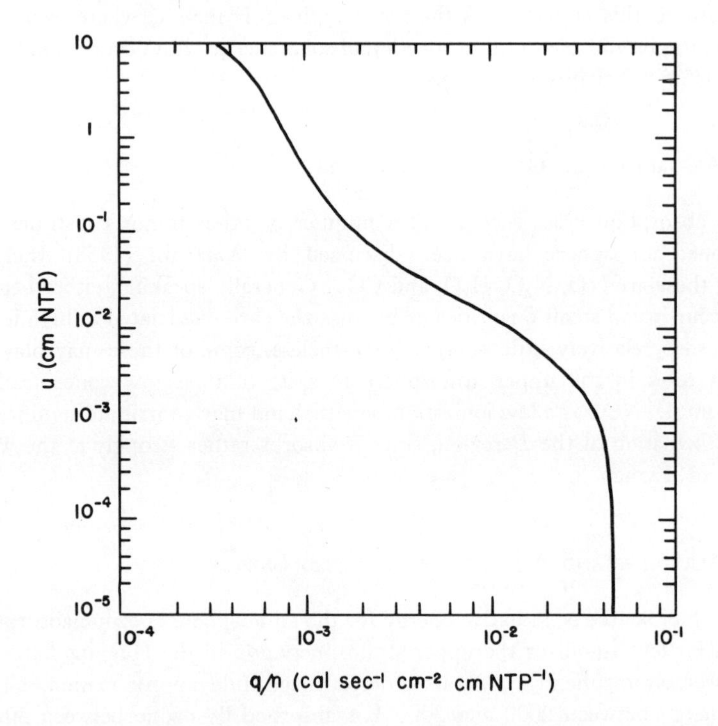

FIG. 4.19. Ozone absorption as a function of path length. The abscissa is energy absorbed per unit time and volume divided by ozone density in cm NTP cm^{-1} for a column of unit area. The ordinate is the amount of ozone in a slant path of 1 cm^2 area between the absorbing volume and the sun. This curve is based on the newer coefficients of Vigroux (1953) and on the solar flux measurements reported by Johnson (1954), according to unpublished calculations of Mr. Philip Sticksel of Florida State University.

noting that the true heating rate in the upper stratosphere involves also some absorption by molecular oxygen in the Herzberg continuum, but this is of small direct effect (although this absorption is crucial in the formation of ozone; see Chapter 5).

Both Pressman (1955) and Murgatroyd (1957) have computed heating rates in the upper stratosphere for representative ozone distributions and for various latitudes and seasons, taking into account the appropriate variations of solar zenith angle and duration of insolation. Although their results differ somewhat in detail, both find an appreciable heat source with maximum heating rate of up to 15°C per day near the summer pole near 45 to 50 km. At the equinoxes, the maximum is less intense and occurs in low latitudes. (Note that the maximum heating rate occurs at a noticeably higher altitude than the maximum rate of energy absorption because the air density decreases with height so rapidly.)

The mesosphere, by contrast, absorbs very little solar radiation. Only the energy in part of the Schumann–Runge bands, perhaps that between 1750 and 1950 A, some of the Lyman-α radiation, and some of the X-ray radiation is involved. Altogether, this is on the order of 10^2 erg cm^{-2} sec^{-1} for each (centimeter)^2 column in the direction of the solar beam. According to Murgatroyd's calculations, the heating rate decreases rapidly with altitude through the mesosphere and at 70 km is on the order of a few degrees per day even at the summer pole.

All energy short of 1750 A (with the exceptions noted above) is absorbed in the thermosphere. This can hardly exceed about 30 erg cm^{-2} sec^{-1} at normal incidence, must of which is absorbed by molecular oxygen in the Schumann–Runge continuum. The altitude of this absorbed energy depends on the vertical distribution of O_2 but must be mostly in the 100–125 km region. It is important to note, however, that in the thermosphere the altitude at which energy is absorbed need not in general correspond to the altitude at which it is converted into thermal energy. In the case of E-region* absorption, Nicolet (1961) estimated that much of the energy is carried downward in chemical form by diffusion of the dissociated oxygen atoms, which recombine mainly at levels below 100 km (see Chapter 7).

The vertical distribution of the absorption of the few ergs (centimeter)$^{-2}$ (second)$^{-1}$ available at wavelengths below 1000 A is of crucial importance in theories of ionospheric formation and thermospheric structure and is rather difficult to establish. Absorption cross sections for the atmospheric gases are only poorly known and the vertical distribution of those gases above 100 km is still somewhat uncertain. According to Hinteregger and Watanabe (1962) at wavelengths between 1027 and 911 A energy penetrates to the 100–120 km region. Shorter wavelengths and soft X-rays combine to give total absorption

---

* See introduction to Section 9.2 for a discussion of ionospheric nomenclature.

that is relatively constant with height up to about 160–200 km, above which absorption decreases.

REFERENCES

ABOUD, A. A., CURTIS, J. P., MERCURE, R., and RENSE, W. A. (1955). Oxygen gas continuous absorption in the extreme ultraviolet. *J. Opt. Soc. Amer.* **45**, 767–768.

ASTOIN, N., and GRANIER, J. (1957). Sur le spectre d'absorption de l'azote dans l'ultraviolet extrème. *Comptes Rendus Acad. Sci. Paris* **244**, 1350–1353.

ATHAY, R. G., and WARWICK, C. S. (1961). Indices of solar activity. *Advances in Geoph.* **8**, 1–83.

BATES, D. R., and SEATON, M. J. (1949). The quantal theory of continuous absorption of radiation by various atoms in their ground states. II, Further calculations on oxygen, nitrogen and carbon. *Mon. Not. Roy. Ast. Soc.* **109**, 698–704.

BLAKE, R. L., CHUBB, T. A., FRIEDMAN, H., and UNZICKER, A. E. (1963). Interpretation of X-ray photograph of the sun. *Ap. J.* **137**, 3–15.

BRIX, P., and HERZBERG, G. (1954). Fine structure of the Schumann-Runge bands near the convergence limit and the dissociation energy of the oxygen molecule. *Canad. J. Phys.* **32**, 110–135.

BUISSON, H., JAUSSERAN, G., and ROUARD, P. (1933). La transparence de la basse atmosphère. *Rev. Opt.* **12**, 70–80.

BYRAM, E. T., CHUBB, T. A., and FRIEDMAN, H. (1956). The solar X-ray spectrum and the density of the upper atmosphere. *J. Geoph. Res.* **61**, 251–263.

CARROLL, P. K. (1959). Predissociation in the Schumann-Runge bands of oxygen. *Ap. J.* **129**, 794–800.

CHANDRASEKHAR, S. (1950). "Radiative Transfer." Oxford Univ. Press, (Clarendon), London and New York. (Reprinted, 1960, Dover, New York.)

CHAPMAN, S. (1931a). The absorption and dissociative or ionizing effect of monochromatic radiation in an atmosphere on a rotating earth. *Proc. Phys. Soc.* **43**, 26–45.

CHAPMAN, S. (1931b). The absorption and dissociative or ionizing effect of monochromatic radiation in an atmosphere on a rotating earth. II, Grazing incidence. *Proc. Phys. Soc.* **43**, 483–501.

CHAPMAN, S. (1939). The atmospheric height distribution of band-absorbed solar radiation. *Proc. Phys. Soc.* **51**, 93–109.

CHAPMAN, S. (1951). Photochemical processes in the upper atmosphere and resultant composition. *In* "Compendium of Meteorology" (T. F. Malone, ed.), pp. 262–274. American Meteorological Society, Boston, Massachusetts.

CONDON, E. U., and SHORTLEY, G. H. (1951). "The Theory of Atomic Spectra." Cambridge Univ. Press, London and New York.

CRAIG, R. A. (1950). The observations and photochemistry of atmospheric ozone and their meteorological significance. *Meteor. Monogr.* **1**, No. 2, 1-50.

CRAIG, R. A. (1951). Radiative temperature changes in the ozone layer. *In* "Compendium of Meteorology" (T. F. Malone, ed.), pp. 292–302. American Meteorological Society, Boston, Massachusetts.

CURTIS, J. P. (1954). Absorption coefficients of air and nitrogen for the extreme ultraviolet. *Phys. Rev.* **94**, 908–910.

DALGARNO, A., and PARKINSON, D. (1960). Photoionization of atomic oxygen and atomic nitrogen. *J. Atmos. Terr. Phys.* **18**, 335–337.

DETWILER, C. R., GARRETT, D. L., PURCELL, J. D., and TOUSEY, R. (1961). The intensity distribution in the ultraviolet solar spectrum. *Ann. Geoph.* **17**, 263–272.

Ditchburn, R. W., and Young, P. A. (1962). The absorption of molecular oxygen between 1850 and 2500 A. *J. Atmos. Terr. Phys.* **24**, 127–139.

Ditchburn, R. W., Bradley, J. E. S., Cannon, C. G., and Munday, G. (1954). Absorption cross-sections for Lyman α and neighboring lines. *In* "Rocket Exploration of the Upper Atmosphere" (R. L. F. Boyd and M. J. Seaton, eds.), pp. 327-334. Pergamon Press, New York.

Elsasser, W. M., (1960). Atmospheric radiation tables. *Meteor. Monogr.*, **4**, No. 23, 1-43.

Friedman, H. (1959). Rocket observations of the ionosphere. *Proc. Inst. Rad. Eng.* **47**, 272–280.

Friedman, H. (1960). The sun's ionizing radiations. *In* "Physics of the Upper Atmosphere" (J. A. Ratcliffe, ed.), pp. 133–218. Academic Press, New York.

Friedman, H. (1961a). X-ray and ultraviolet radiation measurements from rockets. *In* "Space Astrophysics" (W. Liller, ed.), pp. 107–120. McGraw-Hill, New York.

Friedman, H. (1961b). Lyman-α radiation. *Ann. Geoph.* **17**, 245–248.

Gledhill, J. A., and Szendrei, M. E. (1950). Theory of the production of an ionized layer in a non-isothermal atmosphere neglecting the earth's curvature, and its application to experimental results. *Proc. Phys. Soc.* **B63**, 427–445.

Götz, F. W. P., and Maier-Leibnitz, H. (1933). Zur ultraviolett Absorption bodennaher Luftschichten. *Z. Geoph.* **9**, 253–260.

Heilpern, W. (1941). Die Absorption des Lichtes durch Sauerstoff bei der Wellenlänge $\lambda = 2144$ A Abhängigkeit vom Druck. *Helv. Phys. Acta* **14**, 329–354.

Herzberg, G. (1944). "Atomic Spectra and Atomic Structure," 2nd ed. Dover, New York.

Herzberg, G. (1950). "Molecular Spectra and Molecular Structure. I, Spectra of Diatomic Molecules," 2nd ed. Van Nostrand, Princeton, New Jersey.

Herzberg, G. (1952). Forbidden transitions in diatomic molecules. II, The ${}^3\Sigma_u^+ \leftarrow {}^3\Sigma_g^-$ absorption bands of the oxygen molecule. *Canad. J. Phys.* **30**, 185–210.

Herzberg, G. (1953). Forbidden transitions in diatomic molecules. III, New ${}^1\Sigma_u^- \leftarrow {}^3\Sigma_g^-$ and ${}^3\Delta_u \leftarrow {}^3\Sigma_g^-$ absorption bands of the oxygen molecule. *Canad. J. Phys.* **31**, 657–669.

Hinteregger, H. E. (1961a). Telemetering monochromator measurements of extreme ultraviolet radiation. *In* "Space Astrophysics" (W. Liller, ed.), pp. 34–95. McGraw-Hill, New York.

Hinteregger, H. E. (1961b). Preliminary data on solar extreme ultraviolet radiation in the upper atmosphere. *J. Geoph. Res.* **66**, 2367–2380.

Hinteregger, H. E., and Watanabe, K. (1962). Photoionization rates in the *E* and *F* regions, 2. *J. Geoph. Res.* **67**, 3373–3392.

Huschke, R. E., ed. (1959). "Glossary of Meteorology." American Meteorological Society, Boston, Massachusetts.

Inn, E. C. Y., and Tanaka, Y. (1953). Absorption coefficient of ozone in the ultraviolet and visible regions. *J. Opt. Soc. Amer.* **43**, 870–873.

Johnson, F. S. (1954). The solar constant. *J. Meteor.* **11**, 431–439.

Kato, S. (1954). On the solar Lyman-beta radiation and the ionosphere. *J. Geomag. Geoelect. Japan* **4**, 153–156.

Kreplin, R. W. (1961). Solar X-rays. *Ann. Geoph.* **17**, 151–161.

Kreplin, R. W., Chubb, T. A., and Friedman, H. (1962). X-ray and Lyman- alpha emission from the sun as measured from the NRL SR-1 satellite. *J. Geoph. Res.* **67**, 2231–2253.

Ladenburg, R., and Van Voorhis, C. C. (1933). The continuous absorption of oxygen between 1750 and 1300 A and its bearing upon the dispersion. *Phys. Rev.* **43**, 315–321.

Lee, P. (1955). Photodissociation and photoionization of oxygen (O_2) as inferred from measured absorption coefficients. *J. Opt. Soc. Amer.* **45**, 703–709.

MALITSON, H. H., PURCELL, J. D., and TOUSEY, R. (1960). The solar spectrum from 2635 to 2085 A. *Ap. J.* **132**, 746–766.

MITRA, S. K. (1952). "The Upper Atmosphere," 2nd ed. Asiatic Society, Calcutta.

MURGATROYD, R. J. (1957). Winds and temperatures between 20 km and 100 km—a review. *Quart. J. Roy. Meteor. Soc.* **83**, 417–458.

NICOLET, M. (1951). Effects of the atmospheric scale height gradient on the variation of ionization and short wave absorption. *J. Atmos. Terr. Phys.* **1**, 141–146.

NICOLET, M. (1961). Structure of the thermosphere. *Plan. Space Sci.* **5**, 1–32.

NICOLET, M., and BOSSY, L. (1949). Sur l'absorption des ondes courtes dans l'ionosphère. *Ann. Geoph.* **5**, 275–292.

OGAWA, M., and TANAKA, Y. (1962). Rydberg absorption series of N_2. *Canad. J. Phys.* **40**, 1593–1607.

PAGEL, B. E. J. (1963). Ultraviolet emission from the sun. *Plan. Space Sci.* **11**, 333-353.

PRESSMAN, J. (1955). Seasonal and latitudinal temperature changes in the ozonosphere. *J. Meteor.* **12**, 87–89.

PRESTON, W. M. (1940). The origin of radio fade-outs and the absorption coefficient of gases for light of wavelength 1215.7 A. *Phys. Rev.* **57**, 887–894.

PURCELL, J. D., and TOUSEY, R. (1960). The profile of solar hydrogen- Lyman-α. *J. Geoph. Res.* **65**, 370–372.

PURCELL, J. D., PACKER, D. M., and TOUSEY, R. (1960). Photographing the sun in Lyman Alpha. *In* "Space Research I" (H. Kallmann Bijl, ed.), pp. 594-598. North-Holland Publ. Co., Amsterdam.

RENSE, W. A. (1953). Intensity of Lyman-alpha line in the solar spectrum. *Phys. Rev.* **91**, 299–302.

RENSE, W. A. (1961). Solar ultraviolet research. *In* "Space Astrophysics" (W. Liller, ed.), pp. 17–33. McGraw-Hill, New York.

SCHNEIDER, E. G. (1937). The effect of foreign molecules on the absorption coefficient of oxygen in the Schumann region. *J. Chem. Phys.* **5**, 106–107.

STACEY, D. S., STITH, G. A., NIDEY, R. A., and PIETENPOL, W. B. (1954). Rocket-borne servo tracks the sun. *Electronics* **27**, 149–151.

TANAKA, Y. (1955). Absorption spectrum of nitrogen in the region from 1075 to 1650 A. *J. Opt. Soc. Amer.* **45**, 663–664.

TANAKA, Y., INN, E. C. Y., and WATANABE, K. (1953). Absorption coefficients of gases in the vacuum ultraviolet. Part IV. Ozone. *J. Chem. Phys.* **21**, 1651–1653.

TOUSEY, R. (1961). Ultraviolet spectroscopy of the sun. *In* "Space Astrophysics" (W. Liller, ed.), pp. 1–16. McGraw-Hill, New York.

TOUSEY, R. (1962). In a scientific discussion reported in *Trans. Int. Ast. Union* **B11**, 194.

TOUSEY, R. (1963). The extreme ultraviolet spectrum of the sun. *Space Sci. Rev.* **2**, 3–69.

VIGROUX, E. (1953). Contribution à l'étude expérimentale de l'absorption de l'ozone. *Ann. Phys.* [6] **8**, 709–762.

VIOLETT, T., and RENSE, W. A. (1959). Solar emission lines in the extreme ultraviolet. *Ap. J.* **130**, 954–960.

WATANABE, K. (1958). Ultraviolet absorption processes in the upper atmosphere. *Advances in Geoph.* **5**, 153–221.

WATANABE, K., and HINTEREGGER, H. E. (1962). Photoionization rates in the *E* and *F* regions. *J. Geoph. Res.* **67**, 999–1006.

WATANABE, K., and MARMO, F. F. (1956). Photoionization and total absorption cross section of gases. II. O_2 and N_2 in the region 850–1500 A. *J. Chem. Phys.* **25**, 965–971.

WATANABE, K., ZELIKOFF, M., and INN, E. C. Y. (1953). Absorption coefficients of several atmospheric gases. *Geoph. Res. Papers* No. 21, 1-80.

WEISSLER, G. L., LEE, P., and MOHR, E. I. (1952). Absolute absorption coefficients of nitrogen in the vacuum ultraviolet. *J. Opt. Soc. Amer.* **42**, 84–90.

WILKINSON, P. G., and MULLIKEN, R. S. (1957). Dissociation process in oxygen above 1750 A. *Ap. J.* **125**, 594–600.

WILSON, N. L., TOUSEY, R., PURCELL, J. D., JOHNSON, F. S., and MOORE, C. E. (1954). A revised analysis of the solar spectrum from 2990 to 2635 A. *Ap. J.* **119**, 590–612.

CHAPTER 5

Composition of the Stratosphere and Mesosphere; Atmospheric Ozone

Above the tropopause, composition becomes a matter of great concern and considerable complexity. Two factors which are absent in the troposphere gradually come into play as one considers the higher levels. The first is the production of different atomic and molecular species as a net result of photodissociation and chemical reactions. The second, which becomes important only in the thermosphere and is discussed in Chapter 6, is the tendency toward a diffusion (rather than a mixing) equilibrium (Sections 1.1 and 1.2).

Photochemical processes in the stratosphere and mesosphere do not appreciably affect the mean molecular weight of air. On the other hand, some of the gases that are formed play very important roles in upper-atmospheric phenomena despite their small relative abundances. Absorption, emission, and ionization of the upper atmosphere often depend critically on these trace constituents.

Ozone is a particularly important trace constituent. Formed principally above 25 km by photochemical processes, it nevertheless is carried to the lower stratosphere (and troposphere) by mixing processes. Owing to its very strong absorption in the ultraviolet, it is responsible, despite its small abundance, for the depletion of solar radiation between 2000 and 3000 A. The energy so absorbed is the principal source of thermal energy in the stratosphere and is responsible for the increase of temperature with height in the upper stratosphere. Ozone also plays a significant role in the infrared transfer of energy, especially in the upper stratosphere. In the lower stratosphere, where ozone is effectively shielded from photochemical destruction once it is transported there, ozone assumes great meteorological importance as a tracer to indicate atmospheric circulation.

For this reason and also because the observations and theory of ozone are more advanced than are those of other trace constituents, a large part of this chapter is devoted to ozone, particularly to ozone in the stratosphere. Section 5.1 is concerned with the methods and results of extensive measurements of the total amount of ozone (in a vertical column of unit area). Section 5.2 takes up the more difficult problem and more meager results of measurements of the vertical distribution of ozone density. In Sections 5.3 and 5.4, the explanation of these observations in terms of photochemical processes (Section 5.3) and transport processes (Section 5.4) is considered.

Ozone is present in the mesosphere, but its abundance and geophysical significance are less than in the stratosphere. Atomic oxygen, for example, becomes relatively more abundant than ozone in the mesosphere. In the mesosphere, photochemical problems are more complex because reactions involving hydrogen and nitrogen must be taken into consideration. Among the trace constituents resulting from these reactions are hydroxyl (OH), emission from which is an important part of the airglow spectrum (Section 9.3), and nitric oxide (NO), which, because of its low ionization potential, is generally believed to be responsible for the ionization at those levels (Section 9.2). The airglow spectrum also reveals the presence of sodium, emission from which originates principally in the mesosphere and lower thermosphere. Section 5.5 is an introduction to all these problems.

Finally, Section 5.6 is concerned with the morphology and explanation of noctilucent clouds which are observed to occur in the vicinity of the mesopause.

5.1 Observations of the Total Amount of Ozone

The earliest and still the most numerous measurements of atmospheric ozone yield only the integrated amount of ozone in a vertical column above the earth's surface, usually referred to as the *total amount* of ozone. This quantity is usually expressed in the unit (10^{-3} cm NTP), sometimes called a Dobson Unit (D.U.). In the older literature, the symbol x was often used for total amount of ozone. Recently Godson (1962) has suggested the symbol Ω, a suggestion adopted here.

5.1.1 Method of Observation

The basic method of measurement was first applied quantitatively by Fabry and Buisson (1913, 1921). The first systematic program of extensive measurements was carried out by Dobson (1930) and his collaborators (1926, 1927, 1929). Dobson was also responsible for developing an appropriate instrument (Dobson, 1931, 1957) which is now standard in many parts of the world and which is briefly described later in this section. In the spectral region of the Huggins bands (3000–3500 A; see Subsection 4.6.4) solar radiation is measurably depleted by ozone absorption but nevertheless reaches the earth's surface in sufficient strength to be detected and measured. The flux of solar radiation reaching the surface varies with zenith angle and wavelength, and observations of these variations, together with values of ozone absorption coefficients determined in the laboratory, allow a determination of the total amount of ozone.

Let $F(\lambda)$ represent the integrated solar flux in a narrow spectral interval centered at wavelength λ. Between z and $(z + dz)$ the fractional change in

$F(\lambda)$ due to ozone absorption, molecular scattering, and dust scattering can be expressed as

$$dF(\lambda)/F(\lambda) = +k(\lambda)\rho_3 \sec\psi\, dz + K(\lambda)\rho \sec\psi\, dz + \delta_z \sec\psi\, dz \tag{5.1}$$

where $k(\lambda)$ = ozone absorption coefficient at wavelength λ

ρ_3 = ozone density at height z

$K(\lambda)$ = molecular scattering coefficient at wavelength λ

ρ = air density at height z

δ_z = fraction scattered (per unit distance) by particulate matter at height z and is to a first approximation independent of wavelength

ψ = zenith angle at height z, and is not in general independent of height because of the sphericity of the earth and atmospheric refraction

Integration of (5.1) from the earth's surface to the outer edge of the atmosphere gives

$$\ln[F(\lambda)/F_\infty(\lambda)] = -k(\lambda)\int_0^\infty \rho_3 \sec\psi\, dz - K(\lambda)\int_0^\infty \rho \sec\psi\, dz - \int_0^\infty \delta_z \sec\psi\, dz \tag{5.2}$$

where $F_\infty(\lambda)$ is the flux outside the atmosphere.

The integrals on the right side of (5.2) are in practice approximated in different ways. In the first, $\sec\psi$ is considered to have the constant value $\sec\psi_h$ where ψ_h is the zenith angle at the level of maximum ozone density (taken for this purpose to be 22 km). The second term is written as (βm) where β is the value of the term for vertical incidence and can be computed from the Rayleigh theory of molecular scattering; and m is the so-called "air mass," a function of zenith angle at the point of observation and taken in practice to be given by Bemporad's function (tabulated by List, 1958). In the third integral, $\sec\psi$ is considered to have the constant value $\sec Z$, where Z is the zenith angle at the ground, since most particulate scattering occurs in the troposphere. Thus,

$$\ln[F(\lambda)/F_\infty(\lambda)] = -k(\lambda) \sec\psi_h \Omega - \beta(\lambda)m - \delta \sec Z \tag{5.3}$$

where δ is $\int_0^\infty \delta_z\, dz$ and is in the first approximation assumed to be independent of wavelength.

In order to minimize the effect of particulate scattering, and to allow a relative measurement of fluxes, observations are made at two wavelengths λ_1 and λ_2. If (5.3) is written separately for the two wavelengths, if the equation for λ_2 is subtracted from the equation for λ_1, and if the resulting equation is solved for Ω, one gets

$$\Omega = \frac{\ln[F_\infty(\lambda_1)/F_\infty(\lambda_2)] - \ln[F(\lambda_1)/F(\lambda_2)] - m[\beta(\lambda_1) - \beta(\lambda_2)]}{\sec\psi_h\, [k(\lambda_1) - k(\lambda_2)]} \tag{5.4}$$

Of the quantities on the right side of (5.4), $\ln[F(\lambda_1)/F(\lambda_2)]$ is the quantity that is measured by the ozone spectrophotometer as is discussed below. The quantities ψ_h and m are functions of the zenith angle at the surface and therefore depend in a known manner on the location and time of observation. All the other quantities are essentially constant for a given pair of wavelengths and have been given by Dobson (1957) for certain standard pairs. The quantity $[\beta(\lambda_1) - \beta(\lambda_2)]$ is computed from the Rayleigh theory of molecular scattering; the quantity $[k(\lambda_1) - k(\lambda_2)]$ is determined from laboratory measurements; and the ratio $[F_\infty(\lambda_1)/F_\infty(\lambda_2)]$ is determined once and for all (and assumed to remain constant) by making a series of measurements over a range of zenith angles and extrapolating the results to the outer edge of the atmosphere.

It should be pointed out that (5.4) is correct only if the dust scattering term δ in (5.3) is independent of wavelength. This is true only for large (compared with the wavelength of the radiation) scattering particles. In practice, not all atmospheric aerosols meet this criterion and the effect of particulate scattering can introduce serious errors if (5.4) is used under conditions where scattering by small particles is appreciable. To minimize this effect, it is preferable to measure two *pairs* of wavelengths. Two equations similar to (5.4) can be written, one for each wavelength pair, and Ω determined by subtracting one of these equations from the other. Then if $[\delta(\lambda_1) - \delta(\lambda_2)]$ is not exactly zero, as is assumed in deriving (5.4), it can be assumed in the second approximation to have the same value for the two wavelength pairs.

It should be pointed out also that the method described above applies to measurements on direct sunlight. However, total amounts of ozone can be determined from measurements on the clear or cloudy zenith sky, measurements which are related empirically to the ozone amount.

The standard instrument for measuring the total amount of ozone is called the Dobson spectrophotometer or ozone spectrophotometer (Dobson, 1931, 1957). It consists essentially of a double monochromator for isolating a pair of narrow spectral intervals, two optical wedges to adjust the fluxes so that a null measurement can be made, and a photomultiplier and amplifier. The wedges are controlled by a calibrated dial and adjusted to reduce the energy at the longer wavelength until the two produce the same current in the photomultiplier and no output current from the amplifier. The instrument and details of its use have been fully described by Dobson (1957).

5.1.2 Results of Observations

Measurements of total amount of ozone were first gathered in a systematic manner (with a different instrument than the one described above, which, however, is now standard) by Dobson (1930) and his collaborators (1926, 1927,

1929). These and essentially all other published observations up to 1950 were summarized by Craig (1950).*

Since 1950 and particularly during the IGY large numbers of additional data have become available. The principal climatological aspects of the variation of total amount of ozone are illustrated in Fig. 5.1, after Godson (1960). This shows the latitudinal and seasonal changes in the Northern Hemisphere, smoothed by averaging large numbers of observations over time and longitude. The total amount of ozone is a maximum in spring and a minimum in autumn with the largest amplitude of variation at high latitudes. Near the equator there is very little

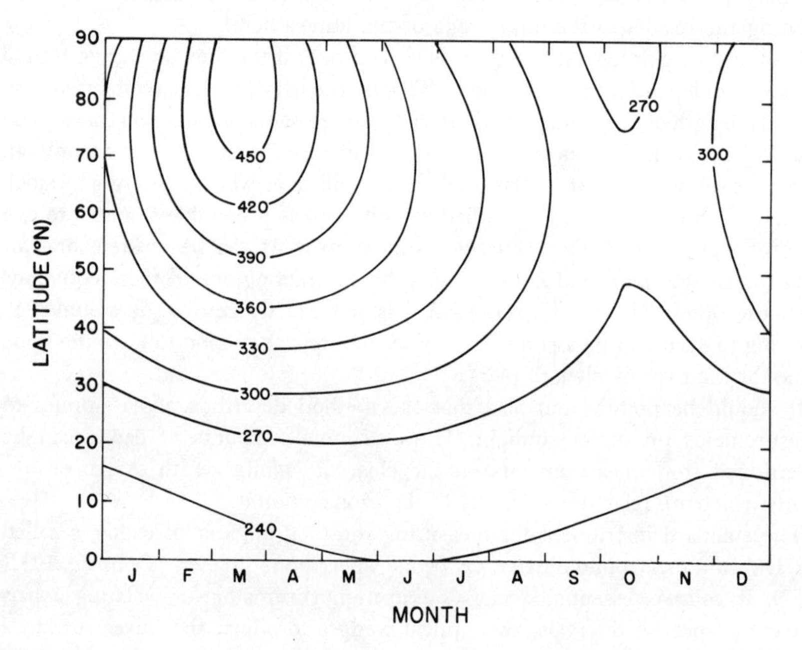

FIG. 5.1. Mean distribution of total ozone (in 10^{-3} cm NTP) as a function of month and latitude in the Northern Hemisphere. (After Godson, 1960.)

* These and other data published prior to July 1957 should be increased by a factor of 1.35 to be consistent with the more recent data. This is because the absorption coefficients appearing in the denominator of (5.4) are now assigned smaller values than before on the basis of the more recent laboratory measurements of Vigroux (1953). As a matter of fact, there is still some question about the absolute scale of the ozone values, because values of Ω determined nearly simultaneously from different wavelength pairs, or different pairs of wavelength pairs, differ among themselves by something like 10 per cent. All are standardized so that the final results are homogeneous, but the disagreement leaves some doubt about the absolute accuracy of the standardized scale. However, all ozone values quoted in this book, unless specifically stated otherwise, are expressed in the new Vigroux scale.

seasonal variation, with a slight maximum in the late spring or early summer. The total amount of ozone increases from the equator to a maximum at about 60°N during most of the year and near or at the pole in the spring. These characteristics of the seasonal and latitudinal variation are illustrated in another way in Fig. 5.2, after London (1962).

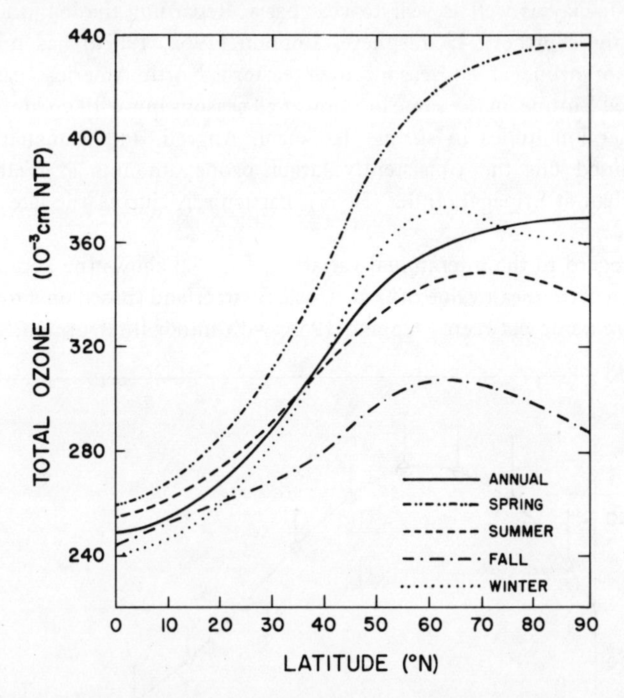

FIG. 5.2. Average variation of total ozone with latitude at different times in the Northern Hemisphere. (After London, 1962.)

Observations from the Southern Hemisphere are relatively less numerous, but have increased substantially during and since the IGY. Kulkarni (1962) has discussed the seasonal and latitudinal variations of total ozone in the Southern Hemisphere compared with those in the Northern Hemisphere. It appears from these few years of data that between the equator and about 55° the seasonal and latitudinal variations are quite similar in the two hemispheres (annual variation being referred to the same seasons, not the same calendar months). There is, however, in the middle latitudes somewhat more ozone in the Southern Hemisphere than in the Northern Hemisphere during the late summer, autumn, and winter. The amplitude of the spring maximum in these latitudes is about the same in the two hemispheres. Poleward of 55°, there is a striking difference during the spring months: maximum total ozone in the Southern Hemisphere occurs later and with smaller magnitude than in the Northern Hemisphere. Since the Southern Hemisphere has more ozone at 55° during summer, autumn, and winter and less ozone poleward of 55° during spring, the latitudinal variation there exhibits a more pronounced maximum at about 55° at all seasons.

There are, in addition, longitudinal and other temporal variations, the latter on a day-to-day as well as year-to-year basis. Regarding the longitudinal variations in the Northern Hemisphere, London (1962, 1963a) has noted largest amounts of ozone at all seasons over eastern North America, eastern Asia, and central Europe, in the same location at all seasons but with greatest difference from other longitudes in spring. Kulkarni, Angreji, and Ramanathan (1959) have pointed out the consistently larger ozone amounts at Tateno, Japan (36°N), than at Srinagar, India (34°N), particularly during the late winter and spring.

With regard to the interannual variation, Fig. 5.3 shows the annual variation of the long-term mean value of Ω at Arosa, Switzerland (based on a unique series of measurements between 1926 and 1958, with minor interruption, which were

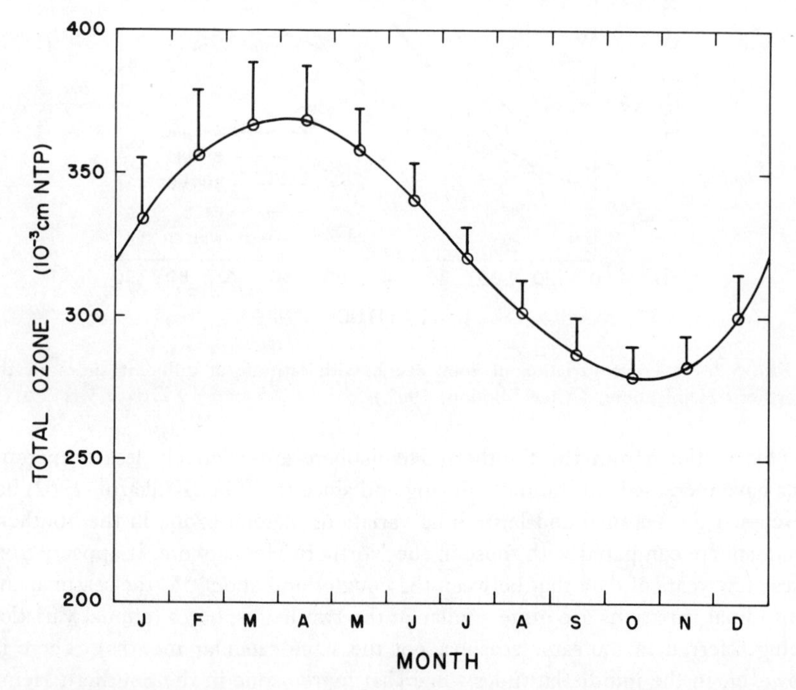

FIG. 5.3. Annual variation of total amount of ozone at Arosa, Switzerland. The circled points show the average values for each calendar month, based on observations covering most of the period 1926–1958. The vertical line for each calendar month has a length (measured along the ordinate scale) equal to the root-mean-square value of the deviations of monthly means for individual years from the long-term average.

due largely to the efforts of the late F. W. P. Götz and have been published by Perl and Dütsch, 1959). The vertical lines extending upward from the curve show the root-mean-square values of the deviations of individual monthly means from the long-term mean for the month. These deviations can be rather sizable, attaining the value of 23×10^{-3} cm NTP in February, but are smaller ($10–12 \times 10^{-3}$ cm NTP) from June through November.

For the Southern Hemisphere, there is no such long series of data on which to base a study of interannual variations. Funk and Garnham (1962) have pointed out a peculiar 24-month variation which occurred at Melbourne (38°S) and Brisbane (27°S) during seven years of observations. The sense of this variation was such that the spring maximum alternated from a relatively high value one year to a relatively low value the next. The autumn minima showed no regular variation. No such systematic interannual variation has been observed in the Northern Hemisphere.

Deviations of individual daily values from the monthly mean are quite large and striking. For example, one can compute from the Arosa data the deviation of each daily value from the mean for the month in which it occurs. The root-mean-square value of such deviations from 1949 through 1958 was 33×10^{-3} cm NTP in March and 16×10^{-3} cm NTP in October. Such a statistic varies markedly from year to year. It is not unusual to note individual daily values of Ω in the autumn that are somewhat larger than certain individual daily values during the period of spring maximum. The specific figures quoted above, of course, apply only to Arosa, but the statements apply qualitatively to all locations not too near the equator. Variations are even more pronounced at higher latitudes than at Arosa (47°N).

Interdiurnal variations also occur in the Southern Hemisphere (Kulkarni, 1962; Funk and Garnham, 1962), where they are also larger in winter and spring than in summer and autumn. However, for corresponding latitudes and seasons, they appear to be smaller in the Southern than in the Northern Hemisphere.

There have been many studies of the correlation of total amounts of ozone with various meteorological parameters, but a discussion of these is deferred until we have considered the vertical distribution of ozone and its photochemical formation and destruction.

5.2 Observations of the Vertical Distribution of Ozone

There is no single method for measuring the vertical distribution of ozone that is comparable in economy, convenience, and accuracy with the Dobson method of determining the total amount. Several techniques have been devised and applied over the years and we shall discuss in turn (a) the Götz *Umkehr* method, (b) the infrared method, (c) the balloon-borne (or rocket-borne) optical method, (d) the chemical method, and (e) the chemiluminescent method. Other methods have been suggested but not widely applied and are not discussed in detail here. These include observation from a satellite of diffusely reflected ultraviolet radiation (Sekera and Dave, 1961a; Twomey, 1961) and photometry of a satellite (Venkateswaran *et al.*, 1961) or balloon (Pittock, 1963).

It will be useful at this point to discuss briefly some of the units that are used to express ozone density. Ozone density, ρ_3, is, of course, expressed in gram (centimeter)$^{-3}$ in the cgs system. However, this unit is seldom used. An older unit, still frequently used, is (10^{-3} cm NTP) (kilometer)$^{-1}$. A unit which is coming more and more into use is the microgram (meter)$^{-3}$ (sometimes called the "gamma"). Still another, which is particulary useful in photochemical studics, is molecules (centimeter)$^{-3}$ (the number density). Some useful numerical relationships are

$$\rho_3\,(\mu\text{g m}^{-3}) = 21.4\,\rho_3\,(10^{-3}\text{ cm NTP km}^{-1}) = 7.97 \times 10^{-11}\rho_3\,(\text{molecules cm}^{-3})$$

The ozone mixing ratio, r_3, is defined as ρ_3/ρ, the ratio by mass of ozone to air in a given volume. It is, of course, dimensionless. A convenient unit is 10^{-6}, sometimes referred to as "parts per million" or "micrograms per gram" (μg g^{-1}).

$$r_3\ \ (\mu\text{g g}^{-1}) = \frac{2.87 \times 10^{-3}\rho_3\,(\mu\text{g m}^{-3})\ T\,(^\circ\text{K})}{p\,(\text{mb})}$$

The partial pressure of ozone, p_3, is given by $(n_3/n)p$ where n_3 is the number density of ozone, n the number density of air, and p the total pressure. Since the gram-molecular weight of ozone is 48, this can be written as $p_3 = 0.603\ r_3 p$. A convenient unit is the micromillibar (μmb).

$$p_3(\mu\text{mb}) = 0.603 r_3\,(\mu\text{g g}^{-1})\,p\,(\text{mb}) = 1.73 \times 10^{-3}\,\rho_3\,(\mu\text{g m}^{-3})\ T(^\circ\text{K})$$

5.2.1 The Götz *Umkehr* Method

The Götz *Umkehr* method, or simply *Umkehr* method, makes use of measurements (for example, by the Dobson spectrophotometer) at large zenith angles of the ratio of scattered light from the zenith sky at two wavelengths in the Huggins bands (Götz, 1931; Götz *et al.*, 1934). The manner in which this ratio varies with zenith angle depends on the vertical distribution of ozone and can be used to deduce with reasonable accuracy (but without great detail) that vertical distribution. Of all the various methods of deducing the vertical distribution of ozone, the *Umkehr* method has been most widely used. The basic equation, which is derived in Appendix E, is

$$\frac{I(\lambda_1)}{I(\lambda_2)} = \frac{K(\lambda_1)F_\infty(\lambda_1)\int_0^\infty \rho\tau_{0,z}(\lambda_1)\tau_{z,\infty}(\lambda_1)\,dz}{K(\lambda_2)F_\infty(\lambda_2)\int_0^\infty \rho\tau_{0,z}(\lambda_2)\tau_{z,\infty}(\lambda_2)\,dz} \tag{5.5}$$

where $I(\lambda)$ is the specific intensity at the surface integrated over a narrow wavelength interval centered at λ, $K(\lambda)$ is the scattering coefficient for Rayleigh scattering at wavelength λ and is inversely proportional to λ^4, $F_\infty(\lambda)$ is the solar flux at the outer edge of the atmosphere integrated over a narrow wavelength interval centered at λ, ρ is the air density, and the τ's are transmission functions defined below.

The transmissivity of air in a vertical column between the surface and altitude z is designated by $\tau_{0,z}(\lambda)$ and takes account of both ozone absorption and Rayleigh scattering. It is given by

$$\tau_{0,z}(\lambda) = \exp\left[-k(\lambda)\int_0^z \rho_3\, dz - K(\lambda)\int_0^z \rho\, dz\right] \tag{5.6}$$

The transmissivity of a slant air column (in the direction toward the sun) between altitude z and the outer edge of the atmosphere is designated by $\tau_{z,\infty}(\lambda)$ and is given by

$$\tau_{z,\infty}(\lambda) = \exp\left[-k(\lambda)\int_z^\infty \rho_3 \sec\psi\, dz - K(\lambda)\int_z^\infty \rho \sec\psi\, dz\right] \tag{5.7}$$

where ψ is the angle the solar beam makes with the (local) vertical at the height z. *Umkehr* measurements are made at very large zenith angles, and it is not possible

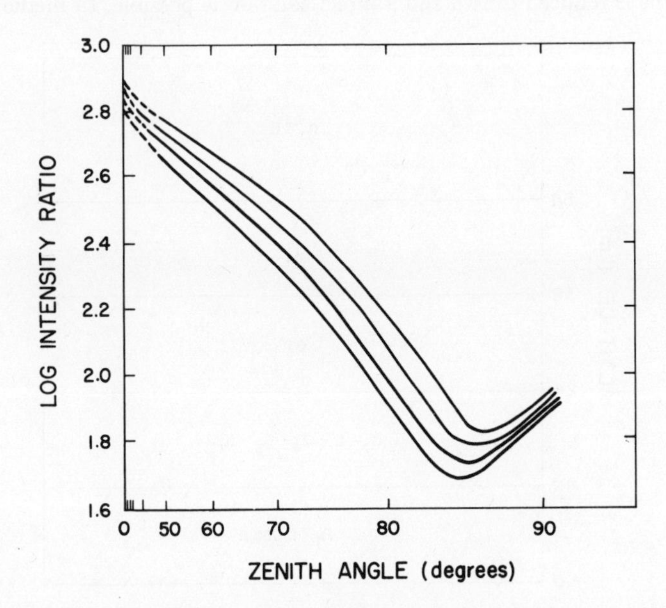

FIG. 5.4. Typical *Umkehr* curves for different total amounts at Delhi, India. These total amounts are (from top curve to bottom curve) 0.155, 0.175, 0.200, and 0.217 cm NTP in the old scale. The abscissa is linear for Z^4. The ordinate is $\log[I(\lambda_1)/I(\lambda_2)]$ plus a constant. (After Karandikar and Ramanathan, 1949.)

to assume ψ constant as solar radiation passes through the atmosphere (see Appendix D). In practice, the Chapman function (Appendix D) is used in the evaluation of the scattering integral, and the absorption integral is solved numerically.

Thus the ratio $I(\lambda_1)/I(\lambda_2)$ in (5.5) depends in a very complicated manner, for a given pair of wavelengths λ_1 and λ_2, on both the zenith angle and the vertical distribution of ozone. Figure 5.4 shows some measurements of this ratio by Karandikar and Ramanathan (1949). In this case $\lambda_1 = 3110$ A and $\lambda_2 = 3300$ A. The ratio decreases as the zenith angle increases until the latter reaches about 85° to 87°; then the ratio increases. The different curves, with different shapes, correspond to different vertical distributions of ozone. The typical minimum in the curves led to the name *Umkehr* (turning back, reversal).

To derive a vertical distribution of ozone from an *Umkehr* curve is a rather complicated matter and standardized approximate methods have been developed. The two methods most commonly employed are called method A and method B. In method A, the atmosphere is divided into five layers for the purpose of approximating numerically the integrals in Eqs. (5.5)–(5.7). It is desired to determine the (assumed constant) ozone density within each layer. A direct solution of (5.5) with even five unknowns is impractical. By further assumptions and a measurement of the total amount of ozone Ω (see Fig. 5.5) the number of unknowns is reduced to two and a direct solution is possible. In method B, the

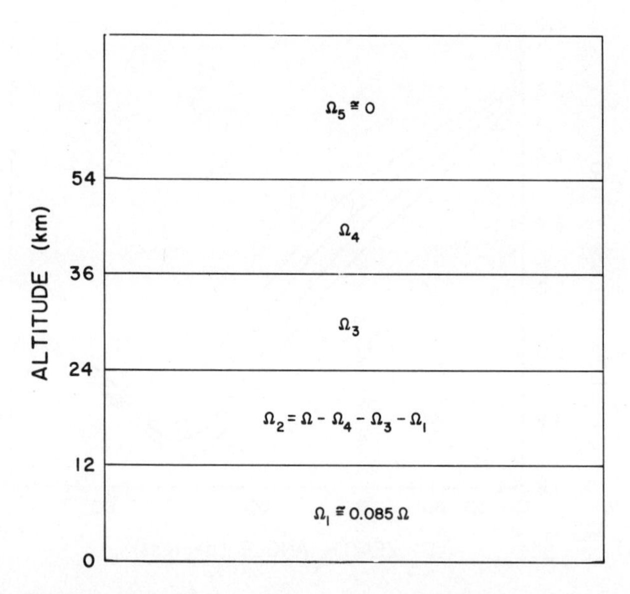

FIG. 5.5. Diagram illustrating a division of the atmosphere into five layers for reduction of *Umkehr* curves by method A. Ω_n refers to the mass of ozone (per unit area column) in the nth layer. Assumptions about Ω_1 and Ω_5 and a measurement of the total amount Ω leave two variables to be determined from the *Umkehr* curve.

atmosphere is divided into nine 6-km layers, plus the region above 54 km where the ozone density is assumed to be zero, and a trial-and-error method is used to determine a vertical distribution that will fit the observed *Umkehr* curve. In general, the final result is not a unique solution, because there are always several distributions that will fit a given *Umkehr* curve within the accuracy of the observations. These, however, differ from one another not grossly but only in detail. Detailed numerical procedures for applying these two methods have been given by Walton (1957) in the case of method A, and by Ramanathan and Dave (1957) in the case of method B. Dütsch (1959) and Mateer (1960) have described variations of method B that are suitable for use on electronic computers.

In closing this discussion, it is necessary to point out that the assumption of primary molecular scattering only is rather a stringent one. In the standard method of reduction, rough corrections are made for the effect of secondary molecular scattering (see Ramanathan and Dave, 1957). This question has been discussed further by Sekera and Dave (1961b). No corrections are made for particulate scattering and *Umkehr* observations should be made in clear air as free as possible from particulate matter.

A very large number of *Umkehr* observations have been made. For example, Fig. 5.6 shows five vertical distributions, for different seasons, computed by

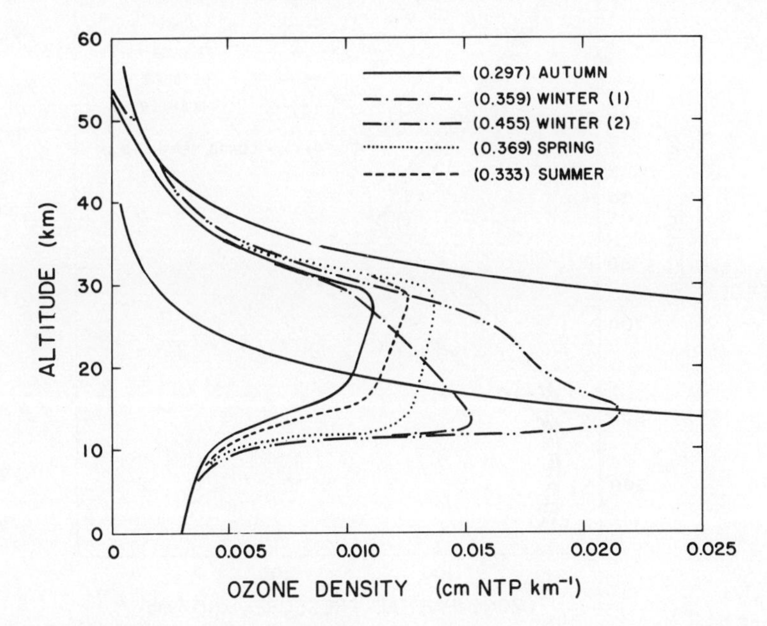

FIG. 5.6. Average vertical distributions of ozone at Moosonee and Edmonton, Canada, for different seasons of the year. The two smooth curves running from upper left to lower right are lines of constant ozone mixing ratio, the upper for 21×10^{-6}, the lower for 2.1×10^{-6}. (After Mateer and Godson, 1960.)

Mateer and Godson (1960) from observations at Moosonee and Edmonton, Canada. These illustrate the principal features of the vertical ozone distribution at high latitudes (although, as pointed out above, the *Umkehr* method is incapable of showing great detail and smooths out small-scale features). The maximum density of ozone occurs in the lower stratosphere, somewhere between about 15 and 30 km. Above 30 km, the ozone density decreases rapidly with height. Below the tropopause, the ozone density is small, but not negligibly so. Note particularly that variations of the total amount of ozone represent mainly variations in the lower stratosphere. Variations in the upper stratosphere and troposphere contribute much less to the variations of the total amount.

Figure 5.7 shows similar seasonal curves for a lower-latitude station, Arosa (47°N), after Dütsch (1962a). The vertical distribution and its seasonal variation is qualitatively the same as described above, but seasonal variations in the lower stratosphere are smaller.

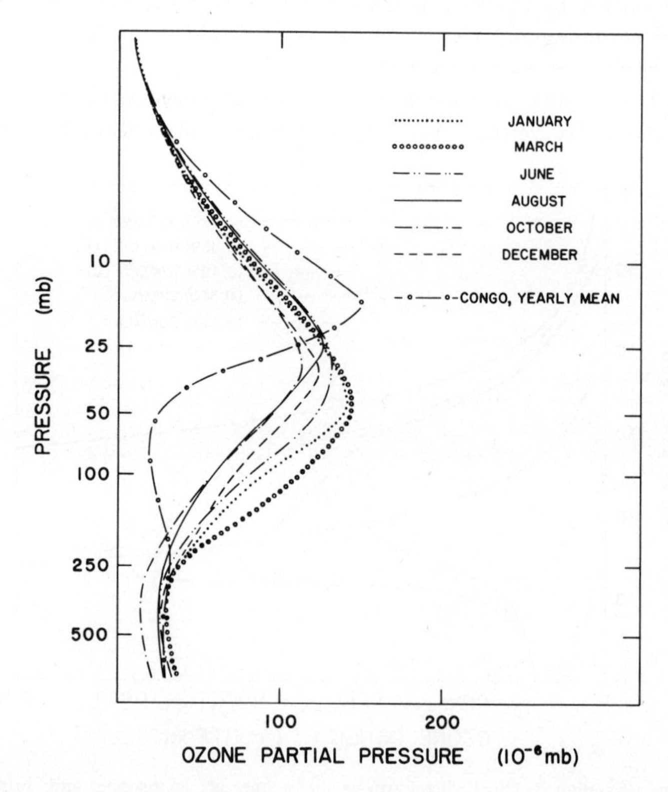

FIG. 5.7. Average vertical distributions of ozone at Arosa for different months of the year. One curve is shown for Leopoldville, Congo, where there is very little seasonal variation. (After Dütsch, 1962a.)

On the other hand, the vertical distribution near the equator seems to be qualitatively different. One such distribution is shown in Fig. 5.7, based on *Umkehr* measurements at Leopoldville in the Congo. The maximum density occurs at a higher altitude and there is little seasonal variation at any altitude.

5.2.2 The Infrared Method

Another method that has been used to deduce the vertical distribution of ozone from ground-based measurements makes use of observations in the infrared. Ozone has an absorption band at 9.6 μ (1043 cm^{-1}) in a region where other atmospheric gases are relatively transparent (see Section 7.1). Measurements of this band in absorption and in emission, combined with a simultaneous determination of the total amount of ozone, suffice to give an approximate (not detailed) picture of the vertical distribution. We shall briefly formulate the relationships between the ozone distribution and these three observations.

The ozone density ρ_3, is related simply to the total amount Ω by

$$\Omega = \int_0^\infty \rho_3 \, dz \tag{5.8}$$

With regard to the absorption measurement, the solar flux measured at the ground (at a frequency within the 9.6 μ band) is related to the solar flux outside the atmosphere by

$$F_\nu / F_{\nu\infty} = \tau_\nu \tag{5.9}$$

where τ_ν is the transmissivity. The transmissivity of the band $\tau(\Delta\nu)$ is given by

$$\tau(\Delta\nu) = \frac{\int_{\Delta\nu} F_\nu \, d\nu}{\int_{\Delta\nu} F_{\nu\infty} \, d\nu} = \frac{\int_{\Delta\nu} F_{\nu\infty} \tau_\nu \, d\nu}{\int_{\Delta\nu} F_{\nu\infty} \, d\nu} = \frac{\int_{\Delta\nu} \tau_\nu \, d\nu}{\Delta\nu} \tag{5.10}$$

The last step in (5.10) is justified because $F_{\nu\infty}$ varies relatively little with frequency over the spectral interval considered and a mean value may be taken outside both integrals.

The transmissivity τ_ν is the exponential of Beer's law, Eq. (4.36), but because the absorption coefficient k_ν varies so rapidly with frequency in the infrared the integral $\int_{\Delta\nu} \tau_\nu \, d\nu$ in (5.10) is not easily evaluated. This question is discussed at much greater length in Section 7.1. Suffice it to say here that this integral and the quantity $\tau(\Delta\nu) \, \Delta\nu$ which can be deduced from the measurement and to which it is equal are related to the vertical distribution of ozone through laboratory measurements in a manner discussed briefly below. In this discussion the quantity $[1 - \tau(\Delta\nu)] \, \Delta\nu$ is referred to as the *band area*, A_r, which is often expressed in units of wave number.

It has been found in the laboratory (Summerfield, 1941; Walshaw, 1957) that A_r is a complicated function of both the amount of ozone in the path and the total pressure of the mixture in the absorption tube (ozone and dry air, in the case of Walshaw's measurements). Since these measurements were made in an absorption tube where the total pressure was constant (for any one measurement) and measurements in the atmosphere are made over a path where pressure varies widely, there is some uncertainty involved in relating the two. In practice (Goody and Roach, 1956; Vigroux, 1959) the following procedure is employed:

(a) Determine by measurement the band area A_r for the atmospheric path.

(b) Using the known total amount of ozone in the path ($\Omega \sec Z$) and A_r, determine the pressure p_e that is consistent with Walshaw's laboratory data.

(c) Assume, according to the Curtis–Godson approximation (see Subsection 7.1.4) that

$$p_e = \frac{\int_0^\infty \rho_3 p \, dz}{\Omega} \tag{5.11}$$

Equation (5.11) furnishes a second equation, in addition to (5.8), for relating the vertical distribution of ozone to quantities derived from observation, in this case to p_e and Ω.

With regard to the emission measurement, let $I(\Delta\nu)$ be the specific intensity (integrated over a frequency interval $\Delta\nu$ covering the 9.6-μ band) received at the ground. Let s be distance measured along the column of air (not necessarily vertical) in a direction perpendicular to the energy-gathering area of the instrument. Let $dI(\Delta\nu)$ be an increment of $I(\Delta\nu)$ that is received from a volume of unit area and thickness ds in this column. Then

$$dI(\Delta\nu) = \int_{\Delta\nu} L_\nu \tau_\nu \, d\nu \tag{5.12}$$

where L_ν is the radiance of this volume and τ_ν the transmissivity of the air between the instrument and the emitting volume. The geometry leading to (5.12) is the same as that discussed in Appendix E.

This can be written, with the help of Kirchhoff's law [see Eq. (4.37)], as

$$dI(\Delta\nu) = \int_{\Delta\nu} k_\nu \rho_3 L_{\nu B} \, ds \, \tau_\nu \, d\nu \tag{5.13}$$

where k_ν is the absorption coefficient (of ozone), ρ_3 the ozone density, and $L_{\nu B}$ the black-body radiance at frequency ν. Here it is assumed that no gas other than ozone is radiatively active near 9.6 μ, an assumption that is discussed below. With the value of τ_ν given by (4.36), this can be written in terms of the band transmissivity, defined by (5.10), as

$$dI(\Delta\nu) = -\bar{L}_{\nu B} \, \Delta\nu \, d\tau(\Delta\nu) \tag{5.14}$$

where $\bar{L}_{\nu B}$ is a mean value of $L_{\nu B}$ in the frequency interval $\Delta\nu$ and it is understood that the differential $d\tau(\Delta\nu)$ is to be evaluated for a value of s corresponding to the distance between the instrument and the incremental volume being considered. The integrated specific intensity $I(\Delta\nu)$ is given by the contribution of all such volumes and therefore by

$$I(\Delta\nu) = -\Delta\nu \int_0^\infty \bar{L}_{\nu B} \frac{d\tau(\Delta\nu)}{ds}\, ds \tag{5.15}$$

It should be noted that $\bar{L}_{\nu B}$ is a function of temperature at the altitude s and $d\tau(\Delta\nu)/ds$ is a function of the integrated mass of ozone between the instrument and s and also of the effective mean pressure between the instrument and s [Eq. (5.11)] in the Curtis–Godson approximation. Equation (5.15) therefore relates the emission measurement to the vertical distribution of ozone in a rather complicated way.

One could, of course, make several emission measurements, in different parts of the band and at different zenith angles, and in principle obtain additional information. In practice this information is largely redundant; see Goody and Roach (1956) and a discussion by Walshaw (1960).

In one application of the infrared method to the computation of the vertical distribution of ozone, as recommended by the International Ozone Commission in 1956 and described by Goody and Roach (1958), the atmosphere is divided into four layers: (1) surface to 300 mb; (2) 300 to 50 mb; (3) 50 to 10 mb; (4) 10 mb to the top of the atmosphere. The parameter describing the concentration of ozone is taken to be ϵ, the mixing ratio r_3 divided by the acceleration of gravity g, in units of centimeters NTP (millibar)$^{-1}$. In the lowest layer ϵ is assumed to be given by the simple formula $\epsilon = ap^{-1}$, in layer 2 by $\epsilon = bp^{-2}$, in the third layer by $\epsilon = cp^{-0.2}$, and to have a constant known distribution in the top layer. The parameters a, b, and c are determined from the three conditions expressed by Eq. (5.8), (5.11), and (5.15). Thus, for example, Eq. (5.8) for these assumptions gives

$$\Omega = \int_{1000}^{300} ap^{-1}\, dp + \int_{300}^{50} bp^{-2}\, dp + \int_{50}^{10} cp^{-0.2}\, dp$$

$$+ \text{const} = 1.202a + 0.0167b + 20.70c + \text{const}$$

Equation (5.11) gives a similarly simple expression. However, the emission measurement cannot be expressed simply in terms of a, b, and c and trial-and-error procedures are required to determine the a, b, and c that fit this condition.

Note that no continuity is enforced on these various segments at the interfaces of the layers (300, 50, and 10 mb) by this technique. Vigroux (1959) has suggested a more complicated procedure, based still on only the three observations, but providing continuity and, further, allowing the altitude of the interface between layers 2 and 3 (which in his method always represents the maximum in the vertical distribution) to be a variable.

Goody and Roach (1958) have reported the results of a year's observations of this type at Ascot, England, but have analyzed only the deduced ozone content of layer 1 (troposphere), the other results being, according to them, of questionable significance. In the first place, the total amount of ozone below 300 mb averaged only about 18×10^{-3} cm NTP,* about 5 per cent of the total amount in the vertical column. The most surprising result of their analysis was a negative correlation, particularly in spring, between short-term deviations of tropospheric ozone content and of total amount of ozone. Other infrared results have been reported by Epstein *et al.* (1956) and by Vigroux and Debaix (1963).

In closing this subsection, it would be well to point out that the preceding account has omitted or glossed over some rather serious difficulties in the application of this method. The 9.6-μ band contains also a weak carbon dioxide band for which correction must be made. It is also a region of weak continuous absorption, due presumably to water vapor (Roach and Goody, 1958), and this must be considered. There are further sources of error in the radiosonde temperatures and the assumed conditions above 10 mb, as well as in the approximation of the pressure effect along a variable-pressure path. Walshaw (1960) has discussed these in some detail.

5.2.3 The Optical Method

The optical method of determining the vertical distribution of ozone is simple in principle. An instrument, carried aloft by a balloon or rocket, measures the solar flux in certain ultraviolet spectral intervals absorbed by ozone. As the instrument rises, the amount of ozone between it and the sun decreases and the measured flux increases. The rate of increase, with suitable corrections for scattering and with the known absorption coefficients of ozone, gives a measure of the vertical distribution of the ozone. Note that in this method the desired parameter, ozone density at some particular height, is necessarily determined from the rate of change of the observed parameter, total ozone above the instrument. Nevertheless, optical methods have given very valuable results and are capable of determining more detail than the ground-based methods discussed above.

---

* The infrared data used by Goody and Roach were based on an early reduction of Walshaw's laboratory data. This reduction made use of ultraviolet absorption coefficients larger than those of Vigroux (1953), which are now considered to be standard. Total amounts of ozone used by Goody and Roach were also expressed in the old scale. It is not obvious what effect this has on the results, but the number quoted above is 1.35 times the number quoted by Goody and Roach.

The earliest direct measurement of the vertical distribution of ozone to great heights was obtained by E. and V. H. Regener (1934) with a quartz spectrograph sent aloft on an unmanned balloon. This measurement, a milestone in early ozone studies, showed a maximum amount of ozone, about 16×10^{-3} cm NTP km^{-1}, at about 24 km. Other such measurements have been made (V. H. Regener, 1938, 1951, 1956; E. Regener, Paetzold, and Ehmert, 1954; Paetzold, 1955) and have showed, above all, irregularities in the vertical distribution and rather wide fluctuations in the altitude and magnitude of the ozone maximum, as illustrated in Fig. 5.8, after E. Regener *et al.* (1954).

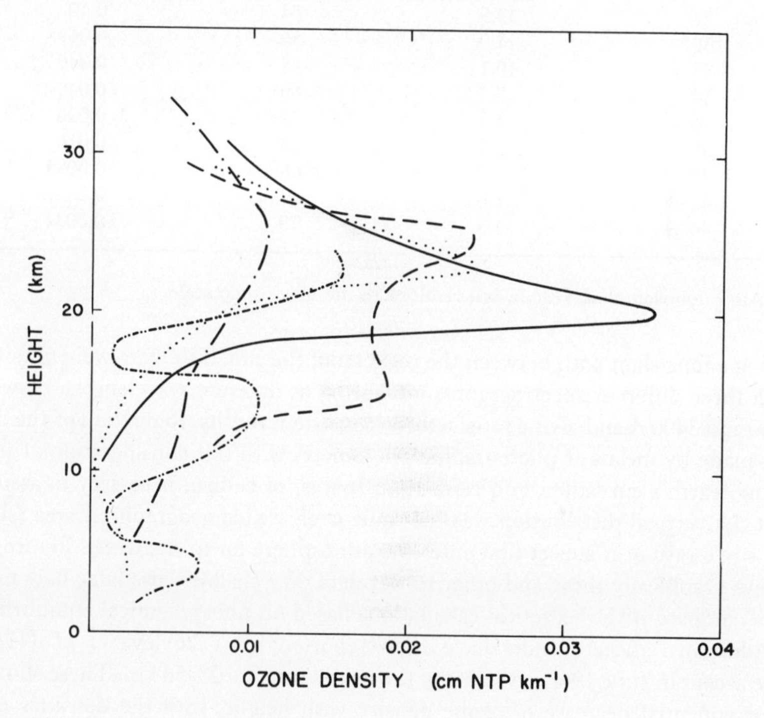

FIG. 5.8. Some vertical distributions of ozone measured by the optical method, illustrating the variability among individual ozone profiles. The ozone densities are in the old scale and must be multiplied by 1.35 to be consistent with other data in this book. (After E. Regener *et al.*, 1954.)

Very valuable results have been obtained by a Naval Research Laboratory group with spectrographs carried aloft on rockets (Johnson *et al.*, 1951, 1952, 1954). Data of this sort give the only direct measurements of the vertical distribution of ozone above the ceiling of balloons. In one instance, 14 June 1949, the vertical distibution was determined between 20 and 70 km and these results are reproduced in Table 5.1. On this flight the sun was at an altitude of 1°,

TABLE 5.1

OZONE DENSITY ABOVE NEW MEXICO, 14 JUNE 1949[a]

Altitude (km)	O_3 density (10^{-3}cm NTP km^{-1})	Altitude (km)	O_3 density (10^{-3}cm NTP km^{-1})
20	12.3	46	0.86
22	13.0	48	0.43
24	13.4	50	0.30
26	14.6	52	0.18
28	13.9	54	0.10
30	12.0	56	0.069
32	10.1	58	0.049
34	7.7	60	0.040
36	5.3	62	0.026
38	3.4	64	0.016
40	2.2	66	0.0088
42	1.6	68	0.0051
44	1.3	70	0.0034

[a] After Johnson *et al.* (1952), but reduced to the new ozone scale.

giving a long slant path between the rocket and the sun. The data were obtained with three different spectrographs, measuring at different wavelengths between 2500 and 3400 A and giving satisfactorily consistent results. Reduction of the data was made by means of photographic photometry with full consideration of scattering, earth's curvature, and refraction. It was, of course, necessary to assume that the vertical distribution was the same over a wide geographical area (since the sun's rays near sunset first enter the atmosphere far to the west). Figure 5.9 shows graphically these and other rocket data. We shall discuss later how these data compare with theoretical calculations based on photochemical equilibrium.

Additional rocket results have been reported by Yakovleva *et al.* (1963). One ascent in June 1960 gave data in the height interval 23-56 km. These showed an exponential decrease of ozone density with height, with the densities only about one third of the densities measured at corresponding levels in the 14 June Naval Research Laboratory flight.

No method of measuring the vertical distribution of ozone can be used on a routine basis unless the instrument is light, compact, rugged, and inexpensive, and unless it telemeters its information so that recovery is not essential. There have been several attempts to utilize the optical method in the development of such a balloon-borne ozonesonde; for example, the pioneering efforts of Coblentz and Stair (1939, 1941) and the more recent instrument of Vassy and Rasool, with which measurements were made at several locations during the IGY (Vassy and Rasool, 1960). The most widely used of the optical devices

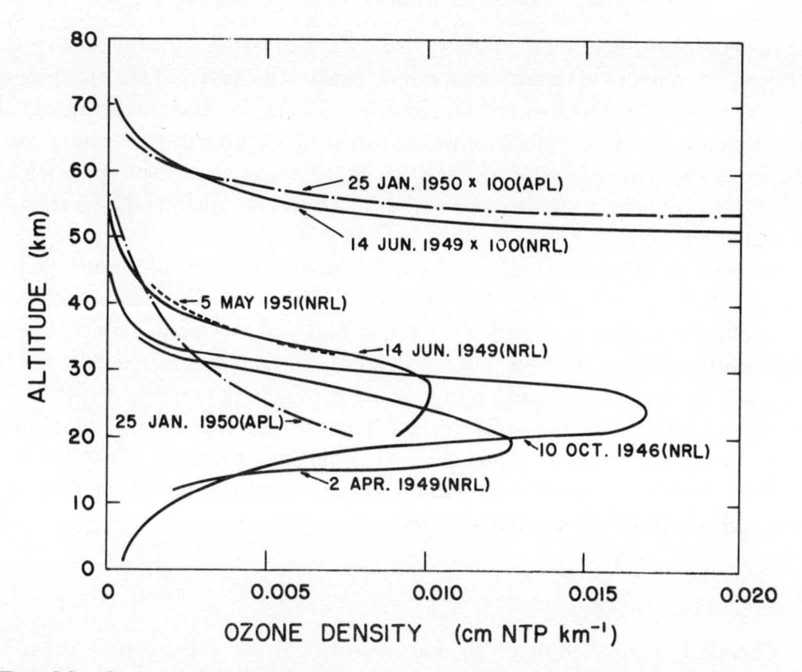

FIG. 5.9. Some vertical distributions of ozone measured by the optical method with rockets. The ozone densities are in the old scale and must by multiplied by 1.35 to be consistent with other data in this book. The data marked APL were obtained by the Applied Physics Laboratory, Johns Hopkins University. Note that the two upper curves are shown on an expanded scale. (After Johnson *et al.*, 1954.)

has been the Paetzold ozonesonde (Paetzold and Piscalar, 1961a), and we shall discuss this only.

The Paetzold ozonesonde employs a quartz ball lined with magnesium oxide as light collector. Two spectral regions of the ultraviolet, one near 3100 A and the other near 3750 A, are passed alternately to a photocell, whose output, after suitable amplification, is telemetered to the ground as a Morse code signal calibrated in terms of the energy reaching the photocell. Pressure is also telemetered. Each spectral region is isolated by means of a combination of two filters, each combination passing a band of about 100-A half-width. Thus, wide spectral regions are involved; in the shorter-wavelength region, ozone absorbs; in the longer one there is no ozone absorption. The variation with pressure of the ratio of the energy in the two intervals gives, after a rather complicated reduction procedure, a measure of the vertical distribution of ozone.

During this reduction (as in any application of the optical method) the curve showing the vertical variation of solar flux must be smoothed because of a certain amount of scatter in the experimental data. Since the ozone density at any altitude follows from the slope of the smoothed curve (in this case from the slope of the curve representing the ratio of two smoothed curves, one for each spectral region), the manner of smoothing is critical (unless the scatter of the experimental data is negligible, which is not usually true). Therefore, one should expect this technique to show up reliably only well-marked features in the vertical-distribution curve. The method works best at high levels, above the level of maximum ozone density, where atmospheric scattering effects are small and ρ_3 is changing rapidly with height.

Paetzold and Piscalar (1961a, b) have discussed the results of soundings made during the IGY at Tromso, Norway (70°N), Weissenau, Germany (48°N), and Leopoldville, the Congo (5°S). They found latitudinal differences in the vertical distribution similar to those discussed earlier in connection with the *Umkehr* method. At Leopoldville, maximum ozone density was found at about 27 km, with very little ozone below 22 km. At Tromso they found much more ozone in the lower stratosphere, especially during the spring months. They point out that the vertical distribution in middle latitudes is sometimes of the polar type and sometimes of the equatorial type.

5.2.4 The Chemical Method

Chemical methods of detecting and measuring ozone usually make use of the reaction of ozone with potassium iodide in solution:

$$O_3 + 2\,KI + H_2O \rightarrow 2\,KOH + O_2 + I_2 \tag{5.16}$$

The iodine is usually detected amperometrically. If a small potential difference is maintained between two platinum electrodes immersed in the iodide solution, the electrodes become polarized and no current flows. When, however, iodine is formed by the ozone reaction, the iodine depolarizes the cathode through $I_2 + 2e \rightarrow 2I^-$ and a current flows. At the same time iodine is reformed at the anode through $2I^- - 2e \rightarrow I_2$.

There are various ways of obtaining a quantitative measure of the amount of ozone. In one (for example, Bowen and Regener, 1951) a known quantity of sodium thiosulfate is added to the solution. Iodine formed by the chemical reaction immediately reacts with the thiosulfate and only when the thiosulfate is used up does the iodine allow a current to flow. The length of time for this to happen, together with the known initial amount of sodium thiosulfate and the metered rate at which air containing ozone is pumped into the reaction chamber, furnishes a measure of the ozone concentration in the air. This method has been used extensively and successfully in measurements of ozone concentration near ground level (V. H. Regener, 1956) but has not been used much for upper-atmospheric soundings.

Brewer and his collaborators (see especially Brewer and Milford, 1960) have developed two types of ozonesondes that make use of the chemical method of detection. In one, the "transmogrifier," the iodide solution is allowed to drip on and flow down a glass rod with a cathode at the top and an anode at the bottom. The iodine formed at the anode drips off the rod so that, in equilibrium, each molecule of ozone entering the reaction chamber produces one molecule of iodine through (5.16) and allows two electrons to flow in the electric circuit. In the other instrument, the "bubbler," the iodine formed at the anode is prevented from recirculating by the use of an anode of silver or mercury. Silver iodide or mercurous iodide is then formed at the anode; because both of these are sufficiently insoluble, iodine is effectively removed from the solution. Air is sucked or blown through the reaction chamber at a metered rate and each molecule of ozone in the air is responsible for the flow of two electrons through the cell.

Brewer's electrochemical instruments have been widely used in aircraft. In England (Kay, 1954) 15 ascents showed ozone densities of 1 to 2 $\times$ 10^{-3} cm NTP km^{-1} in the troposphere. In Norway, Brewer (1957) found larger amounts of 3 to 4 $\times$ 10^{-3} cm NTP km^{-1} through the troposphere, increasing sharply at the tropopause. The instruments have also been adapted for use on balloons in conjunction with the British radiosonde. Brewer and Milford (1960) have given the results of 32 ascents at Liverpool, England, made during the IGY.

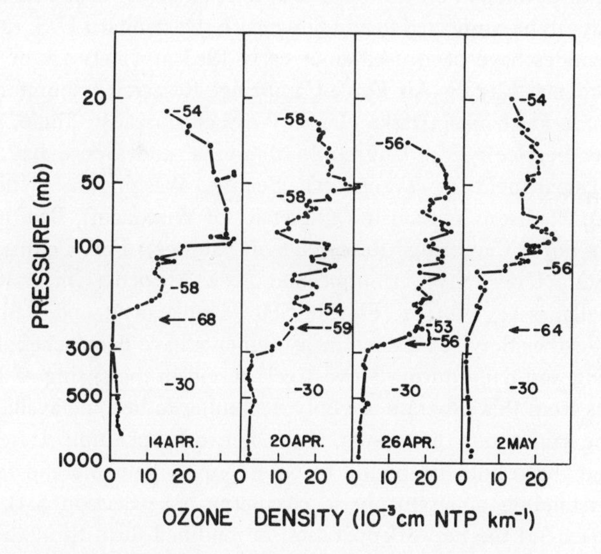

FIG. 5.10. Vertical distributions of ozone measured by Brewer's chemical instrument at Liverpool on the indicated dates in 1958. The numbers give temperatures (in degrees C) and the arrows indicate tropopauses. Note that ozone density starts increasing sometimes at the tropopause, sometimes below it, and sometimes above it. Note the fine detail in the sounding, some of which was reproduced on descent (data not shown). (After Brewer and Milford, 1960.)

Griggs (1963) has reported on a few ascents made at Nairobi in East Africa (2°S).

Chemical measurements show small-scale detail in the ozone profiles that would never be detected by the *Umkehr*, infrared, or optical methods. In cases where measurements made during ascent are compared with measurements made during subsequent descent (by parachute), there is generally good agreement, even with regard to fairly small-scale features. Figure 5.10 shows a few ascents made at Liverpool. The absolute accuracy of the chemical method is estimated (Griggs, 1963) to be 10 to 20 per cent.

5.2.5 The Chemiluminescent Method

V. H. Regener (1960) described a method of measuring the vertical distribution of ozone which is now in extensive use. Air is drawn past a disk coated with a luminescent material (Luminol) suitably bound to the disk. Regener found that the reaction releases light proportional to the amount of ozone in the air sample, that the reduction of the ozone is complete and rapid, and that the Luminol is itself only very slowly affected so that the calibration of light released against ozone amount is a stable one.

Regener has developed on the basis of this reaction an ozone-measuring instrument that can be connected to and flown with the standard U.S. radiosondes. Such ozonesondes have been used since early 1963 at a network of stations in a program organized by the Air Force Cambridge Research Laboratories. These stations include Fairbanks, Alaska (U. S. Weather Bureau); Thule, Greenland (Air Weather Service); Fort Churchill, Manitoba, and Goose Bay, Labrador (Canadian Department of Transport); Seattle, Washington (University of Washington); Madison, Wisconsin (University of Wisconsin); Bedford, Massachusetts (Air Force Cambridge Research Laboratories); Fort Collins, Colorado (Colorado State University); Albuquerque, New Mexico (University of New Mexico); Tallahassee, Florida (Florida State University); and Balboa, Canal Zone (Air Weather Service). Ozone measurements have been scheduled weekly at each station and daily during a two-week period in the spring of 1963. Published results from this program are only beginning to become available, so not much can be said about the results of the network operation. It seems clear, however, that the results are bound to be extremely valuable and to lead to a better understanding of stratospheric circulation (see Section 5.4). Plans are underway to extend the network operation in modified form through the International Quiet Sun Years, 1964 and 1965.

With regard to individual soundings with the chemiluminescent ozonesonde, fine detail comparable with that obtained with the Brewer chemical ozonesonde is revealed and often verified on descent. The instrument is light, compact, and rugged and its relative accuracy on any one ascent appears to be very good.

The absolute accuracy, which depends on a preflight calibration procedure, leaves something to be desired at the present time, and subsequent calibration with a surface measurement of total ozone is advisable.

5.2.6 Intercomparison Tests

Obviously, it is of very great interest and importance to determine whether several different methods, operating on different principles, are capable of giving consistent results. For this purpose several intercomparison tests have been conducted, one of which was at Arosa, Switzerland, from mid-July to mid-August 1958 (Brewer *et al.*, 1960). Vertical distributions were obtained simultaneously on as many days as possible by the Götz *Umkehr* method (Dütsch), the infrared method (Vigroux), the chemical method (Brewer), and the optical method (Paetzold).

Measurements were obtained by all four methods on four days, and a sample of the results is shown in Fig. 5.11. One is struck by the fact that, while all

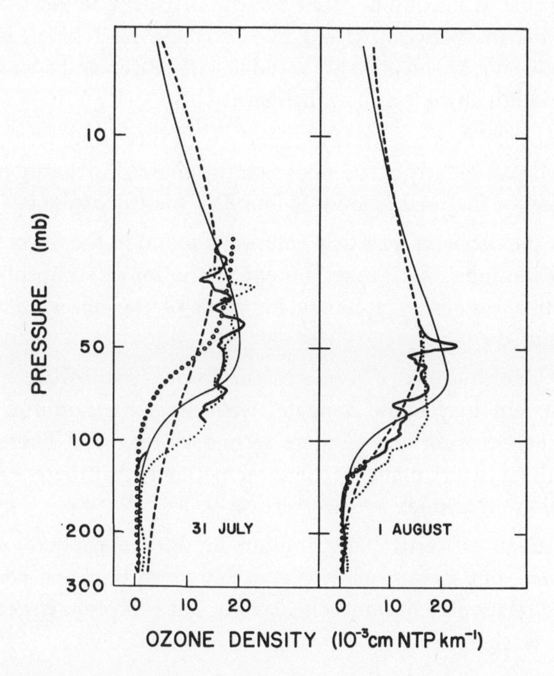

Fig. 5.11. Vertical distributions of ozone at Arosa, Switzerland, on 31 July and 1 August 1958; smooth, full-line curves by the infrared method; rapidly varying full-line curves by the chemical method (ascent of instrument); small-dot curves by the chemical method (descent of instrument); dashed-line curves by the *Umkehr* method; circle curves by the optical method. (After Brewer *et al.*, 1960.)

show more or less the same amount of ozone at more or less the same altitudes, there is not very satisfactory agreement in detail. By and large, the *Umkehr* and optical methods tended to show higher altitudes for maximum ozone density than did the chemical and infrared methods (see mean soundings in original reference).

Although good agreement on such tests would be very reassuring, lack of agreement does not give much positive information. No one of the methods can be judged at this moment to be clearly the most accurate and therefore can be used as a standard for checking the others. (As a matter of fact, probably no one of them is consistently the most accurate at all altitudes.)

5.2.7 Summary of Results

Inasmuch as samples of results have been interspersed throughout the preceding discussion of techniques, it is appropriate to summarize here the principal results bearing on the vertical distribution of ozone:

(1) The altitude of maximum ozone density is usually found between 20 and 30 km, with a density between 10 and 25×10^{-3} cm NTP km^{-1}. In middle and high latitudes this maximum is highly variable with season and over short periods, with respect to both altitude and ozone density.

(2) Ozone density decreases rather rapidly with increasing altitude above the maximum level, particularly in the upper stratosphere. Furthermore, a relatively small proportion of the total amount is found in the troposphere.

(3) Most of the ozone in a vertical column is found in the lower stratosphere. Except at low latitudes, the ozone content of the lower stratosphere is widely variable and this variability accounts for most of the observed variability of the total amount.

(4) Vertical distributions of ozone in the lower stratosphere are subject to wide variation and may show a single, well-defined maximum, a broad flat maximum, or on occasion one or more secondary maxima. There is reason to believe, from the measurements by chemical methods, that there is also considerable stratification on a much smaller vertical scale.

(5) Observations of vertical distribution in the troposphere, as elsewhere, are rather scarce, but seem to show that at any given time and place the ozone mixing ratio is relatively constant with height. An exception is near the surface at night when large-scale vertical mixing is inhibited.

5.3 Photochemical Processes Affecting Ozone

A great variety of photochemical processes take place in the upper atmosphere, of which only those involving the various forms of oxygen are important in determining the amount of ozone (except perhaps in the mesosphere; see Subsection 5.5.2). Accordingly, we discuss in this section the photochemistry of oxygen, with emphasis on the stratosphere. In Section 5.5, the discussion of photochemical processes is extended to higher levels.

It is instructive to consider the implications of a *photochemical equilibrium* for ozone. In photochemical equilibrium, the number of ozone molecules formed exactly equals the number destroyed in unit volume and unit time. The "equilibrium amount" of ozone is just that amount required to maintain photochemical equilibrium under given environmental conditions. The assumption of photochemical equilibrium turns out to represent rather well the situation in the upper stratosphere, but to fail completely in the lower stratosphere, where transport of ozone by various types of atmospheric motion is the important factor affecting the ozone density and its spatial and temporal variations.

The photochemistry of the upper atmosphere has been discussed extensively. Especially interesting with regard to the ozone problem are the classical paper of Chapman (1930), sections of Bates and Nicolet's (1950) paper on the photochemistry of water vapor, and several studies of ozone in particular (Wulf and Deming, 1936a, b, 1937; Dütsch, 1946, 1956; Craig, 1950; Johnson *et al.*, 1952; and others).

Ozone is formed by the three-body collision

$$O + O_2 + M \rightarrow O_3 + M \tag{5.17}$$

where M is any third atom or molecule. Of course, this reaction can take place only after the formation of atomic oxygen, which occurs when the oxygen molecule is dissociated by a quantum of solar energy,

$$O_2 + h\nu \quad (\lambda < 2423\ \text{A}) \rightarrow O + O \tag{5.18}$$

Ozone, on the other hand, is destroyed both by collision and by photodissociation,

$$O_3 + O \rightarrow 2\,O_2 \tag{5.19}$$

$$O_3 + h\nu \quad (\lambda < 11{,}000\ \text{A}) \rightarrow O + O_2 \tag{5.20}$$

Note that reaction (5.20) releases an oxygen atom which may be available to take part in reaction (5.17).

Direct association of the oxygen atoms released by reaction (5.18) or (5.20) may take place by three-body collision or radiative association

$$O + O + M \rightarrow O_2 + M \tag{5.21}$$

$$O + O \rightarrow O_2 + h\nu \tag{5.22}$$

The first of these may be neglected below 50 to 60 km, the second below 100 to 110 km.

Let us fix our attention on a certain volume of air, which may in general be situated at any altitude and location, at some particular time when the sun is at zenith angle Z. Let n_1, n_2, n_3, n_m be the number densities, respectively, of O atoms and of O_2, O_3, and air molecules. The number of ozone molecules formed by reaction (5.17) per unit volume and time is proportional to the product $n_1 n_2 n_m$; let κ_{12} be the proportionality factor, which is called the *rate coefficient*. (Rate coefficients may in general be temperature dependent, although the dependence on temperature of this particular coefficient is not marked). On the other hand, the number of oxygen atoms (or oxygen molecules) lost as a result of the same reaction (5.17), per unit volume and time, is also given by $(\kappa_{12} n_1 n_2 n_m)$. Let κ_{13} and κ_{11} be the rate coefficients for (5.19) and (5.21), respectively. Let q_2 and q_3 be the number of dissociating quanta absorbed by O_2 and O_3, respectively, per unit time and unit volume. With this notation, the rates of change of n_1, n_2, and n_3 in the unit volume, due to the photochemical processes (5.17)–(5.21), are

$$\frac{\partial n_1}{\partial t} = -\kappa_{12} n_1 n_2 n_m + 2q_2 - \kappa_{13} n_1 n_3 + q_3 - 2\kappa_{11} n_1 n_1 n_m \tag{5.23}$$

$$\frac{\partial n_2}{\partial t} = -\kappa_{12} n_1 n_2 n_m - q_2 + 2\kappa_{13} n_1 n_3 + q_3 + \kappa_{11} n_1 n_1 n_m \tag{5.24}$$

$$\frac{\partial n_3}{\partial t} = \kappa_{12} n_1 n_2 n_m - \kappa_{13} n_1 n_3 - q_3 \tag{5.25}$$

(5.17) (5.18) (5.19) (5.20) (5.21)

Each parenthesized number beneath this set of equations indicates the reaction responsible for the terms above it.

It should be noted that q_2 and q_3 are proportional, respectively, to n_2 and n_3. Specifically,

$$q_2 = n_2 \int_0^\infty k_{\nu 2} F_{\nu\infty} \tau_\nu \, d\nu \tag{5.26}$$

$$q_3 = n_3 \int_0^\infty k_{\nu 3} F_{\nu\infty} \tau_\nu \, d\nu \tag{5.27}$$

where $F_{\nu\infty}$ is the solar flux at the outer edge of the atmosphere (in quanta [centimeter]$^{-2}$ [second]$^{-1}$), $k_{\nu 2}$ and $k_{\nu 3}$ are the absorption coefficients per molecule of O_2 and O_3, respectively, and τ_ν is the transmissivity of the atmosphere above the volume under consideration in the direction toward the sun. These transmissivities therefore depend on the overlying mass of atmosphere that absorbs at the frequency in question. The frequency integrations must be carried out over those regions of the spectrum where absorption results in dissociation. For ozone, this includes the Chappuis continuum in the visible and the Hartley continuum in the ultraviolet. For oxygen, in the stratosphere, it includes the Herzberg continuum. In the region of the Schumann–Runge continuum, with its much stronger absorption, solar radiation does not penetrate to the stratosphere, or even to the mesosphere. Note that, between 2000 and 2500 A, both ozone and oxygen absorb significantly and the transmissivities in (5.26) and (5.27) must take into account absorption by both gases.

For photochemical equilibrium, $\partial n_1/\partial t = \partial n_2/\partial t = \partial n_3/\partial t = 0$. If this condition is assumed and applied to (5.23)–(5.25), there result three homogeneous equations, which are not, however, independent. Under the conditions for which these equations were set up, it is evident that the total number of oxygen atoms (bound or free) must remain constant, that is, $n_1 + 2n_2 + 3n_3 =$ constant. However, in the stratosphere (and mesosphere), the very great proportion of oxygen is in diatomic form and photochemical processes have negligible effect on the value of n_2. Therefore (5.24) may be ignored and the values of n_2 and n_m that enter into (5.23) and (5.25) may be considered to be constant, depending on the assumed density of the volume under consideration. With this assumption and for photochemical equilibrium, equilibrium values of n_1 and n_3 can be obtained from (5.23) and (5.25).

If, for the time being, we neglect reaction (5.21), and therefore the last term in (5.23), as is appropriate for the stratosphere, we get

$$n_{3e} = \tilde{n}_3 \frac{q_2}{q_2 + q_3} \tag{5.28}$$

where n_{3e} is the equilibrium number density of ozone and $\tilde{n}_3 \equiv (\kappa_{12}/\kappa_{13}) n_2 n_m$. The latter therefore depends on the density in the volume and, through the rate coefficients, on temperature (see below). Note that q_3 is proportional to n_{3e} [Eq. (5.27)] and that (5.28) is therefore a quadratic equation. This can, however, be solved without ambiguity because only one of the two solutions gives a positive value for n_{3e}. The expression for n_{1e} is

$$n_{1e} = \frac{q_2 + q_3}{\kappa_{12} n_2 n_m} \tag{5.29}$$

In order to make numerical computations with (5.28) and (5.29) it is necessary to specify environmental conditions and related physical parameters. These include

(a) atmospheric structure parameters giving density and temperature as a function of altitude (discussed in Chapters 2 and 3);

(b) solar zenith angle, which enters into the transmissivities in (5.26) and (5.27);

(c) solar flux at the outer edge of the atmosphere (discussed in Subsection 4.5.2);

(d) oxygen and ozone absorption coefficients (discussed in Subsections 4.6.3 and 4.6.4); and

(e) the rate coefficients and their dependence on temperature.

The rate coefficients are not well known. Many authors have used the provisional estimates of Bates and Nicolet (1950), according to which

$$\kappa_{11} = 8 \times 10^{-33} (T/256)^{\frac{1}{2}} \quad \text{cm}^6 \text{ molecule}^{-2} \text{ sec}^{-1}$$

$$\kappa_{12} = 8 \times 10^{-35} (T/256)^{\frac{1}{2}} \quad \text{cm}^6 \text{ molecule}^{-2} \text{ sec}^{-1}$$

$$\kappa_{13} = 2.4 \times 10^{-10} (T/256)^{\frac{1}{2}} \exp(-3070/T) \quad \text{cm}^3 \text{ molecule}^{-1} \text{ sec}^{-1}$$

According to measurements of Reeves, Mannella, and Harteck (1960), the value of κ_{11} is

$$\kappa_{11} = 2.7 \times 10^{-33} \quad \text{cm}^6 \text{ molecule}^{-2} \text{ sec}^{-1}$$

and according to the measurements of Kaufman and Kelso (1961), it is

$$\kappa_{11} = 3 \times 10^{-33} \quad \text{cm}^6 \text{ molecule}^{-2} \text{ sec}^{-1}$$

With regard to the value of κ_{12} Benson and Axworthy (1957) found

$$\kappa_{12} = 1.6 \times 10^{-34} \exp(302/T) \quad \text{cm}^6 \text{ molecule}^{-2} \text{ sec}^{-1}$$

which corresponds to a value of 1.9×10^{-34} at 256°K. The measurements of Kaufman and Kelso (1961) give

$$\kappa_{12} = 1.2 \times 10^{-34} \quad \text{cm}^6 \text{ molecule}^{-2} \text{ sec}^{-1}$$

The rate coefficient κ_{13} is strongly temperature dependent. Benson and Axworthy found

$$\kappa_{13} = 5 \times 10^{-11} \exp(-3020/T) \quad \text{cm}^3 \text{ molecule}^{-1} \text{ sec}^{-1}$$

The ratio $(\kappa_{12}/\kappa_{13})$, which enters into the expression for equilibrium ozone density in the stratosphere [Eq. (5.28)], was measured by Eucken and Patat (1936) and was used in most of the early ozone studies (Dütsch, 1946; Craig, 1950). Bates and Nicolet (1950) chose their values of κ_{12} and κ_{13} so that the ratio would agree with these measurements, which could be represented [after a correction to account for the efficiency of nitrogen molecules in the three-body reaction (5.17)] by

$$\kappa_{12}/\kappa_{13} = 3.3 \times 10^{-25} \exp(-3070/T) \quad \text{cm}^3 \text{ molecule}^{-1}$$

The corresponding expression from Benson and Axworthy's measurements is

$$\kappa_{12}/\kappa_{13} = 3.2 \times 10^{-24} \exp(-3322/T) \quad \text{cm}^3 \text{ molecule}^{-1}$$

There have been many computations of the vertical distribution of ozone under photochemical equilibrium (see, for example, the references cited at the beginning of this section). These have encompassed many of the possible assumptions about atmospheric and related physical factors, one of the most vexing of which is the question of the pressure dependence of O_2 absorption (see the end of Subsection 4.6.3). Figure 5.12 gives as examples two results of Craig (1950) both for vertical incidence of the sun, one for no pressure dependence (I), the other for a rather arbitrary assumption about pressure dependence near 2000 A (II). These curves probably give too low values near the stratopause, because the assumed temperatures at that level were too high (affecting the ratio κ_{12}/κ_{13}), and too high values at and above 60 km because the computations were made without regard to reaction (5.21).

Without going into detail about the many calculations that have been made, we shall point out the most important limiting factors on the results. According to Eq. (5.28), the equilibrium amount of ozone should decrease very rapidly with altitude because of the factor $\tilde{n}_3$ which depends critically on air density. This accounts for the behavior above the maxima of the curves in Fig. 5.12. On the other hand, the factor q_2 in Eq. (5.28) is vanishingly small at an altitude low enough so that all the dissociating quanta have already been absorbed at higher levels. This accounts for the behavior below the maxima of the curves in Fig. 5.12.

Typically, calculations of the photochemical-equilibrium amounts of ozone compare with observations as follows:

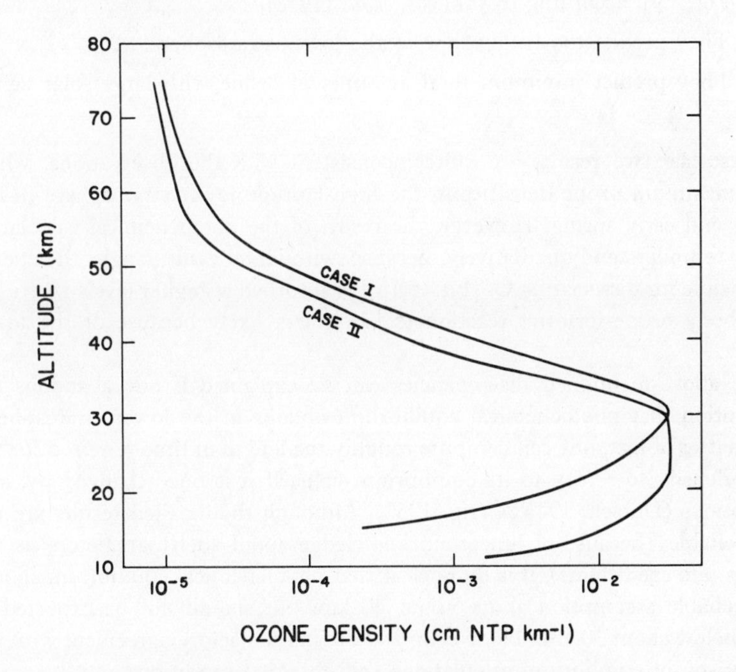

FIG. 5.12. The vertical distribution of ozone computed for photochemical equilibrium with an overhead sun. Cases I and II correspond to two different assumptions about the pressure dependence of O_2 absorption coefficients. (After Craig, 1950.)

(a) The calculations predict approximately the correct amount of ozone at levels above 30 to 40 km. This is illustrated in Table 5.2, which compares

TABLE 5.2

CALCULATED AND MEASURED OZONE DENSITIES[a]

Altitude (km)	Observed n_3 (cm^{-3})	Calculated n_3 (cm^{-3})
70	6×10^8	1.2×10^9
60	7.3×10^9	7.0×10^9
50	5.5×10^{10}	4.3×10^{10}
40	4.6×10^{11}	4.8×10^{11}
30	2.5×10^{12}	6.7×10^{12}

[a] Observations were reduced and calculations were made with the Ny and Choong absorption coefficients. After Johnson *et al.* (1952).

calculated ozone densities with those measured during the rocket flight of 14 June 1949 (Fig. 5.9), according to Johnson *et al.* (1952).

(b) They predict too little ozone in the lower stratosphere.

(c) They predict minimum total amounts of ozone with large solar zenith angle.

These last two results are quite inconsistent with the observations, which show maximum ozone densities in the high-latitude lower stratosphere in late winter and early spring. However, the result of the photochemical calculation is easy to understand qualitatively, because, with larger zenith angle, the energy responsible for dissociating O_2 [Eq. (5.18)] is absorbed at higher levels where the three-body ozone-forming reaction (5.17) is less likely because of the lower density.

The above-mentioned discrepancies can be explained if one abandons the assumption that photochemical equilibrium obtains in the lower stratosphere. As a matter of fact, one can compute roughly the length of time required for the ozone density to return to its equilibrium value if it is once changed by, say, air motions (Dütsch, 1946; Craig, 1950). Although the detailed results are not very reliable (because of uncertain knowledge about such parameters as the various rate coefficients), it is quite clear that photochemical equilibrium should be a reliable assumption above about 40 km, but should not be expected to occur below about 30 km. These estimates are in satisfactory agreement with the comparison of equilibrium calculations and observed conditions.

Where, then, does the ozone observed in the lower stratosphere and troposphere come from? It seems quite clear that the great bulk of it must originate in the upper stratosphere or upper part of the lower stratosphere and be transferred by one atmospheric process or another to the lower levels. At these lower levels, it is protected (on balance, see below) from photochemical destruction and becomes a conservative property of the air. The ozone that is transported by various atmospheric mixing processes into the troposphere and toward the earth's surface is eventually destroyed by chemical reactions with particulate constituents of the troposphere or at the ground. The details of the atmospheric processes that transfer ozone from upper stratosphere to lower stratosphere to troposphere are not clearly understood and we shall discuss them in Section 5.4.

It is often said that ozone in the lower stratosphere is "protected" from photochemical destruction by the overlying mass of atmosphere. This is only a manner of speaking, because ozone is actually dissociated by radiation in the near ultraviolet and visible at all levels of the atmosphere down to the earth's surface. What actually happens, however, is that the oxygen atom liberated in the photodissociation process (5.20) very quickly associates with O_2 to form a new ozone molecule according to the reaction (5.17). Of course, it is possible for reaction (5.19) to occur, but this is enormously less likely in the lower stratosphere, as is easy to demonstrate.

Given a certain number of oxygen atoms, n_1, per unit volume, as a result of reaction (5.20), then the number of ozone molecules formed through (5.17) per unit volume and time is $\kappa_{12} n_1 n_2 n_m$ and the number destroyed through (5.19) is $\kappa_{13} n_1 n_3$. The ratio of these is $(\kappa_{12}/\kappa_{13})(n_2 n_m/n_3)$. Around 20 km, $\kappa_{12}/\kappa_{13} \cong 10^{-18}$ cm^3 molecule^{-1}, $n_2 n_m \cong 10^{36}$ $(\text{molecules})^2$ cm^{-6}, and n_3 (for maximum observed values, not calculations) $\cong 10^{13}$ molecules cm^{-3}. Therefore, the ratio is about 10^5; only about 1 in 10^5 dissociations of ozone actually results in a *net* destruction of ozone. A more detailed analysis shows that this can affect the ozone content of an air parcel appreciably only after a very long period of time, measured in years.

Thus, there is a sort of equilibrium between ozone destruction and formation in the lower stratosphere. This, however, is not a "photochemical equilibrium" in the usual meaning of the term because the ozone involved could not be formed *in situ* but only *maintained* after transport from higher levels.

Another question of interest is the possibility of diurnal variation of ozone density. During the night when sunlight is absent, Eqs. (5.23) and (5.25) reduce to

$$\frac{\partial n_1}{\partial t} = -\kappa_{13}(\tilde{n}_3 + n_3)n_1 \tag{5.30}$$

$$\frac{\partial n_3}{\partial t} = \kappa_{13}(\tilde{n}_3 - n_3)n_1 \tag{5.31}$$

Up to the stratospause [above which, in any case, Eq. (5.23) is incomplete because it neglects reaction (5.21)] $\tilde{n}_3 \gg n_3$, which is another way of saying that reaction

(5.17) is more likely to occur than (5.19). If n_{30} and n_{10} are the numbers of O_3 molecules and O atoms at sunset, then integration of Eqs. (5.30) and (5.31), with n_3 neglected, gives

$$n_1 = n_{10} \exp(-\kappa_{13}\tilde{n}_3 t) \tag{5.32}$$

$$n_3 = n_{30} + n_{10} - n_1 \tag{5.33}$$

where t is the time measured from sunset. Now the value of κ_{13} is not known very accurately, but for any reasonable estimate it is clear that n_1 tends toward zero very rapidly after sunset. For example, at 40 km, $\tilde{n}_3 \simeq 5 \times 10^{13}$ molecules cm^{-3} and if κ_{13} is taken to be 10^{-15} cm^3 $(molecule)^{-1}$, then n_1 is reduced to $n_{10}/100$ in less than two minutes. At the same time, a new molecule of ozone is formed for each oxygen atom lost. Thus, almost immediately after sunset, the atomic oxygen disappears and ozone is formed. However, up to the stratopause the concentration of oxygen atoms is so small that the effect of this process on the number of ozone molecules per unit volume is quite small. Certainly the integrated effect throughout the entire column, which would have to be detected in terms of measurements of "total amount" of ozone, is very small. No significant diurnal variation of total amount of ozone should be expected and none is observed.

5.4 Meteorological Processes Affecting Ozone

Although it is clear that the observed vertical distribution of ozone and the seasonal and latitudinal variations of total ozone are intimately connected with the circulation of the stratosphere, a detailed understanding of the relationships is not yet achieved. This is because meteorological measurements up through the stratosphere have been available only in recent years and over limited geographical areas, and also because regular measurements of the vertical distribution of ozone on a synoptic basis are still not available. Therefore, many studies of the problem have been limited to comparisons of total amounts of ozone with meteorological information at relatively low levels in the stratosphere.

For purposes of discussion, the problem of ozone variations as they are related to meteorological processes can be divided into two parts: the question of short-period (day-to-day or week-to-week) fluctuations at any given location in middle and high latitudes; and the question of long-period (seasonal) variations. Although it is convenient to discuss the problem in this way, the two may turn out to be intimately related when a full understanding is achieved.

5.4.1 Short-Period Ozone Variations

Superimposed on the seasonal variations of ozone are short-period fluctuations of ozone amount at any given location. These are mentioned in Section 5.1

where some specific examples are given in connection with the observations at Arosa of the total amount of ozone. These variations are too large and rapid to be explained by photochemical processes and must reflect mechanisms that redistribute ozone from one location to another, apart from any net formation or destruction. Observations of the total amount of ozone, of course, reveal only the integrated effect of such mechanisms, which might conceivably include increases at some altitudes and decreases at others in the same air column. However, to a first approximation, changes of total amount appear to represent changes in the lower stratosphere and it is natural, in view of the lack of detailed vertical distributions, to look for correlations with meteorological conditions there.

Until recently, most studies of the relationships between ozone and meteorological conditions have involved either tropospheric or low-stratospheric (up to 100 mb) meteorological parameters correlated on a day-to-day basis with the total amount of ozone. These studies show a moderately consistent and moderately significant relationship between large total amounts of ozone and

(a) cold troposphere,

(b) low tropopause,

(c) warm lower stratosphere,

(d) a trough in the upper troposphere and low stratosphere.

To illustrate these statements, we show some results of Normand (1951)

TABLE 5.3

Correlation Coefficients between Total Amount of Ozone at Oxford and the Indicated Meteorological Parameters, during the Indicated Time Periods[a]

Meteorological parameter	13 Nov. to 8 Dec. 1950	9 Dec. to 3 Jan. 1951	4 Jan. to 29 Jan. 1951	30 Jan. to 24 Feb. 1951
Surface pressure	− .46	− .69	− .32	− .46
Thickness 1000/700 mb	− .81	− .47	− .44	− .56
Thickness 700/500 mb	− .72	− .65	− .32	− .38
Thickness 500/300 mb	− .89	− .71	− .27	− .67
Temperature 250 mb	+ .33	− .09	+ .09	+ .50
Tropopause height	− .94	− .65	− .34	− .78
Tropopause temperature	+ .68	+ .35	+ .14	+ .38
Tropopause pressure	+ .74	+ .53	+ .41	+ .77
Temperature 200 mb	+ .53	+ .40	+ .08	+ .67
Temperature 150 mb	+ .62	+ .28	+ .17	+ .60
Temperature 100 mb	+ .60	+ .52	+ .53	+ .59
Temperature 70 mb	+ .59	+ .55	+ .53	+ .53

[a] According to Normand (1951).

in Table 5.3. These results are consistent with the earlier results of Dobson *et al.* (1927) and Meetham (1937), and generally with those of more recent studies, for example, Ohring and Muench (1960). The latter used only 100-mb temperature (of those variables listed in Table 5.3), but computed correlation coefficients separately for each of several different months at each of several different European stations. Although the coefficients fluctuate rather widely, even occasionally being negative for some months at some stations, they are on the whole positive with average values around 0.5, no consistent annual variation, and apparently largest values near 50°N compared with more northerly or more southerly locations. Ohring and Muench also found generally negative correlations (of somewhat smaller magnitude) between total ozone and 100-mb height.

One other aspect of Table 5.3 requires comment. The negative correlation coefficient between total ozone and surface pressure reflects the well-known tendency at some locations for large amounts of ozone to be associated with surface low-pressure systems and small amounts of ozone with surface high-pressure systems. However, this is not universally true. For example, in China (Lejay, 1937), large amounts of ozone are associated with extensions of the Siberian anticyclone. The inconsistency, however, disappears if one considers conditions in the upper troposphere and lower stratosphere: both the well-developed cyclonic systems of western Europe and the cold anticyclones of the Asian continent are usually associated with warm low tropopause and a trough aloft.

While the relationships that have been revealed by these studies explain only part of the variance of the total ozone, they are undoubtedly real and represent the systematic operation of certain meteorological factors. These include both advection and divergence. In connection with the large amounts of ozone associated with troughs, consider the following simplified model.

Associated with a well-developed trough in the upper troposphere, there is generally a low, warm tropopause and a warm lower stratosphere. Immediately above the tropopause, the trough, being associated with air warmer than the air around it, gradually decreases in amplitude with increasing altitude. To the west of the trough, the air is from the north and subsiding, with the magnitude of the downward motion decreasing upward. Because ozone is a conservative quantity, ozone mixing ratio (not density) is conserved in the subsiding air. Inasmuch as ozone mixing ratio typically increases with height in this part of the atmosphere, a hypothetical air column approaching the trough undergoes a gradual increase in its ozone content. Figure 5.13 shows an example of this process computed by Reed (1950). The initial ozone distribution plotted in terms of mixing ratio need simply be displaced an appropriate amount to correspond to the net effect of vertical motion at each level over a period of time and the new mixing ratio curve converted back to the more standard units of centimeters NTP (kilometer)$^{-1}$. In this particular case there is a net gain of

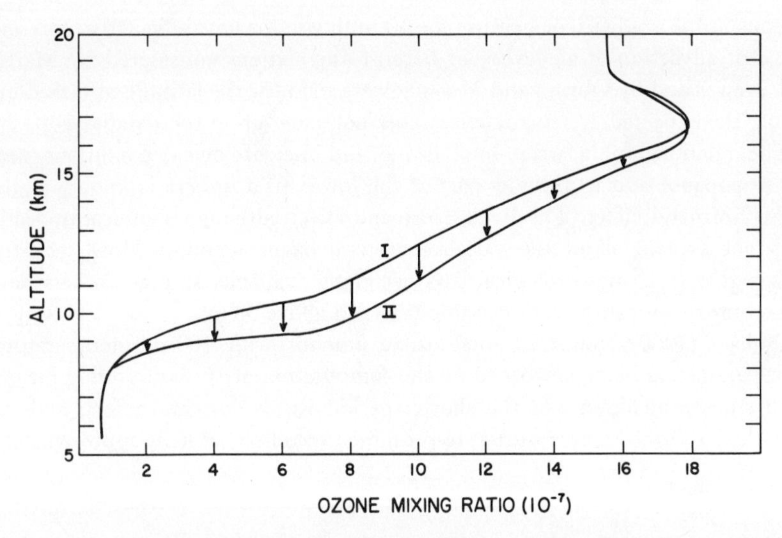

FIG. 5.13. Diagram illustrating the change of ozone mixing ratio due to subsidence. An air column with initial distribution marked I has the distribution marked II after subsiding variable distances indicated by the lengths of the arrows. (After Reed, 1950.)

0.024 cm NTP of ozone in the part of the column that undergoes subsidence of the type assumed. There would be a net loss of ozone if the sign of the vertical motions were reversed.

At first thought, it might appear that the air column pictured in Fig. 5.13 is getting "something for nothing." However, if one realizes that there is vertical divergence and hence horizontal convergence above the level of maximum downward motion, and vertical convergence and hence horizontal divergence below that level, then one can think of air relatively richer in ozone being added to the upper region at the same time that air with less ozone is leaving the bottom part. The net increase of ozone really comes at the expense of the surroundings.

Along with this effect, there must, of course, be effects of horizontal advection. It is natural to think in terms of the mean latitudinal ozone gradient and to connect north winds with ozone increases at all latitudes south of the high-latitude ozone maximum. Miyake and Kawamura (1956) have found such a correlation for Tokyo, but Martin and Brewer (1959) and Ohring and Muench (1960) working with data in the European area found larger total amounts of ozone connected with trajectories (in the first case) or winds (in the second) from the south at 100 mb. There is no question that ozone, being a conservative quantity, is advected with the wind; undoubtedly if daily ozone maps for particular levels were available from a synoptic network of stations measuring vertical distribution, they would show well-marked advective effects. However, either because the mean latitudinal gradient is not sufficiently representative of instantaneous ozone gradient, or because the 100-mb wind is not sufficiently representative of advection at all levels, or because the stations considered by Martin and Brewer and by Ohring and Muench were too near the latitude of maximum ozone, the expected systematic effect does not show up in their statistics.

The relationship between total ozone and meteorological conditions near the tropopause and in the low part of the lower stratosphere is known as the Reed–Normand effect. The Reed–Normand effect, although significant, clearly does not explain all of the variance of total ozone amount. More recently, with higher-level meteorological data becoming available, several workers have looked for relationships with conditions at and above 50 mb.

Godson (1960) compared total ozone amount with 100-mb temperature, both parameters being smoothed by the computation of 10-day running means. Such smoothing filters out the shorter-period Reed–Normand effect, and the smoothed 100-mb temperatures are presumed to reflect, at least approximately,

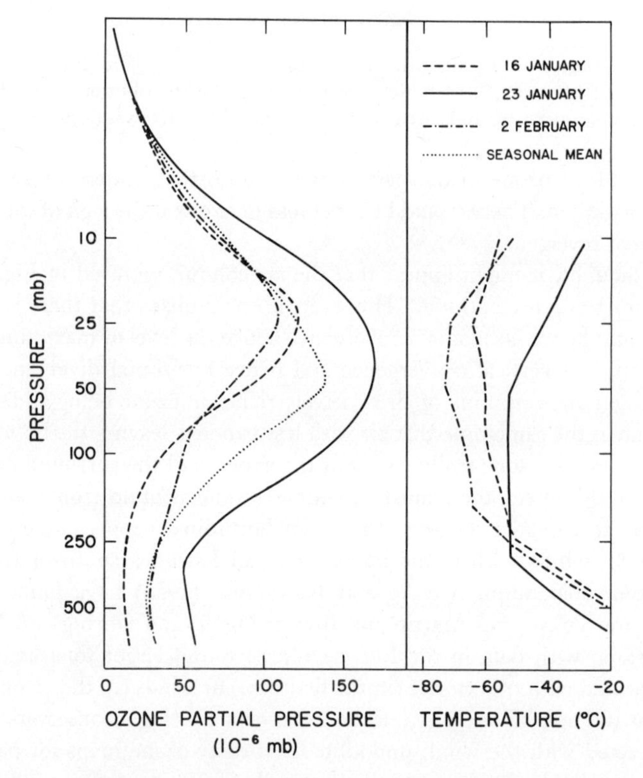

FIG. 5.14. Vertical distributions of ozone and temperature measured at Arosa at the indicated times in 1958. Note the large ozone densities at all levels on 23 January 1958 during a major stratospheric warming. (After Dütsch, 1962b.)

the longer-period temperature variations characteristic of higher levels in the lower stratosphere (say, 50 mb). (The variations of 100-mb temperature, smoothed in this manner, are discussed in Section 2.5.) Godson's study of conditions at different locations during several winters reveals a correspondence between the two parameters which is not very marked in detail but which is notable for the simultaneous large rise of (smoothed) temperature and (smoothed) total ozone at the time of major stratospheric warmings.

Allington, Boville, and Hare (1960) found a marked correspondence between total ozone amount and the thickness of the 100- to 25-mb layer at Edmonton and Moosonee, Canada, for two months in the winter of 1958-1959. The curves show a consistently positive correlation, ranging in the four cases (two stations, two months each) from 0.25 to 0.73. Boville and Hare (1961) showed two interesting case studies in which exceptionally small total amounts of ozone in middle latitudes were associated with unusual southward extensions of cold troughs associated with the disturbed polar-night circulation at about 25 mb.

Particularly interesting results have come from case studies of the major stratospheric warming of 1958, which originated over western Europe in late January. Dütsch (1962b) has shown the changes in vertical distribution of ozone at Arosa, as revealed by *Umkehr* observations, during the time of rapid warming. As revealed by Fig. 5.14, ozone increased at all levels well up into the upper stratosphere during the period of rapid temperature rise, 16 to 23 January. Dütsch comments especially on the increase at the upper levels:

> The really outstanding, and in our observation series unequalled, feature is the sudden and very pronounced increase in the layer between 23 and 35 km. It amounts to 30 to 50% at a height where normally the ozone fluctuations are already rather small (standard deviation around 10%). At the 10 mb level the absolute maximum of our observation series was reached during these days though the seasonal maximum occurs in late spring or early summer!

Dütsch (see also London, 1963b) also found a good correspondence between large (total) amounts of ozone and the location of the warm center as it subsequently moved northwestward.

Rather little attention has been given to the ozone content of the troposphere and its fluctuations. Junge (1962) has suggested that variations of tropospheric ozone are rather small, and, in any case, the mean value of tropospheric ozone is a relatively small proportion of the total amount. One would not expect therefore that variations of the total amount of ozone would reflect to any great extent variations in tropospheric values. Goody and Roach (1958), in fact, found a negative correlation.

It thus appears from these studies that the variations of ozone within a given season result largely from the superposition of meteorological effects near and just above the tropopause and other effects higher in the stratosphere. These effects seem to have no consistent simple correlation with each other (Allington *et al.*, 1960). Furthermore, it appears that particularly large rises of total ozone leading to the spring maximum are associated with major stratospheric warmings.

5.4.2 Seasonal Ozone Variations

Although there is some uncertainty, and even argument, about the details of the meteorological processes that lead to the large-scale ozone variations, there is little doubt about what the net effect of these processes must be. The total ozone content of the Northern Hemisphere increases by a significant factor during the period November through March. The most important contribution to this over-all increase is a rapid rise of the ozone content of the high-latitude lower stratosphere, where the additional ozone is most certainly not formed by photochemical processes. During the remainder of the year, from about April through October, the hemispheric ozone content gradually decays. From the point of view, then, of the large-scale changes, one must postulate a circulation mechanism that transfers ozone from the middle- (or low-) latitude upper stratosphere to the high-latitude lower stratosphere during the winter months. This ozone is protected and "stored" while new ozone is formed by photochemical processes in the source region. The transfer process operates during the entire winter, and leads to the early spring maximum, but must be missing or very weak during the remainder of the year. The ozone thus stored during the winter is gradually lost during the remainder of the year, either by a return to high levels and photochemical destruction, or by mixing down into the troposphere and chemical destruction.

Meridional transport of atmospheric quantities, such as ozone, can be accomplished by meridional cells or by quasi-horizontal, large-scale eddy processes. Mathematical expressions for these processes are formulated and discussed in Appendix F. A large part of the uncertainty about transport mechanisms in the upper atmosphere centers around the question of the relative importance of these two processes under various conditions.

Several scientists have postulated slow circulations in vertical-meridional planes to explain ozone transport and other phenomena. Brewer (1949) pointed out that the dryness of the lower stratosphere over England (Subsection 2.3.1) could be explained if the air there had passed at some previous time through the cold tropical tropopause. He suggested a circulation with air rising through the tropical tropopause, moving northward and subsiding in the temperate- and high-latitude stratosphere. Dobson (1956) discussed the same type of air circulation as an explanation of ozone changes as well as water-vapor observations. Goldie (1950) suggested three cells in each hemisphere, vertically aligned. The lowest in the troposphere corresponds to the well-known Hadley cell: up near the equator, poleward in the vicinity of the tropopause, down in lower middle latitudes, equatorward near the ground. The next of Goldie's cells has the opposite sense of circulation, with downward motion in the stratosphere over the equator from about 24 km down to the tropopause and poleward motion near the tropopause. The upper cell is again in the same direction as the bottom one; up over the equator from 24 to 40 km and poleward at high levels. Libby and Palmer (1960), in order to explain certain features of radioactive fallout, have recently suggested that the upper cell of the Goldie model is usually based much lower than 24 km.

It must be remembered that such circulations represent small net vertical and meridional components that would remain when velocity is averaged over longitude and time (see Appendix F). They would be superimposed on the very much larger quasi-horizontal, quasi-geostrophic motions observed in the atmosphere. Wind data in the stratosphere are totally inadequate to demonstrate their existence directly.

It is now well known from the work of Starr and his collaborators over the past dozen years that the meridional transport of various quantities (heat, momentum, water vapor) in the extratropical troposphere can be explained in terms of the large-scale, quasi-horizontal mixing processes connected with ridges and troughs. In view of the presently emerging picture of the winter arctic stratosphere as a region containing meteorologically active disturbances (for example, Hare, 1960) it is natural to consider whether the same may be true there.

Ozone data do not exist in sufficient quantity or detail to calculate quantitatively the transport of ozone by standing or transient eddies (Appendix F). For this purpose, one would like to have values of ozone density at a particular level, or several particular levels, at a network of stations evenly and closely spaced around a latitude circle. Most ozone observations are of total amount and do not give directly the ozone amount at a given level. However, we have seen that the total amount of ozone is rather closely related to the ozone content of the lower stratosphere. Newell (1963) has calculated the contribution of transient and standing eddies at 100 and 50 mb on the assumption that total amounts are representative of the ozone densities at those levels. His results seem to show that, particularly during the January–March period, both transient and standing eddies accomplish a northward transport of ozone in the lower stratosphere.

The connection of total ozone amount with meteorological disturbances at and above 50 mb during the winter months also seems to favor the importance of large-scale mixing processes. Godson (1960) has argued strongly for this point of view, which was advanced earlier by Wexler (1958) and Wexler and Moreland (1958). Although these arguments, coupled with the work of Newell, are quite compelling, it must be remembered that they do not yield a very reliable quantitative assessment and do not rule out the simultaneous operation of transport by meridional cells. Both mechanisms may be important, and their relative importance may vary with altitude, latitude, and season. Calculations based on presently available data are not able to settle this question definitively.

Recently Prabhakara (1963) (see also London and Prabhakara, 1963) has constructed a rather elaborate numerical model to investigate the seasonal and latitudinal variation of ozone in the Northern Hemisphere. The essence of the scheme is to incorporate in the expression for the equilibrium amount of ozone [Eq. (5.28)] a term representing transport processes, in a manner first suggested by Dütsch (1946). Such a transport term can include both meridional and eddy processes, with the latter being parameterized in terms of the traditional *Austausch* (large-scale eddy diffusion) coefficients. There results a partial-differential equation, with y and z as independent variables and equilibrium* ozone mixing ratio as dependent variable. This can be solved numerically with appropriate assumptions about the transport processes and the boundary conditions. It is possible, with reasonable assumptions, to reproduce the observed ozone observations rather well. It is impossible to discuss the details of the scheme here, and the reader is referred to the original references.

The ozone content of the troposphere, according to Junge (1962), is rather uniform geographically and seasonally, but exhibits a flat winter minimum and a spring maximum, the latter being delayed a month or two after the time of maximum total amount. According to present ideas, the troposphere gains ozone from the stratosphere† through large-scale mixing processes near the tropopause. This is a subject of considerable complexity and great concern, because the same processes that transfer ozone must also transfer radioactive debris deposited in the stratosphere by nuclear explosions. This subject is discussed more fully and in a broader context in Chapter 10. The sink for atmospheric ozone is near or at the surface, through chemical reactions, although in certain situations there may be large local additions of ozone as a result of chemical reactions in polluted air (Los Angeles). The relatively uniform distribution of tropospheric ozone, in the vertical (see Subsection 5.2.7) and with latitude and season (Junge, 1962), indicates that the characteristic time of tropospheric mixing is small compared with the characteristic times of injection and destruction. Tropospheric ozone is a general subject that has been relatively neglected but will undoubtedly receive much more attention as time goes on.

---

* This is an equilibrium value with respect to both photochemical and transport processes.

† It has been suggested (Kroening and Ney, 1962) that lightning discharges produce significant amounts of ozone in the troposphere. Although some ozone is undoubtedly produced in this manner, it is difficult to accept the idea that this is a significant source, when the seasonal and latitudinal distributions of ozone and thunderstorms are compared.

5.5 Composition of the Mesosphere

In the mesosphere there are many more possible photochemical reactions to be considered than in the stratosphere. There may well be reactions involving various oxides of nitrogen, and it is clear from the presence of hydroxyl bands in the spectrum of the airglow (Section 9.3) that hydrogen is also present and chemically active. Nevertheless, since these other reactions introduce tremendous complexity, and since many of the relevant rate coefficients are only poorly known, it is useful to continue consideration of a pure oxygen atmosphere (Subsection 5.5.1) before we turn to the consideration of nitrogen (Subsection 5.5.2) and hydrogen (Subsection 5.5.3). The spectrum of the airglow also reveals the presence of sodium in the upper atmosphere. Its possible origin and its vertical distribution are considered briefly in Subsection 5.5.4.

5.5.1 An Oxygen Atmosphere

The photochemical processes affecting the oxygen species remain the same as at lower levels, except that reaction (5.21), the three-body association of atomic oxygen, must be included. Assuming still that the value of n_2 is sensibly unaffected by the photochemical processes, we rewrite Eqs. (5.23) and (5.25), with one change in notation: Let $q_2 = n_2 J_2$ and $q_3 = n_3 J_3$, J_2 and J_3 being defined by Eqs. (5.26) and (5.27), respectively.

$$\frac{\partial n_1}{\partial t} = -\kappa_{12} n_1 n_2 n_m + 2n_2 J_2 - \kappa_{13} n_1 n_3 + n_3 J_3 - 2\kappa_{11} n_1{}^2 n_m \tag{5.34}$$

$$\frac{\partial n_3}{\partial t} = \kappa_{12} n_1 n_2 n_m - \kappa_{13} n_1 n_3 - n_3 J_3 \tag{5.35}$$

We can proceed to determine the photochemical-equilibrium conditions by expressing n_{3e} in terms of n_{1e} from (5.35):

$$n_{3e} = \frac{\kappa_{12} n_2 n_m n_{1e}}{J_3 + \kappa_{13} n_{1e}} \tag{5.36}$$

Substitution of this expression in (5.34) gives a cubic equation in n_{1e}:

$$n_{1e}^3 + n_{1e}^2 \left[\frac{J_3}{\kappa_{13}} + \frac{\kappa_{12} n_2}{\kappa_{11}} \right] - n_{1e} \frac{n_2 J_2}{\kappa_{11} n_m} - \frac{n_2 J_2 J_3}{\kappa_{11} n_m \kappa_{13}} = 0 \tag{5.37}$$

In order to make calculations with (5.36) and (5.37), it is necessary to estimate the values of the rate coefficients κ_{11}, κ_{12}, κ_{13} and of J_2 and J_3. The situation with respect to the rate coefficients was discussed in Section 5.3, and it is clear that there is considerable uncertainty in their values. The value of J_3 depends rather little on absorption in the visible and mainly on absorption in the Hartley continuum. In the mesosphere, J_3 is essentially independent of altitude, absorption being so small as to leave the solar flux sensibly unchanged. The value of J_2 does not depend on absorption in the Schumann-Runge continuum, because energy in that spectral region is already absorbed at higher levels, in the thermosphere. It clearly does include absorption in the Herzberg continuum, where its value is somewhat uncertain owing to the pressure effect on the oxygen absorption coefficients (Subsection 4.6.3). It probably also includes some of the absorption in the Schumann–Runge bands, as a result of predissociation (Subsection 4.6.3). It is clear that many uncertainties are involved but for illustrative purposes some numerical results are given in Table 5.4, which is constructed on the following assumptions:

(a) Values of n_2, n_m, and T are those given by Nicolet (1958) for a certain model atmosphere.

(b) Values of κ_{11}, κ_{12}, and κ_{13} are those given by Bates and Nicolet (1950) for the temperatures specified.

(c) J_3 is considered to have the constant value 3.6×10^{-3} sec^{-1}, which is an approximate average of the slowly varying values given by Horiuchi (1961) (computed for overhead sun and with Vigroux's absorption coefficients).

(d) J_2 is considered to have the constant value 10^{-9} sec^{-1} [computed for overhead sun, considering the Herzberg continuum only, and with pressure effects on the absorption coefficients neglected, by Horiuchi (1961)].

When the values tabulated in Table 5.4 are used in the coefficients of the cubic equation (5.37), it turns out that the latter has three real roots, only one of which is positive. This root can be approximated closely by

$$n_{1e}^2 = \frac{n_2 J_2}{\kappa_{11} n_m + \kappa_{12}\kappa_{13} n_2 n_m / J_3} \tag{5.38}$$

Values of n_{1e} and n_{3e} in Table 5.4 have been computed from (5.38) and (5.36). In view of the uncertainties involved these values may be incorrect by as much as an order of magnitude. Horiuchi (1961) has given results taking into consideration absorption in the Schumann–Runge bands.

The problem of day-time nonequilibrium conditions in this part of the atmosphere is a very complicated one. Any attempt to solve the simultaneous (nonlinear) differential equations (5.34) and (5.35) by eliminating one or the other of the variables n_1 or n_3 leads to an equation of great complexity. However, by computing instantaneous rates of change for various values of n_1 and n_3, one can gain considerable insight into the problem. For this purpose it is convenient to replace the variables n_1 and n_3 by $f = n_1/n_{1e}$ and $g = n_3/n_{3e}$. Here n_{1e} and

TABLE 5.4

NUMERICAL VALUES OF QUANTITIES BEARING ON THE PHOTOCHEMISTRY OF OXYGEN IN THE MESOSPHERE

z (km)	n_m (cm^{-3})	n_2 (cm^{-3})	T (°K)	κ_{11} (cm^6 sec^{-1})	κ_{12} (cm^6 sec^{-1})	κ_{13} (cm^3 sec^{-1})	n_{1e} (cm^{-3})	n_{3e} (cm^{-3})
50	2.35×10^{16}	4.94×10^{15}	274	8.28×10^{-33}	8.28×10^{-35}	3.37×10^{-15}	2.32×10^{10}	6.06×10^{10}
55	1.28×10^{16}	2.69×10^{15}	274	8.28×10^{-33}	8.28×10^{-35}	3.37×10^{-15}	3.11×10^{10}	2.39×10^{10}
60	7.31×10^{15}	1.54×10^{15}	253	7.96×10^{-33}	7.96×10^{-35}	6.54×10^{-16}	8.33×10^{10}	2.05×10^{10}
65	3.99×10^{15}	8.38×10^{14}	232	7.62×10^{-33}	7.62×10^{-35}	2.10×10^{-16}	1.36×10^{11}	9.56×10^{9}
70	2.06×10^{15}	4.33×10^{14}	210	7.24×10^{-33}	7.24×10^{-35}	4.93×10^{-17}	1.66×10^{11}	2.96×10^{9}
75	1.01×10^{15}	2.12×10^{14}	183	6.77×10^{-33}	6.77×10^{-35}	5.11×10^{-18}	1.76×10^{11}	7.08×10^{8}
80	4.45×10^{14}	9.34×10^{13}	156	6.24×10^{-33}	6.24×10^{-35}	2.57×10^{-19}	1.83×10^{11}	1.32×10^{8}

n_{3e} are given by (5.36) and (5.38); f and g are dimensionless variables, always positive (or zero). Equations (5.35) and (5.34) take the form (the latter approximately)

$$\frac{\partial f}{\partial t} = J_3(g - f) + g(1 - f)\kappa_{13}n_{1e} \tag{5.39}$$

$$\frac{\partial g}{\partial t} = 2\kappa_{11}n_m n_{1e}(1 - g^2) + \frac{\kappa_{12}\kappa_{13}n_2 n_m n_{1e}}{J_3}(2 - g - gf) - \kappa_{12}n_2 n_m(g - f) \tag{5.40}$$

Let us consider first conditions in the lower mesosphere, say at 50 km. At 50 km, for the data in Table 5.4,

$$\frac{\partial f}{\partial t} = 3.6 \times 10^{-3}(g - f) + 7.8 \times 10^{-5}g(1 - f) \tag{5.39a}$$

$$\frac{\partial g}{\partial t} = 9.0 \times 10^{-6}(1 - g^2) + 2.1 \times 10^{-4}(2 - g - gf) - 9.6 \times 10^{-3}(g - f) \tag{5.40a}$$

where all the coefficients are in units of $(\text{second})^{-1}$. It is clear that the terms involving $(g - f)$ are numerically the only important ones as long as g and f are not nearly equal. For initial conditions such that $g > 1$ and $f < 1$, $(g - f)$ is positive, f increases very rapidly, g decreases very rapidly, and equilibrium values are approached within a matter of minutes. By a similar argument, equilibrium conditions are also approached very rapidly if initial conditions are such that $g > 1$ and $f < 1$. On the other hand, if initial conditions are such that $g < 1$, $f < 1$ or that $g > 1$, $f > 1$, then the dominant terms act in such a way that the condition $g = f$ is approached rather rapidly. Further changes then proceed according to the rates associated with the smaller terms and a return to equilibrium may take several hours. It is clear that initial conditions are rather important in determining the characteristic times of photochemical processes in this part of the atmosphere.

In the upper mesosphere, conditions are quite different. At 80 km, for example, according to the data in Table 5.4,

$$\frac{\partial f}{\partial t} = 3.6 \times 10^{-3}(g - f) + 4.7 \times 10^{-8}g(1 - f) \tag{5.39b}$$

$$\frac{\partial g}{\partial t} = 1.0 \times 10^{-6}(1 - g^2) + 3.4 \times 10^{-10}(2 - g - gf) - 2.6 \times 10^{-6}(g - f) \tag{5.40b}$$

For any initial conditions, the condition $g = f$ is approached rapidly with f changing rapidly and g varying hardly at all. Further changes toward equilibrium can be expected to be quite slow, in view of the coefficients of the terms not involving $(g - f)$. Photochemical equilibrium near the mesopause is not a good assumption for either O_3 or O.

At night in the mesosphere, if n_1 and n_3 at sunset have nearly their day-time equilibrium values, n_1 decreases and n_3 increases as in the stratosphere (see Section 5.3). The rate of decrease of n_1 is considerably slower than in the stratosphere; nevertheless, because the proportion of oxygen atoms with respect to ozone molecules is much larger than at lower levels, there is a greater proportional increase of the latter. In the mesosphere there should be a large diurnal variation of n_3 (see Nicolet, 1955b).

5.5.2 A Nitrogen-Oxygen Atmosphere

The complexities introduced by the presence of nitrogen have been discussed extensively, for example, by Bates (1952, 1954), Nicolet (1954, 1955a,b, 1957), Harteck (1957), and Horiuchi (1961). Most of the rate coefficients are only very poorly known, in some cases even the order of magnitude being quite uncertain. Of the large number of possible reactions (see Harteck, 1957, table 1) the following appear to be the most important:

$$N + O_2 \rightarrow NO + O \tag{5.41a}$$

$$N + O + M \rightarrow NO + M \tag{5.41b}$$

$$N + O_3 \rightarrow NO + O_2 \tag{5.41c}$$

$$N + O_2 + M \rightarrow NO_2 + M \tag{5.41d}$$

$$NO + O \rightarrow NO_2 + h\nu \tag{5.41e}$$

$$NO + O_3 \rightarrow NO_2 + O_2 \tag{5.41f}$$

$$NO + N \rightarrow N_2 + O \tag{5.41g}$$

$$NO + h\nu \rightarrow N + O \tag{5.41h}$$

$$NO_2 + O \rightarrow NO + O_2 \tag{5.41i}$$

$$NO_2 + h\nu \rightarrow NO + O \tag{5.41j}$$

Nitrogen in atomic form is necessary for the formation of nitric oxide by (5.41a), (5.41b), or (5.41c), of which (5.41a) is probably the most important, and nitric oxide in turn leads to the formation of nitrogen peroxide. No reactions involving nitrous oxide, N_2O, are listed because this molecule is generally believed to play an unimportant role in the upper atmosphere (Bates and Witherspoon, 1952).

As mentioned earlier (Subsection 4.6.2), no dissociation continua have been observed for the nitrogen molecule; however, predissociation in the $^1\Pi_g$ state could lead to the production of nitrogen atoms in the upper atmosphere (Herzberg and Herzberg, 1948). Absorption producing this lies in certain of the

Lyman–Birge–Hopfield bands between 1200 and 1250 A and Bates (1954) has estimated a rate coefficient of 10^{-12} sec^{-1} at zero optical depth for this process. However, this energy is also absorbed by molecular oxygen and presumably not much of it can reach the mesosphere. Atomic nitrogen can also be produced in the upper atmosphere by ionization of N_2 (in the ionization continua short of 796 A or by X-ray radiation) followed by dissociative recombination [Eq. (4.16)], but this occurs in the thermosphere ordinarily.

Neither of these processes can produce large amounts of atomic nitrogen in the mesosphere, and it might appear at first sight that the nitrogen reactions could be neglected there. As a matter of fact, there is no direct experimental evidence for the presence of N, NO, or NO_2 in the mesosphere. However, the most attractive hypothesis explaining ionization in the mesosphere (*D* region), originally suggested by Nicolet (1945), and generally accepted for lack of any more reasonable alternative, is that NO is ionized by Lyman-α radiation. This is the only known or plausible constituent of the region to have a low enough ionization potential. Higher-energy (shorter-wavelength) radiation does not reach the *D* region from the quiet sun.

As a matter of fact, atomic nitrogen need not be produced *in situ*. Either nitric oxide or atomic nitrogen could well have a long enough lifetime to be mixed down into the mesosphere from production regions in the low thermosphere, just as ozone is produced primarily in the upper stratosphere but appears in important quantities in the lower stratosphere. There is good reason to believe (Nicolet, 1955a) that nitric oxide under mesospheric conditions would have a long life; in the mesosphere it is probably distributed according to a mixing equilibrium, rather than according to a photochemical equilibrium.

Nicolet has estimated (Nicolet and Aikin, 1960) that the proportion of NO atoms to the total number of molecules near 85 km may be on the order of 10^{-9} to 10^{-10}. This is not inconsistent with the upper limit of 10^8 cm^{-3} at any level between 63 and 87 km, estimated by Jursa, Tanaka, and LeBlanc (1959) from their failure to observe nitric oxide absorption bands in the solar spectrum near 1910 A from an ascending rocket.

5.5.3 A Hydrogen-Oxygen Atmosphere

The presence of hydrogen leads to many more possibilities for photochemical changes and reactions. Given a source of hydrogen from the dissociation of water vapor, presumed to be mixed upward from the earth's surface, and in the presence of the various forms of oxygen, one would expect to find atomic hydrogen, molecular hydrogen, the hydroxyl and perhydroxyl radicals, and hydrogen peroxide. The problem was first treated in some detail by Bates and Nicolet (1950), although their results were admittedly based on very rough estimates of some of the rate coefficients involved. Recently Wallace (1962) has extended this study, but many of the uncertainties involving the rate coefficients still remain.

It does not seem worth while here to list all the possible reactions and enter into a detailed discussion of the photochemical-equilibrium conditions. Figure 5.15 gives the results of Bates and Nicolet for the likely vertical distributions of the various constituents between 60 and 95 km. There are several points of interest in these results:

(a) The presence of hydrogen reduces the equilibrium concentrations of both O and O_3 relative to their values as computed without regard for hydrogen. Thus the values of Table 5.4 are overestimates (in relation to this one factor and apart from all the other uncertainties involved).

(b) Water-vapor density decreases with altitude much more rapidly than it would in the absence of dissociation and with complete mixing (constant mixing ratio).

(c) Above 70 to 75 km a substantial amount of hydrogen is expected to be in atomic form, rather than a member of heavier molecules, and will tend to escape by diffusion upward into interplanetary space.

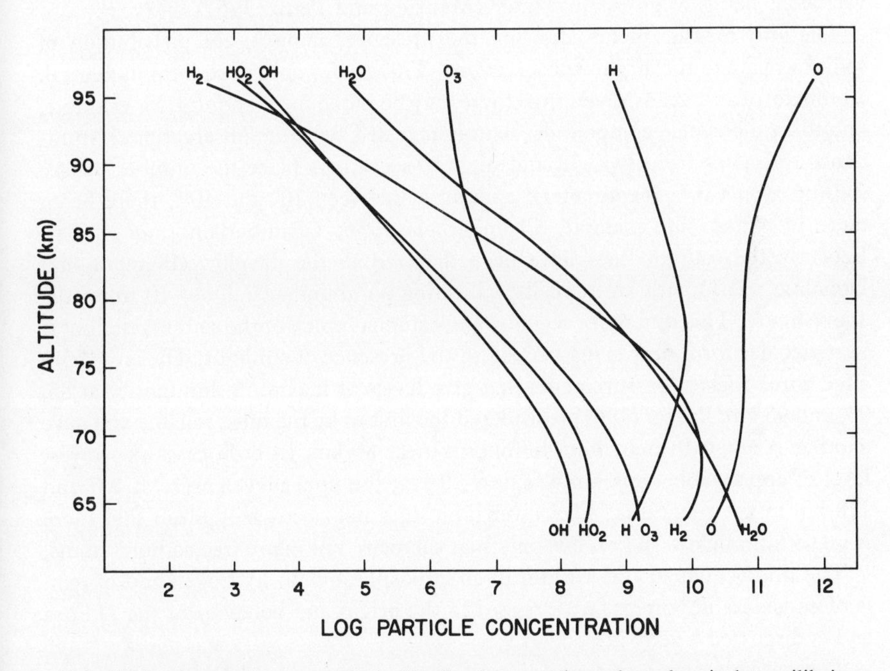

FIG. 5.15. Approximate vertical distributions under photochemical equilibrium of some constituents of the upper atmosphere. (After Bates and Nicolet, 1950.)

(d) The presence of the hydroxyl radical, known from observations of the airglow spectrum, is explained.

It must be remembered that these are only rough estimates of equilibrium conditions and, even to the extent that they are correct, may not represent actual conditions very well if photochemical equilibrium is only slowly approached.

5.5.4 Sodium in the Upper Atmosphere

One of the features of the airglow spectrum is the D doublet of sodium in the yellow at 5890 and 5896 A. This was first noted by Slipher (1929), but it was not until 1938 that the identification was established (Cabannes *et al.*, 1938; Bernard, 1939). The D doublet arises from the transition from the lowest excited term of sodium $3p\ ^2P^o$ with $J = \frac{3}{2}$ and $J = \frac{1}{2}$ to the ground level $3s\ ^2S_{\frac{1}{2}}$.

These lines flash out brightly near sunset and sunrise, as a result of scattering of sunlight. During the remainder of the night they remain at a more or less steady, lower intensity, becoming somewhat brighter as the night progresses (Pettit *et al.*, 1954). Their intensity also undergoes a rather marked seasonal variation, being brightest in winter (Manring and Pettit, 1957).

The only measurements showing the presence, amount, and distribution of sodium relate to the airglow observations. These refer only to sodium in neutral, atomic form and at any given time there may be much more sodium, for example, bound in molecular compounds, than is revealed by these observations. Abundance estimates from twilight and night observations place the number of free sodium atoms in a (centimeter)^2 column at between 10^9 and 10^{10}, there being more in winter (for example, Chamberlain, 1956; Chamberlain *et al.*, 1958). Recently the sodium lines have been detected in the dayglow (Blamont and Donahue, 1961) with an intensity indicating an abundance about 10 times the above figure. This difference seems to indicate that much of the sodium at night is in molecular form, but is dissociated in the presence of sunlight. Tousey (1958) cited three rocket measurements that give levels of maximum luminosity at 85, 93, and 95 km. Packer (1961) considered the first to be the most reliable and gave another result with maximum luminosity near 89 km. In both cases the lowest level of appreciable emission was near 70 km; the upper levels were at 107 and 118 km in the two cases. The lower limit does not necessarily mean that there is no sodium below 70 km, but only that there are not many free sodium atoms.

The photochemistry of sodium in an atmosphere with hydrogen and oxygen is of considerable interest with regard to the origin and behavior of the D lines and is discussed briefly in Section 9.3.

The possible origin of the sodium in the mesosphere and low thermosphere has received considerable attention and several suggestions have been made.

The question has been reviewed and discussed critically by Junge, Oldenberg, and Wasson (1962), whose paper contains a number of references to earlier work. The source may be terrestrial (sea salt, volcanic, and continental dust) or extraterrestrial (meteors, interplanetary dust, interstellar dust, material ejected from the sun). These authors favor meteoric influx as the principal source.

5.6 Noctilucent Clouds

We shall close this chapter with a brief discussion of luminous night clouds or *noctilucent clouds*. These were first observed, or at least recognized as a distinctive phenomenon, in 1885. During the succeeding decade, they were evidently exceptionally frequent and bright. The observations of this period were analyzed by Jesse (1896), who was able to indicate at that early date many of the essential observational facts about the clouds. Others who have reported or summarized noctilucent cloud observations include Vestine (1934), Störmer (1935), Khvostikov (1952), Paton (1949, 1954), Ludlam (1957), and Witt (1957). Ludlam's paper contains an excellent critical survey of the subject.

Noctilucent clouds when observed are within 10° or so of the horizon when the sun is 10°–20° below the horizon. The light which renders them visible is sunlight scattered by particles comparable in size to the wavelength of visible light (Deirmendjian and Vestine, 1959). They are generally whitish in appearance, but may take on a yellowish tint near the horizon or a silvery hue at the higher elevations. Their brightness varies from occurrence to occurrence and they are sometimes bright enough to cast a shadow at the earth's surface. Triangulation on distinguishable features by vaiious observers (for example, Störmer, 1935) has given rather consistent altitudes near 82 km; individual determinations seldom vary much from this. In visual appearance, the clouds often contain a series of bands (long bright streaks, sometimes hundreds of kilometers in length) against a fainter background. In addition, there is often a series of "billows," alternate darker and brighter areas often but not always oriented more or less across the direction of the bands and with a wavelength of about 10 km. Some striking photographs of noctilucent clouds have been presented by Witt (1962).

The clouds have been observed only within a month or two of the summer solstice and only in latitudes between 50° and 70°N. This is the time and location when the geometry of the sun-earth system gives the longest duration of conditions suitable for illumination of the altitude in question. Nevertheless, Vestine (1934) and Ludlam (1957) have both argued that the clouds should occasionally have been observed at other times of year in high latitudes and at other latitudes if they were indeed present. Furthermore, they have been observed only very rarely in Canada, most occurrences having been reported from Scandinavia and the U.S.S.R. Baker and Currie (1961), in reporting one such observation in Canada, have suggested that lack of suitable observational programs and confusion with weak auroral occurrences may be responsible for this.

It must not be supposed that the clouds appear every night during the season and location of their occurrence. For example, during the summers of 1954 and 1955 Ludlam (1957) observed altogether two faint, three moderate, and five bright displays from a location in Sweden near 63°N. Although he did not keep watch every night during the entire time of possible occurrence (roughly 2100 to 0300 local time at that latitude and time of year) there were clearly times of observation when the clouds were not present.

The question of the nature of the particles that constitute noctilucent clouds has never been definitively settled. It has been variously suggested that they are ice crystals, or that they are minute solid particles of either volcanic or extra-terrestrial origin.

The ice-crystal theory, first advanced by Humphreys (1933), seems to require either more water vapor than is generally believed to be present or lower temperature than is generally observed at the mesopause. Ludlam (1957), for example, argued that if the mixing ratio of about 10^{-6} indicated in the low stratosphere by the British frost-point hygrometer measurements (see Section 2.3) were to remain constant with altitude, the vapor density would be about 3×10^{-14} g cm^{-3} near 80 km and this would require a temperature less than 145°K for saturation, considerably lower than has been observed. The difficulty with this argument is that any estimate of the amount of water vapor at the altitude in question is subject to considerable uncertainty. Hesstvedt (1961, 1962) has argued that a mixing ratio of 10^{-4} (consistent with other measurements in the upper part of the lower stratosphere; see Section 2.3) together with the low mesopause temperatures measured by the rocket-grenade experiments at Fort Churchill in summer (see Figs. 3.10 and 3.11) does indicate the possibility of saturation with respect to ice at 80 km. Apart from the difficulty of extrapolating widely conflicting estimates of mixing ratio at 10 to 30 km up to 80 km, there is in addition the question of photochemical destruction of water vapor in the mesosphere (Subsection 5.5.3). Bates and Nicolet (1950), although concluding that in *photochemical equilibrium* water vapor at 80 km would be largely dissociated, nevertheless mentioned specifically that they did not consider this fatal to the ice-crystal theory of noctilucent clouds because large departures from equilibrium might sometimes exist. It is evidently not possible to disprove the ice-crystal theory by this line of argument. As a matter of fact, the Churchill measurements of low mesopause temperatures during the season and at the latitude where noctilucent clouds occur, taken together with the infrequent and irregular appearances of the clouds (indicating that somewhat unusual conditions are required), make the theory more attractive than it was a few years ago.*

Most investigators, though, do not favor an ice-crystal theory. Ludlam (1957), for example, concluded that water probably plays no role. A few pertinent sentences of his conclusions follow:

> It appears very unlikely that the condensation of water vapour plays any part in the formation of noctilucent clouds. They are probably composed of small solid particles, which may enter the stratosphere from below during major volcanic eruptions, or which may be produced in the stratosphere by the condensation of gases in the wakes of meteors. During the high latitude summer steep lapse rates develop in the upper stratosphere because of the prolonged heating of the ozone layer, and convective motions, possibly aided by solar warming of particle clouds, distribute the particles throughout a layer extending up to the 80 km inversion. Irregularities and wave disturbances in the top of the dusty layer produce visible clouds.

Several investigators have attempted to gain support for either a terrestrial or extraterrestrial origin of such particles by correlating unusual occurrences of noctilucent clouds with either large volcanic eruptions or meteor showers. Ludlam has discussed these studies and concluded that neither attempt has yielded conclusive results.

It might be remarked in conclusion, as Ludlam has pointed out, that the presence of small solid particles is almost certainly a necessary, even if it may not be a sufficient, condition for noctilucent clouds. For, even should water play a role, it is likely that condensation nuclei are required.

REFERENCES

Allington, K., Boville, B. W., and Hare, F. K. (1960). Midwinter ozone variations and stratospheric flow over Canada, 1958–1959. *Tellus* **12**, 266–273.

Baker, K. D., and Currie, B. W. (1961). An observation of noctilucent clouds in western Canada. *Canad. J. Phys.* **39**, 1515.

Bates, D. R. (1952). Some reactions occurring in the earth's upper atmosphere. *Ann. Geoph.* **8**, 194–204.

* A very recent announcement in the *Bulletin of the American Meteorological Society* (**44**, 806, December 1963) concerns rocket-grenade measurements in Sweden by scientists from the University of Stockholm. According to this preliminary announcement, temperatures at the mesopause at the time of occurrence of noctilucent clouds were determined to be 130°K. When noctilucent clouds were not present, temperatures of about 150° K were measured. See also *Bulletin of the American Meteorological Society* **45**, 431, July 1964 for evidence that water may sometimes be present.

BATES, D. R. (1954). The physics of the upper atmosphere. *In* "The Earth as a Planet" (G. P. Kuiper ed.), pp. 576–643. Univ. of Chicago Press, Chicago, Illinois.

BATES, D. R., and NICOLET, M. (1950). The photochemistry of atmospheric water vapor. *J. Geoph. Res.* **55**, 301–327.

BATES, D. R., and WITHERSPOON, A. E. (1952). The photo-chemistry of some minor constituents of the earth's atmosphere (CO_2, CO, CH_2, N_2O). *Mon. Not. Roy. Ast. Soc.* **112**, 101–124.

BENSON, S. W., and AXWORTHY, A. E., Jr. (1957). Mechanism of the gas phase, thermal decomposition of ozone. *J. Chem. Phys.* **26**, 1718–1726.

BERNARD, R. (1939). The identification and the origin of atmospheric sodium. *Ap. J.* **89**, 133–135.

BLAMONT, J. E., and DONAHUE, T. M. (1961). The dayglow of the sodium *D* lines. *J. Geoph. Res.* **66**, 1407–1423.

BOVILLE, B. W., and HARE, F. K. (1961). Total ozone and perturbations in the middle stratosphere. *Quart. J. Roy. Meteor. Soc.* **87**, 490–501.

BOWEN, I. G., and REGENER, V. H. (1951). On the automatic chemical determination of atmospheric ozone. *J. Geoph. Res.* **56**, 307–324.

BREWER, A. W. (1949). Evidence for a world circulation provided by the measurements of helium and water vapor distribution in the stratosphere. *Quart. J. Roy. Meteor. Soc.* **75**, 351–363.

BREWER, A. W. (1957). Ozone-concentration measurements from an aircraft in N. Norway. *Quart. J. Roy. Meteor. Soc.* **83**, 266–268.

BREWER, A. W., and MILFORD, J. R. (1960). The Oxford-Kew ozone sonde. *Proc. Roy. Soc.* **A256**, 470–495.

BREWER, A. W., DÜTSCH, H. U., MILFORD, J. R., MIGEOTTE, M., PAETZOLD, H. K., PISCALAR, F., and VIGROUX, E. (1960). Distribution verticale de l'ozone atmosphérique. Comparaison de diverses méthodes. *Ann. Geoph.* **16**, 196–222.

CABANNES, J., DUFAY, J., and GAUZIT, J. (1938). Sodium in the upper atmosphere. *Ap. J.* **88**, 164–172.

CHAMBERLAIN, J. W. (1956). Resonance scattering by atmospheric sodium. I. Theory of the intensity plateau in the twilight airglow. *J. Atmos. Terr. Phys.* **9**, 73–89.

CHAMBERLAIN, J. W., HUNTEN, D. M., and MACK, J. E. (1958). Resonance scattering by atmospheric sodium. 4. Abundance of sodium in twilight. *J. Atmos. Terr. Phys.* **12**, 153–165.

CHAPMAN, S. (1930). A theory of upper-atmospheric ozone. *Mem. Roy. Meteor. Soc.* **3**, 103–125.

COBLENTZ, W. W., and STAIR, R. (1939). Distribution of ozone in the stratosphere. *J. Res. Nat. Bur. Stand.* **22**, 573–606.

COBLENTZ, W. W., and STAIR, R. (1941). Distribution of ozone in the stratosphere: measurements of 1939 and 1940. *J. Res. Nat. Bur. Stand.* **26**, 161–174.

CRAIG, R. A. (1950). The observations and photochemistry of atmospheric ozone and their meteorological significance. *Meteor. Monogr.* **1**, No. 2, 1-50.

DEIRMENDJIAN, D., and VESTINE, E. H. (1959). Some remarks on the nature and origin of noctilucent cloud particles. *Plan. Space Sci.* **1**, 146–153.

DOBSON, G. M. B. (1930). Observations of the amount of ozone in the earth's atmosphere, and its relation to other geophysical conditions, part IV. *Proc. Roy. Soc.* **A129**, 411–433.

DOBSON, G. M. B. (1931). A photoelectric spectrophotometer for measuring the amount of atmospheric ozone. *Proc. Phys. Soc.* **43**, 324–339.

DOBSON, G. M. B. (1956). Origin and distribution of polyatomic molecules in the atmosphere. *Proc. Roy. Soc.* **A236**, 187–193.

DOBSON, G. M. B. (1957). Observers' handbook for the ozone spectrophotometer. *Ann. I.G.Y.* **5**, 46–89.

DOBSON, G. M. B., and HARRISON, D. N. (1926). Measurements of the amount of ozone in the earth's atmosphere and its relation to other geophysical conditions. *Proc. Roy. Soc.* **A110**, 660–693.

DOBSON, G. M. B., HARRISON, D. N., and LAWRENCE, J. (1927). Measurements of the amount of ozone in the earth's atmosphere and its relation to other geophysical conditions, part II. *Proc. Roy. Soc.* **A114**, 521–541.

DOBSON, G. M. B., HARRISON, D. N., and LAWRENCE, J. (1929). Measurements of the amount of ozone in the earth's atmosphere and its relation to other geophysical conditions, part III. *Proc. Roy. Soc.* **A122**, 456–486.

DÜTSCH, H. U. (1946). Photochemische Theorie des atmosphärischen Ozons unter Berücksichtigung von Nichtgleichgewichts-zuständen und Luftbewegungen. Ph. D. Dissertation, Zürich.

DÜTSCH, H. U. (1956). Das atmosphärische Ozon als Indikator für Strömungen in der Stratosphäre. *Archiv Meteor., Geoph., Biokl.* **A9**, 87–119.

DÜTSCH, H. U. (1959). Vertical ozone distribution from Umkehr observations. *Archiv Meteor., Geoph., Biokl.* **A11**, 240–251.

DÜTSCH, H. U. (1962a). Mittelwerte und wetterhafte Schwankungen des atmosphärischen Ozongehaltes in verschiedenen Höhen über Arosa. *Archiv Meteor., Geoph., Biokl.* **A13**, 167–185.

DÜTSCH, H. U. (1962b). Ozone distribution and stratospheric temperature field over Europe during the sudden warming in January/February 1958. *Beitr. Phys. Atmos.* **35**, 87–107.

EPSTEIN, E. S., OSTERBERG, C., and ADEL, A. (1956). A new method for the determination of the vertical distribution of ozone from a ground station. *J. Meteor.* **13**, 319–334.

EUCKEN, A., and PATAT, F. (1936). Die Temperaturabhängigkeit der photochemischen Ozonbildung. *Z. Phys. Chem.* **B33**, 459–474.

FABRY, C., and BUISSON, M. (1913). L'absorption de l'ultraviolet par l'ozone et la limite du spectre solaire. *J. Phys. Rad.* [5] **3**, 196–206.

FABRY, C., and BUISSON, M. (1921). Étude de l'extrémité ultraviolette du spectre solaire. *J. Phys. Rad.* [6] **2**, 197–226.

FUNK, J. P., and GARNHAM, G. L. (1962). Australian ozone observations and a suggested 24 month cycle. *Tellus* **14**, 378–382.

GODSON, W. L. (1960). Total ozone and the middle stratosphere over arctic and sub-arctic areas in winter and spring. *Quart. J. Roy. Meteor. Soc.* **86**, 301–317.

GODSON, W. L. (1962). The representation and analysis of vertical distributions of ozone. *Quart. J. Roy. Meteor. Soc.* **88**, 220–232.

GOLDIE, A. H. R. (1950). The average planetary circulation in vertical meridian planes. *Cent. Proc. Roy. Meteor. Soc.* pp. 175–180.

GOODY, R. M., and ROACH, W. T. (1956). Determination of the vertical distribution of ozone from emission spectra. *Quart. J. Roy. Meteor. Soc.* **82**, 217–221.

GOODY, R. M., and ROACH, W. T. (1958). The determination of tropospheric ozone from infra-red emission spectra. *Quart. J. Roy. Meteor. Soc.* **84**, 108–117.

GÖTZ, F. W. P. (1931). Zum Strahlungsklima des Spitzbergensommers. Strahlungs- und Ozonmessungen in der Königsbucht, 1929. *Gerl. Beitr. Geoph.* **31**, 119–154.

GÖTZ, F. W. P., MEETHAM, A. R., and DOBSON, G. M. B. (1934). The vertical distribution of ozone in the atmosphere. *Proc. Roy. Soc.* **A145**, 416–446.

GRIGGS, M. (1963). Measurements of the vertical distribution of atmospheric ozone at Nairobi. *Quart. J. Roy. Meteor. Soc.* **89**, 284–286.

HARE, F. K. (1960). The disturbed circulation of the arctic stratosphere. *J. Meteor.* **17**, 36–51.

HARTECK, P. (1957). A discussion of the reactions of nitrogen and nitrogen oxides in the upper atmosphere. *In* "The Threshold of Space" (M. Zelikoff, ed.), pp. 32–39. Pergamon Press, New York.

HERZBERG, G., and HERZBERG, L. (1948). Production of nitrogen atoms in the upper atmosphere. *Nature* **161**, 283.

HESSTVEDT, E. (1961). Note on the nature of noctilucent clouds. *J. Geoph. Res.* **66**, 1985–1987.

HESSTVEDT, E. (1962). On the possibility of ice cloud formation at the mesopause. *Tellus* **14**, 290–296.

HORIUCHI, G. (1961). Odd oxygen in the mesosphere and some meteorological considerations. *Geoph. Magaz.* **30**, 439–520.

HUMPHREYS, W. J. (1933). Nacreous and noctilucent clouds. *Mon. Wea. Rev.* **61**, 228–229.

JESSE, O. (1896). Die Höhe der leuchtenden Nachtwolken. *Ast. Nachr.* **140**, 161–168.

JOHNSON, F. S., PURCELL, J. D., and TOUSEY, R. (1951). Measurements of the vertical distribution of atmospheric ozone from rockets. *J. Geoph. Res.* **56**, 583–594.

JOHNSON, F. S., PURCELL, J. D., TOUSEY, R., and WATANABE, K. (1952). Direct measurements of the vertical distribution of atmospheric ozone to 70 kilometers altitude. *J. Geoph. Res.* **57**, 157–176.

JOHNSON, F. S., PURCELL, J. D., and TOUSEY, R. (1954). Studies of the ozone layer above New Mexico. *In* "Rocket Exploration of the Upper Atmosphere" (R. L. F. Boyd and M. J. Seaton, eds.), pp. 189–199. Pergamon Press, New York.

JUNGE, C. E. (1962). Global ozone budget and exchange between stratosphere and troposphere. *Tellus* **14**, 363–377.

JUNGE, C. E., OLDENBERG, O., and WASSON, J. T. (1962). On the origin of the sodium present in the upper atmosphere. *J. Geoph. Res.* **67**, 1027–1039.

JURSA, A. S., TANAKA, Y., and LEBLANC, F. (1959). Nitric oxide and molecular oxygen in the earth's upper atmosphere. *Plan. Space Sci.* **1**, 161–172.

KARANDIKAR, R. V., and RAMANATHAN, K. R. (1949). Vertical distribution of atmospheric ozone in low latitudes. *Proc. Ind. Acad. Sci.* **29**, 330–348.

KAUFMAN, F., and KELSO, J. R. (1961). The homogeneous recombination of atomic oxygen. *In* "Chemical Reactions in the Lower and Upper Atmosphere," p. 255. Wiley (Interscience), New York.

KAY, R. H. (1954). The measurement of ozone vertical distribution by a chemical method to heights of 12 km from aircraft. *In* "Rocket Exploration of the Upper Atmosphere" (R. L. F. Boyd and M. J. Seaton, eds.), pp. 208–211. Pergamon Press, New York.

KHVOSTIKOV, I. A. (1952). Silvery clouds. *Priroda, Moscow* **5**, 49–59.

KROENING, J. L., and NEY, E. P. (1962). Atmospheric ozone. *J. Geoph. Res.* **67**, 1867–1875.

KULKARNI, R. N. (1962). Comparison of ozone variations and of its distribution with height over middle latitudes of the two hemispheres. *Quart. J. Roy. Meteor. Soc.* **88**, 522–534.

KULKARNI, R. N., ANGREJI, P. D., and RAMANATHAN, K. R. (1959). Comparison of ozone amounts measured at Delhi (28½°N), Srinagar (34°N) and Tateno (36°N) in 1957–1958. *Pap. Meteor. Geoph., Meteor. Res. Inst., Japan* **10**, 85–92.

LEJAY, P. (1937). Mesures de la quantité d'ozone contenue dans l'atmosphère à l'observatoire de Zô-Sé, 1934-1935-1936; les variations de l'ozone et les situations météorologiques. *Notes Météor. Phys., Obs. de Zi-Ka-Wei, Fasc.* **7**, 1-16.

LIBBY, W. F., and PALMER, C. E. (1960). Stratospheric mixing from radioactive fallout. *J. Geoph. Res.* **65**, 3307–3317.

List, R. J. (1958). "Smithsonian Meteorological Tables," 6th ed., p. 422. Smithsonian Institution, Washington, D.C.

London, J. (1962). The distribution of total ozone over the Northern Hemisphere. *Sun at Work* **7**, No. 2, 11–12.

London, J. (1963a). The distribution of total ozone in the Northern Hemisphere. *Beitr. Phys. Atmos.* **36**, 254–263.

London, J. (1963b). Ozone variations and their relation to stratospheric warmings. *Meteor. Abhandlungen, Univ. Berlin* **36**, 299–310.

London, J., and Prabhakara, C. (1963). The effect of stratospheric transport processes on the ozone distribution. *Meteor. Abhandlungen, Univ. Berlin* **36**, 291–297.

Ludlam, F. H. (1957). Noctilucent clouds. *Tellus* **9**, 341–364.

Manring, E. R., and Pettit, H. B. (1957). A study of the airglow emissions at 5577, 5890 and 6300 A with a photometer of high spectral purity. *In* "The Threshold of Space" (M. Zelikoff, ed.), pp. 58-64. Pergamon Press, New York.

Martin, D. W., and Brewer, A. W. (1959). A synoptic study of day-to-day changes of ozone over the British Isles. *Quart. J. Roy. Meteor. Soc.* **85**, 393–403.

Mateer, C. L. (1960). A rapid technique for estimating the vertical distribution of ozone from Umkehr observations. Circ. No. 3291, Meteor. Branch, Dept. Transport, Toronto.

Mateer, C. L., and Godson, W. L. (1960). The vertical distribution of atmospheric ozone over Canadian stations from Umkehr observations. *Quart. J. Roy. Meteor. Soc.* **86**, 512–518.

Meetham, A. R. (1937). The correlation of the amount of ozone with other characteristics of the atmosphere. *Quart. J. Roy. Meteor. Soc.* **63**, 289–307.

Miyake, Y., and Kawamura, K. (1956). Studies on atmospheric ozone at Tokyo. *Sci. Proc. Int. Assoc. Meteor., Rome, 1954* pp. 172–176.

Newell, R. E. (1963). Transfer through the tropopause and within the stratosphere. *Quart. J. Roy. Meteor. Soc.* **89**, 167–204.

Nicolet, M. (1945). Contribution à l'étude de la structure de l'ionosphère. *Mem. Inst. Meteor. Belg.* **19**, 1-162.

Nicolet, M. (1954). Dynamic effects in the high atmosphere. *In* "The Earth as a Planet" (G. P. Kuiper, ed.), pp. 644–712. Univ. of Chicago Press, Chicago, Illinois.

Nicolet, M. (1955a). The aeronomic problem of nitrogen oxides. *J. Atmos. Terr. Phys.* **7**, 152–169.

Nicolet, M. (1955b). Nitrogen oxides and the airglow. *J. Atmos. Terr. Phys.* **7**, 297–309.

Nicolet, M. (1957). Nitrogen oxides and the airglow. *In* "The Threshold of Space" (M. Zelikoff, ed.), pp. 40–57. Pergamon Press, New York.

Nicolet, M. (1958). Aeronomic conditions in the mesosphere and lower thermosphere. Sci. Rep. 102, AF 19(604)–1304. Penn. State Univ., University Park, Pennsylvania.

Nicolet M., and Aikin, A. C. (1960). The formation of the *D* region of the ionosphere. *J. Geoph. Res.* **65**, 1469–1483.

Normand, Sir Charles (1951). Some recent work on ozone. *Quart. J. Roy. Meteor. Soc.* **77**, 474–478.

Ohring, G., and Muench, H. S. (1960). Relationships between ozone and meteorological parameters in the lower stratosphere. *J. Meteor.* **17**, 195–206.

Packer, D. M. (1961). Altitudes of the night airglow radiations. *Ann. Geoph.* **17**, 67–75.

Paetzold, H. K. (1955). New experimental and theoretical investigations on the atmospheric ozone layer. *J. Atmos. Terr. Phys.* **7**, 128–140.

Paetzold, H. K., and Piscalar, F. (1961a). Die Messung der vertikalen Ozonverteilung mittels einer optischen Radiosonde. *Beitr. Phys. Atmos.* **34**, 53–68.

Paetzold, H. K., and Piscalar, F. (1961b). Meridionale Ozonverteilung und stratosphärische Zirkulation. *Naturwiss.* **48**, 474.

Paton, J. (1949). Luminous night clouds. *Meteor. Magaz.* **78**, 354–357.

Paton, J. (1954). Direct evidence of vertical motion in the atmosphere at a height of about 80 km provided by photographs of noctilucent clouds. *Proc. Toronto Meteor. Conf., 1953* pp. 31–33. Roy. Meteor. Soc., London.

Perl, G., and Dütsch, H. (1959). Die 30-jährige Aroser Ozonmessreihe. *Ann. Schweiz. Meteor. Zentralamstalt, 1958* **8**.

Pettit, H. B., Roach, F. E., St. Amand, P., and Williams, D. R. (1954). A comprehensive study of atomic emissions in the nightglow. *Ann. Geoph.* **10**, 326–347.

Pittock, A. B. (1963). Determinations of the vertical distribution of ozone by twilight balloon photometry. *J. Geoph. Res.* **68**, 5143–5155.

Prabhakara, C. (1963). Effects of non-photochemical processes on the meridional distribution and total amount of ozone in the atmosphere. *Mon. Wea. Rev.* **91**, 411–431.

Ramanathan, K. R., and Dave, J. V. (1957). The calculation of the vertical distribution of ozone by Götz Umkehr-effect (Method B). *Ann. I.G.Y.* **5**, 23–45.

Reed, R. J. (1950). The role of vertical motions in ozone-weather relationships. *J. Meteor.* **7**, 263–267.

Reeves, R. R., Mannella, G., and Harteck, P. (1960). Rate of recombination of oxygen atoms. *J. Chem. Phys.* **32**, 632–633.

Regener, E., and Regener, V. H. (1934). Aufnahmen des ultravioletten Sonnenspektrums in der Stratosphäre und vertikale Ozonverteilung. *Phys. Z.* **35**, 788–793.

Regener, E., Paetzold, H. K., and Ehmert, A. (1954). Further investigations on the ozone layer. *In* "Rocket Exploration of the Upper Atmosphere" (R. L. F. Boyd and M. J. Seaton, eds.), pp. 202–207. Pergamon Press, New York.

Regener, V. H. (1938). Neue Messungen der vertikalen Verteilung des Ozons in der Atmosphäre. *Z. Phys.* **109**, 642–670.

Regener, V. H. (1951). Vertical distribution of atmospheric ozone. *Nature* **167**, 276–277.

Regener, V. H. (1956). New experimental results on atmospheric ozone. *Sci. Proc. Int. Assoc. Meteor., Rome, 1954* pp. 181–188.

Regener, V. H. (1960). On a sensitive method for the recording of atmospheric ozone. *J. Geoph. Res.* **65**, 3975–3977.

Roach, W. T., and Goody, R. M. (1958). Absorption and emission in the atmospheric window from 770 to 1,250 cm^{-1}. *Quart. J. Roy. Meteor. Soc.* **84**, 319–333.

Sekera, Z., and Dave, J. V. (1961a). Determination of the vertical distribution of ozone from the measurement of diffusely reflected ultra-violet solar-radiation. *Plan. Space Sci.* **5**, 122–136.

Sekera, Z., and Dave, J. V. (1961b). Diffuse transmission of solar ultraviolet radiation in the presence of ozone. *Ap. J.* **133**, 210–227.

Slipher, V. M. (1929). Emissions in the spectrum of the light of the night sky. *Publ. Ast. Soc. Pacific* **41**, 262–263.

Störmer, C. (1935). Measurements of luminous night clouds in Norway, 1933 and 1934. *Astroph. Norv.* **1**, 87–114.

Summerfield, M. (1941). Pressure dependence of the absorption in the 9.6 micron band of ozone. Ph. D. Thesis, Calif. Inst. Tech.

Tousey, R. (1958). Rocket measurements of the night airglow. *Ann. Geoph.* **14**, 186–195.

Twomey, S. (1961). On the deduction of the vertical distribution of ozone by ultraviolet spectral measurements from a satellite. *J. Geoph. Res.* **66**, 2153–2162.

Vassy, A., and Rasool, I. (1960). Répartition verticale de l'ozone atmosphérique à différentes latitudes. *Ann. Geoph.* **16**, 262–263.

Venkateswaran, S. V., Moore, J. G., and Krueger, A. J. (1961). Determination of the vertical distribution of ozone by satellite photometry. *J. Geoph. Res.* **66**, 1751–1771.

Vestine, E. H. (1934). Noctilucent clouds. *J. Roy. Ast. Soc. Canada* **28**, 249.

Vigroux, E. (1953). Contribution à l'étude expérimentale de l'absorption de l'ozone. *Ann. Phys.* [6] **8**, 709–762.

Vigroux, E. (1959). Distribution verticale de l'ozone atmosphérique d'après les observations de la bande 9.6 μ. *Ann. Geoph.* **15**, 516–538.

Vigroux, E., and Debaix, A. (1963). Résultats d'observations de l'ozone atmosphérique par la méthode infra-rouge. *Ann. Geoph.* **19**, 31–42.

Wallace, L. (1962). The OH nightglow emission. *J. Atmos. Sci.* **19**, 1–16.

Walshaw, C. D. (1957). Integrated absorption by the 9.6 μ band of ozone. *Quart. J. Roy. Meteor. Soc.* **83**, 315–321.

Walshaw, C. D. (1960). The accuracy of determination of the vertical distribution of atmospheric ozone from emission spectrophotometry in the 1043 cm^{-1} band at high resolution. *Quart. J. Roy. Meteor. Soc.* **86**, 519–529.

Walton, G. F. (1957). The calculation of the vertical distribution of ozone by Götz Umkehr-effect (Method A). *Ann. I.G.Y.* **5**, 9–22.

Wexler, H. (1958). A meteorologist looks at the upper atmosphere. *In* "Atmospheric Explorations" (H. G. Houghton, ed.), pp. 79–100. The Technology Press, Cambridge, Massachusetts, and Wiley, New York.

Wexler, H., and Moreland, W. B. (1958). Winds and temperatures in the arctic stratosphere. *Proc. Polar Atm. Symp.* Part I., pp. 71–84. Pergamon Press, New York.

Witt, G. (1957). Noctilucent cloud observations. *Tellus* **9**, 365–371.

Witt, G. (1962). Height, structure and displacements of noctilucent clouds. *Tellus* **14**, 1–18.

Wulf, O. R., and Deming, L. S. (1936a). The theoretical calculation of the distribution of photochemically-formed ozone in the atmosphere. *Terr. Magn. Atmos. Elect.* **41**, 299–310.

Wulf, O. R., and Deming, L. S. (1936b). The effect of visible solar radiation on the calculated distribution of atmospheric ozone. *Terr. Magn. Atmos. Elect.* **41**, 375–378.

Wulf, O. R., and Deming, L. S. (1937). The distribution of atmospheric ozone in equilibrium with solar radiation and the rate of maintenance of the distribution. *Terr. Magn Atmos. Elect.* **42**, 195–202.

Yakovleva, A. V., Kudryavtseva, L. A., Britaev, A. S., Gerasev, V. F., Kachalov, V. P., Kuznetsov, A. P., Pavlenko, N. A., and Iozenas, V. A. (1963). Spectrometric investigation of the ozone layer up to a height of 60 km. *Plan. Space Sci.* **11**, 709–721.

CHAPTER 6

Composition and Structure of the Thermosphere

Below the thermosphere, problems of composition involve certain trace constituents which, although they may have very important physical and chemical effects, do not change the mean molecular weight of air appreciably. On the other hand, in the thermosphere such factors as dissociation of oxygen and the action of diffusion bring into question the composition of air in its gross aspects. Composition is, of course, a matter of great interest in itself. In addition, information about composition is necessary to interpret pressure and density measurements (from rockets or satellites) in terms of kinetic temperature. Therefore it is necessary to consider composition as a variable parameter interrelated with pressure, temperature, and density. In this chapter we are concerned with the neutral components of air and defer (to Section 9.2) a discussion of the ionized components, except insofar as they affect the main topic. Neither do we at this time consider air motions (see Chapter 8).

It should be made clear at the outset that some important questions about thermospheric composition and structure have not yet been definitively answered. The answers are at this time slowly emerging from a steadily increasing number of measurements and from considerations of the interrelated physical processes involved. It should not surprise anybody if the true situation differs in some respects from the present estimates.

In this chapter we start by discussing the photochemical (Section 6.1) and diffusive processes (Section 6.2) that affect composition. Section 6.3 reviews the composition measurements that are available. In Section 6.4, the methods and results of pressure and density observations are discussed. In Section 6.5 the interpretation of the latter in terms of composition and kinetic temperature is considered. The emphasis is on the atmosphere up to about 300 km. However, measurements of density above 300 km contribute greatly to the understanding of atmospheric structure and composition at much lower levels. There is, therefore, in Sections 6.4 and 6.5 considerable reference to structure above 300 km, despite the fact that this book is primarily concerned with lower levels.

6.1 Photochemical Processes in the Thermosphere

In the present context, the photochemical processes of overwhelming interest are those that might lead to a dissociation of the main constituents of air, N_2 and O_2. It is generally believed that nitrogen occurs predominantly in molecular form throughout the thermosphere, but that oxygen is almost entirely dissociated at high enough levels. Therefore, the emphasis in this section is on oxygen.

As noted in Chapter 4, the nitrogen molecule has no observed dissociation continuum. Furthermore, as is discussed in Section 6.3, there is no observational evidence for the presence of substantial amounts of atomic nitrogen in the upper atmosphere. These are good reasons to neglect the presence of atomic nitrogen in considerations of mean molecular weight. Nevertheless, some atomic nitrogen undoubtedly exists and plays an important role in certain processes. It may be formed, for example, by dissociative recombination (Section 4.2),

$$N_2^+ + e \rightarrow N + N \tag{6.1}$$

by ion-atom interchange (Section 4.2),

$$O^+ + N_2 \rightarrow NO^+ + N \tag{6.2}$$

or by a predissociation process (Section 4.1). Lines of atomic nitrogen are observed in the spectrum of the aurora and airglow. As discussed in Subsection 5.5.2, Nicolet (1955) has suggested that atomic nitrogen formed in the thermosphere by one of these processes may be transported into the mesosphere and play an important role in the photochemistry of that region.

Oxygen, on the other hand, has a very pronounced dissociation continuum (Schumann–Runge) with cross section appropriate to absorption of solar energy at thermospheric levels. Chapman (1930) first pointed out that oxygen must be almost completely dissociated at high enough levels. The presence of pronounced lines of atomic oxygen in the spectra of the aurora and airglow as well as more recent direct measurements verify its presence in important quantities.

Although there is really no question in principle about the important role played by oxygen dissociation, there is unfortunately still considerable doubt about the quantitative aspects of the problem. A definitive and reliable determination of the vertical distribution of the ratio $n(O)/n(O_2)$ (and possible temporal and spatial variations thereof) would go a long way toward clearing up questions of thermospheric composition and structure, at least up to 200–300 km.

The principal photochemical reactions affecting oxygen in the thermosphere are

$$O_2 + h\nu \rightarrow O + O \qquad (\lambda < 1751 \text{ A}) \tag{6.3}$$

$$O + O + M \rightarrow O_2 + M \tag{6.4}$$

$$O + O \rightarrow O_2 + h\nu \tag{6.5}$$

The process (6.5) must become the predominant one at sufficiently high levels where low density renders three-body collisions highly unlikely. However, three-body association may be the predominant process in the lower thermosphere. Because the rate coefficients for these processes are not well known there is some uncertainty about their relative importance at intermediate levels. In either case, the equation governing oxygen changes (which neglects for the time being any transport processes) is

$$\partial n(O_2)/\partial t = -n(O_2)J_2 + \kappa[n(O)]^2 \tag{6.6}$$

where J_2 represents the number of dissociating quanta absorbed per molecule and unit time, as used in Subsection 5.5.1. In the thermosphere, only absorption in the Schumann–Runge continuum is important. The rate coefficient κ is the rate coefficient for radiative association if (6.5) is the predominant process and is equal to $\kappa_{11}n(\mathrm{m})$ (see Section 5.3) if three-body association is the predominant process.

Until about 1954, in the absence of direct reliable rocket measurements, attempts to study the oxygen problem postulated a photochemical equilibrium, so that

$$n(O) = \left[\frac{J_2 n(O_2)}{\kappa}\right]^{1/2} \tag{6.7}$$

In photochemical equilibrium studies for the stratosphere and mesosphere we assume that the density of one of the constituents (O_2) is so little affected by the photochemical processes that its vertical distribution can be specified by an atmospheric model based on density or pressure measurements. For the thermosphere, where a transition region from O_2 at lower levels to O at higher levels is to be anticipated, no such simplification is possible. It is therefore necessary to assume some other (rather arbitrary) condition that will lead to another relation between $n(O)$ and $n(O_2)$ in addition to that given by (6.7). Furthermore, in the thermosphere at low enough levels, the optical depth of the overlying atmosphere is large for Schumann–Runge radiation and J_2 at a given level depends on the integrated amount of O_2 between that level and the sun.

Several investigators (Rakshit, 1947; Penndorf, 1949; Moses and Wu, 1951, 1952) have carried out photochemical equilibrium calculations for the thermosphere under various assumptions. Despite the different assumptions, all calculations of this type lead to a rather sharp transition region near 100 km. As an example, Penndorf's methods and results are briefly reviewed.

Penndorf (1949) obtained another relation between $n(O_2)$ and $n(O)$ by assuming that the ratio by mass of oxygen (O plus O_2) to nitrogen (N_2) remains the same (approximately 1 : 4) at any level regardless of the degree of dissociation. Thus

$$n(O_2)\mu(O_2) + n(O)\mu(O) = \tfrac{1}{4}n(N_2)\mu(N_2) \tag{6.8}$$

He computed $n(N_2)$ as a function of height hydrostatically for an assumed reference pressure and a specified temperature distribution, assuming a mean molecular mass μ appropriate to an O_2–N_2 atmosphere (before dissociation).

Starting high enough in the atmosphere that J_2 could be assumed to be independent of overhead O_2 (in practice 130 km) he then worked down, computing $n(O_2)$ and $n(O)$ from the equivalent of Eqs. (6.7) and (6.8), each time modifying J_2 to correspond to values of $n(O_2)$ already computed for higher levels. He made computations for various zenith angles and various black-body temperatures of the sun, assuming the three-body association process. He found a very sharp and narrow transition region, with the oxygen mostly in molecular form around 90 km and mostly in atomic form around 105 km. Table 6.1 gives a typical result. Rakshit (1947) earlier had obtained a similar result (using the two-body radiative association process and a 6000°K sun) with the transition region a few kilometers higher than Penndorf's.

TABLE 6.1

CONCENTRATION BY VOLUME (IN PER CENT) OF THE PRINCIPAL CONSTITUENTS UNDER PHOTOCHEMICAL EQUILIBRIUM WITH ZENITH ANGLE 0° AND A BLACK-BODY TEMPERATURE FOR THE SUN OF 5300°K[a]

z (km)	N_2	O_2	O
85	82.0	18.0	0
90	81.6	17.3	1.1
95	81.2	16.2	2.6
100	76.9	10.9	12.2
105	69.9	0.4	29.7
110	69.4	0	30.6

[a] The atmosphere was assumed to have 82 per cent N_2 and 18 per cent O_2 in the absence of photochemical processes. After Penndorf (1950).

However, in the early 1950's, it became apparent from rocket measurements that the assumption of photochemical equilibrium (and the resulting sharp transition from an O_2 atmosphere to an O atmosphere just above 100 km) was an untenable one. In the first place, measurements of O_2 concentration in the thermosphere (see Section 6.3) showed many more molecules than were predicted by the equilibrium theory; in the second place, measurements of solar energy in the Schumann–Runge continuum, while not very quantitative, showed that equivalent black-body temperatures of 5000° to 6000°K were too high.

Nicolet and Mange pointed out the implications of these rocket results in two important papers (Nicolet and Mange, 1954; Nicolet, 1954). Nicolet and

Mange (1954) interpreted the earliest rocket results (an early measurement of O_2 reported by Friedman, Lichtman, and Byram, 1951) to mean that oxygen could not be considered to be in photochemical equilibrium. Their principal result is shown in Fig. 6.1. In this figure the dot-dash curve shows the observed variation of the O_2 optical depth with height. The dashed-line curve represents Nicolet and Mange's modification of these results to correct for some instrumental uncertainty of unknown origin, a modification necessary to allow a reasonable interpretation at low levels. The solid-line curve shows the computed optical depth of molecular oxygen as a function of altitude according to a certain model atmosphere in which $n(O_2)$ maintains a constant proportion to the total number density. The details of this model are unimportant, but it is important that these data are incompatible with the concept of photochemical equilibrium.

Nicolet and Mange also showed that photochemical equilibrium is not a reasonable working hypothesis for the high levels, in view of the observational evidence that solar energy in the Schumann–Runge continuum is much less than would correspond to the black-body temperatures (5000–6000°K) assumed by the earlier workers. This is a critical factor in the maintenance of photochemical equilibrium. Consider Eq. (6.6). At a high enough level, $n(O_2) \ll n(O)$ and $n(O)$ may be considered constant insofar as it affects $n(O_2)$. Furthermore, at this high level, the optical depth of the overlying atmosphere may be considered

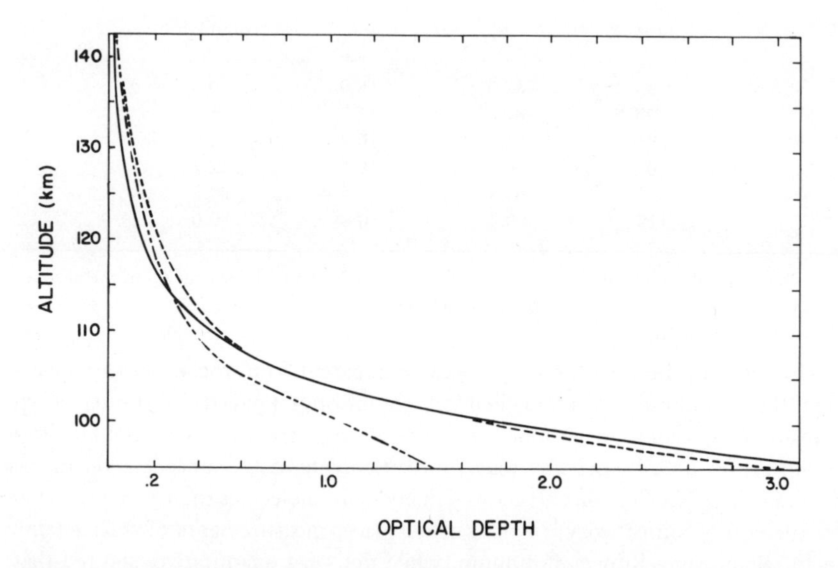

FIG. 6.1. Variation with altitude of the optical depth (referred to 1500 A) of molecular oxygen. The dot-dash curve was measured; the dashed curve represents a corrected version of the measurements; the full curve represents the calculated variation for a certain mixed atmosphere. (After Nicolet and Mange, 1954.)

to be so small that J_2 has the value appropriate to zero optical depth. Then (6.6) can be integrated to give

$$n(O_2) = n_0(O_2)e^{-J_2 t} + \kappa \frac{[n(O)]^2}{J_2}(1 - e^{-J_2 t}) \tag{6.9}$$

where $n_0(O_2)$ is an initial value of $n(O_2)$. The time required to go halfway to photochemical equilibrium ($\kappa[n(O)]^2/J_2$) from any initial value $n_0(O_2)$ (the dissociation time) is simply $t_D = 0.7/J_2$. Table 6.2 shows the value of t_D for various values of the sun's black-body temperature. The effect of radiation temperature is striking; the assumption of photochemical equilibrium at high levels was quite justified by the earlier workers, but is unjustified in the light of later indications of much smaller solar emission in the Schumann–Runge continuum. At great heights with a 4500° sun, excess amounts of O_2 would be 50 per cent dissociated only after about 10 days (each with a 12-hour duration of sunlight). At lower levels, where the optical depth of O_2 is not negligible and to which less radiation penetrates, the times would be correspondingly larger.

TABLE 6.2

DISSOCIATION TIME (t_D) OF MOLECULAR OXYGEN IN THE THERMOSPHERE FOR VARIOUS BLACK-BODY TEMPERATURES AT ZERO OPTICAL DEPTH[a]

T (°K)	t_D
4000	65 days
4500	5 days
5000	15 hours
5500	3 hours
6000	40 minutes

[a] Computed from data of Nicolet and Mange (1954).

Furthermore, such oxygen atoms as are produced at high levels can be shown to have very long lifetimes, so long that most of them must be transported downward by convection or diffusion before they have an opportunity to recombine. Nicolet (1960b) has estimated that at altitudes above 120 km the "recombination time" of oxygen atoms (the time required for one-half the oxygen atoms to recombine by three-body collision) is more than a year; at 100 km it is still more than a month.

In a study of additional rocket data on O_2 distribution (the 1 December flight of Table 6.3; see Section 6.3) Nicolet (1954) showed that these data also could not be fitted by an assumption of photochemical equilibrium. In this case, the best fit was found for an atmosphere in which O_2 (assumed to be a minor constituent at this level) was in diffusive equilibrium above 110 km in a main mixed atmosphere of N_2 and O (see Section 6.2). (The next section is concerned with the question of diffusion.) Again, the significance of this is not so much in the particular model chosen to fit the data as in the demonstration that photochemical equilibrium is definitely not involved.

The realization that oxygen is not in photochemical equilibrium in the thermosphere has many implications for the interpretation of upper-atmospheric phenomena. In the first place, solar radiation absorbed at high levels from the Schumann–Runge continuum is not all realized as heat energy at those levels; instead, some of the dissociated atoms may move downward and recombine below 100 km where the energy is finally converted to heat or to radiation at longer wavelengths, as visible in the airglow. In the second place, the knowledge that molecular oxygen can exist in much greater quantities at high levels than previously imagined has had important consequences for theories of ionization in the E and F regions and for the interpretation of thermospheric structure measurements.

On the other hand, it should be pointed out that the latest estimate of solar flux in the Schumann–Runge continuum (see Section 4.5) indicates energy equivalent to that from a black body at about 4700°K. Earlier estimates had placed the figure between 4000°K and 4500°K. According to Table 6.2, the assumed value of this temperature in the range 4500° $\pm$ 500°K is a rather critical factor in any assessment of the importance of photochemical processes. It may be that this importance has been underestimated in recent years, just as it was overestimated prior to 1954. If solar energy in this spectral region should turn out to be significantly variable (which would be rather unexpected) the thermospheric problem would be further complicated.

6.2 Diffusion in the Thermosphere

An understanding of the structure of the thermosphere must take into account the role of diffusive separation of the gases. Contrary to the case of a mixed atmosphere, where the mixture of gases is distributed vertically in a hydrostatic equilibrium that depends on a mean molecular mass, in the case of diffusive equilibrium each individual gas is distributed according to its own molecular mass. Lighter gases tend to spread out over a greater vertical extent and at high enough levels predominate over the heavier ones.

It has long been known that the effect of mixing successfully opposes this diffusive tendency up to a certain level, above which diffusive effects become important. The altitude of this "certain level" is a matter of extreme interest. According to the measurements reported in Section 6.3, it now seems to be established at about 110 $\pm$ 10 km. The purpose of the present section is to discuss from an elementary point of view some of the expected effects of diffusion on structure.

For a given vertical distribution of temperature and given boundary conditions at the base of an atmosphere in diffusive equilibrium, it is easy enough to specify the vertical distribution of the gases. However, when other processes such as mixing or photochemical changes operate to oppose such an equilibrium, one is faced with a time-dependent problem that is very formidable. Consequently, theoretical studies of diffusion are highly idealized.

Typically, they investigate the diffusion of a minor constituent in a stable main atmosphere that is and remains thoroughly mixed. The minor constituent is presumed to have some idealized vertical distribution at an initial moment (for example, that corresponding to thorough mixing with the other gases), and the theory allows one to calculate the distribution of the minor constituent at later times due to diffusion only. The length of time required for the minor constituent to approach a diffusive-equilibrium distribution (the other gases remaining mixed) is presumed to be a measure of the tendency of the real atmosphere to be in diffusive equilibrium. It is not possible to calculate quantitatively the effects of mixing, or to discover the diffusion times of main constituents such as N_2. Nevertheless, there are interesting applications of the theory, for example to show that the photochemical-equilibrium distribution of molecular oxygen at high enough levels could not persist because of diffusion.

Mange (1955, 1957, 1961a) has done considerable work on the problem of thermospheric diffusion. In Mange's model, the main atmosphere is assumed to have a mean molecular mass μ independent of altitude and time, and a constant gradient of scale height, $dH/dz = \beta$. In this case, as pointed out in Subsection 1.1.2,

$$\frac{p}{p_0} = \left(\frac{H}{H_0}\right)^{-1/\beta} \tag{6.10}$$

$$\frac{n}{n_0} = \left(\frac{H}{H_0}\right)^{-(1+\beta)/\beta} \tag{6.11}$$

where H_0, p_0, and n_0 are scale height, pressure, and number density at a reference level where $z = 0$.

Now consider a minor constituent* (indicated by the subscript 1). If it is mixed with the main atmosphere, p_1/p_{10} and n_1/n_{10} are given by the right-hand sides of Eqs. (6.10) and (6.11). On the other hand, in diffusive equilibrium, a constant gradient β_1 of the scale height H_1 of the minor constituent gives

$$\frac{p_1}{p_{10}} = \left(\frac{H_1}{H_{10}}\right)^{-1/\beta_1} \tag{6.12}$$

$$\frac{n_1}{n_{10}} = \left(\frac{H_1}{H_{10}}\right)^{-(1+\beta_1)/\beta_1} \tag{6.13}$$

* A gas may be taken to be a "minor constituent" if $n_1 \ll n$ and further $n_1\mu_1 \ll n\mu$. Therefore the values of n and μ of the main atmosphere are not appreciably changed by any possible redistribution of the minor constituent.

Inasmuch as $H_1/H = \beta_1/\beta = \mu/\mu_1$, it is clear that

$$\frac{p_1}{p_{10}} = \left(\frac{H}{H_0}\right)^{-(\mu_1/\mu)(1/\beta)} \tag{6.14}$$

$$\frac{n_1}{n_{10}} = \left(\frac{H}{H_0}\right)^{-[(\mu_1/\mu)+\beta]/\beta} \tag{6.15}$$

A convenient bit of notation introduced by Nicolet is the "vertical-distribution factor" X, defined so that one may write the expression for n_1/n_{10} in the form

$$\frac{n_1}{n_{10}} = \left(\frac{H}{H_0}\right)^{-X[(1+\beta)/\beta]} \tag{6.16}$$

It is clear that for a mixing equilibrium, when (6.11) applies, $X = 1$ and for a diffusive equilibrium, when (6.15) applies, $X = [(\mu_1/\mu) + \beta]/(1 + \beta)$.

To illustrate the equilibrium vertical distributions of number densities of various constituents according to Eq. (6.16), Fig. 6.2 shows the results of Mange (1955). In the model for which these computations were made, all constituents are assumed to be uniformly mixed up to 110 km, above which the main mixed atmosphere consists of N_2 and O with a mean molecular weight of 24 and $\beta = 0.2$. If all gases continued to be uniformly mixed, all would follow the curve $X = 1$. If they were in diffusive equilibrium above 110 km, their distributions would follow the curves as labeled.

If a minor constituent is not initially in diffusive equilibrium, the equation governing its redistribution is the equation of continuity,

$$\frac{\partial n_1}{\partial t} + \frac{\partial}{\partial z}(n_1 w_1) = 0 \tag{6.17}$$

where w_1 is the mean velocity in the vertical direction of all the molecules of type 1, and horizontal gradients of mass transport are neglected. The effects of any sources or sinks (such as photochemical changes) are neglected for the time being. Nicolet (1960b) has shown that the expression for w_1, for the special case being considered here and in place of the general expression of Chapman and Cowling (1952), is

$$w_1 = -D\left[\frac{1}{n_1}\frac{\partial n_1}{\partial z} + \frac{1}{H_1} + (1 + \alpha)\frac{1}{T}\frac{\partial T}{\partial z}\right] \tag{6.18}$$

Here D is the diffusion coefficient, whose form is discussed below, and α is the thermal diffusion factor, which is very nearly constant for a given gas and may be neglected for the heavier gases. It is left out hereafter, although it may be of importance for hydrogen and helium. See Mange (1955) and Nicolet (1960b) for the equations with its inclusion.

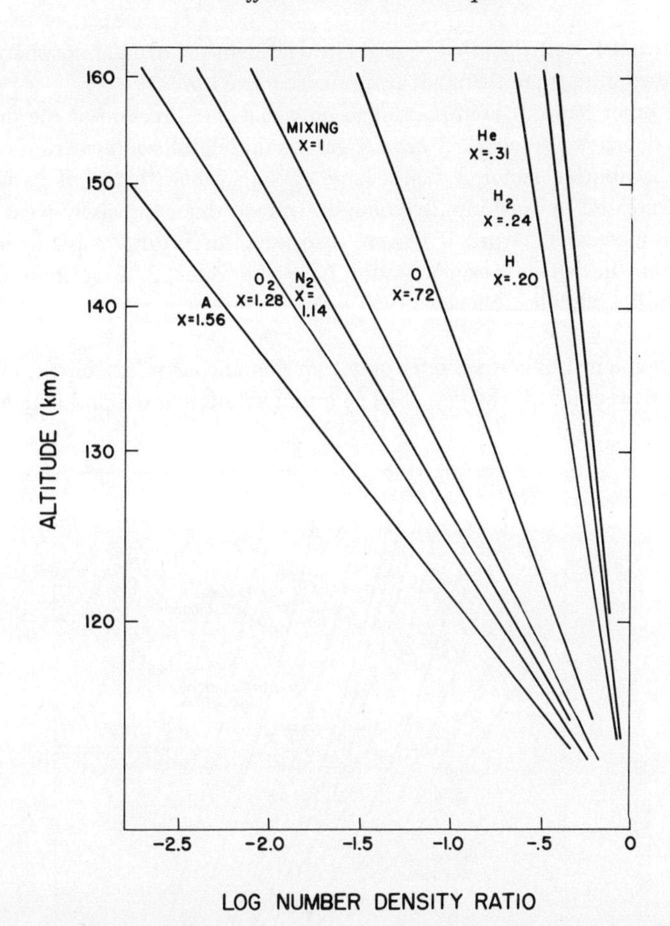

FIG. 6.2. Vertical distributions of minor constituents assumed to be in diffusive equilibrium above 110 km in a main mixed atmosphere of N_2 and O. (Curves for N_2 and O are simply illustrative.) The abscissa gives ratio of number density at height z to number density at 110 km. The ordinate is linear in scale height. (After Mange, 1955.)

It follows easily that, if the main atmosphere has the constant scale-height gradient β so that $(1/T)(\partial T/\partial z) = \beta/H$ and if the minor constituent has the distribution factor X, so that $(1/n_1)(\partial n_1/\partial z) = -X(1 + \beta)/H$, then

$$w_1 = \frac{D}{H}\left[X(1 + \beta) - \left(\frac{\mu_1}{\mu} + \beta\right)\right] \tag{6.19}$$

This form of the equation for w_1, discussed for example by Mange (1955), gives immediate insight into the sign of w_1. If the minor constituent is initially mixed with the main atmosphere ($X = 1$), w_1 is positive (upward) if $1 > (\mu_1/\mu)$

and negative if $1 < (\mu_1/\mu)$; that is, gases heavier than the main atmosphere move down, those lighter than the main atmosphere move up.

On the other hand, it is important to note that the direction of the diffusive transport depends also on X. A *heavier* gas would be diffused *upward* if at some time its distribution factor X had a large enough value (that is, if its number density decreased upward rapidly enough). In fact, this is precisely what would happen to molecular oxygen if it were distributed according to photochemical equilibrium, in which case X would have the value 2.92 at high enough levels with $\beta = 0.2$ (see Nicolet, 1954). We show some results for this case later.

Such arguments, of course, refer only to instantaneous values of w_1, because, as diffusion proceeds, X changes. The complete solution of the problem depends

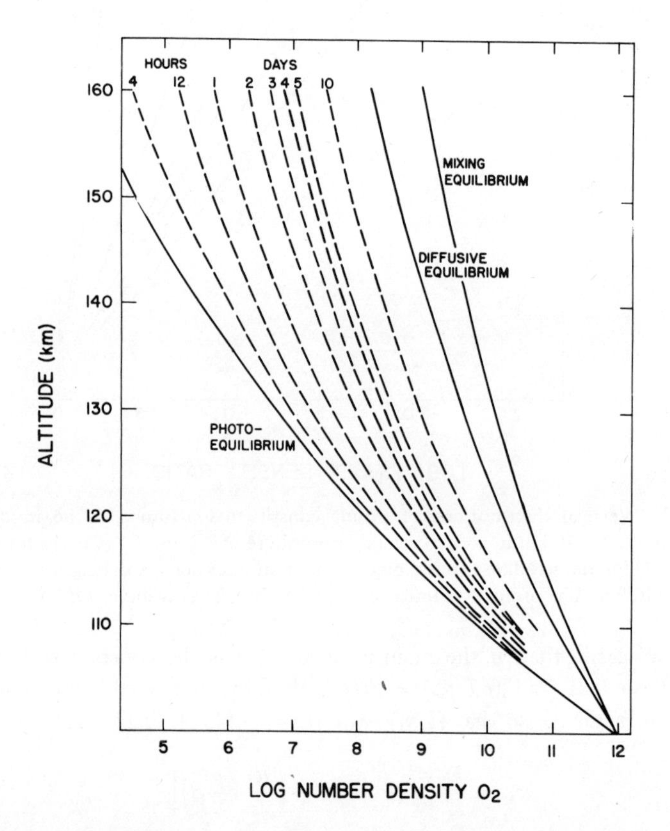

FIG. 6.3. Time required for molecular oxygen, initially in photochemical equilibrium, to assume indicated vertical distributions as a result of diffusion, even though dissociations continue to occur. Oxygen is assumed to be a minor constituent in a main mixed atmosphere of nitrogen and atomic oxygen. (After Mange, 1955.)

on solving (6.17) subject to suitable boundary conditions, when w_1 is given by (6.18), and D is given by

$$D = \frac{3}{8}\left(\frac{g}{2\pi}\right)^{\frac{1}{2}}\left[1 + \frac{\mu}{\mu_1}\right]^{\frac{1}{2}}\frac{H^{\frac{1}{2}}}{n\sigma^2} \tag{6.20}$$

where σ is the collision diameter of the molecules. The resulting differential equation has been discussed by Epstein (1932), Sutton (1943), and Mange (1957). Here we show only some of the results of Mange.

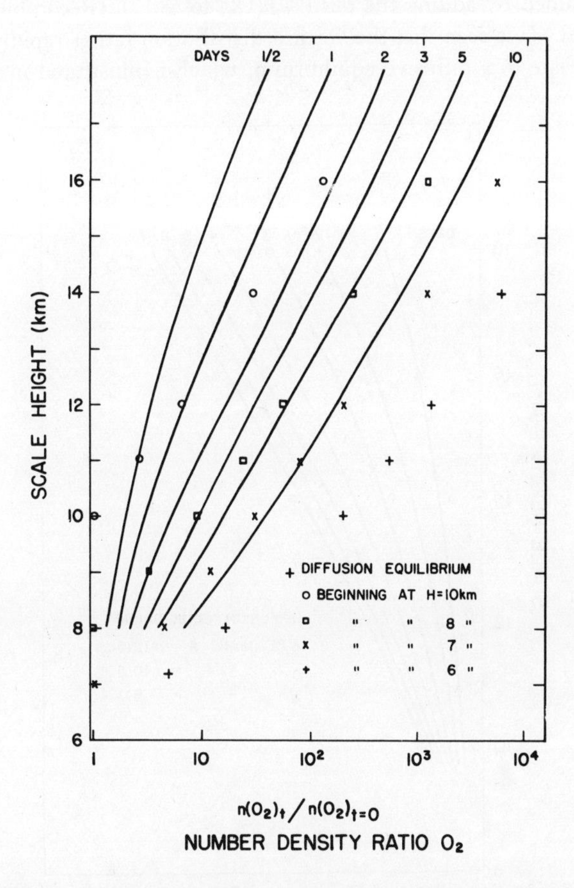

FIG. 6.4. Effect of diffusion on an initial photochemical-equilibrium distribution of molecular oxygen. In the abscissa label, $n(O_2)_t$ refers to the number density of molecular oxygen at time t after the beginning of diffusion. The ordinate is altitude expressed in terms of scale height. The curves labeled in days tell the time necessary to achieve the conditions represented by the coordinates of the plotted points. Calculations are made for different levels (as indicated) taken to represent the base of the diffusion layer. (After Nicolet, 1960b.)

A particularly interesting calculation of Mange (1955) is shown in Fig. 6.3. Here molecular oxygen is assumed to be *initially* in photochemical equilibrium and above 100 km to be a minor constituent in a main mixed atmosphere of N_2 and O, with a distribution as shown in Fig. 6.3. The scale height H is taken to be 8 km at 110 km and to vary linearly at all levels with $\beta = 0.2$. The dash-line curves in Fig. 6.3 show the calculated vertical distributions of O_2 at various indicated times after the initial time, as a result of the action of diffusion (no mixing) above 100 km as well as dissociation. Mathematically, the effect of the latter is included by adding the term $J_2 n(O_2)$ to (6.17). (Recombination effects are neglected.) It is seen that the vertical distribution rather rapidly approaches that appropriate to a diffusive equilibrium, which is illustrated in the diagram.

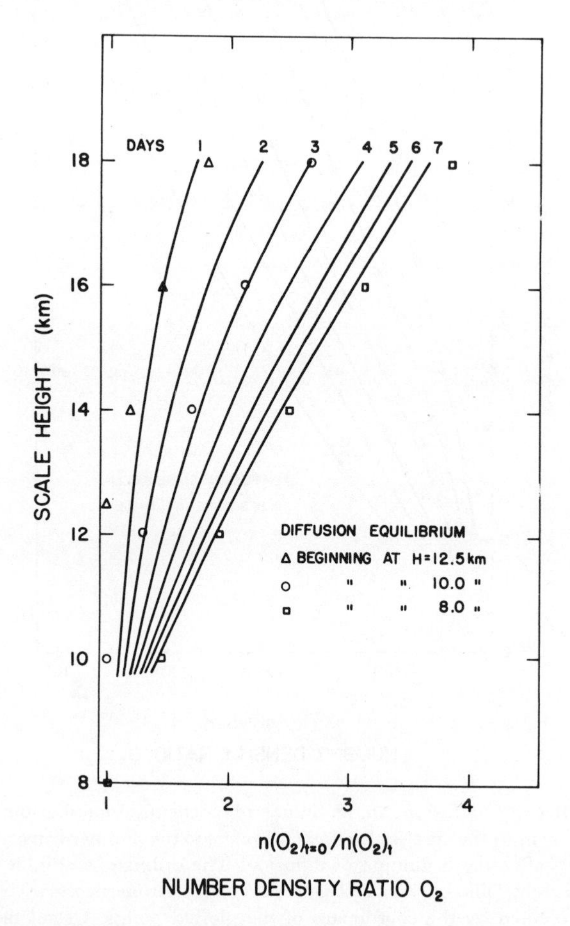

FIG. 6.5. Same as Fig. 6.4, but for oxygen distributed initially according to a mixing equilibrium. (After Nicolet, 1960b.)

In this problem, the (unknown) effects of mixing can only assist the diffusive processes in destroying the photochemical distribution and the departure from the latter would be more rapid than shown. Mange (1961a) has pointed out that if β is larger than 0.2, the diffusion time is considerably shortened.

Actually, the choice of reference level (here 100 km, $H_0 = 6$ km) above which diffusion is assumed to act is arbitrary. Choice of a higher level, such as 105 or 110 km, would give more rapid changes at higher levels. Calculations by Nicolet (1960b) for a similar problem but with various choices for the lower level are illustrated in Fig. 6.4. In both Mange's and Nicolet's calculations solar radiation is assumed to be present for 12 hours per day.

Similar calculations are possible if the initial distribution corresponds to a *mixing* equilibrium rather than a photochemical equilibrium. Figure 6.5 shows these calculations of Nicolet. In this case, of course, the effect of mixing counteracts the diffusive effects to an unknown degree so that the times shown are minimum.

One more example, for argon from an initial mixing equilibrium, is shown in Fig. 6.6. Note the much greater ratio $n(A)_{t=0}/n(A)_t$ than for O_2 ; this is because the molecular mass of A differs more from the mean molecular mass of the main atmosphere than does that of O_2, so the difference between mixing and diffusive densities, particularly at high levels, is much greater.

On the basis of calculations such as these, Nicolet (1960b) has concluded:

> In the light of these results, it is safe to consider that in the thermosphere above 110 km, all *minor* constituents which are sufficiently

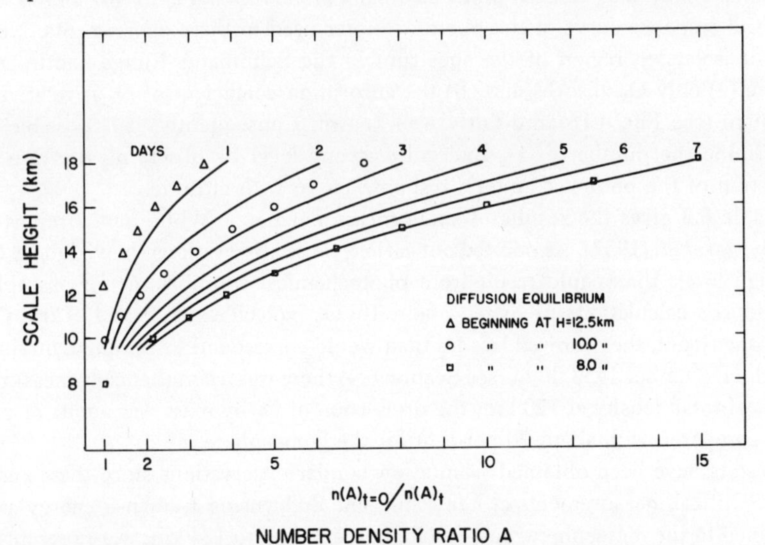

FIG. 6.6. Same as Fig. 6.4 but for argon. (After Nicolet, 1960b.)

inert and neutral to be free from rapid chemical reactions are distributed according to a height gradient corresponding to a diffusion equilibrium.

6.3 Composition Measurements

Direct measurements of thermospheric composition are not available in sufficient number to define the composition very satisfactorily. The problem is a difficult one and even the measurements that are available are subject to various uncertainties and in some cases conflict with one another.

In this section, we review some of these measurements. The earliest information was derived from optical determinations of O_2 concentration, which were mentioned in Section 6.1. These are treated in Subsection 6.3.1. In Subsection 6.3.2, we consider measurements specifically aimed at establishing a level above which diffusive effects become important. The meager evidence on the height dependence of mean molecular weight and of the $n(O)/n(O_2)$ ratio is covered in Subsection 6.3.3.

6.3.1 The Vertical Distribution of Molecular Oxygen

An optical method, similar in principle to that described in Subsection 5.2.3 in connection with ozone, can be used to determine the vertical distribution of molecular oxygen in the thermosphere. The earliest results of this type were reported by Byram, Chubb, and Friedman (1955, 1957). Figure 6.7 shows the spectral-response curve of the photon counter used in these experiments. Note that it isolates a region of the spectrum in the Schumann–Runge continuum where (a) only O_2 absorbs and (b) the absorption coefficient of O_2 is relatively constant (see Fig. 4.16) and fairly well known. Consequently, it is possible to determine the number of O_2 molecules at any level by observing the rate of variation of the photon counter's response curve with altitude.

Table 6.3 gives the results of three measurements at White Sands reported by Byram *et al.* (1957). As pointed out earlier, these observations showed more O_2 at high levels than could result from photochemical equilibrium; for example, Penndorf's calculations predicted about 10^7 O_2 molecules cm^{-3} at 130 km. On the other hand, they showed less O_2 than would correspond to complete mixing. On the December 1953 flight (see Section 6.4) there was a simultaneous measurement of total density at 120 km; the proportion of O_2 by mass was about 11 per cent compared with about 20 per cent for the homosphere.

Results have been obtained from a few similar observations since these early ones. Optical measurements at Churchill, one flight using Lyman-α energy and confined to the mesosphere, and one flight extending to 124 km, were presented by Kupperian *et al.* (1958). Jursa *et al.* (1963) discussed three flights at White

Sands, one of which gave quantitative results for the thermosphere (between 124 and 164 km).

Hinteregger (1962) has discussed in considerable detail the potentialities of the optical method for measuring the vertical distribution of not only molecular oxygen, but also atomic oxygen and molecular nitrogen. Measurements of

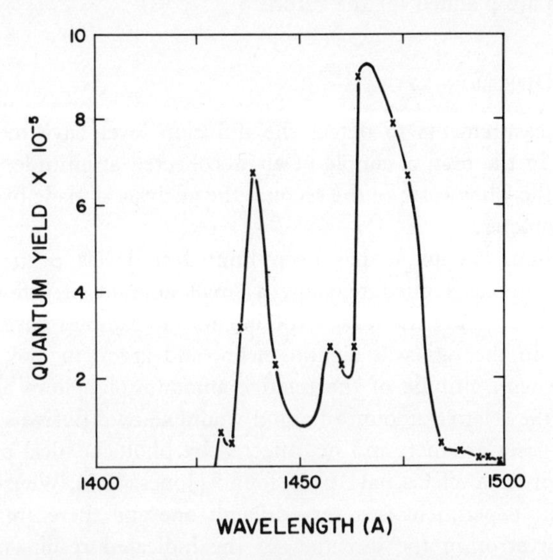

FIG. 6.7. Spectral response curve of photon counter used in optical measurements of O_2 distribution. (After Byram, Chubb, and Friedman, 1957.)

TABLE 6.3

NUMBER DENSITY (CM^{-3}) OF MOLECULAR OXYGEN AT WHITE SANDS[a]

Altitude (km)	Date		
	1 Dec. 1953	18 Oct. 1955	21 Oct. 1955
110	16×10^{10}	—	—
114	7.5×10^{10}	12×10^{10}	—
116	5.4×10^{10}	9×10^{10}	—
118	4×10^{10}	2.3×10^{10}	—
120	3×10^{10}	1.8×10^{10}	—
130	1×10^{10}	—	—
150	—	—	4×10^{9}
160	—	—	1.4×10^{9}
170	—	—	8×10^{8}
180	—	—	4×10^{8}

[a] According to three measurements of Byram *et al.* (1957).

extreme ultraviolet radiation already obtained by Hinteregger and his collaborators (Subsection 4.5.2) were not designed with this purpose in mind, but enabled Hinteregger to obtain some preliminary and illustrative results. He concludes that measurements in the extreme ultraviolet planned especially for the purpose of determining composition would have a useful accuracy and such experiments are planned for the future.

6.3.2 The Diffusion Level

Rocket measurements to detect the diffusion level have made use of two techniques: in the first, a sample of air is collected at altitude, recovered, and analyzed in the laboratory; in the second, the analysis is made by a rocket-borne mass spectrometer.

The sampling technique has been limited to levels primarily below the mesopause and has tended to show a small amount of diffusive separation between 60 and 90 km. In these experiments the samples have been analyzed for changes in the ratios of helium, neon, and argon to molecular nitrogen. An increase with altitude of the relative amounts of helium and neon and a decrease of the relative amount of argon would show diffusive separation, since all of these gases are inert and not affected by photochemical processes. Small indications of such effects have been found (Jones, 1954; Wenzel *et al.*, 1958). However, the experiment is a very difficult one and there are possibilities of experimental error in the direction of the indicated results (G. R. Martin, 1954). There are also theoretical difficulties in explaining lack of mixing in the mesosphere. In any case, the departures from the appropriate tropospheric ratios have been small and it is generally accepted that diffusive separation below the mesopause, if any, is negligibly small.

Analyses by rocket-borne mass spectrometers gave the first definite evidence of diffusive separation. The instrument used was a Bennett mass spectrometer (Bennett, 1950) flown on Aerobee-Hi rockets and particular emphasis was placed on the ratio of the densities of argon and molecular nitrogen. It is presumed that nitrogen as well as argon is sensibly unaffected by photochemical processes in the altitudes studied. The earliest flight for which results were reported (Townsend *et al.*, 1954) indicated no diffusive separation between 96 and 137 km over White Sands, New Mexico, on 12 February 1953. According to C. Y. Johnson (1961), this early flight did not give consistent enough data to justify a reliable conclusion that diffusive separation was really absent. Three later experiments over Churchill, Canada, on 20 November 1956, 21 February 1958, and 23 March 1958, while differing from each other in detail, showed clear evidence of diffusive separation in the altitude range 112–160 km. These have been described by Meadows and Townsend (1958, 1960). Figure 6.8 shows the results of the three experiments. Some explanation is necessary before this figure is discussed.

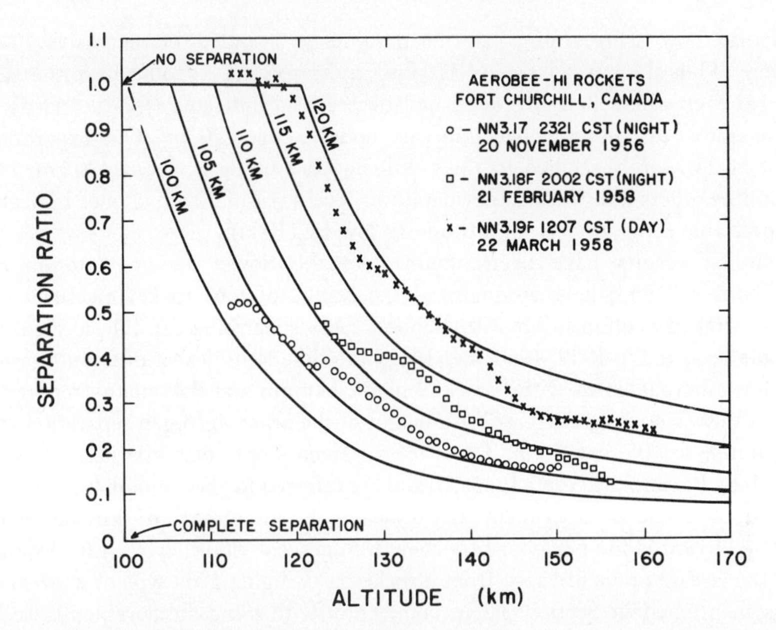

FIG. 6.8. Separation ratio, argon to nitrogen, measured on dates indicated. Full curves refer to distributions calculated on the assumption of complete mixing below and diffusive equilibrium above the indicated altitudes. (After Newell, 1960.)

In an isothermal region with diffusive equilibrium, the variation of number density of an individual gas with altitude is

$$n_i = n_{i0} \exp(-g\mu_i z/kT) \tag{6.21}$$

where an appropriate mean value of g must, of course, be used. If this equation is written for argon and also for molecular nitrogen and a ratio is formed, then

$$\frac{n(\mathrm{A})/n(\mathrm{N_2})}{n_0(\mathrm{A})/n_0(\mathrm{N_2})} = \exp\left[-\frac{gz[\mu(\mathrm{A}) - \mu(\mathrm{N_2})]}{kT}\right] \tag{6.22}$$

The left side of (6.22) is referred to as the *separation ratio*, r. With no diffusion, r would retain the value of 1, because both gases would have a scale height that depended on the mean molecular mass. With diffusion, r would decrease steadily with altitude, because the molecular weight of argon is about 40 and that of molecular nitrogen is about 28.

Figure 6.8 has r as ordinate and altitude as abscissa and the reference ratio $[n_0(\mathrm{A})/n_0(\mathrm{N_2})]$ is the same as in the troposphere. The continuous curves in Fig. 6.8 give the expected values of r according to Eq. (6.22) for different choices of a level below which r is assumed everywhere to be 1 (complete mixing) and above which a complete diffusion equilibrium is postulated. These curves were calculated by using appropriate mean values of g and T in successive 10-km layers. Thus the curve labeled "120 km" indicates that r should have the value of 1.0 everywhere below 120 km and decrease as indicated (for the particular temperature distribution chosen in this model) above 120 km. One experiment, that of 23 March, seemed to show diffusion beginning at about 120 km. Data from the others indicate a somewhat lower level, that of 20 November indicating appreciable separation at an altitude as low as 110 km.

Similar results have been obtained in the Soviet Union. Istomin and Pokhunkov (1963) have summarized the results of four rocket measurements of neutral gas composition with Bennett mass spectrometers. These measurements occurred in 1959, 1960, and 1961 in middle latitudes, at different seasons, and at different times of the day and night. Istomin and Pokhunkov report that all of these showed a decrease with height of the argon-nitrogen separation ratio beginning at 105 to 110 km. These measurements have been discussed in more detail by Pokhunkov (1960, 1963a, b) and are referred to again in Subsection 6.3.3.

Another type of experiment that indicates the importance of diffusion above 110 ± 10 km is the behavior of a chemiluminescent cloud, created for example by the release of alkali vapor from a rocket at twilight. This type of experiment was mentioned in Section 3.3 in connection with wind measurements. In the present context it is of importance to note that the character of the cloud is usually observed to change at a certain level presumed to be close to the diffusion level. According to Blamont and de Jager (1962), below this level the trail shows "globular distortions in the form of elements having an average diameter of 0.5 km; above that height the trail is completely smooth, though highly curved, showing no small irregularities." Various observations of Blamont and de Jager (1961) have indicated the altitude of this level to lie between 96 and 112 km (five observations). They find that it is possible to account for the expansion of the trail above the critical level on the theory of molecular diffusion. From similar observations Groves (1960) has reported a diffusion level of 102 km, Rees (1961) a level of 103 km, and Manring *et al.* (1961) a level of about 111 km.

In the radio observations of ionized meteor trials, it is possible to account for the expansion of the ionized trails on the theory of molecular diffusion (for example, Greenhow and Hall, 1961).

6.3.3 Mean Molecular Weight; the Ratio $n(O)/n(O_2)$

The same mass-spectrometer measurements that gave the $n(A)/n(N_2)$ ratio at Churchill also detected the presence of other atmospheric constituents in the range of atomic mass numbers 10 to 50, for example N, O, and O_2. From the ion currents associated with these it is possible in principle to determine the mean molecular weight of air as a function of altitude.

Meadows and Townsend (1960) in this manner deduced that the mean molecular weight of air over Churchill decreases surprisingly slowly, still having a value between 27 and 28 in the vicinity of 200 km, then decreasing rather rapidly above there. As might be expected, since the ratio $n(O)/n(O_2)$ is the principal variable affecting the mean molecular weight in this altitude interval, this result implies a rather small amount of O_2 dissociation (less than would be inferred from the O_2 measurements discussed in Subsection 6.3.1).

C. Y. Johnson (1961) criticized this conclusion on the grounds that laboratory calibrations for mass discrimination or nonlinear response of the spectrometer were not applied [as they were in the special case of the $n(A)/n(N_2)$ ratio]. There is also a very serious uncertainty in the case of the oxygen species about the effect of recombination of atomic oxygen within the instrument; such recombination would, of course, reduce the measured value of $n(O)/n(O_2)$ relative to the ambient value, perhaps seriously.

A more recent mass-spectrometer measurement at Wallops Island, Virginia (Schaefer, 1963), was designed insofar as possible to minimize the uncertainty resulting from possible recombination of atomic oxygen in the spectrometer. Preliminary results indicate a much higher $n(O)/n(O_2)$ ratio than do the Churchill data. In the vicinity of 130 km this ratio was found to be between 2 and 3 rather than the value of less than 0.1 indicated by Meadows and Townsend.

From one of the Soviet mass-spectrometer flights, Pokhunkov (1963a, b, c) has deduced mean molecular weights and concentrations of the principal constituents up to 210 km. The data apply to a September midnight (in 1960) at "moderate latitudes of the European part of the U.S.S.R." He found a mean molecular weight decreasing approximately linearly from about 28 at 100 km to about 24 at 210 km. He estimated the probable error as 2 per cent at the lower level and $\pm$ 4 per cent at the higher level. These results show an atmosphere in which $n(O) = n(O_2)$ at about 130 km and in which at 210 km more than 50 per cent of the particles are N_2, less than 10 per cent O_2.

On the other hand, Hinteregger's preliminary analysis of his extreme-ultraviolet absorption data seem to indicate much less O_2 and much more O than do these results. According to this analysis (Hinteregger, 1962) over White Sands in daytime in August 1961, the atmosphere at 200 km consisted mostly of atomic oxygen and at 130 km $n(O)$ was about 10 times $n(O_2)$ and even exceeded $n(N_2)$.

To what extent these wide differences may result from space and time variations and to what extent they may result from experimental uncertainties await further elucidation.

Both the Churchill and the U.S.S.R. mass-spectrometer measurements showed only small amounts of atomic nitrogen representing no more than 1 to 2 per cent dissociation of the nitrogen molecules, according to Istomin and Pokhunkov (1963). One of the Churchill measurements and all of the Soviet measurements also were able to detect in the range of very small atomic mass number but detected no significant amounts of hydrogen or helium. The atmosphere between 100 and 200 km seems to consist of N_2, O_2, and O, but in just what proportions under different conditions is not yet experimentally established.

6.4 Structure Measurements

In this section we consider methods and results of measuring thermospheric structure—most of which apply to density. Only a few data above 100 km were available in the pre-IGY period; these were supplemented during and especially after the IGY by additional rocket results and especially by observations of the orbits of artificial satellites.

In the 100–200-km interval, structure information comes primarily from observations from moving rockets. Above 200 km, observations are mostly made with the use of satellites. It turns out that the latter data are far more copious. The stıucture of the thermosphere above 200 km is rather well defined now, even to the details of its variability, which is pronounced. On the other hand, in the transition region 100–200 km observations are scanty. Although the large-scale features are now probably established, more structure (as well as composition) data are needed for this region to fill in the picture.

In Subsection 6.4.1, we discuss the rocket measurements in the 100–200-km region. Subsection 6.4.2 takes up the methods used and the general results achieved from satellite observations above 200 km. Subsection 6.4.3 is concerned with the fascinating patterns of density variation that occur in the thermosphere above 200 km. This section is confined primarily to a consideration of density, the variable that is usually directly determined, and leaves for Section 6.5 the question of interpretation of these data in terms of scale height and temperature.

It should be pointed out that kinetic temperature can be deduced roughly from the characteristics of airglow and auroral spectra. (See Hunten, 1961). In addition a few temperature determinations have been made from the Doppler widths of the emission lines of sodium clouds released from rockets (for example, Blamont *et al.*, 1961). In principle, temperatures can also be inferred from the vertical variation of the number densities of individual constituents, as revealed by mass-spectrometer and optical measurements, if a complete diffusive equilibrium is assumed to prevail. Pokhunkov (1963a, b) has given some temperatures deduced in this manner.

6.4.1 Pressure and Density Measurements, 100–200 km

Most of the few available measurements of atmospheric structure in the 100–200-km region come from pressure-gauge measurements on moving rockets. The interpretation of these experimental data in terms of ambient pressures and densities is rather complicated and has been discussed in some detail by Horowitz and LaGow (1957) and by Ainsworth, Fox, and LaGow(1961). Suffice it to say here that usually ambient pressure is determined between about 100 and 115 km and ambient density above that level. Densities for the lower levels can be deduced from the slope of the pressure-height curve through the hydrostatic equation.

Density can also be determined by the optical method from measurements of atmospheric absorption of X-rays in the spectral region near 50 A. Within reasonable limits of atmospheric composition, the absorption coefficient of air is a function of density only at this wavelength. In the measurements that have been made (Byram *et al.*, 1956; Chubb *et al.*, 1958) the photon counter used (with a mylar window) responded in the spectral region 44–60 A, and it was therefore necessary to assume something about the relative spectral variation of solar energy in this interval. A gray-body distribution at a temperature of 7×10^{5}°K was used, but as a matter of fact, other reasonable assumptions did not change the results significantly. This spectral interval is on the flat, long-wave side of peak emission for black-body curves at any reasonable coronal temperature.

One or two measurements have been made (Faucher, Procunier, and Sherman, 1963) with an adaptation of the falling-sphere experiment used at lower levels (Subsection 3.3.3). An inflatable sphere is ejected at an altitude between 80 and 100 km during the ascent of a rocket (Aerobee). The sphere continues to high levels on a free-fall trajectory modified by drag effects. Accelerations are measured by accelerometers positioned at the center of the sphere. One source of uncertainty in this method is the somewhat doubtful value of drag coefficient under conditions obtaining in the lower thermosphere.

Densities can also be inferred from the rate of growth of artificial sodium clouds (Blamont, 1959) or of meteor trails (Greenhow and Hall, 1960). With particular reference to an altitude of 120 km, Whitehead (1963) has summarized some of these data, as well as some of the rocket data and data based on collision frequencies deduced from ionospheric measurements in the *E* region.

Table 6.4 gives the results of most of the available rocket data in terms of the derived densities. The columns in Table 6.4 are headed by letters A, B, C, etc. and it is convenient to describe here the time, location, instrumentation, and published source appropriate to the data in each column.

A: See Horowitz and LaGow (1957). The data were obtained 7 August 1951, daytime, at White Sands, New Mexico, with an ionization gauge mounted in the side of the nose cone of a Viking rocket.

B, C, D, E: See Horowitz and LaGow (1958); Horowitz, LaGow, and Giuliani (1959); LaGow, Horowitz, and Ainsworth (1960); Ainsworth, Fox,

TABLE 6.4

DENSITIES IN THE THERMOSPHERE FROM ROCKET MEASUREMENTS[a]

Altitude (km)	Density (g cm^{-3})											
	A	B	C	D	E	F	G	H	I	J	K	L
80	—	2.2(8)	1.4(8)	—	—	—	—	—	—	—	—	—
90	—	—	3.4(9)	—	—	—	—	—	—	—	2.9(9)	—
100	2.5(10)	7.2(10)	6.7(10)	—	—	—	—	—	2.9(10)	—	3.7(10)	4.8(10)
110	5.0(11)	1.3(10)	1.5(10)	—	—	1.0(10)	—	4(11)	6.2(11)	—	5.9(11)	9.5(11)
120	1.2(11)	2.6(11)	2.5(11)	—	—	3.0(11)	4(11)	1.1(11)	1.5(11)	—	1.5(11)	2.4(11)
130	3.3(12)	6.4(12)	6.7(12)	—	—	6(12)	1.2(11)	3.8(12)	—	1.1(11)	6.0(12)	7.0(12)
140	1.2(12)	3.0(12)	3.0(12)	—	—	1.4(12)	5(12)	1.2(12)	—	6(12)	3.0(12)	3.1(12)
150	6.6(13)	1.9(12)	1.8(12)	—	—	5(13)	2.0(12)	5(13)	—	3.1(12)	1.8(12)	1.7(12)
160	4.3(13)	1.4(12)	1.2(12)	—	—	1.9(13)	8(13)	3.0(13)	—	—	1.1(12)	1.1(12)
170	3.0(13)	1.1(12)	8.6(13)	—	—	8(14)	4(13)	2.0(13)	—	—	—	8.3(13)
180	2.3(13)	8.9(13)	6.4(13)	—	—	6(14)	2.7(13)	1.3(13)	—	—	—	6.6(13)
190	1.8(13)	7.9(13)	—	—	—	4(14)	1.9(13)	—	—	—	—	4.7(13)
200	1.4(13)	6.7(13)	—	3.5(13)	1.3(13)	3.5(14)	1.3(13)	—	—	—	—	3.6(13)
210	1.1(13)	6.2(13)	—	—	—	3.0(14)	1.1(13)	—	—	—	—	—
220	9.0(14)	—	—	—	—	—	—	—	—	—	—	—

[a] See text for details about the sources of the data. In some cases, data were read from published graphs by the present author. A number in parentheses gives the power of 10 by which the preceding number must be multiplied (with minus sign omitted); thus "(10)" means "$\times 10^{-10}$."

and LaGow (1961). These data were obtained at Churchill on, respectively, 29 July 1957, daytime; 31 October 1958, daytime; 17 November 1956, daytime; 24 February 1958, nighttime. The measurements were made with Aerobee-Hi rockets especially instrumented with pressure gauges for high-altitude structure measurements during the IGY by the Naval Research Laboratory. Some data were also obtained at lower levels during these flights (see Subsection 3.3.2). In addition to the pressure sensors, each rocket carried equipment for telemetering, determination of rocket aspect, and determination of rocket trajectory. See LaGow, Horowitz, and Ainsworth (1960) for a general description of the equipment, and the other references for details of individual flights.

F, G, H: See Townsend and Meadows (1958); Meadows and Townsend (1960). These data were obtained at Churchill on, respectively, 20 November 1956, nighttime; 21 February 1958, nighttime; 22 March 1958, nighttime. They were obtained with equipment (Bennett mass spectrometers) primarily designed to measure atmospheric composition, but used in this application as a pressure gauge.

I, J: See Byram, Chubb, and Friedman (1956); Chubb, Byram, Friedman, and Kupperian (1958). These data were obtained at White Sands on, respectively, 1 December 1953 and 18 October 1955. These data were obtained from X-ray absorption measurements.

K: See Faucher, Procunier, and Sherman (1963). These data were obtained by the falling-sphere method at Eglin Gulf Test Range in Florida on 7 December 1961, evening.

L: Data from the International Reference Atmosphere of the Committee on Space Research (see Subsection 6.5.1 and Table 6.6) are shown here for direct comparison with the measurements.

Table 6.4 illustrates two things: First, there are not many published structure measurements for the region from the mesopause to 200 km (a few of the satellite measurements of density, discussed in Subsection 6.4.2, refer to the upper part of this region but are not included in Table 6.4); second, what data there are exhibit considerable variability. It is difficult to tell to what extent this variability reflects real variability in the atmosphere and to what extent it results from experimental errors.

Consider the situation at 100 km. The reported densities differ by a factor of about 3, and the implied pressures (since temperature at 100 km cannot vary much percentually) by about the same. The lowest value refers to White Sands in the summer of 1951 and the highest value to Churchill in the summer of 1957. If this difference is considered to be real and is ascribed to a difference in latitude, then one finds inconsistencies with presumed conditions at lower levels. Since the 70-km pressures in summer at White Sands and Churchill certainly do not differ by this factor, the difference implies much lower average summer temperature between 70 and 100 km at White Sands than at Churchill; but this is not consistent with the deep temperature minimum near 80 to 90 km at Churchill in summer, as revealed by the rocket-grenade experiment (Figs. 3.10 and 3.11). There remains the possibility of a systematic difference associated with the sunspot cycle, but such a large difference so low in the thermosphere hardly seems likely.

The density at 200 km, according to these data, varies by a factor of about 20 (6, if one low value in column F is not included). The variations appear not to be simply related to latitude, season, and time of day. According to satellite measurements and current theories of thermospheric structure (Subsections 6.4.2 and 6.5.3) one would not expect such density variations at this elevation.

6.4.2 Density above 200 km

Density above 200 km has been inferred primarily from observations of the variable orbits of artificial satellites. The satellite experiences a retarding force caused by the drag of the atmosphere through which it is moving. The resulting effect on the orbit is most easily measured in terms of the slowly decreasing period of revolution, although, of course, other orbital elements are also affected.

The drag acceleration may be expressed in the form*

$$a_D = \frac{\rho V^2 A_n c_D}{2M} \tag{6.23}$$

where ρ is the atmospheric density, A_n the satellite's area projected on a plane normal to the direction of motion, V its velocity relative to the center of the earth, M its mass, and c_D the drag coefficient with a value at these altitudes in the neighborhood of 2.2 (see, for example, Stirton, 1960).

There are several difficulties in the way of deducing densities from (6.23), apart from the somewhat uncertain value of the drag coefficient. Unless the satellite is spherical, A_n varies as the satellite rolls and tumbles during its orbit, and it is necessary to take some appropriate average value that depends on the shape. For some of the Soviet satellites the shape has not been well known. Furthermore, since satellites typically have eccentric orbits, the air density encountered during one revolution is not at all constant, and it is the integrated effect of the drag during the entire orbit that causes the orbital variations. It is

* More exactly the right side of this expression should be multiplied by a factor usually between 0.9 and 1 to take account of the earth's rotation (Sterne, 1959; King-Hele and Walker, 1961b). This is because the atmosphere (apart from any "wind" relative to the earth's surface) rotates with the earth, introducing a velocity relative to the satellite that is not included in the V of (6.23).

therefore customary to assume some functional relationship between air density and altitude in order to calculate this integrated effect. Of course, since air density decreases approximately exponentially with altitude, the higher densities near perigee have much the greatest effect, and the results apply to density at and immediately above perigee.

A common assumption about vertical density variation is that the density above perigee may be represented by

$$\rho = \rho_\pi \exp[-(z - z_\pi)/H_\rho] \tag{6.24}$$

where ρ_π is perigee density, z_π is perigee altitude, and H_ρ is the density scale height,* assumed here to be vertically constant. As long as only one satellite is considered, this is not too bad an approximation, because only the higher densities near perigee are very effective in slowing down the satellite. If the drag force is included in the equation of motion of the satellite, with ρ expressed as a function of height according to (6.24), then it is possible to derive the following approximate expression for ρ_π (Sterne, 1958; King-Hele, 1959):

$$\rho_\pi = -\frac{dP}{dt}\frac{M}{3A_n c_D}\left(\frac{2e}{\pi a H_\rho}\right)^{\frac{1}{2}}\left[1 - 2e - \frac{H_\rho}{8ae}\right] \tag{6.25}$$

where some second-order terms in e and (H_ρ/ae) are left out of the bracket. This expression is applicable only when $0.02 < e < 0.2$, which is usually the case. For an expression applicable when $e < 0.02$, see King-Hele and Walker (1961b). Here P is the satellite's period, M its mass, e the eccentricity of the orbit, and a the semimajor axis of the orbit. Perigee distance r_π (measured from the center of the earth) is related to eccentricity and the semimajor axis by $r_\pi = a(1 - e)$. Note that even if the orbital elements and time change of the period, as well as the characteristics of the satellite, are known accurately, this equation gives ρ_π only if H_ρ is known or estimated. To put it another way, the satellite information gives not ρ_π directly but $\rho_\pi H_\rho^{\frac{1}{2}}$ (if we neglect the small term involving H_ρ in the brackets, because an error of estimating H_ρ in this term is not very important).

King-Hele (1959) (see also Whitney, 1959) introduced an interesting device to get around the uncertain value of H_ρ in Eq. (6.25). Suppose the density above perigee is distributed according to Eq. (6.24), and H_ρ' is a "best estimate" of the

* Chapman (1961) defines the density scale length with respect to the vertical coordinate as $(d \ln \rho/dz)^{-1}$. What is here called the density scale height, H_ρ, is the negative of this. In some applications of (6.24), H_ρ has been identified with scale height H. However, this is true only in an atmosphere where H is vertically constant. It can be shown easily from (6.11) that in an atmosphere where $H = H_0 + \beta z$ the relationship between the two is $H = (1 + \beta)H_\rho$.

actual density scale height H_ρ. Then the density at the altitude $z = z_\pi + H_\rho'/2$ is given by

$$\rho(\text{at } z_\pi + H_\rho'/2) = \rho_\pi \exp(-H_\rho'/2H_\rho) \tag{6.26}$$

Substituting the value of ρ_π from Eq. (6.25), we get

$$= -\frac{dP}{dt}\frac{M}{3A_n c_D}\left(\frac{2e}{\pi a H_\rho}\right)^{\frac{1}{2}} \exp\left(-\frac{H_\rho'}{2H_\rho}\right)\left[1 - 2e - \frac{H_\rho}{8ae}\right]$$

$$= -\frac{dP}{dt}\frac{M}{3A_n c_D}\left(\frac{2e}{\pi a H_\rho'}\right)^{\frac{1}{2}} \left[\left(\frac{H_\rho'}{H_\rho}\right)^{\frac{1}{2}} \exp\left(-\frac{H_\rho'}{2H_\rho}\right)\right]\left[1 - 2e - \frac{H_\rho}{8ae}\right] \tag{6.27}$$

This expression gives the density, not at perigee, but at the distance $H_\rho'/2$ above perigee, where H_ρ' is the estimated, not necessarily correct, density scale height. The first quantity in square brackets depends on the (unknown) ratio H_ρ'/H_ρ. However, it is rather insensitive to this ratio, having the value 0.55 when $H_\rho'/H_\rho = 0.5$, the value 0.61 when $H_\rho'/H_\rho = 1$, and the value 0.58 when $H_\rho'/H_\rho = 1.5$. Therefore one may take this quantity to have a numerical value of, say, 0.59 with confidence that the error in ρ will be only a few per cent. The H_ρ that appears in the second set of square brackets appears in a small term and may be incorrectly estimated with only small penalty.

Jacchia (1960a) discussed the effect of a vertically variable density scale height and King-Hele (1963) gave formulae taking into account such a vertical variation.

There is now a very large and rapidly growing number of contributions related to the subject of density determinations from the changing orbits of artificial satellites. Most of the early studies were referenced by Kallmann Bijl (1961) and Nicolet (1961a). The earliest results generally applied to values of density derived from changes of period averaged over very many orbits. Therefore, they tended to average out short-period density fluctuations that are now known to be very important. In this part of Section 6.4, we discuss only the average vertical distribution of density between 200 and 700 km and the large changes therein from day to night. In Subsection 6.4.3, the question of other variations, associated with solar activity of one kind or another, is taken up. In these discussions, the references are intended to be representative but by no means exhaustive.

Jacchia (1959a) first pointed out the importance of diurnal variations. The location of perigee in space moves slowly relative to the sun-earth line and a given satellite may have its perigee point on the sunlit side of the earth for a few months, then in the earth's shadow for a few months. Furthermore, the altitude of perigee usually decreases rather slowly, and often remains essentially constant while the perigee point moves from sunlit to dark hemisphere. In this circumstance, if the orbital period is observed to decrease more rapidly when perigee occurs during the day than when it occurs during the night, this is an indication of higher density during the day at essentially the same altitude. This is exactly what is observed and is illustrated for Explorer 7 (1959ι 1 and 2)* in Fig. 6.9, which shows a smoothed graph of this satellite's period *versus* time. The marked changes in the slope of this curve as perigee passes from day to night and back to day conditions illustrates the point. In this case perigee was in the vicinity of 550 to 600 km.

By way of quantifying this effect, the results of King-Hele and Walker (1961b) based on 29 satellites launched through 1960 are shown in Table 6.5. Below 300 km no day-night variation is shown. Above 300 km two sets of data

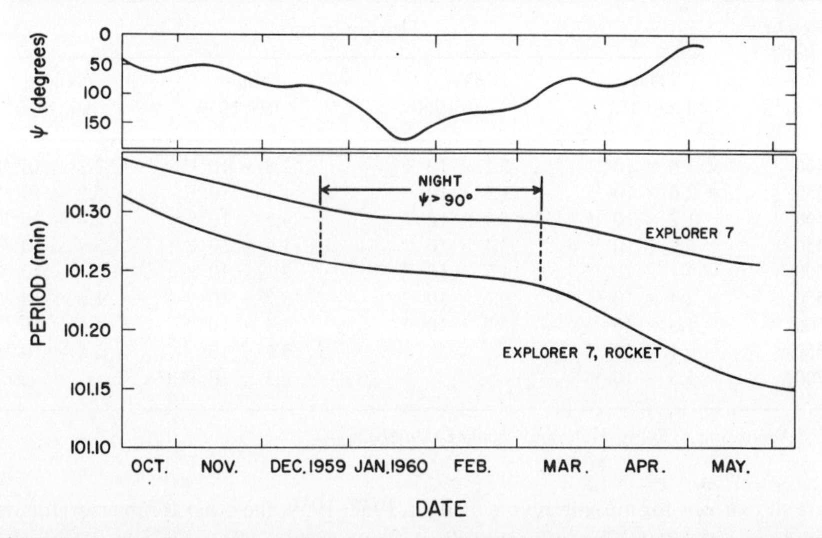

FIG. 6.9. Density variation from night to day, as revealed by the variation of the period of Explorer 7. In the ordinate label, ψ is the geocentric angle between the sun and the perigee point. The period decreases because of atmospheric drag, but less rapidly when perigee occurs in the dark hemisphere. This indicates lower nighttime density near the altitude of perigee (in this case 550–600 km). (After King-Hele and Walker, 1961a.)

---

* Prior to 1963, satellites were designated by the year they were launched, followed by a lower-case Greek letter in the order of launching for the year. In case more than one object went into orbit from a single launch (for example, rocket carrier and capsule), a number suffix distinguished them in order of brightness. Effective in 1963, satellites are designated by the year of launch, followed by a number in the order of launching for the year. In case more than one object goes into orbit from a single launch, a letter suffix distinguishes them: A for the component carrying the principal scientific payload; B, C, etc. for components carrying secondary payloads; subsequent letters for inert components in the order of brightness. For more details, see National Academy of Sciences (1962).

TABLE 6.5

DENSITY AS A FUNCTION OF ALTITUDE[a]

Height (km)	Density (g cm^{-3})
180	6.2×10^{-13}
200	3.9×10^{-13}
220	2.8×10^{-13}
240	1.8×10^{-13}
260	9.8×10^{-14}

Height (km)	Density (g cm^{-3})			
	Day, late 1958	Day, late 1960	Night, late 1959	Night, late 1960
300	3.6×10^{-14}	2.5×10^{-14}	2.8×10^{-14}	2.3×10^{-14}
350	2.0×10^{-14}	1.1×10^{-14}	1.1×10^{-14}	6.2×10^{-15}
400	1.2×10^{-14}	5.7×10^{-15}	4.5×10^{-15}	2.4×10^{-15}
450	6.9×10^{-15}	3.1×10^{-15}	1.9×10^{-15}	9.4×10^{-16}
500	4.1×10^{-15}	1.7×10^{-15}	8.1×10^{-16}	3.9×10^{-16}
550	2.4×10^{-15}	9.6×10^{-16}	3.3×10^{-16}	1.6×10^{-16}
600	1.5×10^{-15}	5.5×10^{-16}	1.4×10^{-16}	6.6×10^{-17}
650	9.1×10^{-16}		5.7×10^{-17}	2.6×10^{-17}
700	5.5×10^{-16}		2.3×10^{-17}	

[a] According to King-Hele and Walker (1961b).

are shown, one for measurements made in 1958–1959, the other for measurements made in late 1960. The difference illustrates a secular effect associated with the sunspot cycle (see Subsection 6.4.3). Considering the 1960 data, we note a variation from night to day by factors of about 1.1 at 300 km, 2.4 at 400 km, 4.4 at 500 km, and 8.3 at 600 km. These are very large differences and imply very large temperature differences between day and night in the upper thermosphere. Other investigators have also studied this effect. Paetzold and Zschörner (1960) reported factors of 1.07 at 220 km, 2.0 at 365 km, and 6.0 at 655 km.

Within the limits of the available data, the diurnal variation can be represented as a function of local time only, as noted by H. A. Martin *et al.* (1961) and by King-Hele and Walker (1961b). According to Paetzold and Zschörner (1960), the maximum occurs at about 1400 local time. According to H. A. Martin *et al.* (1961), the density reaches a minimum at 0500 local time, rises sharply to a maximum at about 1400 local time, then decreases nearly as rapidly until 1800, then decreases rather slowly until the 0500 minimum.

Jacchia (1960b, 1963) has pictured the diurnal variation as a "diurnal bulge" with maximum density (at a given altitude) occurring at the same latitude but about 30° (two hours) east of the subsolar point. It seems reasonable to expect that density is a function of the geometric angular distance from this point and not simply of local time, which depends only on the longitudinal displacement from the point. Jacchia (1963) has searched for evidence of such a dependence on the basis of accurately determined densities from a few selected satellites, but has concluded that the data are insufficient to show more than the dependence on local time.

All of the results discussed above depend on the effect of density on the orbit of a satellite. A few observations have been made more directly by instruments carried on satellites, giving densities from ram-pressure measurements (Pokhunkov, 1960; Sharp, Hanson, and McKibbin, 1962). They confirm the results deduced from the orbital data within satisfactory limits.

6.4.3 Density Variations and Solar Activity

Some very interesting correlations between density and various indices of solar activity have been detected in the first few years of satellite observations. Some of these involve indices that are believed to represent variations of solar emission in the extreme ultraviolet (200–1000 A). The others involve indices of solar corpuscular radiation. These are discussed in turn.

In 1959 Jacchia (1959a, b) and Priester (1959) established a correlation between density variations and variable solar radiation. The density variations in question, quasi-periodic with a period of about 27 days, were shown to correlate closely with changes in solar radio emission in the decimeter range (10.7 cm in the case of Jacchia's correlation, 20 cm in the case of Priester's). These radio emissions originate in the corona and are believed to be highly correlated with the emission of extreme ultraviolet radiation (Section 4.5). Priester (1961) showed that if ρ is assumed to be proportional to F^m, where F is the solar decimetric flux and m is to be determined empirically, then m increases with height above 300 km and is larger at night than in the daytime. Jacchia (1961a) showed that this behavior can be explained if the atmosphere above 200 to 300 km is assumed to be vertically isothermal (see Section 6.5) at a temperature T and if T is linearly correlated with the decimetric flux. According to Jacchia (1963) the best result is obtained if $\partial T/\partial F_{10} = 2.5°\text{K}$ where F_{10} is the solar flux at 10.7 cm expressed in units of 10^{-22} watt m^{-2} $(\text{cycles/sec})^{-1}$. Figure 6.10 shows an example of this effect and the correction, after Jacchia and Slowey (1963a).

Since 1958, during the years of declining solar activity, observations have shown gradually decreasing densities above 200 to 300 km. This is illustrated in Table 6.5 and is shown also in Fig. 6.11, after King-Hele and Rees (1963). The effect is striking. For example, at 500 km during the day the density decreased between 1958 and 1962 by a factor of more than 10. During a period

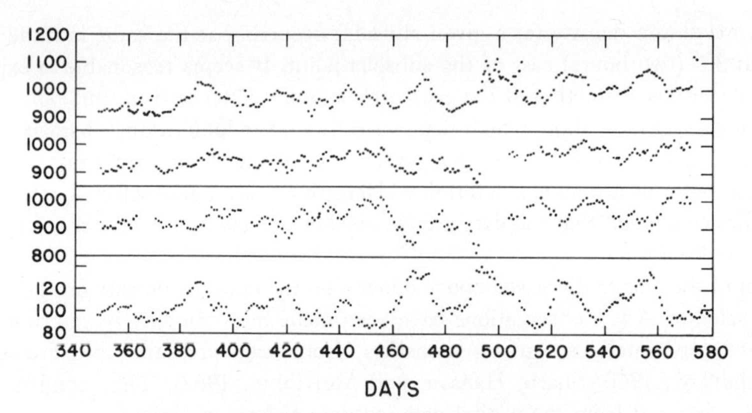

FIG. 6.10. An example showing the correlation between temperature in the thermosphere and 10.7-cm solar radiation. The abscissa is time in mean Julian days, with an initial "37" left off each number. The ordinate for the bottom curve is 10.7-cm solar flux (F_{10}) in units of 10^{-22} watt m^{-2} (cycles/sec)$^{-1}$. The ordinate for the top three curves is temperature in degrees K, as inferred from satellite drag effects (see text). The top curve gives the uncorrected temperature, which is evidently correlated with F_{10}. The second curve from the top gives the temperature corrected according to $\partial T/\partial F_{10} = 2.5°K$. The third curve from the top gives the temperature corrected according to $\partial T/\partial F_{10} = 4.0°K$, which gives an overcorrection. (After Jacchia and Slowey, 1963a.)

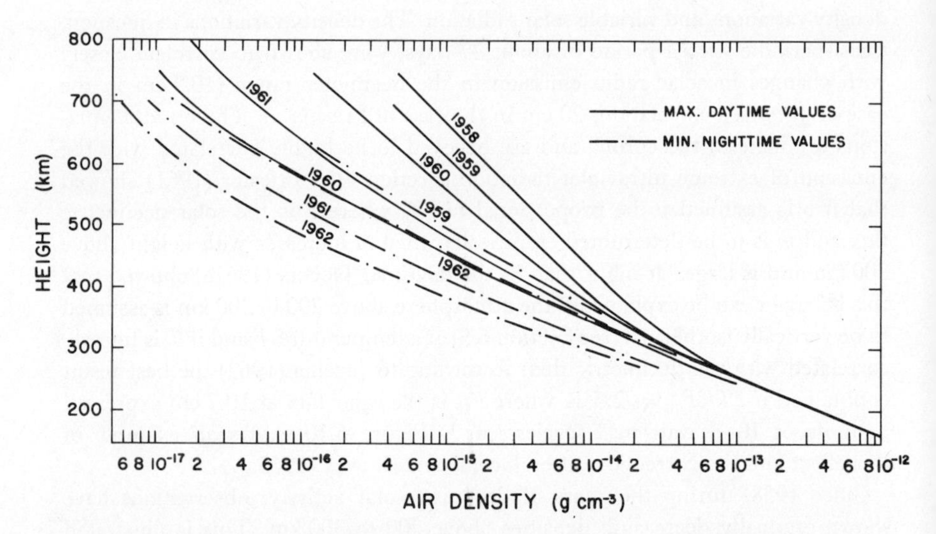

FIG. 6.11. Variation of thermospheric density with phase of the sunspot cycle. Densities in 1958, near sunspot maximum, were higher than in succeeding years. The curves are intended to represent conditions near the middle of the indicated years. (After King-Hele and Rees, 1963.)

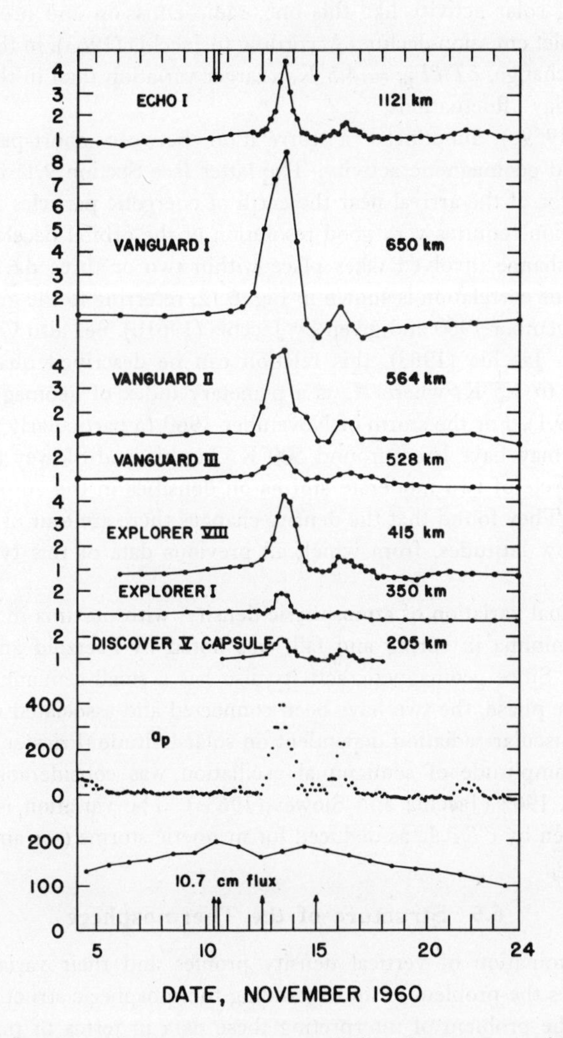

FIG. 6.12. Density variations during a magnetic disturbance, as revealed by variations of the periods of several satellites during magnetic disturbances in November 1960. The ordinate for each of the top curves is the ratio of instantaneous acceleration to the mean acceleration for the period. Altitudes of perigee are given on the right. The bottom curve shows that the large variations of acceleration observed 13–14 November and the smaller variations observed 16–17 November were not correlated with 10.7-cm flux variations. The second curve gives the variation of an index of geomagnetic activity (Section 9.1), which does show a marked correlation with the acceleration curves. The arrows indicate occurrences of solar flares. (After Jacchia, 1961b.)

of decreasing solar activity like this one, radio emission and presumably also solar ultraviolet emission decline. According to Jacchia (1963), in the case of this long-period change, $\partial T/\partial F_{10} = 4.5°K$, a larger variation than in the case of the shorter "27-day" fluctuations.

Jacchia (1959c) announced a correlation between short-period density variations and geomagnetic activity. The latter (see Section 9.1) is believed to be an indicator of the arrival near the earth of energetic particles from the sun. This correlation requires very good resolution of the orbital deceleration, since the density change involved takes place within two or three days. A striking example of this correlation is shown in Fig. 6.12, referring to the great magnetic storm of November 1960 and given by Jacchia (1961b). See also Groves (1961). According to Jacchia (1963), this relation can be described quantitatively if $\partial T/\partial A_p = 1$ to 1.5°K, where A_p is a planetary index of geomagnetic activity (see Section 9.1). For the storm of November 1960 (a particularly intense one), the increase may have been around 500°K. Jacchia and Slowey (1963b) have noted the effects of two moderate storms on densities in the auroral zone (see Section 9.3). They found that the density changes there are four or five times as great as at low latitudes, from which all previous data of this type had been derived.

A semiannual variation of atmospheric density, with maxima in summer and winter and minima in spring and fall, was noted by Paetzold and Zschörner (1960, 1961). Since geomagnetic activity also has a small semiannual variation with the same phase, the two have been connected and associated with a model of solar corpuscular radiation dependent on solar latitude (Priester and Cattani, 1962). The amplitude of semiannual oscillation was considerably greater in 1958 than in 1962 (Jacchia and Slowey, 1963a). The variation is larger than would be given by $\partial T/\partial A_p$ as deduced for magnetic storms (see above).

6.5 Structure of the Thermosphere

The determination of vertical density profiles and their variations by no means satisfies the problem of understanding thermospheric structure. There is in addition the problem of interpreting these data in terms of pressure, temperature, and mean molecular mass. From hydrostatic considerations, only the ratio of the latter two can be determined, a ratio which is usually expressed in terms of the scale height, H, or the molecular-scale temperature T_m (Subsection 1.1.2). The values of kinetic temperature and mean molecular mass can be determined from this ratio only by (a) satisfactory observational knowledge of the vertical vaiiation of one or the other or (b) an appeal to physical reasoning to suggest the most likely interpretation in the light of other evidence. We have already seen that satisfactory observations are lacking: kinetic temperatures are obtainable only from scanty spectral data, and composition measurements are few and contradictory. We must therefore have recourse to certain physical arguments that have led to an interpretation of the density data, an interpretation which is consistent and undoubtedly correct in its broad outline but which still leaves many details unsettled.

In Subsection 6.5.1 we discuss some determinations of pressure and scale height that have been made and present certain results from some recent "model" atmospheres. Subsection 6.5.2 considers briefly some of the physical factors that influence the resolution of the temperature and mean molecular weight variations. These include the vertical distributions of constituents in a diffusive equilibrium, heat sources as they might relate to possible temperature and constituent distributions, and the molecular conduction of heat in the thermosphere. In Subsection 6.5.3 some model atmospheres that are consistent with both density observations and other considerations are discussed.

6.5.1 Vertical Distributions of Pressure and Scale Height

To each vertical distribution of density there correspond vertical distributions of pressure and scale height that, except for a constant of integration, are specified by the hydrostatic equation and the gas law. Although this constant of integration is troublesome and leads to uncertainties near the ceiling of the density determinations, at a level below this ceiling where the pressure and density are, say, 10–100 times their values at the top, little uncertainty is involved. There are various specific procedures that one might adopt to go from a density profile to pressure and scale-height profiles (see, for example, Subsection 3.3.3), but they do not differ in principle. Inasmuch as some density data are available well above 1000 km, we shall not consider that structure parameters up to 600–700 km suffer seriously from this difficulty.

On the other hand, the data are no better than the density data from which they are derived and details of the results may certainly be suspect because of density uncertainties. Particularly in the 100–200-km region, the few observations available are contradictory in some respects (Table 6.4).

Two model or reference atmospheres have become available during the past few years. One, prepared by the Committee on Space Research (COSPAR) of the International Council of Scientific Unions, was released in 1961 (COSPAR, 1961). It is referred to as the COSPAR International Reference Atmosphere (CIRA). Another, prepared by the U. S. Committee on Extension of the Standard Atmosphere, is referred to as the U. S. Standard Atmosphere, 1962 (Sissenwine *et al.*, 1962). Figures 1.1 and 1.2 were based on the latter. The extension above 90 km is labeled "speculative" since it cannot be considered "standard" to the same degree as the results for the well-explored troposphere.

Table 6.6 gives values of density, scale height, temperature, and molecular weight for selected levels from the CIRA "average" atmosphere and the U. S.

TABLE 6.6

THERMOSPHERIC STRUCTURE AND COMPOSITION, 100–600 KM[a]

Altitude (km)	Density ($g\ cm^{-3}$)		Scale height (km)		Temperature (°K)		Mean molecular weight		Acc. of gravity ($cm\ sec^{-2}$)
	CIRA	USSA	CIRA	USSA	CIRA	USSA	CIRA	USSA	
100	4.78(−10)	4.97(−10)	6.43	6.36	212.10	210.02	28.85	28.88	950.52
110	9.49(−11)	9.83(−11)	8.10	7.90	265.02	257.00	28.72	28.56	947.59
120	2.44(−11)	2.44(−11)	10.55	10.96	342.96	349.49	28.60	28.07	944.66
130	6.95(−12)	—	17.68	—	569.44	—	28.43	—	941.76
140	3.07(−12)	—	25.05	—	799.13	—	28.25	—	938.86
150	1.69(−12)	1.84(−12)	32.08	29.46	1014.53	892.79	28.09	26.92	935.97
160	1.11(−12)	1.16(−12)	36.90	34.17	1155.26	1022.20	27.90	26.66	933.10
170	8.26(−13)	8.03(−13)	38.96	37.36	1176.53	1103.40	27.70	26.40	930.24
180	6.59(−13)	—	38.94	—	1193.17	—	27.47	—	927.40
190	4.73(−13)	4.34(−13)	39.95	41.93	1210.51	1205.40	27.25	25.85	924.57
200	3.61(−13)	—	40.99	—	1226.77	—	27.00	—	921.75
230	1.67(−13)	1.56(−13)	44.28	48.73	1273.57	1322.30	26.18	24.70	913.37
250	1.03(−13)	—	46.65	—	1301.45	—	25.55	—	907.85
300	1.34(−14)	3.58(−14)	53.21	58.76	1358.51	1432.10	23.74	22.66	894.27
350	1.23(−14)	—	60.67	—	1401.28	—	21.80	—	880.98
400	5.09(−15)	6.49(−15)	68.79	71.45	1436.16	1487.40	20.00	19.94	867.99
450	2.33(−15)	—	76.83	—	1465.97	—	18.55	—	855.29
500	1.17(−15)	1.58(−15)	83.20	82.44	1474.15	1499.20	17.48	17.94	842.86
550	6.15(−16)	—	90.46	—	1474.15	—	16.31	—	830.70
600	3.45(−16)	4.64(−16)	95.46	90.91	1474.15	1506.10	15.70	16.84	818.00

[a] According to the COSPAR International Reference Atmosphere, 1961 (CIRA) and to the U. S. Standard Atmosphere, 1962 (USSA). The notation $(-x)$ means the preceding number must be multiplied by 10^{-x}.

Standard Atmosphere. Table 6.7 gives the "minimum" and "maximum" values of these quantities at certain levels above 200 km. These refer not to absolute maxima and minima during a solar cycle, but to day maxima and night minima during a period of moderate solar activity (1958–1960). Pressures are not explicitly given but may, of course, be obtained from $p = \rho g H$.

For the present, the reader's attention is directed to the values of scale height in Table 6.6. We shall discuss temperature and molecular weight later. Although the details differ in the two versions and other details of interpretation might be consistent with the data, two facts stand out. The first is the clear tendency for scale height to increase steadily from 100 km to the ceiling of the data given. The second, indicated by the rocket data, is the large scale-height gradient between 120 km and 160 km, with the gradient smaller above there.

TABLE 6.7

THERMOSPHERIC STRUCTURE AND COMPOSITION, 200–600 KM[a]

Altitude (km)	Density ($g\ cm^{-3}$)		Scale height (km)		Temperature (°K)		Mean molecular weight	
	Min	Max	Min	Max	Min	Max	Min	Max
200	3.83(−13)	4.09(−13)	39.97	49.86	1186.01	1492.44	26.77	27.00
250	1.08(−13)	1.43(−13)	43.12	55.42	1186.01	1546.16	25.19	25.55
300	3.27(−14)	5.26(−14)	48.03	67.40	1186.01	1720.95	22.96	23.74
350	1.10(−14)	2.35(−14)	54.70	76.32	1186.01	1762.93	20.46	21.80
400	4.21(−15)	1.16(−14)	61.07	84.90	1186.01	1773.10	18.60	20.00
450	1.80(−15)	6.06(−15)	66.05	93.60	1186.01	1785.96	17.46	18.55
500	8.24(−16)	3.37(−15)	70.53	102.62	1186.01	1811.11	16.59	17.41
550	3.95(−16)	1.98(−15)	75.03	110.68	1186.01	1830.02	15.82	16.55
600	1.99(−16)	1.23(−15)	78.99	117.11	1186.01	1833.70	15.25	15.90

[a] According to CIRA, 1961, minimum and maximum values. The notation $(-x)$ means that the preceding number should be multiplied by 10^{-x}.

An increase of scale height with altitude is explained by some combination of temperature increase, molecular weight decrease, and gravity decrease. The effect of the last is well known and is much too small to play much role except in the uppermost regions. The relative effects of the first two at different altitudes are difficult to unravel. In their attempts to do this aeronomers have been aided by certain lines of reasoning which will be pursued in the following paragraphs.

6.5.2 Diffusion, Energy Sources, and Conduction

It is now reasonably certain (Section 6.3) that inert minor constituents are in diffusive equilibrium above 110 km and it may be inferred that the distribution of major constituents is essentially determined by diffusion above a somewhat higher level—say 120 to 150 km. It is interesting to note that if one could specify (a) a level above which all constituents are in diffusive equilibrium, (b) physical conditions at that level, including number densities of all constituents, and (c) a certain vertical variation of scale height at and above that level, then the problem would be uniquely solved. A particular temperature distribution would define the vertical distributions of the individual constituents, from which could be computed the vertical distribution of the mean molecular mass; only one such temperature distribution and its resulting distribution of molecular mass would be consistent with the given vertical distribution of scale height. However, it turns out that the resulting answer would depend rather critically on the boundary conditions at the bottom and the details of the scale-height variation—to a degree beyond our present power of resolution of these conditions.

For example, Mange (1961b) considered all constituents to be in diffusive equilibrium above 120 km and carried out two computations under identical conditions except for the assumed mean molecular mass at 120 km. When he took this to be 27, he found a mean molecular mass of 23.84 and a temperature of 1145°K at 200 km. When he took it to be 28, he found a mean molecular mass of 26.09 and a temperature of 1252°K at 200 km. The difference between 27 and 28 at 120 km is much less than our present experimental uncertainty about this factor; indeed it might conceivably have larger natural variations. In this connection it is pertinent to note that, under conditions of diffusive equilibrium, the decrease of mean molecular mass with height is slower the higher the temperature. In a warmer atmosphere, all constituents are spread out over a thicker layer and the lighter ones do not predominate so markedly until a greater geometric altitude is reached. Naturally, in Mange's computation, with a specified scale height at 120 km, the temperature at that level was higher when the mean molecular mass was assumed to be higher.

The consideration of energy sources and sinks and of heat transfer can give valuable clues to the structure of the thermosphere. Although these are discussed in considerably more detail and with appropriate references in Section 7.3, a short qualitative discussion is necessary here in connection with the present question.

In the first place, the thermosphere above 120 to 130 km suffers no important radiative heat losses. No polyatomic molecule can persist in the presence of the dissociating radiations; O_2 and N_2 have no dipole moment and therefore no important infrared bands; other diatomic molecules such as CO and NO are present only in small quantities. Instead energy must be lost by molecular conduction of heat downward. The vertical flux of heat downward is proportional to the magnitude of the vertical temperature gradient, so that one expects the vertical temperature gradient to be controlled by the strength and distribution of the heat sources.

In the second place, as we have seen (Section 4.5), absorption of solar extreme ultraviolet radiation is an important source of heat for the thermosphere and probably the principal source. Very little of this is absorbed above 200 km, where the optical depth of the atmosphere is extremely small, and we should therefore expect the large temperature gradients necessary for conductive heat transfer to occur at levels below 200 km. There remains the likelihood that some heat is provided by conduction from an extended solar corona or by interaction of some kind with solar corpuscular radiation but at the present time this is thought to be a secondary source.

In the third place, at high enough levels conductive heat transfer is very rapid and no significant temperature gradients are to be expected except in the immediate presence of a strong heat source and energy flow. Thus, for example, if the thermosphere is heated primarily by solar electromagnetic radiation, the atmosphere above a level where most of the energy is absorbed can be expected to be nearly isothermal in the vertical at a given time and place.

6.5.3 Vertical Distribution of Temperature and Molecular Mass

With these considerations in mind, let us return to the question of interpreting observed density and scale-height variations in terms of temperature and mean molecular mass.

With reference to the large vertical gradient of scale height observed between 120 and 160 km it is clear that at one extreme one might explain this by a rapid increase of temperature with rather little decrease of molecular weight, even in the presence of diffusion. Mange's calculation cited above shows that it is quite possible for the molecular weight factor to play only a small role in this height interval under certain conditions. On the other hand, some of Nicolet's calculations (1960b) show that the observed rapid increase of scale height could be explained by only moderate temperature increases accompanied by a change to an almost entirely atomic-oxygen atmosphere at 150 to 200 km. Naturally, conditions in the real atmosphere can lie between these extremes, but it is convenient for discussion to consider these two cases and to refer to them, respectively, as Models I and II.

Several considerations now indicate that real conditions are much closer to Model I than to Model II. In the first place, the existence of a temperature gradient of the required magnitude in the 100–200-km region is entirely consistent with what is known about the absorption of solar electromagnetic radiation and the magnitude of heat conduction. This was shown quantitatively by F. S. Johnson (1958). In the second place, the adoption of Model II leads to grave difficulties at levels above 200 km: if the atmosphere is already atomic at that level, the observed further increases of scale height require* temperature increases that are inconsistent with the known heat sources and the conductive processes. In the third place, an atomic oxygen atmosphere in the region of the F_1 layer near 180 to 200 km is difficult to reconcile with ionospheric observations. Observed electron concentrations are much less than would be expected to occur in a slowly recombining atomic-oxygen atmosphere; a natural explanation is that much of the ionizing solar radiation is absorbed by N_2, in which case recombination occurs very rapidly through the dissociative recombination process. This explanation, however, requires substantial amounts of N_2 at that level, in contradiction to Model II.

The present picture then is of a thermosphere heated primarily by extreme ultraviolet electromagnetic radiation in the 120- to 200-km region, leading to

* Further decreases of mean molecular weight due to the presence of helium and hydrogen become important only at greater elevations.

large vertical temperature gradients in this region. At 200 km, the atmosphere still contains an important number of molecules, although present data do not allow any very precise determination of the exact proportion. Above 200 km, the atmosphere is quasi-isothermal, although during the day the temperature may continue to increase with height (less rapidly) up to a higher base of the quasi-isothermal layer. Temperature and density (and their variations) in the thermosphere above the level of electromagnetic heating are influenced primarily by conditions at the lower levels. Continued increase of scale height in the quasi-isothermal region results from a decrease of mean molecular mass (and of the acceleration of gravity) in an atmosphere in diffusive equilibrium.

The ability of this kind of atmosphere to explain the satellite density observations is illustrated by the calculated models of Nicolet. Nicolet (1960a, b, 1961a) recognized that these density data were consistent with the idea of an atmosphere in diffusive equilibrium and essentially isothermal above 200 to 300 km at any one time and place. An early demonstration of this is illustrated in Fig. 6.13. According to these ideas, time variations of density at high levels correspond to

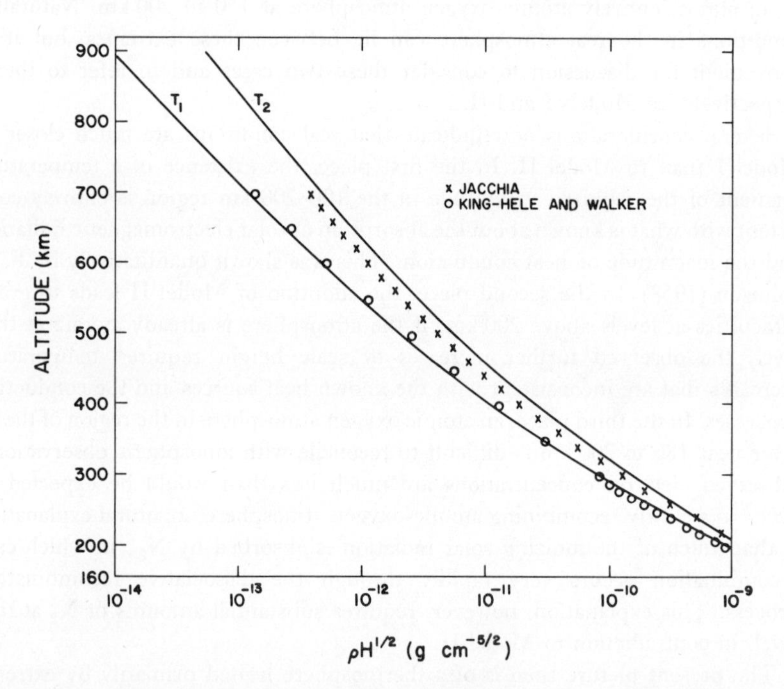

FIG. 6.13. Computations illustrating the ability of isothermal atmospheres in diffusive equilibrium to represent satellite observations above 200 km. The left curve is constructed for a temperature $T_1 = 1190°K$, and fits some results of King-Hele and Walker. The right curve is constructed for a temperature $T_2 = 1425°K$ and fits some results of Jacchia. (After Nicolet, 1961a.)

time variations of the temperature of the quasi-isothermal layer, and we have already referred in Subsection 6.4.3 to Jacchia's correlations of this temperature (derived from Nicolet's models; see below) with solar decimetric flux and with indices of geomagnetic activity. Nicolet further recognized that for specified atmospheric conditions at a low level in the thermosphere, say 120 km, and for an atmosphere in diffusive equilibrium, the vertical gradient of temperature between 120 km and 150 km is the essential parameter that fixes conditions at high levels. For given lower boundary conditions, variations of this temperature gradient over a wide range lead to practically no density variation at 200 km, as required by the satellite observations, but to increasingly great density variations at high altitudes, also as required by the satellite observations.

Table 6.8 is an illustrative selection of conditions in three of the early models published by Nicolet (1961b). These show how different temperature gradients in the 120- to 150-km region influence conditions at the high levels, when the temperature gradient decreases with height in a manner consistent with the heat-conduction equation. Note especially that the density at 200 km is hardly

TABLE 6.8

Conditions in Some Early Model Atmospheres of Nicolet (1961b)[a]

	$\Delta T = 552°K$	$\Delta T = 452°K$	$\Delta T = 374°K$
ρ_{200}	4.07×10^{-13}	3.98×10^{-13}	3.03×10^{-13}
H_{200}	56.0	39.0	32.8
m_{200}	24.8	24.3	23.8
T_{200}	1540	1051	867
ρ_{300}	6.86×10^{-14}	3.45×10^{-14}	1.71×10^{-14}
H_{300}	81.3	52.0	43.3
m_{300}	22.6	20.4	19.4
T_{300}	1975	1150	903
ρ_{400}	1.93×10^{-14}	5.16×10^{-15}	1.82×10^{-15}
H_{400}	96.9	61.5	52.1
m_{400}	20.6	18.0	16.6
T_{400}	2086	1155	903
ρ_{520}	5.45×10^{-15}	7.76×10^{-16}	1.92×10^{-16}
H_{520}	113.4	72.3	63.6
m_{520}	18.6	15.8	14.1
T_{520}	2123	1155	903

[a] All refer to an atmosphere where at 120 km $\rho = 3.5 \times 10^{-11}$ g cm^{-3}, $H = 10.37$ km, $m = 27.4$ g mole^{-1}, $T = 325°K$. ΔT refers to the temperature at 150 km minus the temperature at 120 km; ρ, H, m, T are given at higher levels in the same units as above.

affected, but the density at higher levels shows wide variations. Boundary conditions at 120 km are also important in fixing the conditions at high levels. Thus, as observed in the introduction to this chapter, satellite measurements at very high levels have played an important role in interpreting observations in the 100- to 200-km region.

It should be emphasized that all the model atmospheres discussed here are subject to revision as time goes on. This is a vigorously expanding area of research and present opinions and conclusions are subject to the tests of further observations. We have seen in the correlation between density and geomagnetic activity some evidence of another heat source whose magnitude and distribution are only poorly understood. We have seen in the latest composition measurements some evidence that the mean molecular weight may decrease more rapidly with height than is envisioned in some of these models, especially the CIRA atmosphere. We have seen evidence of large changes during the sunspot cycle and continued measurements over a full cycle are bound to make important contributions to the understanding of thermospheric structure. Until the thermosphere is thoroughly explored under all solar conditions, there will continue to be serious uncertainties.

REFERENCES

AINSWORTH, J. E., FOX, D. F., and LAGOW, H. E. (1961). Upper-atmosphere structure measurement made with the pitot-static tube. *J. Geoph. Res.* **66**, 3191–3212.

BENNETT, W. H. (1950). Radiofrequency mass spectrometers. *J. Appl. Phys.* **21**, 143–149.

BLAMONT, J. E. (1959). Nuages artificiels de sodium. Vitesse du vent, turbulence et densité de la haute atmosphère. *Comptes Rendus Acad. Sci. Paris* **249**, 1248–1250.

BLAMONT, J. E., and DE JAGER, C. (1961). Upper atmospheric turbulence near the 100 km level. *Ann. Geoph.* **17**, 134–144.

BLAMONT, J. E. and DE JAGER, C. (1962). Upper atmospheric turbulence determined by means of rockets. *J. Geoph. Res.* **67**, 3113–3119.

BLAMONT, J. E., LORY, M. L., SCHNEIDER, J. P., and COURTES, G. (1961). Mesure de la temperature de la haute atmosphère à l'altitude de 370 km. *In* "Space Research II" (H. C. van de Hulst, C. de Jager, and A. F. Moore, eds.), pp. 974–980. North-Holland Publ. Co., Amsterdam.

BYRAM, E. T., CHUBB, T. A., and FRIEDMAN, H. (1955). Dissociation of oxygen in the upper atmosphere. *Phys. Rev.* **98**, 1594–1597.

BYRAM, E. T., CHUBB, T. A., and FRIEDMAN, H. (1956). The solar X-ray spectrum and the density of the upper atmosphere. *J. Geoph. Res.* **61**, 251–263.

BYRAM, E. T., CHUBB, T. A., and FRIEDMAN, H. (1957). The dissociation of oxygen at high altitudes. *In* "Threshold of Space" (M. Zelikoff, ed.), pp. 211–216. Pergamon Press, New York.

CHAPMAN, S. (1930). On ozone and atomic oxygen in the upper atmosphere. *Phil. Magaz.* [7] **10**, 369–383.

CHAPMAN, S. (1961). Scale times and scale lengths of variables: with geomagnetic and ionospheric illustrations. *Proc. Phys. Soc.* **77**, 424–432.

CHAPMAN, S., and COWLING, T. G. (1952). "The Mathematical Theory of Non-Uniform Gases." Cambridge Univ. Press, London and New York.

CHUBB, T. A., BYRAM, E. T., FRIEDMAN, H., and KUPPERIAN, J. E., Jr. (1958). The use of radiation absorption and luminescence in upper air density measurements. *Ann. Geoph.* **14**, 109–116.

COSPAR (1961). "COSPAR International Reference Atmosphere, 1961." North-Holland Publ. Co., Amsterdam.

EPSTEIN, P. S. (1932). Über Gasentmischung in der Atmosphäre. *Beitr. Geoph.* **35**, 153–165.

FAUCHER, G. A., PROCUNIER, R. W., and SHERMAN, F. S. (1963). Upper-atmosphere density obtained from measurements of drag on a falling sphere. *J. Geoph. Res.* **68**, 3437–3450.

FRIEDMAN, H., LICHTMAN, S. W., and BYRAM, E. T. (1951). Photon counter measurements of solar X-rays and extreme ultraviolet light. *Phys. Rev.* **83**, 1025–1030.

GREENHOW, J. S., and HALL, J. E. (1960). Diurnal variations of density and scale height in the upper atmosphere. *J. Atmos. Terr. Phys.* **18**, 203–214.

GREENHOW, J. S., and HALL, J. E. (1961). The height variation of the ambipolar diffusion coefficient for meteor trails. *Plan. Space Sci.* **5**, 109–114.

GROVES, G. V. (1960). Wind and temperature results obtained in Skylark experiments. *In* "Space Research I" (H. Kallmann Bijl, ed.), pp. 144–153. North-Holland Publ. Co., Amsterdam.

GROVES, G. V. (1961). Correlation of upper atmosphere air density with geomagnetic activity, November 1960. *In* "Space Research II" (H. C. van de Hulst, C. de Jager, and A. F. Moore, eds.), pp. 751–753. North-Holland Publ. Co., Amsterdam.

HINTEREGGER, H. E. (1962). Absorption spectrometric analysis of the upper atmosphere in the EUV region. *J. Atmos. Sci.* **19**, 351–368.

HOROWITZ, R., and LAGOW, H. E. (1957). Upper air pressure and density measurements from 90 to 220 kilometers with the Viking-7 rocket. *J. Geoph. Res.* **62**, 57–78.

HOROWITZ, R., and LAGOW, H. E. (1958). Summer-day auroral-zone atmospheric-structure measurements from 100 to 210 kilometers. *J. Geoph. Res.* **63**, 757–773.

HOROWITZ, R., LAGOW, H. E., and GIULIANI, J. F. (1959). Fall-day auroral-zone atmospheric structure measurements from 100 to 188 km. *J. Geoph. Res.* **64**, 2287–2295.

HUNTEN, D. M. (1961). Temperatures deduced from aurora and airglow spectra. *Ann. Geoph.* **17**, 249–255.

ISTOMIN, V. G., and POKHUNKOV, A. A. (1963). Mass-spectrometer measurements of atmospheric composition in the USSR. *In* "Space Research III" (W. Priester, ed.), pp. 117–131. North-Holland Publ. Co., Amsterdam.

JACCHIA, L. G. (1959a). Solar effects on the acceleration of artificial satellites. *Smiths. Inst. Ap. Obs. Spec. Rep.* No. 29, pp. 1–15. (Reprinted in *Smiths. Contr. Ap.* **6**, 55–65.)

JACCHIA, L. G. (1959b). Two atmospheric effects in the orbital acceleration of artificial satellites. *Nature* **183**, 526–527.

JACCHIA, L. G. (1959c). Corpuscular radiation and the acceleration of artificial satellites. *Nature* **183**, 1662–1663.

JACCHIA, L. G. (1960a). The effect of a variable scale length on determinations of atmospheric density from satellite accelerations. *Smiths. Inst. Ap. Obs. Spec. Rep.* No. 46, pp. 1–4. (Reprinted in *Smiths. Contr. Ap.* **6**, 77–79.)

JACCHIA, L. G. (1960b). A variable atmospheric-density model from satellite accelerations. *J. Geoph. Res.* **65**, 2775–2782.

JACCHIA, L. G. (1961a). A working model for the upper atmosphere. *Nature* **192**, 1147–1148.

Jacchia, L. G. (1961b). Satellite drag during the events of November 1960. *In* "Space Research II" (H. C. van de Hulst, C. de Jager, and A. F. Moore, eds.), pp. 747–750. North-Holland Publ. Co., Amsterdam.

Jacchia, L. G. (1963). Electromagnetic and corpuscular heating of the upper atmosphere. *In* "Space Research III" (W. Priester, ed.), pp. 3–18. North-Holland Publ. Co., Amsterdam.

Jacchia, L. G., and Slowey, J. (1963a). Accurate drag determinations for eight articial satellites; atmospheric densities and temperatures. *Smiths. Contr. Ap.* **8**, No. 1, 1–99.

Jacchia, L. G., and Slowey, J. (1963b). Atmospheric heating in the auroral zones: a preliminary analysis of the atmospheric drag of the Injun-III satellite. *Smiths. Inst. Ap. Obs. Spec. Rep.* No. 136, 1–18.

Johnson, C. Y. (1961). Aeronomic parameters from mass spectrometry. *Ann. Geoph.* **17**, 100–108.

Johnson, F. S. (1958). Temperatures in the high atmosphere. *Ann. Geoph.* **14**, 94–108.

Jones, L. M. (1954). The measurement of diffusive separation in the upper atmosphere. *In* "Rocket Exploration of the Upper Atmosphere" (R. L. F. Boyd and M. J. Seaton, eds.), pp. 143–156. Pergamon Press, New York.

Jursa, A. S., Nakamura, M., and Tanaka, Y. (1963). Molecular oxygen distribution in the upper atmosphere. *J. Geoph. Res.* **68**, 6145–6155.

Kallmann Bijl, H. K. (1961). Daytime and nighttime atmospheric properties derived from rocket and satellite observations. *J. Geoph. Res.* **66**, 787–795.

King-Hele, D. G. (1959). Density of the atmosphere at heights between 200 km and 400 km, from analysis of artificial satellite orbits. *Nature* **183**, 1224–1227.

King-Hele, D. G. (1963). Improved formulae for determining upper-atmosphere density from the change in a satellite's orbital period. *Plan. Space Sci.* **11**, 261–268.

King-Hele, D. G., and Rees, J. M. (1963). The decrease in upper-atmosphere density between 1957 and 1963, as revealed by satellite orbits. *J. Atmos. Terr. Phys.* **25**, 495–506.

King-Hele, D. G., and Walker, D. M. C. (1961a). Atmospheric densities at heights of 180–700 km. *Ann. Geoph.* **17**, 162–171.

King-Hele, D. G., and Walker, D. M. C. (1961b). Upper-atmosphere density during the years 1957 to 1961, determined from satellite orbits. *In* "Space Research II" (H. C. van de Hulst, C. de Jager, and A. F. Moore, eds.), pp. 918–957. North-Holland Publ. Co., Amsterdam.

Kupperian, J. E., Jr., Byram, E. T., Friedman, H., and Unzicker, A. (1958). Molecular oxygen densities in the mesosphere over Ft. Churchill. *IGY Rocket Rep. Ser.* No. 1, pp. 203–207.

LaGow, H. E., Horowitz, R., and Ainsworth, J. (1960). Results of IGY atmospheric density measurements above Fort Churchill. *In* "Space Research I" (H. Kallmann Bijl, ed.), pp. 164–174. North-Holland Publ. Co., Amsterdam.

Mange, P. (1955). Diffusion processes in the thermosphere. *Ann. Geoph.* **11**, 153–168.

Mange, P. (1957). The theory of molecular diffusion in the atmosphere. *J. Geoph. Res.* **62**, 279–296.

Mange, P. (1961a). Diffusion in the thermosphere. *Ann. Geoph.* **17**, 277–291.

Mange, P. (1961b). The atmospheric mean molecular mass considering diffusion above the 120 km level. *In* "Space Research II" (H. C. van de Hulst, C. de Jager, and A. F. Moore, eds.), pp. 1002–1004. North-Holland Publ. Co., Amsterdam.

Manring, E., Bedinger, J., and Knaflich, H. (1961). Some measurements of winds and of the coefficient of diffusion in the upper atmosphere. *In* "Space Research II" (H. C. van de Hulst, C. de Jager, and A. F. Moore, eds.), pp. 1107–1124. North-Holland Publ. Co., Amsterdam.

MARTIN, G. R. (1954). The composition of the atmosphere above 60 km. *In* "Rocket Exploration of the Upper Atmosphere" (R. L. F. Boyd and M. J. Seaton, eds.), pp. 161–168. Pergamon Press, New York.

MARTIN, H. A., NEVELING, W., PRIESTER, W., and ROEMER, M. (1961). Model of the upper atmosphere from 130 through 1600 km, derived from satellite orbits. *In* "Space Research II" (H. C. van de Hulst, C. de Jager, and A. F. Moore, eds.), pp. 902–917. North-Holland Publ. Co., Amsterdam.

MEADOWS, E. B., and TOWNSEND, J. W., Jr. (1958). Diffusive separation in the winter night time arctic upper atmosphere 112 to 150 km. *Ann. Geoph.* **14**, 80–93.

MEADOWS, E. B., and TOWNSEND, J. W., Jr. (1960). IGY rocket measurements of arctic atmospheric composition above 100 km. *In* "Space Research I" (H. Kallmann Bijl, ed.), pp. 175–198. North-Holland Publ. Co., Amsterdam.

MOSES, H. E., and WU, T. Y. (1951). A self-consistent treatment of the oxygen dissociation region in the upper atmosphere. *Phys. Rev.* **83**, 109–121.

MOSES, H. E., and WU, T. Y. (1952). A self-consistent calculation of the dissociation of oxygen in the upper atmosphere. *Phys. Rev.* **87**, 628–632.

National Academy of Sciences (1962). Space research data interchange (IGY Bulletin 61). *Trans. Amer. Geoph. Union* **43**, 345–350.

NEWELL, H. E., Jr. (1960). The upper atmosphere studied by rockets and satellites. *In* "Physics of the Upper Atmosphere" (J. A. Ratcliffe, ed.), pp. 73–132. Academic Press, New York.

NICOLET, M. (1954). The aeronomic problem of oxygen dissociation. *J. Atmos. Terr. Phys.* **5**, 132–140.

NICOLET, M. (1955). The aeronomic problem of nitrogen oxides. *J. Atmos. Terr. Phys.* **7**, 152–169.

NICOLET, M. (1960a). Les variations de la densité et du transport de chaleur par conduction dans l'atmosphère supérieure. *In* "Space Research I" (H. Kallmann Bijl, ed.), pp. 46–89. North-Holland Publ. Co., Amsterdam.

NICOLET, M. (1960b). The properties and constitution of the upper atmosphere. *In* "Physics of the Upper Atmosphere" (J. A. Ratcliffe, ed.), pp. 17–71. Academic Press, New York.

NICOLET, M. (1961a). Structure of the thermosphere. *Plan. Space Sci.* **5**, 1–32.

NICOLET, M. (1961b). Density of the heterosphere related to temperature. *Smiths. Inst. Ap. Obs. Spec. Rep.* No. 75, pp. 1–30. (Reprinted in *Smiths. Contr. Ap.* **6**, 175–187.)

NICOLET, M., and MANGE, P. (1954). The dissociation of oxygen in the high atmosphere. *J. Geoph. Res.* **59**, 15–45.

PAETZOLD, H. K., and ZSCHÖRNER, H. (1960). Bearings of Sputnik III and the variable acceleration of satellites. *In* "Space Research I" (H. Kallmann Bijl, ed.), pp. 24–36. North-Holland Publ. Co., Amsterdam.

PAETZOLD, H. K., and ZSCHÖRNER, H. (1961). The structure of the upper atmosphere and its variations after satellite observations. *In* "Space Research II" (H. C. van de Hulst, C. de Jager, and A. F. Moore, eds.), pp. 958–973. North-Holland Publ. Co., Amsterdam.

PENNDORF, R. (1949). The vertical distribution of atomic oxygen in the upper atmosphere. *J. Geoph. Res.* **54**, 7–38.

PENNDORF, R. (1950). The distribution of atomic and molecular oxygen in the upper atmosphere. *Phys. Rev.* **77**, 561–562.

POKHUNKOV, A. A. (1960). The study of upper atmosphere neutral composition at altitudes above 100 km. *In* "Space Research I" (H. Kallmann Bijl ed.), pp. 101–106. North-Holland Publ. Co., Amsterdam.

POKHUNKOV, A. A. (1963a). Gravitational separation, composition and structural parameters of the night atmosphere at altitudes between 100 and 210 km. *Plan. Space Sci.* **11**, 441–449.

POKHUNKOV, A. A. (1963b). Gravitational separation, composition and the structural parameters of the atmosphere at altitudes above 100 km. *In* "Space Research III" (W. Priester, ed.), pp. 132–142. North-Holland Publ. Co., Amsterdam.

POKHUNKOV, A. A. (1963c). On the variation in the mean molecular weight of air in the night atmosphere at altitudes of 100 to 210 km from mass spectrometer measurements. *Plan. Space Sci.* **11**, 297–304.

PRIESTER, W. (1959). Sonnenactivität und Abbremsung der Erdsatelliten. *Naturwiss.* **46**, 197–198.

PRIESTER, W. (1961). Solar activity effect and diurnal variation in the upper atmosphere. *J. Geoph. Res.* **66**, 4143–4148.

PRIESTER, W., and CATTANI, D. (1962). On the semiannual variation of geomagnetic activity and its relation to the solar corpuscular radiation. *J. Atmos. Sci.* **19**, 121–126.

RAKSHIT, H. (1947). Distribution of molecular and atomic oxygen in the upper atmosphere. *Ind. J. Phys.* **21**, 57–68.

REES, J. A. (1961). Diffusion coefficients determined from sodium vapor trails. *Plan. Space Sci.* **8**, 35–42.

SCHAEFER, E. J. (1963). The dissociation of oxygen measured by a rocket-borne mass spectrometer. *J. Geoph. Res.* **68**, 1175–1176.

SHARP, G. W., HANSON, W. B., and MCKIBBIN, D. D. (1962). Atmospheric density measurements with a satellite-borne microphone gage. *J. Geoph. Res.* **67**, 1375–1382.

SISSENWINE, N., DUBIN, M., and WEXLER, H. (1962). The U. S. standard atmosphere, 1962. *J. Geoph. Res.* **67**, 3627–3630 .

STERNE, T. E. (1958). Formula for inferring atmospheric density from the motion of artificial earth satellites. *Science* **127**, 1245.

STERNE, T. E. (1959). Effect of the rotation of a planetary atmosphere upon the orbit of a close satellite. *Jet Propulsion* **29**, 777–782.

STIRTON, R. J. (1960). The upper atmosphere and satellite drag. *Smiths. Contr. Ap.* **5**, No. 2, 9–15.

SUTTON, W. G. L. (1943). On the equation of diffusion in a turbulent medium. *Proc. Roy. Soc.* **A182**, 48–75.

TOWNSEND, J. W., Jr., and MEADOWS, E. B. (1958). Density of the winter nighttime arctic upper atmosphere 110 to 170 km. *Ann. Geoph.* **14**, 117–130.

TOWNSEND, J. W., Jr., MEADOWS, E. B., and PRESSLY, E. C. (1954). A mass spectrometric study of the upper atmosphere. *In* "Rocket Exploration of the Upper Atmosphere" (R. L. F. Boyd and M. J. Seaton, eds.), pp. 169–188. Pergamon Press, New York.

WENZEL, E. A., LOH, L. T., NICHOLS, M. H., and JONES, L. M. (1958). The measurement of diffusive separation in the upper atmosphere. *IGY Rocket Rep. Ser.* No. 1, pp. 91–106.

WHITEHEAD, J. D. (1963). The density of the atmosphere in the *E*-region of the ionosphere. *Plan. Space Sci.* **11**, 513–521.

WHITNEY, C. A. (1959). The structure of the high atmosphere. I, Linear models. *Smiths. Inst. Ap. Obs. Spec. Rep.* No. 21, pp. 1–37. (Partially reprinted in *Smiths. Contr. Ap.* **6**, 35–41.)

CHAPTER 7

Radiative Processes and Heat Transfer

We have seen in the previous three chapters some of the effects of solar radiation on the upper atmosphere. It is clear that the composition and thermal structure of the upper atmosphere are profoundly affected by absorption of this radiation.

Another important part of the picture that must be considered is the internal transfer and ultimate disposition of this energy. In the long run, the earth-atmosphere system as a whole must radiate back to space as much energy as it receives. However, a radiative balance need not and in general does not occur during particular time periods at particular locations. The atmosphere has efficient mechanisms for the internal transfer of heat. As a matter of fact, a time-honored approach to the problem of atmospheric circulation is to regard radiative imbalance as the driving force for the circulation. Although the total problem is really a nonlinear one, this is a useful approach because the radiative processes (complicated as they are) are amenable to order-of-magnitude calculations whose results place certain constraints on the much more complicated problem of the circulation.

In the stratosphere and mesosphere radiative transfer results from the infrared radiations of the minor polyatomic constituents water vapor, carbon dioxide, and ozone. In Section 7.1 we discuss the usual meteorological formulation of this problem, with emphasis on the peculiar difficulties that arise in connection with applications to the upper atmosphere. At high enough levels, probably in the upper mesosphere, this usual formulation is inadequate because of the inapplicability of Kirchhoff's law. At still higher levels, probably in the lower thermosphere, polyatomic constituents must disappear in the presence of dissociating radiation; however, other mechanisms of radiative heat loss become important. Section 7.2 considers these general problems. At high enough levels in the thermosphere, the heat gained from solar energy is lost mainly by molecular conduction downward to the lower levels. The problem of thermospheric conduction and the resulting temperature distribution are considered in Section 7.3.

7.1 Infrared Transfer in the Atmosphere

Carbon dioxide, water vapor, and ozone all possess important bands in the part of the spectrum where black-body emission is relatively large at atmospheric temperatures. According to (4.26), the wavelength of maximum emission varies from 14.5 μ when the temperature is 200°K to 9.7 μ when the temperature is 300°K. These temperatures include most of those that occur up to the mesopause.

The most important radiating constituent of the troposphere is water vapor. In the spectral region of interest, this gas has a vibration-rotation band centered near 6.3 μ which is of importance, and an extensive structure of pure rotational lines lying mainly on the long-wave side of 20 μ. Because of its presumably small concentration above the tropopause, the contribution of water vapor to the radiative balance of the stratosphere and mesosphere is believed to be small and is usually neglected.

Carbon dioxide is a symmetrical molecule and has no pure rotational transitions. In the spectral region of interest, it has a very strong vibration-rotation band centered near 15 μ. In the troposphere, this band is usually neglected insofar as flux *divergence* and cooling rates are concerned, because its very strength causes quite thin atmospheric layers to be opaque in the region of the band. In the stratosphere and mesosphere, however, this band assumes great importance.

The ozone molecule has vibration-rotation bands at 9.0 μ and 9.6 μ, the second of which is considerably stronger (Kaplan *et al.*, 1956). Another such band lies near 14 μ and partially overlaps the 15-μ carbon dioxide band. Of these, the absorption and emission near 9 to 10 μ are the more important because in this region of the spectrum the atmosphere is otherwise nearly transparent.

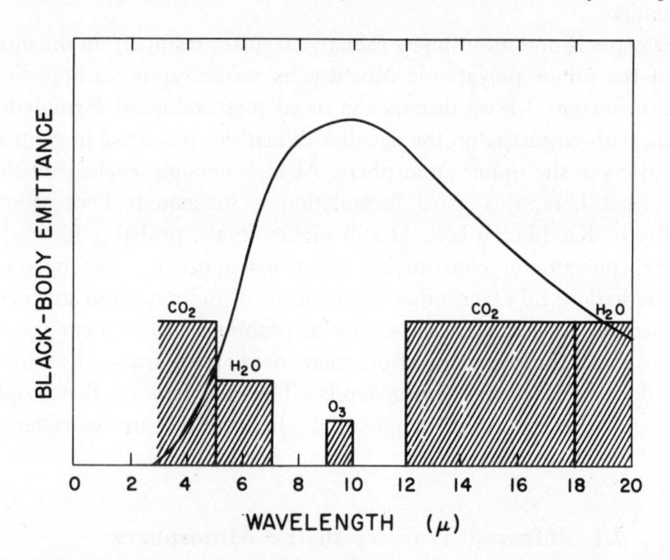

FIG. 7.1. Schematic representation of the infrared spectrum of air. The curve shows black-body emittance at $T = 300°K$ in relative units. The hatched areas represent the spectral regions of appreciable absorption due to CO_2, H_2O, and O_3.

In Fig. 7.1 this brief description of the infrared spectra of atmospheric gases is summarized schematically. The rest of this section is devoted to the mathematical formulation of the atmospheric infrared-transfer problem and to a discussion of numerical results obtained in a few selected investigations of stratosphere and mesosphere. This section extends the discussion of radiative transfer begun in Section 4.3, which should be consulted for definitions of symbols and statements of certain basic relationships.

7.1.1 Assumptions and Solution for Monochromatic Radiation

In the formulation of the infrared-transfer problem for the earth's atmosphere, the following assumptions are usually made:

(a) Kirchhoff's law is obeyed so that the source function is $L_{\nu B}$ and the equation of transfer is given by (4.37).

(b) The atmosphere consists of plane-parallel sheets, each being homogeneous in the horizontal and infinite in extent.

(c) Only one substance is radiatively active at a given wavelength.

The geometry of the problem is illustrated in Fig. 7.2. It is desired to compute the net flux through an elemental area dA, oriented perpendicular to the vertical coordinate z, and located at some "reference level." For convenience, we divide this into two parts, the upward flux and the downward flux, because boundary conditions are different in the two directions. There is always a boundary below the reference level, the ground or a cloud top, either of which we take to be black. There may or may not be a cloud boundary above the reference level.

The formal procedure is to solve the equation of transfer for the given assumptions, integrate over direction, and finally integrate over frequency. If $F\uparrow$ is the upward flux so obtained (counted positive upward) and if $F\downarrow$ is the downward flux (counted positive downward) then the temperature change of a volume of unit area, vertical thickness dz, and total density ρ, due to flux divergence, is

$$\frac{\partial T}{\partial t} = -\frac{1}{\rho c_p} \frac{\partial}{\partial z}(F\uparrow - F\downarrow) \tag{7.1}$$

where c_p is the specific heat at constant pressure of air.

It is convenient to take as a vertical coordinate the integrated mass of the absorbing material, defined by

$$u = \int \rho_a \, dz \tag{7.2}$$

where ρ_a is the density of the radiatively active gas. We take $u = 0$ at the reference level, and count it positive downward during the computation of upward flux and positive upward during the computation of downward flux.

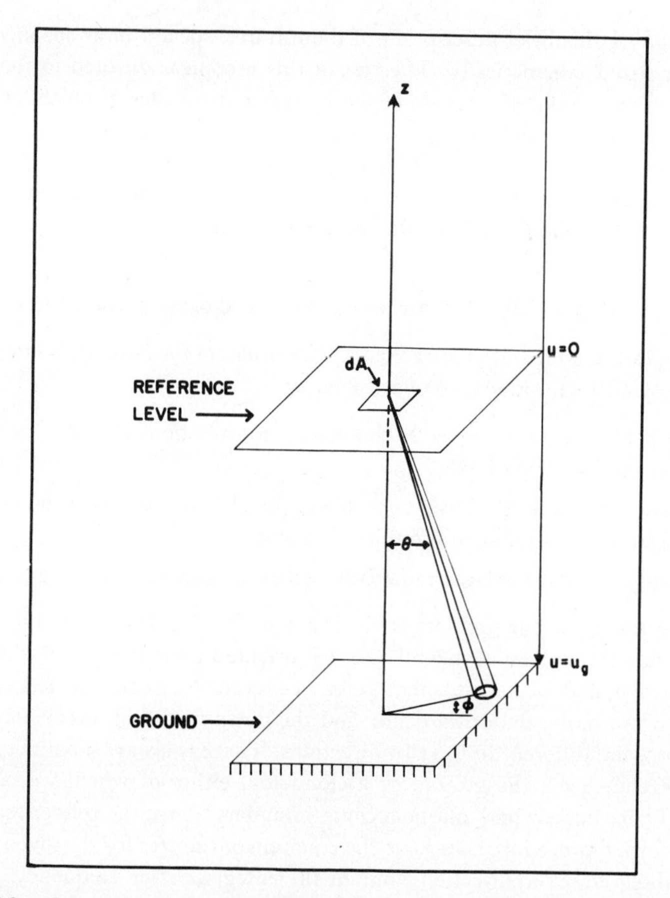

FIG. 7.2. Geometry of the atmospheric infrared-transfer problem. The integrated mass of absorbing material between the reference level and the ground (in a vertical column) is u_g.

The formal solution of the equation of transfer (4.37) gives for the upward specific intensity at $u = 0$ and in a direction θ from the vertical (see Fig. 7.2)

$$I_\nu = L_{\nu B}(u_g) \exp\left(-\sec\theta \int_0^{u_g} k_\nu \, du\right) + \int_0^{u_g} L_{\nu B}(u) \exp\left(-\sec\theta \int_0^{u} k_\nu \, du\right) \sec\theta \, k_\nu \, du \tag{7.3}$$

The physical meaning of this expression is easy to see. The first term represents black-body radiation from the ground (or cloud top) attenuated by absorption of the intervening medium. The second term represents the integrated effect of emission from mass along the path, the emission of each elemental mass being attenuated by absorption between it and the reference level.

The flux is given by (4.28), where the upper integration limit of $\pi/2$ for θ corresponds to our assumed horizontally infinite atmosphere. The integration over θ is accomplished in terms of the exponential integral, which is defined by

$$Ei_n(x) = \int_1^\infty \exp(-x\eta)\,\frac{d\eta}{\eta^n} \tag{7.4}$$

With the substitution $\eta = \sec\theta$, and for isotropic radiation, we get

$$F_\nu = 2\pi L_{\nu B}(u_g) Ei_3\left(\int_0^{u_g} k_\nu\,du\right) + \int_0^{u_g} 2\pi k_\nu L_{\nu B}(u) Ei_2\left(\int_0^u k_\nu\,du\right) du \tag{7.5}$$

The integrated flux is obtained formally by integrating (7.5) over frequency.

Downward flux is given by an expression similar to (7.5) except that if there is no upper boundary (cloud) the first term is missing and the upper integration limit u_g in the second term is replaced by ∞.

At this point the problem is formally "solved." However, there are several practical difficulties in the way of applying equations like (7.5) directly to the

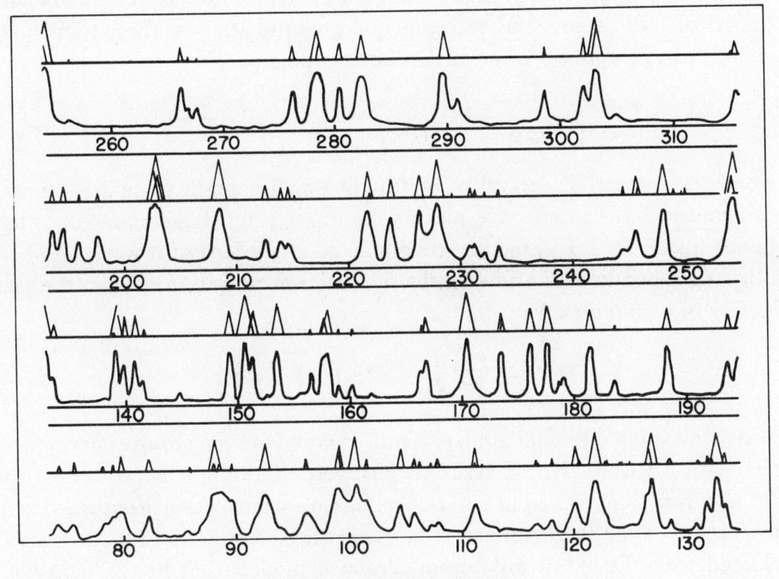

FIG. 7.3. Absorption spectrum of the water-vapor rotational band at high resolution. Triangles above spectrum represent quantum-mechanical computations. (After Randall *et al.*, 1937.)

atmosphere. The chief one is the rapid variation of absorption coefficient with frequency in the presence of the rotational lines of the infrared spectrum. By way of illustration, Fig. 7.3 shows a portion of the water-vapor spectrum in the rotational band, after Randall *et al.* (1937). It is immediately evident that the variation of k_ν with frequency is so rapid that the number of spectral intervals required for a numerical integration over frequency would be prohibitive. In this connection, it should be noted that k_ν varies with temperature and pressure, which in turn vary along a particular atmospheric path. Therefore the integrations involved in the arguments of the exponential integrals would have to be carried out separately for each reference level and furthermore repeated for each new atmosphere, since temperature and pressure in general vary from time to time in their dependence on u. Finally, even if a very large computing machine should make all this practical, the exact values of k_ν and their complicated dependence on temperature and pressure are not known precisely in all parts of the spectrum.

In practice, as atmospheric radiation theory has developed, the way out of this dilemma has been to consider not monochromatic radiation but finite spectral intervals for which one can define by theory or experiment effective transmission functions. These transmission functions are in general much more complicated than the simple exponential that holds for monochromatic radiation [Eq. (4.36)]. Let us now look into this question and see how these transmission functions can be applied to the atmospheric problem.

7.1.2 The Use of Transmission Functions

Consider a spectral interval of width $\Delta\nu$, small enough so that $L_{\nu B}$ may be considered to have an effective mean value $\bar{L}_B(\Delta\nu)$, but large enough so that it includes many lines. In practice, $\Delta\nu$ might be a few tens of a wave number in width (corresponding to a few tenths of a micron near 10 μ). Then the transmissivity of a column u is

$$\tau(\Delta\nu) = \frac{1}{\Delta\nu}\int_{\nu}^{\nu+\Delta\nu} \exp\left[-\int_0^u k_\nu\, du\right] d\nu \tag{7.6}$$

In order to evaluate this integral, one must know how the absorption coefficient varies with ν within the interval. Alternatively, one might hope to determine $\tau(\Delta\nu)$ as a function of u from laboratory measurements. As it has turned out, a combination of the two approaches is required. Transmission functions can be evaluated from Eq. (7.6) for certain idealized models that fit the behavior of atmospheric gases fairly well, but laboratory measurements are useful to verify the models and to establish certain numerical values relating to the transmission. On the other hand laboratory measurements are not sufficient in themselves because it is impossible to duplicate in the laboratory the conditions existing in the atmosphere. One of the most troublesome problems is the variation of k_ν with pressure (and to a lesser extent temperature); in the atmosphere, pressure and temperature vary along the transmission path, while in the laboratory they are constant in the absorption tube for any one measurement. Furthermore, it is impossible to duplicate in the laboratory the extremely small path lengths and low pressures that occur in the upper stratosphere and mesosphere.

Before discussing the models for which τ may be evaluated with Eq. (7.6) or the measurements against which these results may be checked, let us formulate the problem in terms of the transmission function. Suppose that in Eq. (7.3) both sides are integrated over a small spectral interval ν to $\nu + \Delta\nu$. If we let $\bar{I}_i$ be a mean value of I_ν over this interval, and let $\bar{L}_{iB}$ be a mean value of $L_{\nu B}$ over this interval, then the result is

$$\bar{I}_i \,\Delta\nu = \bar{L}_{iB}(u_g) \int_\nu^{\nu+\Delta\nu} \exp\left(-\sec\theta \int_0^{u_g} k_\nu \, du\right) d\nu$$
$$+ \int_0^{u_g} \bar{L}_{iB}(u) \int_\nu^{\nu+\Delta\nu} \exp\left(-\sec\theta \int_0^{u} k_\nu \, du\right) \sec\theta \, k_\nu \, d\nu \, du \tag{7.7}$$

In view of our definition of τ [Eq. (7.6)], this can be written

$$\bar{I}_i = \bar{L}_{iB}(u_g)\tau(u_g \sec\theta) - \int_0^{u_g} \bar{L}_{iB}(u) \frac{d\tau(u \sec\theta)}{du} \, du \tag{7.8}$$

where $\tau(u \sec\theta)$ means that du in (7.6) is to be replaced by $du \sec\theta$.

According to Eq. (4.28) the integrated flux is to be obtained by multiplying both sides of (7.8) by $(\sin\theta \cos\theta \, d\theta \, d\phi)$ and integrating over the angles. The integration over ϕ gives simply 2π, for isotropic radiation. The integration over θ is usually handled formally by defining the *slab transmission function* or *diffuse transmission function* $\tau_F(u)$ as

$$\tau_F(u) = 2 \int_0^{\pi/2} \tau(u \sec\theta) \sin\theta \cos\theta \, d\theta \tag{7.9}$$

With this notation, (7.8) yields

$$\bar{F}_i = \pi \bar{L}_{iB}(u_g)\tau_F(u_g) - \pi \int_0^{u_g} \bar{L}_{iB}(u) \frac{d\tau_F}{du} \, du \tag{7.10}$$

If appropriate transmission functions can be found to represent moderately wide spectral intervals, this preliminary integration over frequency achieves a tremendous simplification of the problem. One very serious difficulty, however, remains: the transmission function for a particular spectral interval is not a unique function of the integrated mass of absorbing material in the path. Because of the pressure and temperature dependence of k_ν, the variation of k_ν with u

depends also on the distributions of pressure and temperature along the path. Thus we might write

$$\int k_\nu \, du = \int k_\nu(T, p) \, du$$

where T and p are functions of u for a particular path in the atmosphere. We shall see later, however, that there are approximations to overcome this difficulty in most if not all practical applications, and we proceed in Subsection 7.1.3 as if we were dealing with a *homogeneous* (constant T and p) path.

7.1.3 Spectral Models and Theoretical Transmission Functions

The line profile usually assumed for studies of infrared transfer in the earth's atmosphere is the Lorentz shape, a simplified result of the theory of pressure broadening of lines. Although this is not an exact expression of the collision effect, being inexact in the wings of the line, it suffices for most atmospheric applications (see discussion by Goody, 1954, p. 155, and by Elsasser, 1960, p. 12). For the Lorentz shape, the absorption coefficient of a line centered at ν_0 is

$$k_\nu = \frac{S\alpha}{\pi} \frac{1}{(\nu - \nu_0)^2 + \alpha^2} \tag{7.11}$$

In (7.11) S is the *intensity* of the line, $S = \int_{-\infty}^{\infty} k_\nu \, d\nu$, and α is the *half-width*, where $k_\nu = \frac{1}{2} k_{\nu 0}$ at $\nu - \nu_0 = \pm \alpha$. The half-width α is to a good approximation proportional to pressure and inversely proportional to the square root of temperature, while the intensity is a function of temperature.

Doppler broadening, caused by thermal motion of the radiating molecules, is negligible in the lower atmosphere. At tropospheric pressures, the half-width α of a Lorentz line is much greater than the half-width α_D of a Doppler-broadened line. However α decreases as the pressure decreases and α_D is pressure independent. Therefore, at high enough levels, the two must be comparable. Goody (1954) has estimated that this occurs for the 15-μ CO_2 band at about 34 km and for the 9.6-μ ozone band at about 33 km.

A Doppler-broadened line centered at ν_0 has the profile

$$k_\nu = \frac{S}{\alpha_D (\pi)^{\frac{1}{2}}} \exp \left[-\left(\frac{\nu - \nu_0}{\alpha_D} \right)^2 \right] \tag{7.12}$$

where S is the intensity of the line and α_D is given by $\alpha_D = (\nu_0/c)(2kT/\mu)^{\frac{1}{2}}$, where c is the speed of light, k is Boltzmann's constant, and μ is the molecular mass. For comparable intensities and half-widths, the Doppler line has more absorption near the center and less in the wings than the Lorentz line. Figure 7.4 illustrates this.

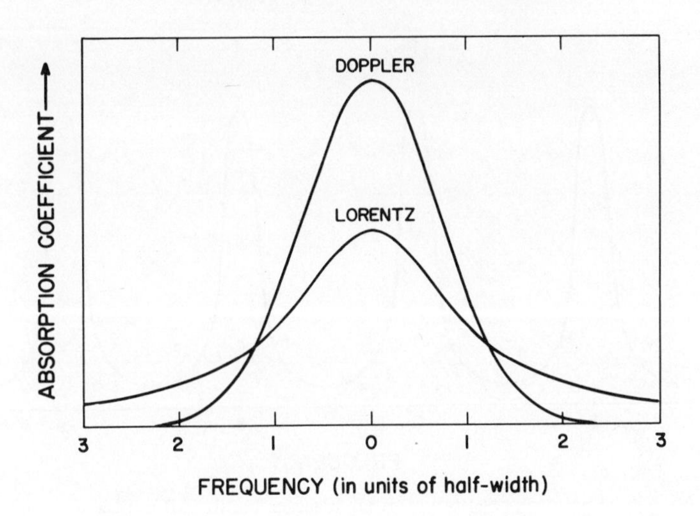

FIG. 7.4. Comparison of line profiles for Doppler broadening and Lorentz broadening for similar intensities and half-widths. (After Goody, 1954.)

When both effects are important, the line shape that results is a combination of the two effects. Plass and Fivel (1953) have discussed the resulting line and its effects on radiative transfer. It appears that until α_D is at least twice as large as α there is rather little difference fıom the Lorentz shape; for strong absorption there is little difference for even greater ratios because the wings of the hybrid line still retain the Lorentz shape. It is probably safe for most purposes to neglect Doppler broadening up to the stratopause.

The spectral interval over which we wish to compute the transmissivity from (7.6) must in general contain many lines. It should be noted that at the pressures existing in most of the atmosphere it is not sufficient to compute the transmissivity due to each line separately without regard to the other lines, because of overlapping effects. The absorption coefficient in general represents contributions from more than one line. Several models describing idealized combinations of lines have been proposed and used. The oldest, which we shall call the *regular* model, is due to Elsasser (1938, 1942, 1960). In the regular model there are several equally spaced Lorentz lines of identical intensity and half-width within the spectral interval. Figure 7.5 shows this periodic pattern schematically.

Elsasser has shown that the transmissivity of the regular model (for a homogeneous path) is given by

$$\tau = \frac{1}{2\pi} \int_{-\pi}^{\pi} \exp\left[-\frac{\beta x \sinh \beta}{\cosh \beta - \cos s}\right] ds \tag{7.13}$$

where $\beta = 2\pi\alpha/\delta$, $x = Su/2\pi\alpha$, $s = 2\pi\nu/\delta$, and δ is the distance between line

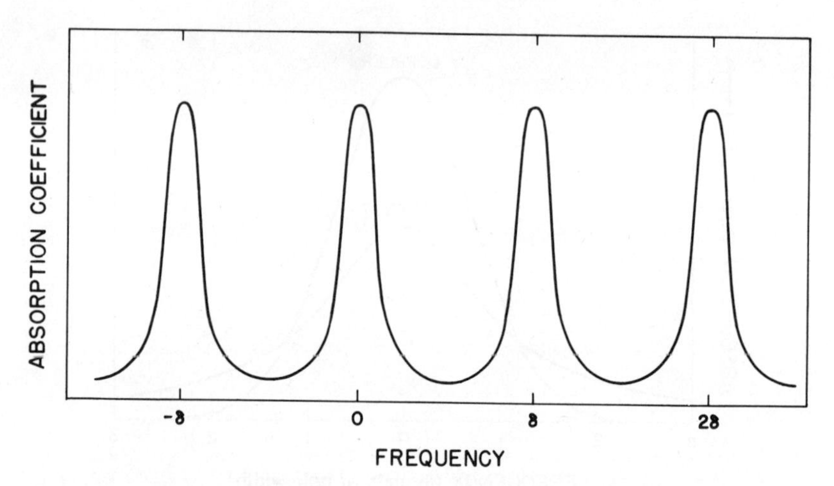

FIG. 7.5. Schematic representation of an Elsasser band. (After Elsasser, 1960.)

centers. This definite integral has been tabulated for $\beta \leqslant 1$ by Kaplan (1953) and by Wark and Wolk (1960).

A model that is in some respects the opposite of this is the *random* model, proposed by Goody (1952a). In this model, lines are considered to be distributed at random within the spectral interval, all lines are taken to have the same half-width, and the distribution of line intensities is specified by some simple function. Goody showed that the transmissivity of the random model is

$$\tau = \exp - \left[\frac{\int_{-\infty}^{\infty} \int_0^{\infty} P(S)\{1 - \exp[-uSf(\nu, \alpha)]\} \, dS \, d\nu}{\delta \int_0^{\infty} P(S) \, dS} \right] \tag{7.14}$$

Here $P(S)\,dS$ is the probability that any line has an intensity between S and $S + dS$; $f(\nu, \alpha)$ is a shape function such that for a line centered at $\nu = 0$ the absorption coefficient of the line at ν is $k = Sf(\nu, \alpha)$; and δ is the mean line spacing. Goody took an exponential distribution of intensities

$$P(S) = (1/S_0) \exp(-S/S_0) \tag{7.15}$$

For this function, it is a straightforward matter to carry out the integration over S in (7.14) and obtain

$$\tau = \exp \left[-\frac{1}{\delta} \int_{-\infty}^{\infty} \frac{uS_0 f(\nu, \alpha)}{1 + uS_0 f(\nu, \alpha)} \, d\nu \right] \tag{7.16}$$

As it stands, (7.16) holds for any line shape. For the Lorentz shape, as noted earlier, $f(\nu, \alpha) = \alpha/[\pi(\nu^2 + \alpha^2)]$ and

$$\tau = \exp \left[-\frac{\beta_0 x_0}{(1 + 2x_0)^{\frac{1}{2}}} \right] \tag{7.17}$$

if, by analogy with the parameters β and x used in (7.13), we define $\beta_0 = 2\pi\alpha/\delta$ and $x_0 = S_0u/2\pi\alpha$.

The transmission functions given above by no means exhaust the possibilities; it is simply beyond the scope of this discussion to go into further detail. For example, a group of Elsasser bands with random relative positions has been proposed (Kaplan, 1954). Also the random model can be generalized to allow for variations of half-width (Kaplan, 1954) and can be used with other probability distributions for the line intensities (Godson, 1955a, b).

Figure 7.6 compares the transmissivities given by (7.13) for $\beta = 1$ and by

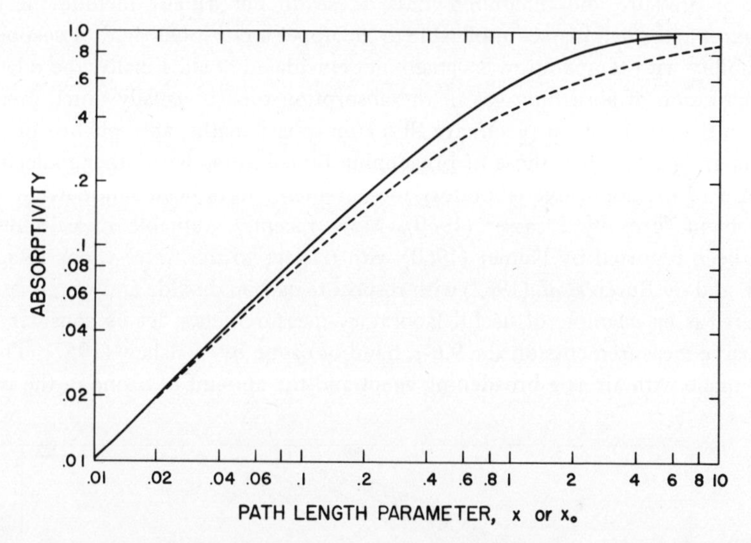

FIG. 7.6. Absorptivities of a regular band (solid curve) and a random band (dashed curve). The curves are constructed for a regular band with $\beta = 1$, and for a random band with an exponential distribution of line intensities and $\beta_0 = 1$.

(7.17) for $\beta_0 = 1$. For small values of x(or x_0), the regular and random bands have nearly the same transmissivity. However, for large x, the random band with an exponential distribution of intensities is more transparent than a regular band (when $S = S_0$). This is also true for a random band with lines of constant intensity S (Godson, 1954).

7.1.4 LABORATORY MEASUREMENTS AND THEIR APPLICATION TO THE ATMOSPHERE

It is possible in principle to approach the infrared problem from the point of view of spectroscopic data on line positions, intensities, and half-widths and their variability with temperature and pressure. As a matter of fact, this would be quite feasible from a computational point of view if the data could be fitted by one of the spectral models that has been studied. In practice, however, laboratory measurements of wide-band transmissivities are often applied to the problem, owing to some uncertainties in the spectroscopic data and also to the failure of any simple model to reproduce exactly the complexities of the line structure.

On the other hand, laboratory data are applied only with difficulty and some uncertainties. They give transmissivities over a parallel, homogeneous path and must be applied to diffuse radiation from a nonhomogeneous slab in the atmosphere. Although measurements are usually made for a wide range of combinations of pressure and integrated mass, these do not usually include the low pressures and small masses applicable to the upper stratosphere and mesosphere. Variability with temperature is usually not elucidated at all. Finally, the relative concentration of absorbing gas in the absorption tube is usually much greater than in the much longer (geometrically) atmospheric paths, and self-broadening effects are greater than those of broadening by collisions with other molecules.

Some of the applicable laboratory measurements have been summarized in a convenient form by Elsasser (1960). More recently, valuable measurements have been reported by Palmer (1960) with respect to the water-vapor rotation band, and by Burch *et al.* (1962) with respect to carbon dioxide and water vapor.

Here, as an example of useful laboratory measurements, let us consider the extensive measurements on the 9.6-μ band of ozone by Walshaw (1957). These were made with air as a broadening agent and the amount of ozone in the tube

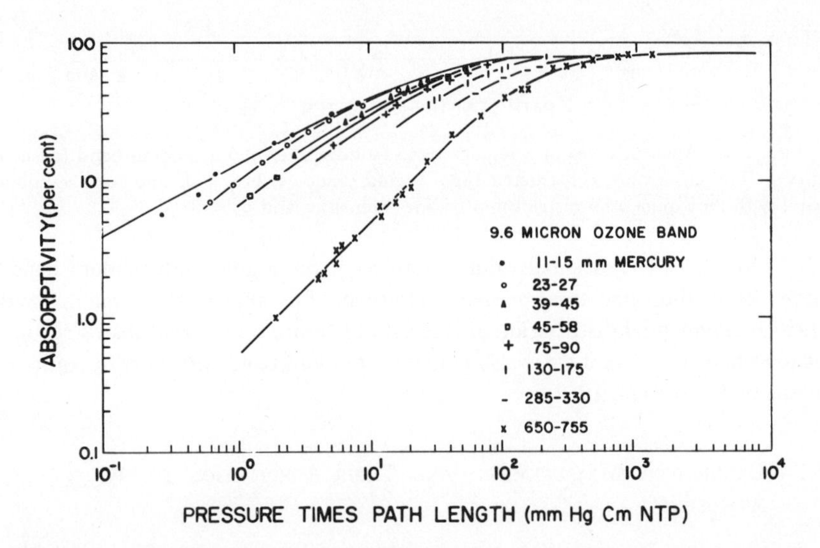

FIG. 7.7. Walshaw's measurements of absorptivity of the 9.6-μ band of ozone. Absorptivities for several nearly constant values of pressure are shown by different symbols. Curves represent theoretical calculations by Plass (see text). (After Plass, 1960.)

was determined from simultaneous measurements of transmissivity in an ultraviolet region with known absorption coefficient (which is pressure independent). Some of these data are shown in Fig. 7.7, according to a representation by Plass (1960), who plotted measured absorptivity as a function of the product pressure times integrated mass. The solid curves in this figure give the theoretical absorptivity for a spectral model based on the random superposition of two regular bands (a model chosen by Plass to give the best fit for the data). Each curve is constructed for a constant pressure, corresponding to the mean pressure for a particular set of experimental data as indicated in the figure.

Among the various problems of applying laboratory data to the atmosphere, the most troublesome is the application of data for homogeneous paths to the nonhomogeneous atmospheric paths. Fortunately, there is a simplified procedure known as the Curtis-Godson approximation which seems to give accurate results in many practical cases. Consider a path in the atmosphere containing integrated mass u_1, where p varies with u and therefore τ is a function of $\int_0^{u_1} f(p)\, du$ (neglecting, for the time being, temperature effects). The Curtis–Godson approximation in its simplest form, as first given by Curtis (Goody, 1952b), states that the transmissivity is numerically equivalent to the transmissivity of a path of the same length u_1 having constant pressure p_e, where p_e is

$$p_e = \frac{\int_0^{u_1} p\, du}{u_1} \tag{7.18}$$

Thus, once the "effective pressure" p_e is determined for a given atmospheric path, the transmissivity measured in the laboratory for constant pressure p_e and integrated mass u_1 is applicable to the atmospheric path.

Godson (1953, 1955a, b) has discussed this question extensively and suggested a more general approximation which accounts also for temperature variability along a nonhomogeneous path. If γ and μ are two functions of temperature only, functions which in principle may be determined from either spectroscopic or laboratory data (Godson, 1962a, b), then the transmission along a nonhomogeneous path of length u_1 is equivalent to transmission along a homogeneous path of length u_e at the constant pressure p_e. The equivalent paths and pressures are defined by

$$u_e = \int_0^{u_1} \gamma\, du \tag{7.19}$$

$$p_e = \frac{\int_0^{u_1} \mu p\, du}{u_e} \tag{7.20}$$

If $\gamma = \mu = 1$, these reduce to (7.18).

It can be shown that the Curtis–Godson approximation is exact in certain limiting cases (see, for example, Kaplan, 1959). For other situations, numerical calculations have established its general accuracy under most circumstances (Godson, 1955a; Kaplan, 1959; Walshaw and Rodgers, 1963). However, the latter authors have questioned its applicability to the computation of heating rates due to radiative transfer in the 9.6-μ ozone band (see also Goody, 1952b).

It should be emphasized that the Curtis–Godson approximation is not the same as the scaling approximation that must be used for radiation diagrams such as the well-known Elsasser diagram. In the latter procedure, the mass of each sublayer is adjusted once and for all, and the adjusted value is used in computations of transmissivity over any path that includes the layer. In applying the Curtis–Godson approximation, on the other hand, a new effective pressure must be found for each and every combination of reference level and radiating level. (However, this same pressure may be used for slant paths between the two levels and therefore in the diffuse transmission function.) Clearly, the Curtis–Godson approximation, although much simpler than an exact solution, still implies a significant computational effort.

Another problem of applying laboratory data to atmospheric problems, particularly the upper atmosphere, is that the range of pressures and path lengths attainable in the laboratory may not encompass the range of interest in the atmosphere. The extremely low pressures and short path lengths of the upper stratosphere and mesosphere, in particular, may not be represented. Extrapolation of laboratory data outside the range of measurements must be done very carefully and with full consideration for the spectral characteristics of the band concerned. For example, a purely empirical representation of laboratory data, no matter how well it fits the data within the range of measurements, should not be relied upon for extrapolation. However, Plass (1958, 1960) and Godson (1962b) have discussed procedures for interpreting laboratory data in terms of spectral models. These procedures consist essentially of determining which spectral model and what parameters best fit the data in the range of measurements (see, for example, Fig. 7.7), and then using these for extrapolation purposes. In extrapolating to very low pressures, of course, account must be taken of Doppler broadening.

7.1.5 Radiation in the Stratosphere and Mesosphere

Shortly after the existence of the tropopause and stratosphere were established by observation, it was recognized (Gold, 1908; Humphreys, 1909) that radiative phenomena must play an important role in their explanation. The earliest studies (Gold, 1909; Humphreys, 1909; Emden, 1913; Milne, 1922) were highly oversimplified. Indeed, even if radiation theory had at that time been developed to its present state, adequate observations of atmospheric composition, of solar flux, and of the radiative characteristics of atmospheric gases would not have been available for its application. Nevertheless, these early studies were extremely interesting and some of their conclusions quite generally valid, and for this reason they are discussed in Appendix G for the interested reader. Some of these results have already been discussed qualitatively in Subsection 2.2.2.

The work of Gowan (1947a, b) was the first that took into account all of the principal processes in the radiation balance of the stratosphere—absorption by ozone and oxygen, and infrared transfer by carbon dioxide, water vapor, and ozone. However, it was impossible at that time, owing to imperfect auxiliary data, to compute these effects very accurately and Gowan's results are not quantitatively correct.

A number of other investigations of various aspects of this problem have since been made. These show considerable variation in technique and assumptions and for the present purposes it does not seem necessary to review all of them. Among them,* the studies of Plass (1956a, b) and of Murgatroyd and Goody (1958) are discussed here.

Plass computed the infrared cooling rates due to ozone (1956a) and to carbon dioxide (1956b). He based his computations on the laboratory measurements of Summerfield in the case of ozone and of Cloud in the case of carbon dioxide, measurements which were unpublished but which are shown in Plass's papers. In applying these to the atmosphere, he took account of the various difficulties mentioned in Subsection 7.1.4, as detailed in his papers and the discussion thereof (Plass, 1957).

Figure 7.8 gives a sample of his results. With reference to the lower stratosphere, it appears that flux divergence due to the 15-μ CO_2 band leads to cooling and that due to the 9.6-μ O_3 band to heating for the temperature distribution used [that of the Rocket Panel (1952) and, at this altitude, essentially the same as that of the standard atmosphere].

Table 7.1 lists some of Plass's results for higher levels. Maximum cooling rate is near the stratopause and of magnitude about 7°K day^{-1}. Ozone contributes less than one-third of this and at other levels, especially above the stratopause, is relatively less important.

Murgatroyd and Goody calculated cooling rates for carbon dioxide by a quite different method (Curtis, 1956), using spectroscopic data as fundamental input and assuming no overlapping of the lines. Their results for this particular temperature distribution are also shown in Table 7.1 for comparison. Considering the complexity of the problem and the possible sources of uncertainty, one must consider the agreement to be good. It seems certain that these calculations establish the correct magnitude of the effect.

Murgatroyd and Goody carried out computations of carbon dioxide cooling for several other vertical distributions of temperature, corresponding to summer

---

* Some of the others are by Oder (1948), Craig (1951), Brooks (1958), Ohring (1958), Manabe and Möller (1961), and Davis (1963).

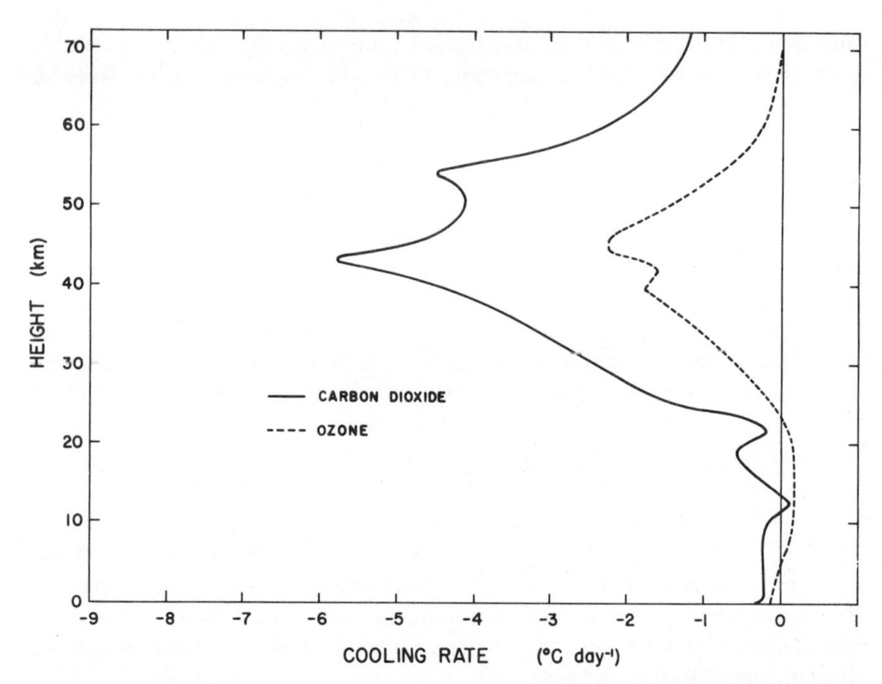

FIG. 7.8. Cooling rates due to the 15-μ carbon dioxide band and the 9.6-μ ozone band, according to calculations of Plass (1956a, b). Both curves are based on the temperature distribution of the Rocket Panel. The carbon dioxide calculations were made for a constant concentration of 0.033 per cent by volume. The ozone calculations were made for an ozone distribution consistent with the NRL rocket measurements of 14 June 1949 (see Fig. 5.9).

TABLE 7.1

COOLING RATES IN THE STRATOSPHERE[a]

Height (km)	Ozone (°K day^{-1})	Carbon dioxide (°K day^{-1})	Total (°K day^{-1})
25	0.1	1.3 (1.7)	1.4
30	0.5	2.4 (2.8)	2.9
35	1.3	3.3 (4.1)	4.6
40	1.8	4.5 (5.2)	6.3
45	2.1	4.9 (6.6)	7.0
50	1.5	4.6 (6.8)	6.1
55	0.8	4.0 (5.0)	4.8
60	0.3	2.3 (3.8)	2.6
65	0.1	1.7	1.8
70	0.0	1.3	1.3

[a] According to Plass (1956a, b). Values of Murgatroyd and Goody (1958) for carbon dioxide are shown in parentheses.

and winter distributions at different latitudes as estimated earlier by Murgatroyd (1957) (see Fig. 3.1). They estimated cooling rates due to ozone by an approximate method of adjusting the results of Plass to fit these temperature distributions. Finally they estimated the heating rates resulting from absorption of solar ultraviolet radiation by ozone and oxygen, taking into account of course the appropriate average zenith angle and length of the day for a given latitude and season. The resulting distribution of net radiative temperature change as a function of latitude and season is shown in Fig. 7.9.

Over a large part of the diagram the net change is quite small, less than $\pm$ 2°K day^{-1}. This is probably no more than the uncertainty inherent in such calculations. The small magnitude of the computed net change (although it could conceivably be accidental) indicates that much of the upper stratosphere and mesosphere is in approximate radiative balance and, furthermore, that the calculations are accurate enough to reveal this. However, poleward of 60° in summer and of 30° in winter radiative imbalances are shown that are likely to be meaningful. Although these results are shown to 90 km, our immediate discussion is confined to the upper stratosphere and low mesosphere, special problems of the mesopause region being reserved for consideration in the next section.

With regard to the summer season, net heating of about 4°K day^{-1} appears in high latitudes near the stratopause. This is caused by ozone absorption during

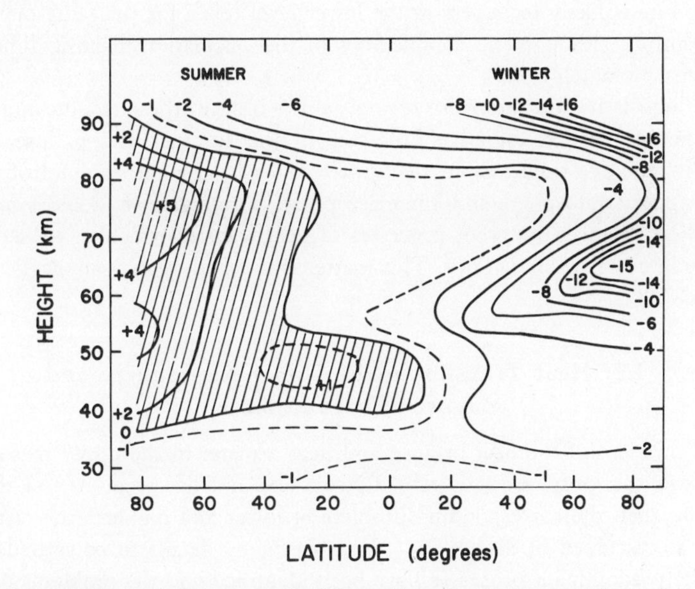

FIG. 7.9. Radiative heating rates (in °K day^{-1}) as the net result of absorption of solar radiation by ozone and oxygen, and of infrared transfer by carbon dioxide and ozone. (After Murgatroyd and Goody, 1958.)

the long summer days, only partially balanced by infrared cooling due to carbon dioxide and ozone. Although the computed rate of cooling at a particular level depends on the entire vertical temperature structure of the model atmosphere under consideration, it is particularly sensitive to the assumed temperature at the level in question. In assessing the significance of such results one must ask whether the assumed temperature is likely to be correct. In this regard, it was pointed out in Section 3.5 that the summer stratopause temperature at Churchill appears from later measurements to be lower than estimated by Murgatroyd. If this is so, then (according to this one factor) the cooling was overestimated by Murgatroyd and Goody and the net heating underestimated. However, the difference is probably rather small. Above the stratopause, the solar heating decreases, but radiative cooling also decreases at such a rate as to keep the net effect approximately constant.

Near the winter pole, net cooling appears throughout the upper stratosphere and mesosphere. Particularly striking is the large cooling rate (15°K day^{-1}) shown at about 65 km. This cooling maximum is associated with the temperature maximum at the stratopause, which appears at 60–65 km in the vertical temperature distribution used for the calculations (see Fig. 3.1; refer to high latitudes in winter). The IGY measurements at Churchill (Figs. 3.8, 3.12, and 3.13) indicate a lower stratopause, near 55 km, in which case the maximum rate of cooling would occur at this lower level. The magnitude of this maximum cooling rate is likely to be less at the lower level (even for the same maximum temperature), although the complexities of the calculation make it difficult to estimate how much less.*

The high-latitude stratopause region, on the basis of these results, appears to gain energy through radiative processes in the summer and to lose energy through radiative processes in the winter. This is to some extent true of the entire upper stratosphere and low mesosphere. These gains and losses must, of course, be balanced by other processes of heat transfer which in all likelihood involve large-scale circulations. This matter is touched upon again at the end of Section 7.2.

7.2 Heat Transfer in the Upper Mesosphere and Lower Thermosphere

The problem of the heat balance and heat transfer in the upper mesosphere and lower thermosphere is a very difficult one and has received less detailed attention than the corresponding problem at lower and higher levels. At lower levels, as discussed in Section 7.1, there are many details to be settled but at least the predominant processes have been identified and the problems defined.

* I am indebted to Professor R. M. Goody for a personal communication relating to the matters discussed in this paragraph.

Similarly, at higher levels, as will be discussed in Section 7.3, the predominant processes have presumably been identified to be absorption of extreme ultraviolet radiation and downward conduction of the heat so gained. Preliminary models based on these two processes agree to a first approximation with present scanty observations. However, in the intermediate region (between, say, the mesopause and 130 km), to which we refer as the "lower thermosphere," the relative importance of the various possible sources and sinks of heat is not known with any certainty as a function of altitude.

In this section, we discuss these problems and what is presently known about them. They include the problems of carbon dioxide, radiative losses by other atmospheric constituents, and heat gains from above through the downward transport (by conduction or otherwise) of sensible or latent heat.

7.2.1 The Problems of Carbon Dioxide Radiation in the Vicinity of the Mesopause

The discussion of the previous section has emphasized the overwhelming importance of the polyatomic molecules, carbon dioxide and, secondarily, ozone, in the radiative heat losses of the upper stratosphere. In the mesosphere, ozone rapidly loses its importance because of its decreasing concentration, leaving carbon dioxide as the sole important radiating constituent. At still higher levels, the importance of carbon dioxide must decrease because of (a) its dissociation at high enough levels and (b) its decreasing ability to convert absorbed radiation to heat energy prior to reradiation.

The photochemistry of carbon dioxide in the earth's atmosphere has been discussed by Bates and Witherspoon (1952). The gas is dissociated by solar radiation at wavelengths less than 1690 A:

$$CO_2 + h\nu \rightarrow CO + O \qquad (\lambda < 1690\ A) \tag{7.21}$$

It is important to note that the dissociating solar energy is strongly absorbed by molecular oxygen in the Schumann–Runge continuum, so that the presence of the latter gas tends to shield carbon dioxide from photodissociation. According to measurements of Wilkinson and Johnston (1950) and of Inn *et al.* (1953), the CO_2 absorption cross section is an order of magnitude less than that of O_2 in the spectral region of interest.

Carbon dioxide is formed mainly through the three-body process

$$CO + O + M \rightarrow CO_2 + M \tag{7.22}$$

Here again the presence of oxygen is a complicating factor because an oxygen atom freed by photodissociation of CO_2 may associate through either of

$$O + O_2 + M \rightarrow O_3 + M \tag{7.23}$$

$$O + O + M \rightarrow O_2 + M \tag{7.24}$$

The second oxygen atom in reaction (7.24) may, of course, have originated in the photodissociation of O_2. According to Bates and Witherspoon, the rate coefficients for processes (7.23) and (7.24) are very much greater than that for process (7.22).

Thus, the photochemistry of carbon dioxide and carbon monoxide is inextricably involved with that of oxygen. Although, nevertheless, estimates could be made of CO and CO_2 concentrations as a function of altitude on the basis of photochemical-equilibrium conditions, we have already seen in Chapters 5 and 6 that diffusion and mixing play a crucial role in the lower thermosphere. Therefore, detailed equilibrium computations are not apt to be very meaningful. It is clear, however, from order-of-magnitude considerations of the cross section involved in (7.21) and the number of available dissociating quanta that the amount of CO_2 is not sensibly affected by photochemical processes up to at least 90 km and perhaps 100 km. Above some higher level, estimated by Bates and Witherspoon to be 110 km (but which may be somewhat higher because they took a 5000°K sun), CO_2 must be essentially absent.

Even if the presence of CO_2 in its normal concentration is accepted up to 90–100 km, the radiative-transfer problem in the mesopause region is not at all straightforward. The formulation of the radiative problem in the lower atmosphere, as discussed in Section 7.1, is based on Kirchhoff's law. The use of this formulation amounts to an assumption that the vibrational (and rotational) energy levels remain populated according to a Boltzmann distribution* determined by the local kinetic temperature. This is possible only if collisions are frequent enough to maintain such a distribution in spite of the radiative processes. A measure of the time required for the establishment of a Boltzmann distribution by collisions is the *relaxation time*. If the relaxation time is short compared with the average radiative lifetime of the excited levels, then a Boltzmann distribution can be maintained. This is the situation in the lower atmosphere but is clearly not true at high levels as the density and collision rate decrease. This problem was noted by Spitzer (1949) and further discussed by Goody (1954). Apparently only Curtis and Goody (1956) have attempted to treat the matter quantitatively.

Goody and also Curtis and Goody pointed out that rotational relaxation times are very much shorter than vibrational relaxation times, while radiative lifetimes of rotational levels are in general longer than those of vibrational levels. Therefore the critical factor in the applicability of Kirchhoff's law is the vibrational relaxation time relative to the vibrational radiative lifetime. The level at which these become equal is sometimes called the level of *vibrational relaxation*. Curtis and Goody estimated the radiative lifetime to be about 0.06 sec for the 6.3-μ water-vapor band, 0.07 sec for the 9.6-μ ozone band, and 0.4 sec for

* See Appendix H.

the 15-μ carbon dioxide band. Vibrational relaxation times were estimated to be about 15×10^{-6} sec for CO_2 and 2×10^{-6} sec for H_2O, both at atmospheric pressure, and to be inversely proportional to total pressure. Accordingly, the level of vibrational relaxation for both bands was placed by Curtis and Goody at 74 ± 10 km, the principal uncertainty arising from their estimates of the relaxation times.

The equation of transfer proposed by Curtis and Goody is based on one simple-harmonic vibrational mode, with the rotational levels populated according to a Boltzmann distribution, and the rotational constants independent of the vibrational excitation. The source function under these conditions is $L_{\nu B}(\Theta + \Lambda X)/(\Theta + \Lambda)$ where Θ is the radiative lifetime of the vibrational state, Λ is the relaxation time for the vibrational mode, and X is defined by $\iint nk_\nu I_\nu \, d\omega \, d\nu / \iint nk_\nu L_{\nu B} \, d\omega \, d\nu$. In the last expression, n is the number density of absorbing molecules and k_ν, the monochromatic absorption coefficient, is in units of $(\text{molecule})^{-1}$; the indicated integrations are over the band and over all directions. The equation of transfer is therefore

$$-\frac{dI_\nu}{ds} = nk_\nu I_\nu - nk L_{\nu B} \frac{\Theta + \Lambda X}{\Theta + \Lambda} \tag{7.25}$$

When $\Theta \gg \Lambda$, Eq. (7.25) reduces to (4.37) as used at lower levels. When $\Lambda \gg \Theta$ at high levels, the net heating rate $-\iint (dI_\nu/ds) \, d\omega \, d\nu$ is zero and the radiation is simply scattered, none of it being converted to translational energy prior to reradiation.

Curtis and Goody used this transfer equation to compute cooling rates due to the 15-μ band. The details of their results are omitted here, because we shall examine some similar results in connection with Fig. 7.9. However, two very important aspects of the computations must be noted. Up to the level of vibrational relaxation, for this band, there is essentially no difference between the cooling rates computed with Kirchhoff's law and those computed with the Curtis–Goody model. Above that level, however, vibrational relaxation becomes extremely important; cooling rates would be very much greater in the absence of vibrational relaxation than estimated by the Curtis–Goody method. For example, at about 100 km (for CO_2 concentration of 3×10^{-4} by volume and the temperature distribution of the Panel Atmosphere) the two computed cooling rates are, respectively, 57°K day^{-1} and 3°K day^{-1}.

Returning now to Fig. 7.9, let us examine the results of Murgatroyd and Goody (1958) at high levels, which took into account vibrational relaxation. Near the high-latitude mesopause in summer the heating of 4° to 5°K day^{-1} is due primarily to infrared transfer. absorption of solar energy by O_2 and O_3 both being relatively unimportant at this altitude. As pointed out earlier (Subsection 7.1.5), the infrared cooling rate, especially at high levels, depends rather sensitively (but not uniquely) on the temperature at the level in question (see also Fig. 3 of Murgatroyd and Goody). The observations at Churchill in summer (Figs. 3.10 and 3.11) indicated somewhat lower mesopause temperatures than used in the calculations (see Fig. 3.1). Considering this factor alone then, the results are not likely to be overstimates of the heating rate and may be underestimates. At higher levels, heating due to O_2 absorption is likely to become important above 90 km, where the diagram ends. However, in view of the many uncertainties at these higher levels, even a semiquantitative assessment of the heat balance seems to be hopeless at the present time.

Near the high-latitude winter mesopause, Fig. 7.9 shows a minimum cooling rate. The minimum is associated with the minimum in the temperature distribution used. As a matter of fact, the basic calculations (see figure 2 of Murgatroyd and Goody) actually showed very slight heating due to carbon dioxide at this level, but this thin layer of heating was smoothed out in the preparation of their figure 5. However, this is of little importance, because the details of such calculations at these levels are closely related to the details of the assumed temperature profile. The adoption of temperature profiles similar to those observed during winter at Churchill during the IGY (Figs. 3.8, 3.12, 3.13) would clearly lead to very large cooling rates near the mesopause. It is not possible to estimate these rates with any certainty without repeating the calculations for the specific temperature profiles concerned. However, there is no doubt that if carbon dioxide is present at these levels in the normal amount and if Curtis and Goody's treatment of vibrational relaxation is even approximately applicable, then radiative cooling near 70 km in the temperature distribution of Fig. 3.12 must be very large indeed—perhaps to be measured in tens of degrees per day.

The large cooling rate shown near 90 km in Fig. 7.9 is very much larger than that calculated by Curtis and Goody (1956) for the same level (5–6°K day^{-1}), presumably by the same method and with essentially the same data, except that they used the temperature distribution of the Panel Atmosphere. In the latter distribution, the temperature at 90 km is some 25°K less than that of Murgatroyd (1957) at 60°N in winter. It would appear that this temperature difference accounts for the difference in computed cooling rate. Since the temperature at this level is not known with any certainty, it is impossible to decide which result is more representative.

According to our present understanding of the behavior of carbon dioxide, the observed temperature structure of the high-latitude region is not to be explained on the basis of carbon dioxide cooling and solar heating alone. This is true of both summer and (especially) winter. Other important means of heat transfer, radiative or otherwise, must be presumed to be operative in this region.

7.2.2 Radiation by Other Thermospheric Constituents

Bates (1951, 1956) has discussed in considerable detail the possible roles of various gases in the radiative budget of the thermosphere.

Molecular oxygen and nitrogen, being homonuclear molecules, have no electric dipole transitions. There is, of course, the possibility of other types of transitions, such as quadrupole transitions. However, from numerical considerations, Bates concluded that these play no important role.

Similar numerical considerations indicate that heteronuclear diatomic molecules, such as CO and NO, probably play no significant role, owing mainly to their presumed low abundance. This might not be true in a planetary atmosphere of different composition. For example, Chamberlain (1962) has called attention to the possibly important role of carbon monoxide in the upper atmosphere of Mars.

Atomic oxygen is an abundant constituent of the thermosphere and its radiations must be considered with care. The red doublet $^3P \leftarrow {}^1D$ (see Fig. 4.5) is radiated from altitudes rather high in the thermosphere, but according to Bates the cooling rate associated with this transition is negligible. On the other hand, Bates (1951) was the first to point out the importance of transitions between the different levels of the ground term ^3P of O. This is now generally accepted as a radiative mechanism of considerable importance in the thermosphere.

The $^3P_0$ and $^3P_1$ levels lie, respectively, only 0.028 and 0.020 ev above the $^3P_2$ level. Such energies are available through collisions from translational energy at thermospheric temperatures. Furthermore, the radiative transitions $^3P_0 \rightarrow {}^3P_1$ and $^3P_1 \rightarrow {}^3P_2$ are forbidden and a Boltzmann distribution is apt to be maintained to very high levels, perhaps 200–300 km. Therefore these transitions, especially $^3P_1 \rightarrow {}^3P_2$ at 62 μ, are apt to be efficient mechanisms for the dissipation of thermal energy.

According to Bates, the heat loss due to the latter transition at high enough levels is given by $10^{-18}n(\mathrm{O})$ (in ergs [centimeter]$^{-3}$ [second]$^{-1}$). This simplified expression is applicable only at high levels where the atmosphere is optically thin for this radiation, perhaps above 150 km. Furthermore, the coefficient of $n(\mathrm{O})$ is temperature dependent, but the dependence is not great.

The concentration of atomic oxygen in the lower thermosphere is not known in detail, but according to Nicolet (1960) it is of the order of 10^{12} cm^{-3} between 90 and 110 km. This would give for an optically thin layer a heat loss of 10^{-6} erg cm^{-3} sec^{-1}, which, with a density of about 5×10^{-10} g cm^{-3} at 100 km, would correspond to a cooling rate of 15°K day^{-1}. However, as emphasized by Chamberlain (1961), the atmosphere at this level is not even approximately optically thin to this radiation and the figure given above must be much larger than the actual heat loss. The radiative-transfer problem has not been treated in detail (see, however, Bauer and Wu, 1954) and the vertical distribution of

n(O) is known only approximately. One can simply conclude that this radiation is apt to play an important role in some parts of the lower thermosphere; however, its importance relative to carbon dioxide radiation is not known as a function of altitude.

Among the other radiations that might be important in the lower thermosphere are those associated with the airglow and known to originate in the low thermosphere (the green-line covariance group; see Section 9.3). These include the green line $^1D \leftarrow {}^1S$ of atomic oxygen, and the Herzberg bands $({}^3\Sigma_u^+ \rightarrow {}^3\Sigma_g^-)$ and atmospheric bands $({}^1\Sigma_g^+ \rightarrow {}^3\Sigma_g^-)$ of O_2. Chamberlain (1961) has estimated the volume emission rates of these radiations to be 10^{-7}-10^{-8} erg cm^{-3} sec^{-1}, mostly from O_2. At 100 km, this would correspond to a cooling rate* on the order of 1°K day^{-1}; but it must be remembered that Chamberlain's figure is only an average one over a large altitude interval and may be considerably higher at some altitudes and lower at others.

Thus in the lower thermosphere, radiative cooling by carbon dioxide, the Bates transition in the ground term of atomic oxygen, and perhaps airglow emission in the green-line covariance group must be taken into account in any detailed study of the radiative budget. The first is undoubtedly the most important in the lowest part of the region and the second in the highest part. The relative importance of the three at intermediate levels, say around 100 km, is not known. Although further radiative-transfer investigations would be useful, any detailed understanding of this problem must await at least some additional observations of the composition and thermal structure of this relatively neglected region of the atmosphere.

7.2.3 Other Heat-Transport Processes in the Lower Thermosphere

We have seen in Subsection 4.6.6 that the lower thermosphere absorbs essentially all of the energy in the Schumann–Runge continuum, which may amount to about 30 erg cm^{-2} sec^{-1} at normal incidence. The average energy input over the earth's surface then is one-fourth of this. Absorption of X-ray radiation is unimportant for this purpose, although it probably plays an important role in the production of ionization (Section 9.2).

The lower thermosphere, as we shall discuss in more detail in Section 7.3, also receives heat from the overlying atmosphere. This heat, absorbed at higher levels, mostly in the 200–1000-A spectral region, is not dissipated by radiation at the higher levels but instead is transported downward by molecular conduction. Some fraction of it is also transported downward by diffusion in chemical

* Strictly speaking, this is not a mechanism for the dissipation of thermal energy, but represents a dissipation of absorbed solar energy prior to its degradation to thermal energy (see Section 9.3). However, it must be considered in a "budget" that includes all absorbed solar radiation as an energy source.

form, that is, in the form of atoms such as O and N that recombine by three-body collision in the lower thermosphere and there convert the original energy of dissociation into thermal energy. Altogether however, the heat flow downward through the 130-km level by these processes can hardly exceed a few ergs $(\text{centimeter})^{-2}$ $(\text{second})^{-1}$ at normal solar incidence and is small in comparison with direct absorption when the average budget of the entire lower thermosphere is under consideration.

However, within the lower thermosphere, any detailed level-by-level study of the heat balance must take into account the vertical transfer of heat through these mechanisms and also, at the lowest levels, by mixing. Nicolet (1960) has estimated that the lifetime of an oxygen atom against three-body recombination is more than a year at 120 km and on the order of a month even at 100 km. Thus, vertical transport processes must play an important role and much of the energy assigned above to the entire lower thermosphere is realized as heat energy only below 100 km or so.

The outstanding problem so far defined by observation and clearly requiring an explanation in terms of heat transport is the observed high temperature of the high-latitude, winter mesopause region and the observed low temperature of the high-latitude, summer mesopause region. The former has received much more attention and is discussed in more detail below. However, the low summer temperatures are equally puzzling and in particular place some constraints on the possible explanations of the high winter temperatures. It appears that any explanation of the wintertime situation must include an explanation of its seasonal character.

Kellogg (1961) has discussed the winter situation extensively and suggested that heating at this altitude takes place as the result of gentle subsidence of air from higher altitudes over the winter polar region. The subsiding air is heated by compression but more significantly by the energy released in the association of oxygen atoms into molecules by the three-body collision process (7.24). The energy so released is approximately 5 ev for each molecule formed and calculations show that this may be a very significant heating factor.

Kellogg assumed a certain vertical profile of atomic oxygen, consistent with what is known about the distribution of this gas, and further assumed that a small net vertical velocity continually transports oxygen atoms to lower altitudes. Oxygen atoms transported downward in this way were assumed to recombine, converting their chemical energy entirely into heat energy, while at the same time horizontal advection from lower latitudes was assumed to maintain the initial atomic-oxygen distribution against the changes due to vertical motion and recombination. He found that rather large amounts of heating could be produced under these assumptions with only a very slight downward motion; for example, $10°\text{K day}^{-1}$ at 95 km with a downward velocity of only 0.05 cm sec^{-1}

In a somewhat more elaborate computation, Young and Epstein (1962) have approached the problem from a different point of view. They assumed that an air parcel with certain initial density and atomic-oxygen concentration starts at 115 km and moves slowly down to the mesosphere without any mixing effects on its composition. They considered the various chemical reactions that might take place within the parcel as its density changes, and by numerical integration computed the changing composition and the heating rate caused by energy released within the parcel. They found that a small vertical velocity of 0.2 cm sec^{-1} leads to heating rates of up to several tens of degrees K day^{-1}, depending on initial conditions.

Although both of these models are quite crude, they illustrate that rates of heating due to the release of stored chemical energy can be quite large near and just above the mesopause. The question remains, of course, of where the energy comes from in the first place. Clearly, no significant amount of oxygen dissociation by sunlight takes place at high latitudes in winter. If the source of the energy is photodissociation, then circulation mechanisms must be at work to transport oxygen atoms from lower latitudes. In the Kellogg model, a slow net meridional circulation is responsible for this transport. Maeda (1963) has proposed that oxygen may be dissociated by auroral particles near 90 km in the polar regions and that this process may provide important amounts of energy. However, even if this process plays a significant role, it is presumably not highly variable with season and one must still postulate some other mechanism, such as downward transport, that is operative principally in the winter months.

At lower levels, say below 75 km, the computations of Kellogg and of Young and Epstein show rather little heating from oxygen recombination. This can be verified by a gross order-of-magnitude computation. Between 60 and 70 km, for example, there is an integrated mass of air of approximately 0.2 g cm^{-2}. To heat this mass at the rate of 10°K day^{-1} would require about 2×10^{18} associations of oxygen atoms into molecules each day, even if all the dissociation energy were converted into heat. This rate of combination would use up all available oxygen atoms between 60 and 110 km in about a day. Some other heating mechanism must be found for the lower mesosphere and stratopause region.

It is also possible that the apparent radiative imbalance near the winter mesopause is due partially to a deficiency of carbon dioxide and therefore less cooling than is implicitly assumed in the search for a heating mechanism. No measurements of carbon dioxide concentration are available at these altitudes. Although it seems reasonable on the basis of photochemical-equilibrium considerations to assume that the gas is present in normal amounts, air originating at higher elevations may be deficient in carbon dioxide, and the time required for its re-formation may be rather long in the presence of the oxygen reactions that compete for available oxygen atoms.

Another possible heating mechanism for the lower thermosphere has been suggested by Hines (1963). As will be discussed in Section 8.3, this region of the atmosphere is the seat of relatively small-scale but large-amplitude motions which show up most commonly in the form of large vertical wind shears (see also Subsection 3.3.5). Hines has suggested that the ultimate degradation of this energy into thermal molecular motion may provide a significant source of heat for the lower thermosphere. The energy presumably originates in the lower atmosphere and is propagated upward. Quantitative estimates are difficult, although Hines (1963) mentions a few degrees per day as an order of magnitude figure. Qualitatively, there is a difficulty in explaining a significant difference between summer and winter; the motions appear to be present at all seasons, although perhaps with somewhat less intensity in summer. In any case, they surely do not represent a cooling mechanism during summer, such as appears to be required by the radiative calculations.

Although all of the factors discussed above may play some role, it seems likely that a large-scale meridional transport of heat in one form or another by the atmosphere is implied. From a meteorological point of view, as discussed in Appendix F, this can result from a mean meridional circulation or from large-scale eddy processes. It is impossible to decide conclusively between these except by calculations from detailed and frequent observations of wind and the property being transported.

The Kellogg hypothesis of a net meridional circulation to explain heat transport near the mesopause, which was advanced in a more general form by Kellogg and Schilling (1951), is now considerably strengthened by consideration of the effect of chemical heating. There are further indications, none of which is conclusive, that it may be preferred over the eddy-transport mechanism. These are (a) the average meridional wind component away from the summer pole and toward the winter pole observed in the few available radar-meteor studies; (b) the difficulty of reconciling the mesospheric wind and temperature distributions in winter under the assumption of geostrophic equilibrium, and (c) the surprising observation that the winter upper mesosphere in high latitudes is not only warmer than it should be from a radiative point of view, but actually warmer than the upper mesosphere at low latitudes (although the eddy process can in principle transport heat against a gradient, rather special conditions are required).

7.3 Heat Conduction and Thermospheric Models

Spitzer (1949) pointed out that molecular conduction of heat would probably be an important mechanism of heat transport in the thermosphere and subsequent studies of the thermosphere have adequately verified this. In fact, the structure of the thermosphere cannot be understood except by consideration of heat conduction.

The coefficient of conductivity K is independent of density but varies with temperature and composition. The expression given by Nicolet (1960) is

$$K = AT^{\frac{1}{2}} \tag{7.26}$$

where

$$A(\mathrm{O}) = 3.6 \times 10^2 \quad \mathrm{erg\ cm^{-1}\ sec^{-1}\ deg^{-\frac{3}{2}}}$$
$$A(\mathrm{O_2}, \mathrm{N_2}) = 1.8 \times 10^2 \quad \mathrm{erg\ cm^{-1}\ sec^{-1}\ deg^{-\frac{3}{2}}}$$

Therefore the vertical flux of heat due to conduction is

$$F_C = -AT^{\frac{1}{2}} \frac{\partial T}{\partial z} \tag{7.27}$$

and the change of heat content per unit volume and unit time due to this process is

$$H_C = \frac{\partial}{\partial z}\left(AT^{\frac{1}{2}} \frac{\partial T}{\partial z}\right) \tag{7.28}$$

One can calculate quickly that downward transport by conduction is an important mechanism for heat loss by the upper thermosphere. For example, at 130 km with the temperature and temperature gradient of the COSPAR International Reference Atmosphere, Eq. (7.27) gives a downward transport of 1 to 2 erg $\mathrm{cm^{-2}\ sec^{-1}}$, the lower figure applying to an (O_2, N_2) atmosphere and the higher to an atomic oxygen atmosphere. This is the same order of magnitude as the amount of solar energy believed to be absorbed and converted into heat at all levels above 130 km (at normal solar incidence). In fact, the adoption of such steep temperature gradients and correspondingly large heat fluxes at 130 to 160 km in the recent standard atmospheres, as discussed in Subsection 6.5.3, is involved in a complicated way with considerations of the heat balance above 130 km as well as with rocket measurements of density and solar energy. In the following paragraphs, we shall try to indicate how these ideas have developed and the reasoning behind the adoption of these steep temperature gradients.

Bates (1951) followed up Spitzer's suggestion with quantitative estimates of the effect of heat conduction in comparison with the estimated or inferred effects of various radiative processes. Rather little was known at that time about the composition and thermal structure of the thermosphere, but radio measurements of the F region implied that the temperature must increase rapidly with height somewhere in the thermosphere. Bates showed that for the inferred vertical temperature gradients, downward transport of heat by conduction would be very rapid indeed and that this transport of heat to lower levels, in conjunction with the radiative loss by the 62-μ radiation, could be balanced only by a far greater energy input than was thought possible at the time. He suggested that this gain of energy occurred though absorption of much more solar radiation in the extreme ultraviolet than was inferred from the observed degree of ionization. The additional ionization resulting from this absorption was thought to disappear very rapidly by dissociative recombination [Eq. (4.16)] and was termed by Bates *unobserved ionization.*

In a later discussion, Bates (1956) pointed out that the model developed in his earlier paper required that the temperature gradient dT/dz must decrease with altitude, at least above a certain level around 160–200 km. At high enough levels, where the radiative heat loss by atomic oxygen can be ignored and where the heat gain by absorption decreases with altitude, the downward transport of heat by conduction and therefore the vertical temperature gradient to which it is proportional, must also decrease with altitude. This conclusion was contrary to the results of model atmospheres that had been proposed up to that time.

More recently, Nicolet (see especially, Nicolet, 1961) has emphasized the importance of conduction as a rapid mechanism of heat transport in the thermosphere. His calculations show, for example, that the atmosphere above 200 km must tend rapidly toward isothermalcy, especially at night. He has also pointed out that, in the presence of conduction, temperature variations with latitude must be very small at high altitudes and that the temperature must vary rather smoothly in the vertical, since conduction would not allow secondary maxima or minima of temperature to exist for very long times.

Johnson (1956, 1958) developed these considerations to the point of computing a steady-state vertical temperature distribution that was consistent with what was known of the energy sources and sinks and also with rocket measurements. Hunt and Van Zandt (1961) repeated the calculations with some variations of procedures and data, and more recently Harris and Priester (1962a) have developed a very elaborate model which considers the temperature variation in the course of a 24-hour day, and to which we shall return after a discussion of the steady-state models.

In a steady state and with no horizontal exchange or vertical mixing

$$H_C + H_A + H_R = 0 \tag{7.29}$$

where H_C is the heat change due to vertical conduction [Eq. (7.28)], H_A is the heat change due to absorption (positive), and H_R is the heat change (negative) due to emission, all per unit volume and time. Integration of this equation from a level z_0 to the outer edge of the atmosphere gives

$$\left(AT^{\frac{1}{2}}\frac{\partial T}{\partial z}\right)_{z_0} = \int_{z_0}^{\infty} (H_A + H_R)\, dz \tag{7.30}$$

if there is no vertical temperature gradient at the outer edge of the atmosphere. The basic idea of the calculation is to estimate the net radiative heat change

$(H_A + H_R)$ as a function of altitude, calculate the right-hand side of Eq. (7.30) as a function of altitude, and then regard Eq. (7.30) as a differential equation from which, with suitable boundary conditions, $T(z)$ can be determined numerically. Actually this is a nonlinear process, principally because the vertical temperature distribution controls the vertical distributions of the absorbing and emitting gases which in turn determine the vertical distributions of H_A and H_R. However, the problem can be solved by successive approximations, starting with an assumed temperature distribution.

The term H_R is due primarily to atomic oxygen radiation at 62 μ and has a maximum value in ergs (centimeter)$^{-3}$ (second)$^{-1}$ of

$$H_R = -\frac{1.67 \times 10^{-18} n(\mathrm{O}) \exp(-228/T)}{1 + 0.6 \exp(-228/T) + 0.2 \exp(-325/T)} \tag{7.31}$$

which may be reduced somewhat at the lowest levels under consideration by radiative-transfer effects. The temperature dependence is not marked in the range of temperatures under consideration. Both Johnson and Hunt and Van Zandt neglected this term, because it is less than 10 per cent of H_A in magnitude in most of the region above 130 km. It is certainly important below 130 km, and depending on the exact vertical distribution of atomic oxygen, which is quite uncertain, may be of importance to altitudes somewhat above 130 km. Figure 7.10 shows the ratio of H_R to H_A as calculated from their final model by Hunt and Van Zandt.

The question of heating and its vertical distribution is a vital one in these models. Johnson, after a discussion of other possibilities and in agreement with Bates, considered only the heat input through absorption of solar radiation in the extreme ultraviolet. There are other possible sources, such as conduction from an extended solar corona (Chapman, 1957) and heating as a net result of solar corpuscular radiation, and we shall see a little later that Harris and Priester (1962a) believe some other such source is involved.

Heat gain through absorption of solar ultraviolet radiation is given (for one kind of particle) by

$$H_A(z) = n(z) \int \epsilon_\nu(z) F_\nu(\infty) \sigma_\nu \exp[-t_\nu(z)]\, d\nu \tag{7.32}$$

where n is the number density of the particle, $F_\nu(\infty)$ the solar flux at the outer edge of the atmosphere, σ_ν the absorption cross section, t_ν the optical depth, and ϵ_ν the efficiency of conversion into heat (thermal efficiency). The last factor arises because most of the energy absorbed in the extreme ultraviolet results in ionization and is not immediately realized as heat energy (thermal motion of the molecules). In the process of dissociative recombination, believed to be the most important one at these altitudes, only a fraction of the energy appears as heat

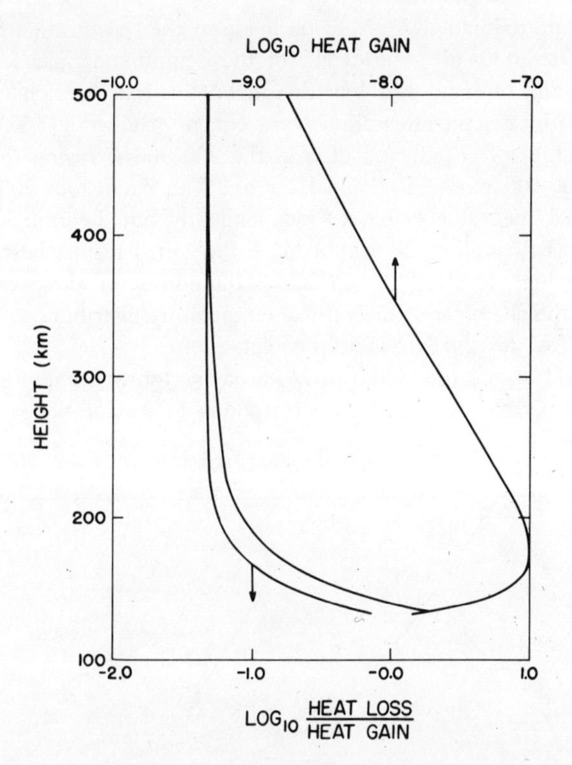

FIG. 7.10. Heat gain (erg [centimeter]$^{-3}$ [second]$^{-1}$) and the ratio of heat loss to heat gain, according to the calculations of Hunt and Van Zandt (1961). The heat gain is computed for an atomic oxygen atmosphere and for $\overline{\epsilon F(\infty)} = 1.1$ erg cm^{-2} sec^{-1} (see text and Fig. 7.13). The heat loss is computed both for an atomic oxygen atmosphere (right curve) and for a certain model atmosphere containing other constituents (left curve). (After Hunt and Van Zandt, 1961.)

and a large part is stored as dissociation energy. If the resulting dissociated neutral particles finally associate in a three-body process, most of the energy appears finally as heat energy; however, three-body association is a very slow process at these low densities and many of the dissociated atoms are likely to be diffused downward and to associate in the lower thermosphere. Probably no more than half, and perhaps much less, of the energy absorbed above 130 km appears as heat at those elevations.

Equation (7.32) represents explicitly the contribution of only one kind of particle; contributions from all of O_2, N_2, and O must be considered. In view of uncertainties about the relative concentrations of these particles at different altitudes, about the solar spectrum at the outer edge of the atmosphere, and about the absorption cross sections and thermal efficiencies involved, it is clear that many approximations have to be made in the specification of $H_A(z)$. We shall not attempt to give the details of these approximations for the models of Johnson and of Hunt and Van Zandt, but instead mention two important points. In both cases the integration over frequency shown in (7.32) was replaced by mean values of $\epsilon F(\infty)$ and of σ in the 200–900-A region. For the latter, Johnson took 500 (cm NTP)$^{-1}$ and Hunt and Van Zandt took 200 (cm NTP)$^{-1}$. Neither fixed the value of $\overline{\epsilon F(\infty)}$ independently but determined it from the calculation as the value that would make their final results best fit measured densities at high levels. Thus an important output of these calculations, in addition to the character of the vertical temperature distribution, is the implied value of $\overline{\epsilon F(\infty)}$ for the extreme ultraviolet.

Figure 7.11 shows the values of H_A used by Johnson and Fig. 7.12 shows his final temperature curve. This corresponds to a value of $\overline{\epsilon F(\infty)}$ of 0.7 erg cm^{-2} sec^{-1}.

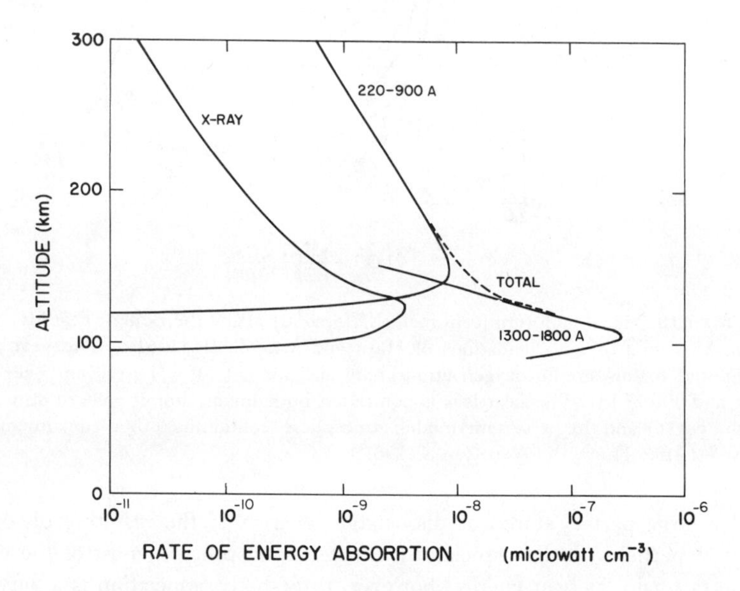

FIG. 7.11. Vertical distribution of the absorption of solar radiation in the thermosphere, as estimated by Johnson (1958). The dashed curve gives the sum of the contributions from different spectral regions. (After Johnson, 1958.)

Figure 7.13 shows the vertical temperature distributions calculated by Hunt and Van Zandt for different values of $\overline{\epsilon F(\infty)}$. It is clear that their calculation for $\overline{\epsilon F(\infty)} = 1.1$ erg cm^{-2} sec^{-1} agrees best with observations.

Both of these models refer to steady-state conditions with vertical incidence. Employing a high-speed computer, Harris and Priester (1962a) have more recently attempted an ambitious study of the time-dependent heat-balance equa-

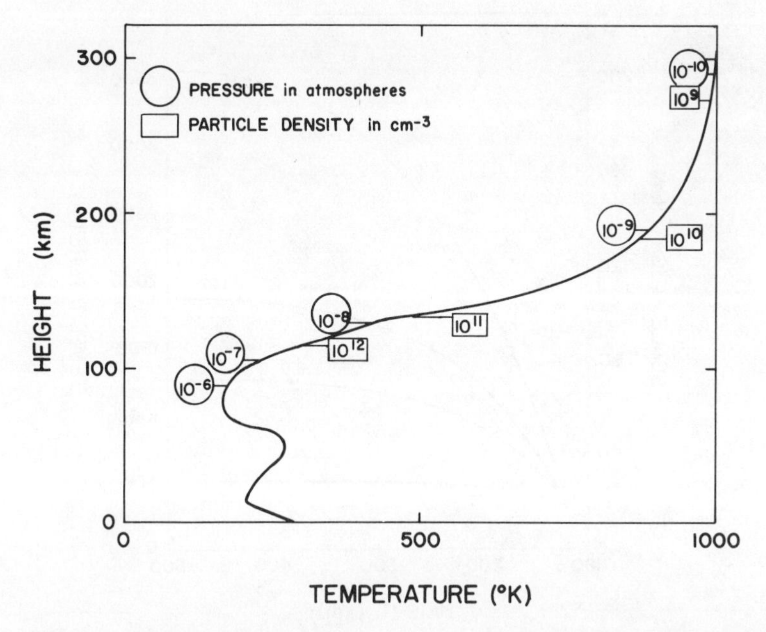

FIG. 7.12. Vertical distribution of temperature in the atmosphere. Temperatures above 130 km were computed by Johnson to give a downward flux of heat by conduction that would balance heat gain by absorption of solar radiation. (After Johnson, 1958.)

tion, essentially Eq. (7.29) with the zero on the right side replaced by $[\rho c_p(\partial T/\partial t)]$. In their model, time-independent boundary conditions are specified at 120 km (those of Nicolet; see Subsection 6.5.3), all constituents are assumed to be in diffusive equilibrium above that level, the radiative loss by atomic oxygen is included, and mean values of σ and $\epsilon F(\infty)$ are used in the extreme ultraviolet region. The numerical solution requires iterations in time (intervals of 0.25 hours were used) and in the vertical (intervals of 1 km were used). The solution was found to converge (that is, give a reproducible diurnal temperature variation at all elevations) after four to five days of real time.

The results of Harris and Priester's calculations, when ultraviolet solar radiation is the sole heat source, disagree with observations in two ways: They require a value of $\epsilon F(\infty)$ in the neighborhood of 2 erg cm^{-2} sec^{-1} at normal incidence, and they predict a high-level temperature and density maximum at 1700 local time rather than 1400 as inferred from satellite observations (see Section 6.4). With regard to the first, the measured values of Hinteregger (Hinteregger, 1961; Watanabe and Hinteregger, 1962; see Table 4.4) give not much more than 2 erg cm^{-2} sec^{-1} for $F(\infty)$ and the mean thermal efficiency is not as high as a comparison of these would require. Incidentally, Harris and Priester's required value of $\epsilon F(\infty)$ is larger than those of Johnson and of Hunt and

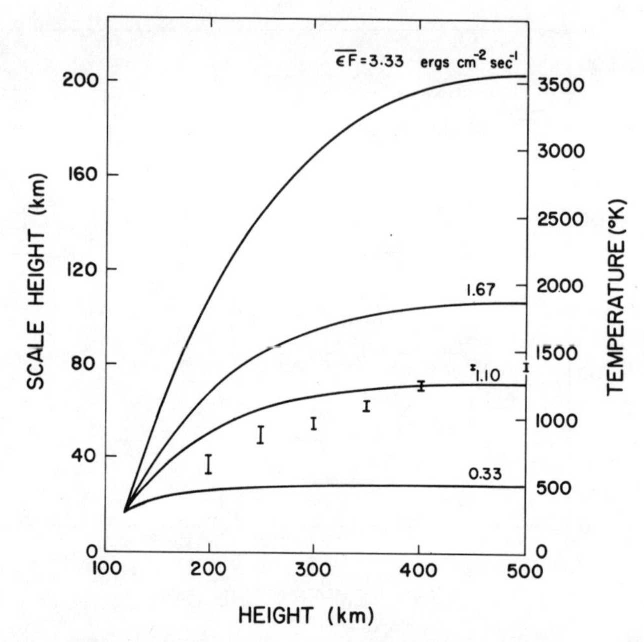

FIG. 7.13. Variations of temperature and scale height with altitude, according to the computations of Hunt and Van Zandt, for different assumed values of absorbed solar energy. The vertical lines show determinations based on satellite measurements. (After Hunt and Van Zandt, 1961.)

Van Zandt primarily because the latter were considering only a time-independent solution with the sun at vertical incidence and Harris and Priester's calculations include conductive losses during the night when the sun is absent.

Harris and Priester found the best agreement with observation (for their model) when they included a second heat source of the type shown in Fig. 7.14 in combination with the ultraviolet heat source also shown in Fig. 7.14. They ascribe this second heat source to energy associated ultimately with solar corpuscular radiation and suggest that the existence of such a source is indicated by density variations associated with geomagnetic disturbances (see also Harris and Priester, 1962b). They give no explanation, however, for the inferred diurnal variation of the energy source and they also have to specify a vertical variation.

These various thermospheric models, with full consideration of molecular conduction, have been very useful in the interpretation and correlation of diverse measurements of composition, density, and solar energy. It must be remembered, however, that the structure of the thermosphere in all its details is still imperfectly understood, and much work, observational and theoretical, remains to be done.

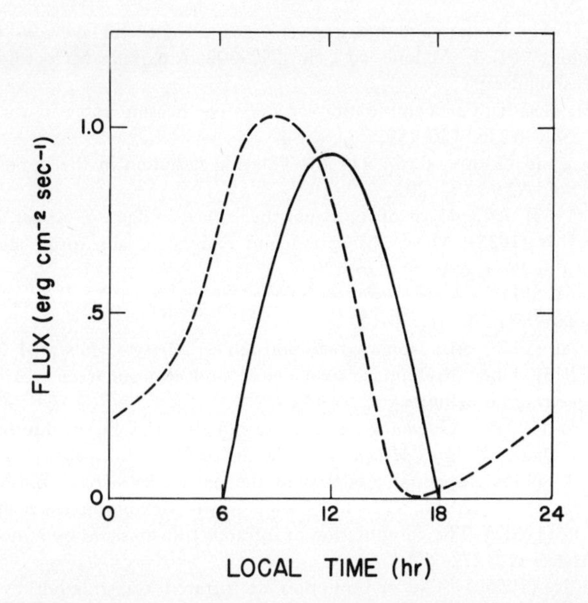

FIG. 7.14. Time-dependent heat sources for the thermosphere above 120 km, as assumed by Harris and Priester. The full curve shows energy due to absorption of solar extreme ultraviolet radiation; the dashed curve shows energy due to a second unspecified source, which is sufficient to reconcile the results of the computations with observations. (After Harris and Priester, 1962a.)

REFERENCES

BATES, D. R. (1951). The temperature of the upper atmosphere. *Proc. Phys. Soc.* **B64**, 805–821.

BATES, D. R. (1956). The thermosphere. *Proc. Roy. Soc.* **A236**, 206–211.

BATES, D. R., and WITHERSPOON, A. E. (1952). The photo-chemistry of some minor constituents of the earth's atmosphere (CO_2, CO, CH_4, N_2O). *Mon. Not. Roy. Ast. Soc.* **112**, 101–124.

BAUER, E., and WU, T.-Y. (1954). The cooling of a gas by radiation. *Proc. Phys. Soc.* **A67**, 741–750.

BROOKS, D. L. (1958). The distribution of carbon dioxide cooling in the lower stratosphere. *J. Meteor.* **15**, 210–219.

BURCH, D. E., GRYVNAK, D., SINGLETON, E. B., FRANCE, W. L., and WILLIAMS, D. (1962). Infrared absorption by carbon dioxide, water vapor, and minor atmospheric constituents. *AFCRL Res. Rep.* No. 62–698, pp. 1–316.

CHAMBERLAIN, J. W. (1961). The energies in the spectra of the airglow and aurora. *Ann. Geoph.* **17**, 90–99.

CHAMBERLAIN, J. W. (1962). Upper atmospheres of the planets. *Ap. J.* **136**, 582–593.

CHAPMAN, S. (1957). Notes on the solar corona and the terrestrial ionosphere. *Smiths. Contr. Ap.* **2**, No. 1, 1–11.

CRAIG, R. A. (1951). Radiative temperature changes in the ozone layer. *In* "Compendium of Meteorology" (T. F. Malone, ed.), pp. 292–302. American Meteorological Society, Boston, Massachusetts.

CURTIS, A. R. (1956). The computation of radiative heating rates in the atmosphere. *Proc. Roy. Soc.* **A236**, 156–159.

CURTIS, A. R., and GOODY, R. M. (1956). Thermal radiation in the upper atmosphere. *Proc. Roy. Soc.* **A236**, 193–205.

DAVIS, P. A. (1963). An analysis of the atmospheric heat budget. *J. Atmos. Sci.* **20**, 5–22.

ELSASSER, W. M. (1938). Mean absorption and equivalent absorption coefficient of a band spectrum. *Phys. Rev.* **54**, 126–129.

ELSASSER, W. M. (1942). Heat transfer by infrared radiation in the atmosphere. *Harvard Meteor. Studies* No. 6, pp. 1–107.

ELSASSER, W. M. (1960). Atmospheric radiation tables. *Meteor. Monogr.* **4**, No. 23, 1–43.

EMDEN, R. (1913). Über Strahlungsgleichgewicht und atmosphärische Strahlung. *S. B. Akad. Wissenschaften, Munich* pp. 55–142.

GODSON, W. L. (1953). The evaluation of infra-red radiative fluxes due to atmospheric water vapor. *Quart. J. Roy. Meteor. Soc.* **79**, 367–379.

GODSON, W. L. (1954). Spectral models and the properties of transmission functions. *Proc. Toronto Meteor. Conf., 1953* pp. 35–42. Royal Meteorological Society, London.

GODSON, W. L. (1955a). The computation of infrared transmission by atmospheric water vapor. *J. Meteor.* **12**, 272–284.

GODSON, W. L. (1955b). The computation of infrared transmission by atmospheric water vapor, II. *J. Meteor.* **12**, 533–535.

GODSON, W. L. (1962a). Infrared transmission by water vapour. *Archiv Meteor., Geoph., Biokl.* **B12**, 1–18.

GODSON, W. L. (1962b). Infrared transmission by water vapour. II, The use of laboratory data. *Archiv Meteor., Geoph., Biokl.* **B12**, 196–223.

GOLD, E. (1908). Contribution to a discussion of "The Isothermal Layer of the Atmosphere." *Nature* **78**, 550–552.

GOLD, E. (1909). The isothermal layer of the atmosphere and atmospheric radiation. *Proc. Roy. Soc.* **A82**, 43–70.

GOODY, R. M. (1952a). A statistical model for water-vapor absorption. *Quart. J. Roy. Meteor. Soc.* **78**, 165–169.

GOODY, R. M. (1952b). Discussion of "Goody (1952a)." *Quart. J. Roy. Meteor. Soc.* **78**, 638–640.

GOODY, R. M. (1954). "The Physics of the Stratosphere." Cambridge Univ. Press, London and New York. (Reprinted, 1958.)

GOWAN, E. H. (1947a). Ozonosphere temperatures under radiation equilibrium. *Proc. Roy. Soc.* **A190**, 219–226.

GOWAN, E. H. (1947b). Night cooling of the ozonosphere. *Proc. Roy. Soc.* **A190**, 227–231.

HARRIS, I., and PRIESTER, W. (1962a). Time-dependent structure of the upper atmosphere. *J. Atmos. Sci.* **19**, 286–301.

HARRIS, I., and PRIESTER, W. (1962b). Theoretical models for the solar-cycle variation of the upper atmosphere. *J. Geoph. Res.* **67**, 4585–4591.

HINES, C. O. (1963). The upper atmosphere in motion. *Quart. J. Roy. Meteor. Soc.* **89**, 1–42.

HINTEREGGER, H. E. (1961). Preliminary data on solar extreme ultraviolet radiation in the upper atmosphere. *J. Geoph. Res.* **66**, 2367–2380.

HUMPHREYS, W. J. (1909). Vertical temperature-gradients of the atmosphere, especially in the region of the upper inversion. *Ap. J.* **29**, 14–32.

Hunt, D. C., and Van Zandt, T. E. (1961). Photoionization heating in the *F* region of the atmosphere. *J. Geoph. Res.* **66**, 1673–1682.

Inn, E. C. Y., Watanabe, K., and Zelikoff, M. (1953). Absorption coefficients of gases in the vacuum ultraviolet. III, CO_2. *J. Chem. Phys.* **21**, 1648–1650.

Johnson, F. S. (1956). Temperature distribution of the ionosphere under control of thermal conductivity. *J. Geoph. Res.* **61**, 71–76.

Johnson, F. S. (1958). Temperatures in the high atmosphere. *Ann. Geoph.* **14**, 94–108.

Kaplan, L. D. (1953). Regions of validity of various absorption-coefficient approximations. *J. Meteor.* **10**, 100–104.

Kaplan, L. D. (1954). A quasi-statistical approach to the calculation of atmospheric transmission. *Proc. Toronto Meteor. Conf., 1953* pp. 43-48. Royal Meteorological Society, London.

Kaplan, L. D. (1959). A method for calculation of infrared flux for use in numerical models of atmospheric motion. *In* "The Atmosphere and the Sea in Motion" (B. Bolin, ed.), pp. 170–177. Rockefeller Inst. Press, New York.

Kaplan, L. D., Migeotte, M. V., and Neven, L. (1956). 9.6–micron band of telluric ozone and its rotational analysis. *J. Chem. Phys.* **24**, 1183–1186.

Kellogg, W. W. (1961). Chemical heating above the polar mesopause in winter. *J. Meteor.* **18**, 373–381.

Kellogg, W. W., and Schilling, G. F. (1951). A proposed model of the circulation in the upper stratosphere. *J. Meteor.* **8**, 222–230.

Maeda, K. (1963). Auroral dissociation of molecular oxygen in the polar mesosphere. *J. Geoph. Res.* **68**, 185–197.

Manabe, S., and Möller, F. (1961). On the radiative equilibrium and heat balance of the atmosphere. *Mon. Wea. Rev.* **89**, 503–532.

Milne, E. A. (1922). Radiative equilibrium: the insolation of an atmosphere. *Phil. Magaz.* [6] **44**, 872–896.

Murgatroyd, R. J. (1957). Winds and temperatures between 20 km and 100 km—a review. *Quart. J. Roy. Meteor. Soc.* **83**, 417–458.

Murgatroyd, R. J., and Goody, R. M. (1958). Sources and sinks of radiative energy from 30 to 90 km. *Quart. J. Roy. Meteor. Soc.* **84**, 225–234.

Nicolet, M. (1960). The properties and constitution of the upper atmosphere. *In* "Physics of the Upper Atmosphere" (J. A. Ratcliffe, ed.), pp. 17–71. Academic Press, New York.

Nicolet, M. (1961). Structure of the thermosphere. *Plan. Space Sci.* **5**, 1–32.

Oder, F. C. E. (1948). The magnitude of radiative heating in the lower stratosphere. *J. Meteor.* **5**, 65–67.

Ohring, G. (1958). The radiation budget of the stratosphere. *J. Meteor.* **15**, 440–451.

Palmer, C. H. (1960). Experimental transmission functions for the pure rotation band of water vapor. *J. Opt. Soc. Amer.* **50**, 1232–1242.

Plass, G. N. (1956a). The influence of the 9.6 micron ozone band on the atmospheric infra-red cooling rate. *Quart. J. Roy. Meteor. Soc.* **82**, 30–44.

Plass, G. N. (1956b). The influence of the 15μ carbon-dioxide band on the atmospheric infra-red cooling rate. *Quart. J. Roy. Meteor. Soc.* **82**, 310–324.

Plass, G. N. (1957). Discussion of "Plass (1956a)" and "Plass (1956b)." *Quart. J. Roy. Meteor. Soc.* **83**, 272–275.

Plass, G. N. (1958). Models for spectral band absorption. *J. Opt. Soc. Amer.* **48**, 690–703.

Plass, G. N. (1960). Useful representations for measurements of spectral band absorption. *J. Opt. Soc. Amer.* **50**, 868–875.

Plass, G. N., and Fivel, D. I. (1953). Influence of Doppler effect and damping on line-absorption coefficient and atmospheric radiation transfer. *Ap. J.* **117**, 225–233.

RANDALL, H. M., DENNISON, D. M., GINSBURG, N., and WEBER, L. R. (1937). The far infrared spectrum of water vapor. *Phys. Rev.* **52**, 160–174.

ROCKET PANEL (1952). Pressures, densities and temperatures in the upper atmosphere. *Phys. Rev.* **88**, 1027–1032.

SPITZER, L., Jr. (1949). The terrestrial atmosphere above 300 km. *In* "The Atmospheres of the Earth and Planets" (G. P. Kuiper, ed.), pp. 211–247. Univ. of Chicago Press, Chicago, Illinois.

WALSHAW, C. D. (1957). Integrated absorption by the 9.6-μ band of ozone. *Quart. J. Roy. Meteor. Soc.* **83**, 315–321.

WALSHAW, C. D., and RODGERS, C. D. (1963). The effect of the Curtis-Godson approximation on the accuracy of radiative heating-rate calculations. *Quart. J. Roy. Meteor. Soc.* **89**, 122–130.

WARK, D. Q., and WOLK, M. (1960). An extension of a table of absorption for Elsasser bands. *Mon. Wea. Rev.* **88**, 249–250.

WATANABE, K., and HINTEREGGER, H. E. (1962). Photoionization rates in the *E* and *F* regions. *J. Geoph. Res.* **67**, 999–1006.

WILKINSON, P. G., and JOHNSTON, H. L. (1950). The absorption spectra of methane, carbon dioxide, water vapor, and ethylene in the vacuum ultraviolet. *J. Chem. Phys.* **18**, 190–193.

YOUNG, C., and EPSTEIN, E. S. (1962). Atomic oxygen in the polar winter mesosphere. *J. Atmos. Sci.* **19**, 435–443.

CHAPTER 8

Atmospheric Tides and Winds in the Lower Thermosphere

To describe motions in the lower thermosphere, it is not enough to consider the seasonally prevailing winds, or even the day-to-day wind changes, which claim the most attention at lower levels. Motions in the lower thermosphere are complicated by the presence of large periodic components of planetary scale, as well as large components of relatively small time and space scales. The former have been suspected ever since Balfour Stewart in 1882 suggested their presence as an explanation of the observed daily variations of the earth's magnetic field. Their explanation in the framework of atmospheric tidal theory has received considerable attention. The importance of the smaller-scale components has become apparent only during the last few years as a result of photographic and radio studies of meteor trails in the 80- to 100-km region. The tidal and small-scale components are extremely important, from both a practical and a theoretical viewpoint.

In Chapter 3, we discussed what little is known of the "prevailing" winds at this altitude. In this chapter we consider the tidal components and the irregular components. Section 8.1 contains a discussion of atmospheric tides, including some of the older observational evidence and the theory which has been built up to explain it. In Section 8.2, we turn to direct observational evidence pertaining to tidal motions in the 80- to 100-km region. This information is derived mainly from radio studies of ionized meteor trails and has become available only within the past five to ten years. Finally, in Section 8.3, the small-scale irregular wind variations are considered.

8.1 Observation and Theory of Atmospheric Tides

The expression *atmospheric tides* is used to refer to atmospheric oscillations whose periods are equal to or submultiples of the solar or lunar day, whether the oscillations are gravitationally or thermally excited. Until the recent accumulation (at a few locations) of sufficient radio-meteor observations to delineate tidal motions at high levels, observations bearing on atmospheric tides have been of two types: detailed studies of their small effects at the surface of the earth, and relatively general inferences from magnetic and ionospheric behavior about their effects in the thermosphere. This evidence is discussed in Subsection 8.1.1. For various reasons, the theory of atmospheric tides has inspired a great deal of interest and effort. This theory is discussed in Subsection 8.1.2 with respect to gravitationally forced oscillations and in Subsection 8.1.3 with respect to thermally forced oscillations.

In many ways, the subject of atmospheric tides is one of the most fascinating in the atmospheric sciences. It involves the atmosphere at both high and low levels and serves to remind us that we are, after all, dealing with only one atmosphere. It involves an interplay between theory and many diverse kinds of observations. Its development has been profoundly stimulated and shaped by two simple ideas, both offered in 1882—Stewart's dynamo theory of magnetic variations and Lord Kelvin's resonance theory of the tides. We can hardly hope to do the subject justice in the pages available, and the reader is referred to the review articles of Wilkes (1949), Chapman (1951, 1961), Kertz (1957), and Siebert (1961) for treatments of various aspects of the problem and for copious references.

8.1.1 Observations Relating to Atmospheric Tides

In discussions of atmospheric tides it is customary to designate an oscillation whose period is l^{-1} of a mean solar day ($l = 1, 2, 3, \ldots$) by S_l; and an oscillation whose period is l^{-1} of a mean lunar day by L_l. The meteorological quantity whose variation is being discussed can be specified in parentheses after this symbol; for example, the solar semidiurnal oscillation of surface pressure can be designated by $S_2(p_0)$. In the following, we shall usually be discussing results pertaining to surface pressure and therefore shall omit this designation. The oscillation at a given location (colatitude θ, longitude λ) can be represented by

$$S_l(\theta, \lambda) = A_l(\theta, \lambda) \sin[lt' + a_l(\theta, \lambda)] \tag{8.1}$$

or, in the case of a lunar oscillation, by

$$L_l(\theta, \lambda) = B_l(\theta, \lambda) \sin[l\tau' + b_l(\theta, \lambda)] \tag{8.2}$$

Here t' is local mean solar time, taken to be zero at local midnight, and τ' is local mean lunar time, taken to be zero at local lower lunar transit, and both t' and τ' are expressed in angular measure at the rate of 2π for each solar (or lunar) day.

At a given location, the amplitude A_l (or B_l) and the phase angle a_l (or b_l) describing the tidal variation of an element can be determined by harmonic analysis* of observations that give the values of the element at different

---

* See Appendix I.

hours of the solar (or lunar) day. It is important that these hourly values be obtained as averages over a sufficiently large number of days so that the effects of other (aperiodic) variations are minimized. The required length of record depends on the amplitude of the tidal oscillation being studied relative to the amplitude of the other variations; criteria for judging the significance of a determination have been discussed in considerable detail by Chapman and Bartels (1940, Chapter 16). They are described briefly in Appendix I.

The amplitudes and phase angles in general vary with both latitude and longitude and these variations may be expressed either functionally or graphically in a complete representation of S_l or L_l. In some instances, the amplitudes and phase angles may have different values and geographical distributions, if the data on which they are based refer to different seasons of the year. However, we do not incorporate this factor in the formal notation and refer to the annual mean oscillations unless otherwise stated.

Numerical investigations of the type outlined above and applied to surface pressure have revealed the relative importance of S_2 among the atmospheric oscillations. In amplitude, it greatly overshadows the lunar oscillations (of which only L_2 has been detected). The solar diurnal oscillation S_1 is by no means negligible in amplitude, but, unlike S_2, varies irregularly over the globe

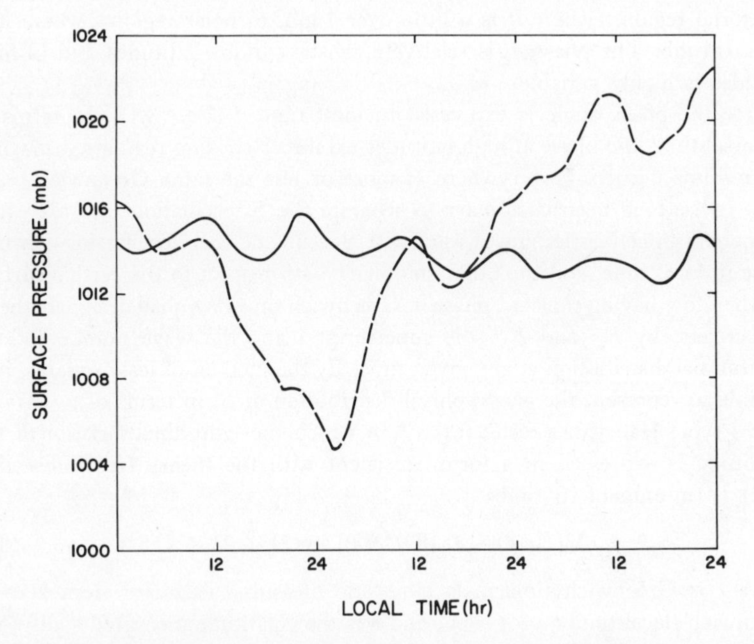

FIG. 8.1. Variation of surface pressure with time at Barbados (13°04′N, solid curve) and at Tallahassee (30°26′N, dashed curve).

and is greatly affected by local weather and topography. The oscillations S_3 and S_4 have been detected but are much smaller in amplitude than S_2. We shall discuss mainly the behavior of S_2.

The importance of S_2 is evident even from inspection of the barograph trace at a low-latitude station. There S_2 has its greatest amplitude and is, furthermore, usually not obscured by large pressure changes connected with traveling meteorological disturbances. Figure 8.1 shows the regular twice-daily rise and fall of the barometer at a low-latitude station, contrasted to a typical larger-amplitude, longer-period variation at a midlatitude station. The pressure maxima of S_2 are observed to occur two to three hours before local noon and midnight, and the range of the variation from maximum to minimum (near the equator) is a bit over 2 mb.

At stations outside low latitudes, S_2 can be determined only from data averaged over many months. Nevertheless, a large number of determinations have been made by various investigators over the years. Some of these were summarized in an early paper by Simpson (1918), which for many years served as the observational basis for discussions of S_2. More recently, Haurwitz (1956) used determinations from 296 stations in a discussion of the geographical distribution and functional representation of S_2. Figures 8.2a and 8.3a show the geographical distributions of A_2 and a_2, respectively. The amplitude A_2 decreases from equatorial regions, where it is a little over 1 mb, to polar regions, where it is about 0.1 mb. The phase a_2 is relatively constant in low latitudes, but in high latitudes is highly variable.

Here the phase angle is expressed in local time t' [Eq. (8.1)]. In terms of universal time, the phase at high latitudes exhibits a striking regularity, maxima and minima occurring everywhere at more or less the same Greenwich mean time. It has long been customary to separate the S_2 oscillation into two wave components, one migrating westward with the sun and therefore having constant phase in local time, and the other stationary with respect to the earth's surface and therefore having constant phase in Greenwich time. We shall designate these, respectively, by S_2^2 and S_2^0, the superscript being the wave number of the longitudinal distribution at any given time. By the method of least squares, it is possible to represent the geographical distribution of S_2 in terms of two waves of this type. Haurwitz's result for S_2^2, in which the latitudinal variation of the amplitude is expressed in a form consistent with the theory (see Subsection 8.1.2), is (in units of 10^{-2} mb)

$$S_2^2 = 123[P_2^2(\theta) - 0.182P_4^2(\theta)] \sin(2t + 2\lambda + 158°) \tag{8.3}$$

where t is Greenwich time, λ is longitude measured eastward from zero at Greenwich (local time $t' = t + \lambda$), and θ is the colatitude measured southward from zero at the North Pole. The functions $P_2^2(\theta)$ and $P_4^2(\theta)$ are seminormalized associated Legendre functions, as tabulated for example by Haurwitz and

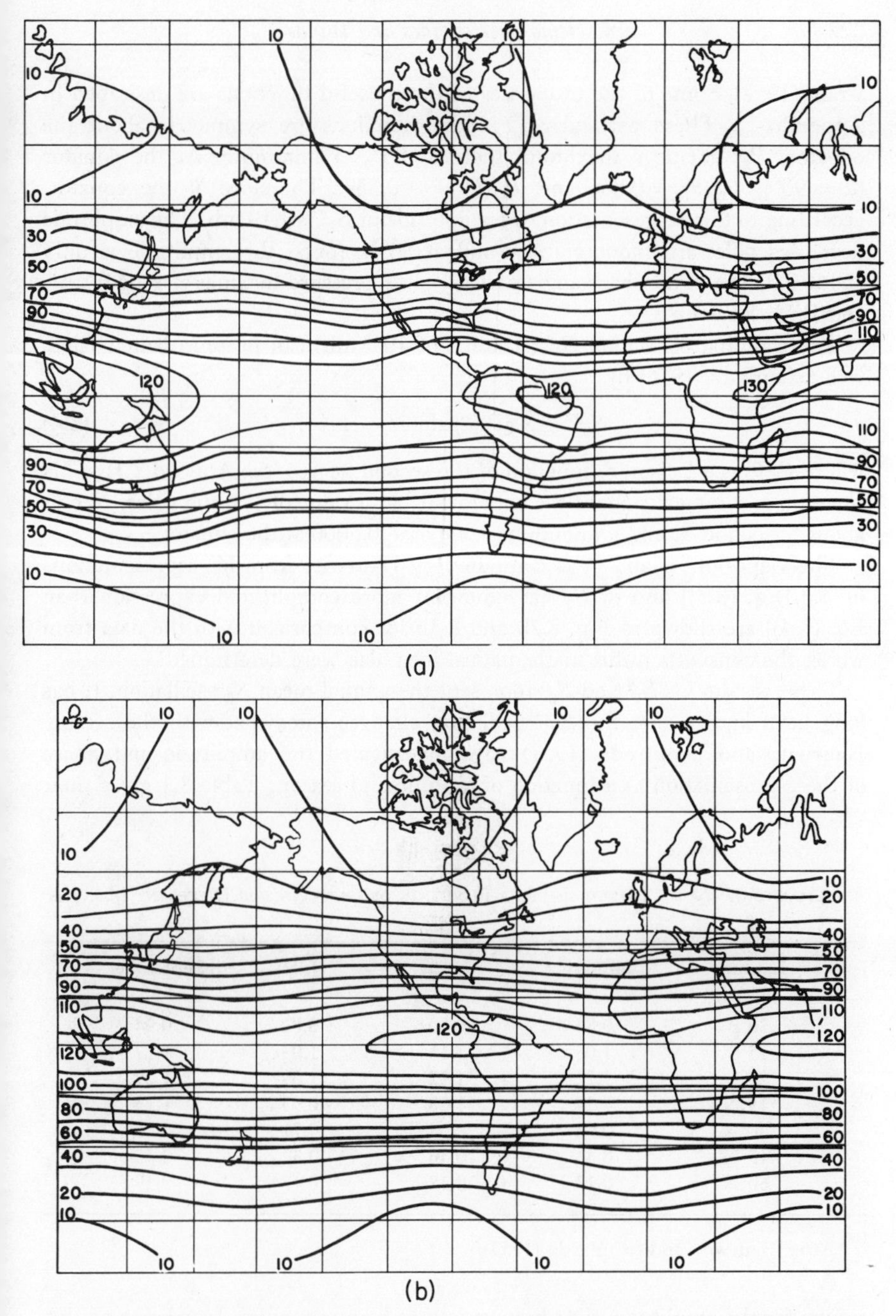

FIG. 8.2. Geographical distribution of A_2, the amplitude of the semidiurnal tide in surface pressure, in units of 10^{-2} mb. (a) An analysis of the results of determinations at 296 stations; (b) based on a mathematical representation of these results (see text). (After Haurwitz, 1956.)

Craig (1952). Some of the properties of these useful functions are described in Appendix J. These particular Legendre functions are symmetric about the equator, P_2^2 having a maximum there and P_4^2 a minimum. At the equator ($\theta = 90°$), $P_2^2(\theta) = 0.866$ and $P_4^2(\theta) = -0.559$. Therefore at the equator, according to this representation, the amplitude of S_2^2 is 1.19 mb. The amplitude decreases poleward monotonically and at 80°N (or S) the amplitude is only about 0.01 mb. The phase angle of 158° gives pressure maxima at 9 : 44 A.M. and P.M. local time.

The standing oscillation S_2^0 (in units of 10^{-2} mb) can be represented fairly well (Haurwitz, 1956) by

$$S_2^0 = 8.5P_2(\theta)\sin(2t + 118°) \tag{8.4}$$

where $P_2(\theta)$ is a Legendre function of the second degree (see Appendix J). This function is symmetric about the equator; it has a value of 1 at the poles, zero at about 55°N and S, and a minimum value of -0.5000 at the equator.

The values of A_2 and a_2 as computed by Haurwitz from his representations of S_2^2 [Eq. (8.3)] and of S_2^0 [a somewhat more complicated expression than Eq. (8.4)] are shown in Fig. 8.2b and 8.3b for comparison with the data from which the constants in his mathematical formulae were determined.

These results for S_2^2 and S_2^0 represent the annual mean S_2 oscillation. It has long been known that the S_2^2 oscillation changes somewhat with the season. Haurwitz and Sepúlveda (1957) have investigated the amplitude and phase of the S_2^2 oscillation as a function of latitude and season. Table 8.1 gives their

TABLE 8.1

Amplitudes of S_2^2 (in Millibars) as a Function of Latitude for Different Months[a]

Latitude	January	March	July	September
30°S	0.85	0.86	0.82	0.94
15°S	1.08	1.16	1.03	1.17
0°	1.22	1.32	1.09	1.29
15°N	1.24	1.28	0.93	1.13
30°N	0.95	0.94	0.71	0.86
45°N	0.43	0.46	0.37	0.45
60°N	0.13	0.18	0.13	0.19

[a] After Haurwitz and Sepúlveda (1957).

results for the amplitude, the phase variations being irregular. Inspection of this table shows that at a given latitude, the amplitude is smallest in July; at some latitudes there is a secondary minimum in December. It will be noted that there

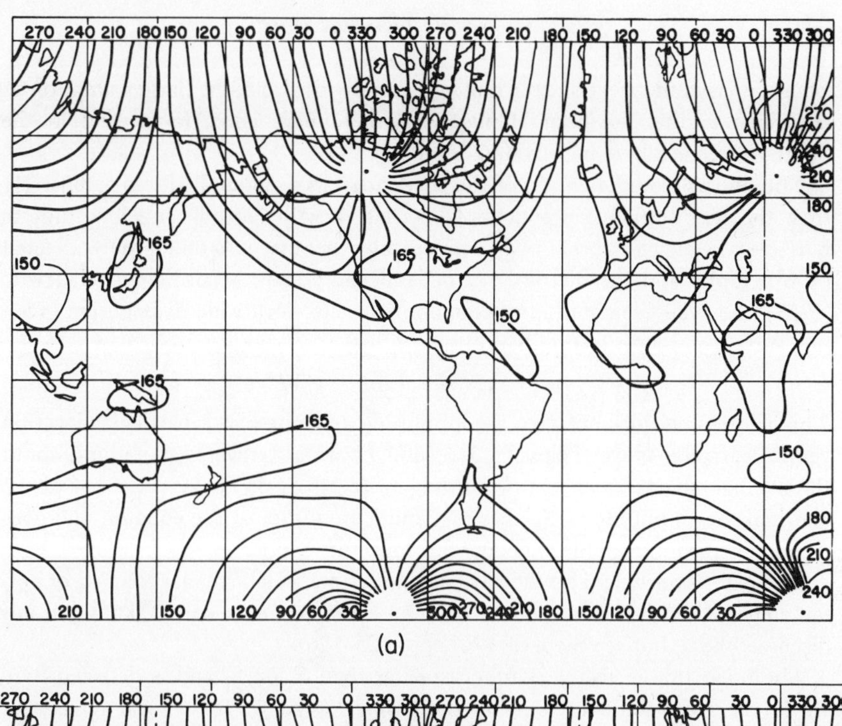

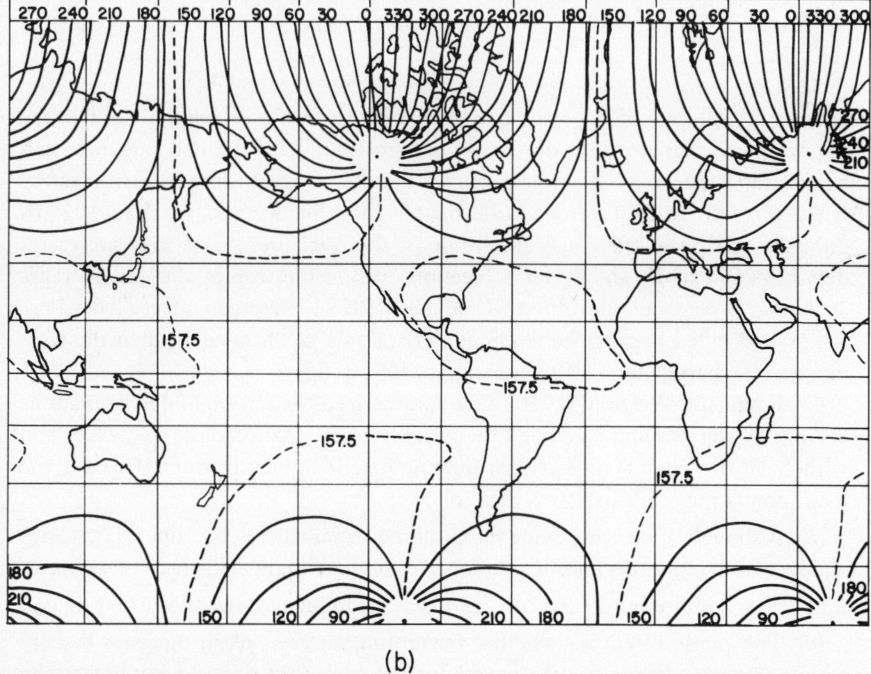

FIG. 8.3. Geographical distribution of a_2, the phase angle of the semidiurnal tide in surface pressure, in degrees (representing local mean time). (a) An analysis of the results of determinations at 296 stations; (b) based on a mathematical representation of these results (see text). (After Haurwitz, 1956.)

is also an asymmetry about the equator in individual months, larger amplitudes occurring in the Northern Hemisphere in January and March, and in the Southern Hemisphere in July and September.

The diurnal component S_1 of the solar tide has been studied less extensively than the semidiurnal component. Its variation with latitude and longitude is much less regular than that of S_2, presumably because it is much more affected by orography and the distribution of land and water. According to Haurwitz (1963), the available data indicate an approximate worldwide distribution given (for surface pressure in millibars) by

$$S_1 = 0.59 \sin^3 \theta \sin(t' + 12°) \tag{8.5}$$

This expression does not give the amplitude and phase at a particular location very accurately. Nevertheless, it is useful because it reveals something about the average character of S_1. According to (8.5), S_1 on the average has about one-half the amplitude of S_2, has maximum amplitude at the equator, and goes through a maximum value at about 0500 local mean time.

The eight-hourly and six-hourly oscillations are much smaller than S_1 or S_2, but may be detected and studied from analyses of long records. They are not discussed here, but see Siebert (1961, pp. 125–126).

The lunar tide is also very much smaller than S_1 or S_2 and is detected only with very great difficulty outside the tropical regions. In tropical regions, a record of about one year's data is required for a reliable determination. Nevertheless, the lunar tide may play an important role at high levels. Furthermore, contrary to the case of the solar tide, for which thermal influences must be very important, the lunar tide is due solely to a gravitational force and this force can be accurately specified. Therefore the lunar tide has great theoretical interest.

Attempts to study the lunar tide have a long history and are fraught with difficulty, owing to the small amplitude of the oscillation. The first successful determination of B_2 and b_2 for a location outside the tropics was for 64 years of data at Greenwich and due to Chapman (1918). Chapman succeeded where others had failed because he confined his analysis to those days when the total range of the barometer was 0.1 inch Hg or less.

Chapman and Westfold (1956) have summarized the results of determinations of L_2 at 69 stations and discussed the geographical variation of L_2 in comparison with S_2 at the same stations. Considerable detail is given in this reference; the main points may be summarized as follows:

(a) Although there are some longitudinal anomalies, B_2, like A_2, varies mainly with latitude, decreasing poleward from a maximum value near the equator.

(b) The amplitude of L_2 is, on the average, about one-twentieth that of S_2.

(c) The phase of L_2 is such that maximum surface pressure occurs usually within an hour after upper (or lower) lunar transit (but occasionally before the transit).

TABLE 8.2

MEAN AMPLITUDES A_2 AND B_2, IN UNITS OF 10^{-3} MB, AND THEIR RATIOS, FOR 10° BELTS OF LATITUDE[a]

Latitude:	N 60–50°	50–40°	40–30°	30–20°	20–10°	N 10–0°	0–10° S	10–20°	20–30°	30–40° S
Number of stations:	6	12(11)	15	7	6	6	6	5	3	3
Mean latitude:	54.4°	44.5°	36.3°	26.3°	14.4°	5.1°N	4.45°S	13.7°	24.8°	37.8°
Mean A_2:	275	399	551	821	1262	1219	1366	1166	876	576
Mean B_2:	11.3	17.8	25.1	45.0	52.4	65.7	83.9	60.6	46.3	26.7
Ratio mean A_2/mean B_2:	24.3	22.4	22.0	18.3	24.1	18.5	16.3	19.2	18.9	21.6

[a] In the latitude belt 50°–40°N, one station included in the computation of the mean A_2 is omitted in the computation of the mean B_2. After Chapman and Westfold (1956).

TABLE 8.3

MEAN PHASE ANGLES a_2 AND b_2 AND THEIR DIFFERENCES FOR 10° BELTS OF LATITUDE[a]

Latitude :	N 60–50°	50–40°	40–30°	30–20°	20–10°	N 10–0°	0–10°S	10–20°	20–30°	30–40°S	All
Number of stations :	5(6)	12(11)	15	7	6	6	6	5	3	3	
Mean a_2 :	152°	152°	149°	157°	157°	149°	154°	149°	157°	163°	153°
Mean b_2 :	93°	74°	65°	61°	67°	74°	71°	73°	96°	80°	72°
Mean a_2 − mean b_2 :	59°	78°	84°	96°	90°	75°	83°	76°	61°	83°	82°

[a] After Chapman and Westfold (1956). (See footnote Table 8.2.)

These points are illustrated in Tables 8.2 and 8.3, taken from Chapman and Westfold (1956).

Chapman (for example, Chapman, 1951) has emphasized the significance of a peculiar annual variation of L_2, such that there is an increased lag of the pressure maximum after lunar transit and a slightly increased amplitude in *both* hemispheres in the months November to February. Thus the annual variation does not depend on the local season, as one might expect. Since, as Chapman points out, there is no annual variation of the lunar tide-producing force, this peculiar behavior must depend on the characteristics of the atmosphere as they determine its response to the tidal force.

In addition to these rather detailed studies of tides at the earth's surface, there has long been less quantitative evidence of tidal motions in ionospheric regions. (More recent direct observations of tidal winds in the lower thermosphere are discussed in Section 8.2.) This evidence consists mainly of the observed daily (solar and lunar) variations of the earth's magnetic field as well as observed daily variations of certain ionospheric parameters. Although, as mentioned in the introduction to this chapter, it is not possible at present to deduce the details of the tidal motions from these observations, certain qualitative deductions have been possible and have greatly stimulated the development of tidal theory. This general subject has been reviewed by Chapman (1961), to whom the reader is referred for a clear and relatively detailed discussion with numerous references.

Values of the elements describing the earth's magnetic field have been recorded regularly at many magnetic observatories for periods of time extending to over 100 years. An outstanding feature of these records is the existence of a small daily oscillation whose form differs from element to element and, for a given element, with latitude (see Section 9.1). These variations show evidence of both diurnal and semidiurnal components. Careful analysis of the records reveals a smaller oscillation, predominantly semidiurnal, related to the lunar day.

The explanation of these small periodic magnetic variations was first suggested by Stewart (1882) and is known as the dynamo theory. According to this theory, periodic tidal motions at high levels carry the ionized air across the lines of force of the earth's main magnetic field and thereby produce a field of induced electromotive force, as in a dynamo. The resulting current flow, mainly due to free electrons, is a rather complicated result of the worldwide distribution of the induced electric field, of the effects of collisions, and of the influence of the magnetic field itself. However, regardless of the details, current systems are expected to result and these (together with currents induced by them in the solid earth) produce magnetic effects that are detected at the earth's surface in the form of the transient, periodic magnetic variations.

It is possible to determine from the observed magnetic variations the strength and worldwide distribution of the upper atmospheric electric current systems that could produce them. These are, indeed, well known and have been for some time (for example, Chapman and Bartels, 1940). To go from the deduced current systems to the periodic air motions presumed to produce them is, however, another matter. The electrical conductivity of the upper atmosphere is a complicated tensor quantity depending on the ionization, collision frequency, and local direction of the magnetic field and therefore varies in space and time. Although this difficulty can in principle be overcome by calculations based on ionospheric and magnetic measurements, a second (unavoidable) difficulty is that the current systems deduced from the magnetic data describe a vertically integrated effect (probably in the general altitude region 90–125 km) and not currents flowing at a particular level. Therefore, if the tidal winds vary significantly with altitude in this region, their net effect is all that can be deduced.

Despite these difficulties, even simple considerations serve to indicate that the tidal wind components in the low thermosphere, if they are to produce the dynamo effect, must be very much larger than those associated with the tides at the surface; and furthermore that the tidal oscillations must have quite different phase than at the surface. The explanation of these two inferences has been one of the goals of tidal theory.

Ionospheric measurements also reveal certain periodic variations related to the solar or lunar day. For example, the equivalent height of reflection of the E layer (see Section 9.2) varies somewhat with a period of one-half the lunar day, as first pointed out by Appleton and Weekes (1939). It is difficult to detect effects of the solar tide from ionospheric data, because the production of ionization is closely related to the solar zenith angle (see Section 4.4), and periodic variations are so closely bound up with that effect as to obscure any effects of periodic air motions. However, certain anomalous effects in the F region (see Section 9.3) are thought to be related to tidal motions (lunar or solar), as discussed extensively by Martyn (see, for example, Martyn, 1955). The theory of the variations at this level is, however, quite complicated, and involves motions that are driven by a motor effect as a result of the primary tidal motions in the E region.

It seems ironic that most of our information about atmospheric tides comes from data that refer to the earth's surface or to ionospheric regions, and not to the atmospheric regions between. Observations of the intervening atmosphere, while plentiful enough for many purposes, are simply not obtained frequently enough for a systematic study of atmospheric tides. However, there have been a few such studies which we shall mention briefly.

Johnson (1955) used 100-mb wind data obtained over the British Isles in 1950–1951 on certain occasions when wind observations were made six hours apart (03, 09, 15, and 21 GMT). He combined these with 150-mb wind data over the British Isles during 1944–1947, also when wind observations were made six hours apart but at 00, 06, 12, and 18 GMT. From these data he computed the first four harmonics of the daily variation and found a very small (amplitude less than 1 knot) semidiurnal variation of wind in phase with that inferred from tidal studies at the earth's surface. He found also a diurnal component of the same magnitude.

Harris, Finger, and Teweles (1962) similarly took advantage of a change in observing times to combine data obtained at 03, 09, 15, and 21 GMT during April 1956–March 1957 with data obtained at 00, 06, 12, and 18 GMT during April 1957–March 1958. Working with data for Lajes Field, Terceira, Azores, they computed the diurnal and semidiurnal components of the daily variation of wind (and also of pressure and temperature) for each calendar month and many pressure levels up to 15 mb. The amplitudes were generally only a few tens of centimeters per second and varied rather irregularly from month to month and from level to level. Their results at 100 mb were in fair agreement with Johnson's, but with rather large phase differences for the diurnal component. It seems clear that tidal oscillations are quite small up to 25–30 km and cannot be studied in great detail with the data available.

There have been one or two attempts to determine tidal components in the upper stratosphere and mesosphere from special observations. Lenhard (1963) took advantage of 23 wind observations made within 24 hours with the Robin sounding system (see Subsection 3.4.2) at Eglin Air Force Base, Florida, and computed the diurnal and semidiurnal components of the wind variation at 35, 43, and 64 km. The amplitudes of these components tended to increase with altitude and were generally between 1 and 10 m sec^{-1}. The phases varied quite irregularly with altitude. However, it is impossible to assess the statistical reliability of tidal determinations from only one day's data (see Appendix I), and these results can be considered to be only suggestive of the size of the tides at those levels.

8.1.2 The Theory of Atmospheric Tides—Gravitational Forces and Resonance

It has long been surprising to theoreticians that S_2 should predominate over both L_2 and S_1 in the earth's atmosphere. The situation was well expressed by Lord Kelvin, in 1882, at which time he also advanced a tentative explanation that has come to be called the *resonance theory*. According to Kelvin (1882),

> The cause of the semi-diurnal variation of barometric pressure cannot be the gravitational tide-generating influence of the sun, because if it were there would be a much larger lunar influence of the same kind, while in reality the lunar barometric tide is insensible, or nearly so. It seems, therefore, certain that the semi-diurnal variation of the barometer is due to temperature. Now, the *diurnal* term, in the harmonic analysis of the variation of *temperature*, is undoubtedly much larger in all, or nearly all, places than the *semi-diurnal*. It is then very remarkable that the *semi-diurnal term of the barometric effect* of the

> variation of temperature should be greater, and so much greater as it is, than the diurnal. The explanation probably is to be found by considering the oscillations of the atmosphere, as a whole, in the light of the very formulas which Laplace gave in his *Mécanique céleste* for the ocean, and which he showed to be also applicable to the atmosphere. When thermal influence is substituted for gravitational, in the tide-generating force reckoned for, and when the modes of oscillation corresponding respectively to the diurnal and semi-diurnal terms of the thermal influence are investigated, it will probably be found that the period of free oscillation of the former agrees much less nearly with 24 hours than does that of the latter with 12 hours; and that, therefore, with comparatively small magnitude of the tide-generating force, the resulting tide is greater in the semi-diurnal term than in the diurnal.

The development of atmospheric tidal theory will not be traced here. See Chapman (1951) and Siebert (1961) for accounts of its interesting history. Instead, we shall outline the present status of the theory with references to past work when they are appropriate in context and especially in connection with the resonance theory.

In atmospheric tidal theory, spherical coordinates r, θ, λ are used, with origin at the center of the earth and with r the radius vector, θ the colatitude positive southward from the North Pole, and λ the longitude positive eastward. The corresponding tidal wind components are w, v_θ, and v_λ respectively.* Undisturbed values of pressure, density, and temperature are denoted by p_0, ρ_0, T_0 and the perturbations of these quantities due to the tidal oscillation by p', ρ', T'. Other symbols will be defined as they appear, if they have not been used regularly in this book. In the discussion immediately following only gravitationally induced tides are considered. The effects of periodic heating are considered in Subsection 8.1.3.

The following assumptions† are made:

(1) T_0 varies with height in an arbitrary way that can be specified, but does not vary with latitude or longitude.

---

* In most treatments of tidal theory, the upward wind speed is designated by w, the southward wind speed v_θ by u, and the eastward wind speed v_λ by v. This usage, although not objectionable in a paper where spherical coordinates are used exclusively, conflicts with the usual meteorological notation used throughout this book, according to which u is the eastward component and v is the northward component.

† For a discussion of the validity of these assumptions and references to studies of their effects, see Siebert (1961, pp. 127-132). There is no reason at present to doubt the general validity of the theory as it applies to most of the atmosphere. However, at high enough levels, nonlinear terms and molecular conductivity and viscosity must become important.

(2) p_0 and ρ_0 are related to T_0 by the hydrostatic equation and the equation of state for an ideal gas.

(3) p', ρ', T', v_θ, v_λ, w are small quantities whoses quares and products may be neglected.

(4) Ellipticity of the earth, vertical acceleration, viscous forces, vertical variation of the radius vector, and vertical variation of g are all neglected.

(5) All changes take place adiabatically (except insofar as a diabatic periodic temperature change may serve as a tide-generating force; see Subsection 8.1.3).

(6) The Coriolis term in the vertical component of the equation of motion and Coriolis terms involving w in the other components of the equation of motion are neglected.

With these assumptions the Eulerian equations of motion for the rotating earth are (see Appendix K)

$$\frac{\partial v_\theta}{\partial t} - 2\omega v_\lambda \cos\theta = -\frac{1}{r_e}\frac{\partial}{\partial\theta}\left(\frac{p'}{\rho_0} + \Omega\right) \tag{8.6}$$

$$\frac{\partial v_\lambda}{\partial t} + 2\omega v_\theta \cos\theta = -\frac{1}{r_e \sin\theta}\frac{\partial}{\partial\lambda}\left(\frac{p'}{\rho_0} + \Omega\right) \tag{8.7}$$

$$\frac{\partial p'}{\partial z} = -g\rho' - \rho_0\frac{\partial\Omega}{\partial z} \tag{8.8}$$

Here ω is the angular velocity of the earth's rotation (7.292×10^{-5} sec^{-1}), r_e is the radius of the earth (6.371×10^8 cm), and Ω is a scalar potential describing the gravitational tide-producing forces (Lamb, 1932). The vertical coordinate z has been substituted for r. It is in the same direction as r, but is zero at the surface of the earth where $r = r_e$.

The equation of continuity is

$$\frac{\partial\rho}{\partial t} + w\frac{\partial\rho_0}{\partial z} + \rho_0\chi = 0 \tag{8.9}$$

where χ is the velocity divergence, which for these assumptions and coordinates is given by

$$\chi = \frac{1}{r_e \sin\theta}\frac{\partial}{\partial\theta}(v_\theta \sin\theta) + \frac{1}{r_e \sin\theta}\frac{\partial v_\lambda}{\partial\lambda} + \frac{\partial w}{\partial z} \tag{8.10}$$

There is, finally, the adiabatic form of the first law of thermodynamics, which can be written

$$\frac{\partial p'}{\partial t} + w\frac{\partial p_0}{\partial z} = \gamma g H_0\left[\frac{\partial\rho'}{\partial t} + w\frac{\partial\rho_0}{\partial z}\right] \tag{8.11}$$

where γ is the ratio of the specific heats (c_p/c_v) and H_0 is the scale height $(RT_0/gm.)$

It is a straightforward, although somewhat tedious, matter to eliminate variables in the five simultaneous, linear differential equations (8.6)–(8.9), (8.11) and arrive at a single partial-differential equation describing one of the variables. Because we are interested only in periodic variations of known period and because the solutions must be single valued in λ, we can first write

$$v_\theta, \quad v_\lambda, \quad w, \quad p', \quad \rho', \quad T' \propto \exp(i\sigma_l t + im\lambda) \tag{8.12}$$

where $\sigma_l = 2\pi l/t_d$, t_d is the length of a mean solar day (for an S_l oscillation) and m is zero for a stationary oscillation or is a positive integer for a westward traveling oscillation. The final partial-differential equation is usually obtained in terms of χ and is*

$$H_0 \frac{\partial^2 \chi}{\partial z^2} + \left(\frac{dH_0}{dz} - 1\right)\frac{d\chi}{dz} - \frac{g}{4r_e^2\omega^2} F\left[\left(\kappa + \frac{dH_0}{dz}\right)\chi\right] = 0 \tag{8.13}$$

where $\kappa = (\gamma - 1)/\gamma$ and F is an operator in the horizontal coordinates given by

$$F = \frac{1}{\sin\theta}\frac{\partial}{\partial\theta}\left(\frac{\sin\theta}{f^2 - \cos^2\theta}\frac{\partial}{\partial\theta}\right) - \frac{m}{f^2 - \cos^2\theta}\left(\frac{m}{\sin^2\theta} + \frac{1}{f}\frac{f^2 + \cos^2\theta}{f^2 - \cos^2\theta}\right) \tag{8.14}$$

In this operator, the period of the oscillation under consideration enters through the parameter f, which is equal to $(\sigma_l/2\omega)$, or, in terms of l, to $(\pi l/\omega t_d)$.

Solutions of (8.13) are sought by the method of separation of variables. Let $\chi(z, \theta) = Z(z)\Theta(\theta)$ and let $1/h$ be the separation constant, where h has the dimension of length. Then (8.13) reduces to the two ordinary differential equations

$$F(\Theta) + \frac{4r_e^2\omega^2}{gh}\Theta = 0 \tag{8.15}$$

$$H_0\frac{d^2 Z}{dz^2} + \left(\frac{dH_0}{dz} - 1\right)\frac{dZ}{dz} + \left(\kappa + \frac{dH_0}{dz}\right)\frac{Z}{h} = 0 \tag{8.16}$$

Equation (8.15) is Laplace's tidal equation. It applies also to the oscillations of an unbounded, homogeneous, and incompressible ocean of uniform depth h surrounding the earth. We will call Eq. (8.16) the radial equation.

The complete solution, since we are dealing with linear equations, is actually a summation of solutions of the type indicated above. Thus, formally,

$$\chi(z, \theta, \lambda, t) = \sum_l \sum_m \sum_n Z_{l,n}^m \Theta_{l,n}^m \exp(i\sigma_l t + im\lambda) \tag{8.17}$$

It is, of course, possible to express the other variables (v_θ, v_λ, w, p', T', ρ') in terms of χ although we shall not write down these formal expressions. Siebert (1961) gives most of them.

To analyze the behavior of a particular tidal oscillation, one can proceed as follows. The oscillation to be studied is characterized by specified values of

* See, for example, Siebert (1961) for details of the derivation. A small term in $(\partial^2\Omega/\partial z^2)$ has been neglected.

l and m; for example $S_2{}^2$ has $l = 2$ and $m = 2$. The frequency σ_l and the factor f that appears in (8.15) are thereby specified. For the given l and m, (8.15) has in general an infinite number of solutions $\Theta_{l,n}^m$ to each of which corresponds a value of h, $h_{l,n}^m$, which is customarily referred to as an *equivalent depth* by analogy with the idealized ocean referred to above. One may speak of oscillations characterized by particular values of l and m as a *wave family*, and refer to one oscillation of this family (a particular value of n) as a *wave type*. For a given wave type, a solution of the radial equation (8.16) is sought, subject to appropriate boundary conditions, with the h in (8.16) having the value $h_{l,n}^m$ that corresponds to the wave type under consideration.

The coefficients of (8.16) contain the factor H_0 and therefore depend on the vertical distribution of temperature assumed and specified for the undisturbed atmosphere under consideration. The amplitude of a wave type, and its variation with altitude, are therefore affected by this vertical temperature distribution. In particular, it may happen that a given wave type has an equivalent depth $h_{l,n}^m$, which, in combination with certain kinds of temperature distribution, leads to a large amplification of that wave type. This is the phenomenon of resonance.

There is a relatively simple procedure to determine, at least qualitatively, whether a given wave type may undergo resonance. In the case of a *free* oscillation ($\Omega = 0$), and for vertical temperature distributions similar to those that occur in our atmosphere, the radial equation has solutions that satisfy the boundary conditions only for special values of h. These values may be called *atmospheric eigenvalues* and denoted by $\bar{h}$. In the case of a *forced* (tidal) oscillation, when $\Omega \neq 0$, the radial equation has solutions that satisfy the boundary conditions for any value of h, for example, for any equivalent depth $h_{l,n}^m$ corresponding to any wave type. This difference arises because one of the boundary conditions is $w = 0$ at $z = 0$; and the expression for w contains a term in Ω in the case of a tidal oscillation, a term which is absent in the case of a free oscillation. However, it turns out that if the equivalent depth of a particular wave type is nearly equal to one of the atmospheric eigenvalues $\bar{h}$, then that wave type is greatly amplified. (Under the assumptions employed in this theory, the amplitude would become infinite if the equivalent depth were exactly equal to one of the $\bar{h}$.) It is therefore convenient to discuss the question of resonance by comparing the atmospheric eigenvalues that correspond to an assumed vertical temperature distribution with the equivalent depths that correspond to different wave types (which equivalent depths are independent of the vertical temperature distribution).*

* Alternatively, one can make this comparison in terms of the periods of the oscillations. If an atmospheric eigenvalue $\bar{h}$ is inserted in Laplace's tidal equation and a value of m is specified, then the solutions of the equation define wave types with specified periods (not, in general, submultiples of the solar or lunar day). These are spoken of as *free periods*, and one may say that we are interested in determining whether our atmosphere has a free period close to 12 hours.

Laplace's tidal equation (8.15) has been studied extensively. Its solution is given in terms of Hough's functions (Hough, 1897, 1898), which may be represented as series of associated Legendre functions. For further reference, the equivalent depths and Hough's functions corresponding to a few important wave types on the earth are given in Table 8.4, after Siebert (1961). For the

TABLE 8.4

EQUIVALENT DEPTHS AND HOUGH'S FUNCTIONS FOR A FEW IMPORTANT WAVE TYPES[a]

Wave family S_2^2:	$l = 2$, $m = 2$, $\sigma = 1.4544 \times 10^{-4}\ \mathrm{sec}^{-1}$, $f = 0.99727$
Wave type $n = 2$:	$h_{2,2}^2 = 7.85$ km
	$\Theta_{2,2}^2 = P_2^2 - 0.339 P_4^2 + \cdots$
Wave type $n = 4$:	$h_{2,4}^2 = 2.11$ km
	$\Theta_{2,4}^2 = 0.202 P_2^2 + P_4^2 - 0.819 P_6^2 + 0.24 P_8^2 - \cdots$
Wave family L_2^2:	$l = 2$, $m = 2$, $\sigma = 1.4052 \times 10^{-4}\ \mathrm{sec}^{-1}$, $f = 0.96350$
Wave type $n = 2$:	$h_{2,2}^2 = 7.07$ km
	$\Theta_{2,2}^2 = P_2^2 - 0.375 P_4^2 + \cdots$
Wave type $n = 4$:	$h_{2,4}^2 = 1.85$ km
	$\Theta_{2,4}^2 = 0.227 P_2^2 + P_4^2 - 0.951 P_6^2 + 0.32 P_8^2 - \cdots$
Wave family S_3^3:	$l = 3$, $m = 3$, $\sigma = 2.1817 \times 10^{-4}\ \mathrm{sec}^{-1}$, $f = 1.4959$
Wave type $n = 3$:	$h_{3,3}^3 = 12.89$ km
	$\Theta_{3,3}^3 = P_3^3 - 0.105 P_5^3 + \cdots$
Wave type $n = 4$:	$h_{3,4}^3 = 7.66$ km
	$\Theta_{3,4}^3 = P_4^3 - 0.164 P_6^3 + \cdots$

[a] Computed for the terrestrial constant $g/(4r_e^2\omega^2) = 1.1349 \times 10^{-2}\ \mathrm{km}^{-1}$. Terms in the Hough's functions with coefficients less than 0.1 are omitted. After Siebert (1961).

important S_2^2 wave family, it can be seen that the wave type $\Theta_{2,2}^2$ has a latitudinal distribution somewhat similar to the observed one as represented, for example, by Haurwitz [Eq. (8.3)]. Figure 8.4 compares these distributions, both normalized to have a value of 1 at the equator. Both show the well-known decrease of S_2^2 from equator to pole, although the observed amplitude is somewhat greater in middle latitudes than that of the wave type $\Theta_{2,2}^2$. Other wave types of the S_2^2 family would have quite different latitudinal distributions and it is clear that, with reference to the theory, only the wave type $\Theta_{2,2}^2$ need be considered in connection with the migrating solar semidiurnal oscillation.

The diurnal oscillations have not been investigated in such detail. However, according to Siebert, the equivalent depths for these oscillations are quite small. For example, Siebert, quoting Kertz, gives a value of $h_{1,1}^1 = 0.63$ km; wave types with larger n would have even smaller values of h.

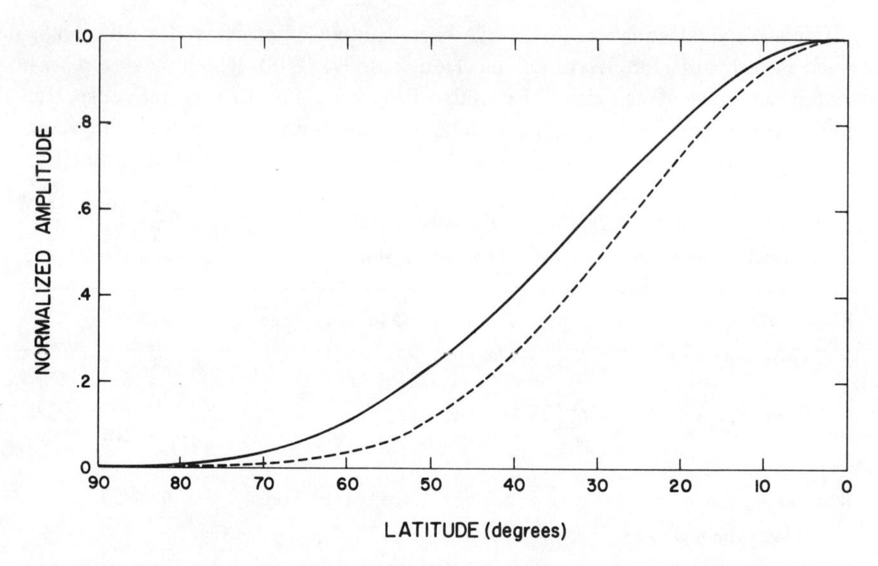

FIG. 8.4. Latitudinal distribution of the theoretical wave type $\Theta_{2,2}^{2}$ (dashed curve) and of the amplitude of $S_2{}^2$ (solid curve) as determined by Haurwitz (1956). Both are normalized to have a value of 1.0 at the equator.

The equivalent depth of $\Theta_{2,2}^{2}$ is 7.85 km and it is pertinent to ask whether the real atmosphere might have an eigenvalue near this value. The apparent answer to this question has oscillated in an interesting way as more and more realistic models of the vertical temperature distribution have been employed. Laplace [see Chapman (1951) for references to Laplace's early, basic work on the theory of tides in atmosphere and ocean] showed that in the case of an atmosphere with isothermal changes of state and an isothermal temperature distribution, $\bar{h} = H_0$. Lamb (1910) demonstrated that, for an atmosphere with adiabatic changes of state and a dry-adiabatic temperature lapse, $\bar{h} = H_0(0)$, the scale height at ground level. This result is, in general, true for any autobarotropic atmosphere (Haurwitz, 1937). For an average surface temperature, the scale height at ground level is not much greater than 7.85 km and the resonance theory received considerable support as a result. However, in 1929, Taylor pointed out that a two-layer atmosphere with constant lapse rate in the lower layer (like the troposphere) and an isothermal top (like the lower stratosphere) has only one eigenvalue, with a value around 10 km. Since this was a more realistic model than the earlier ones, the resonance theory was seriously questioned.

Investigating a suggestion of Taylor (1936) that a more realistic model atmosphere might have more than one eigenvalue, Pekeris (1937) considered a five-layer atmosphere embodying a temperature maximum around 60 km and a minimum around 80 km. He found that for temperature distributions consistent with what was known about the upper atmosphere at that time the atmosphere might indeed have a second eigenvalue very near 7.85 km. Furthermore, Pekeris showed that in an atmosphere of this kind, the $\Theta_{2,2}^2$ wave type would have much larger amplitude and quite different phase at ionospheric levels than at ground level, in agreement with the theory of magnetic variations. Further calculations of this type were carried out by Weekes and Wilkes (1947) and Wilkes (1949), who defined the types of temperature distribution that would support such a conclusion. Figure 8.5 from Wilkes (1949) shows one of their principal results.

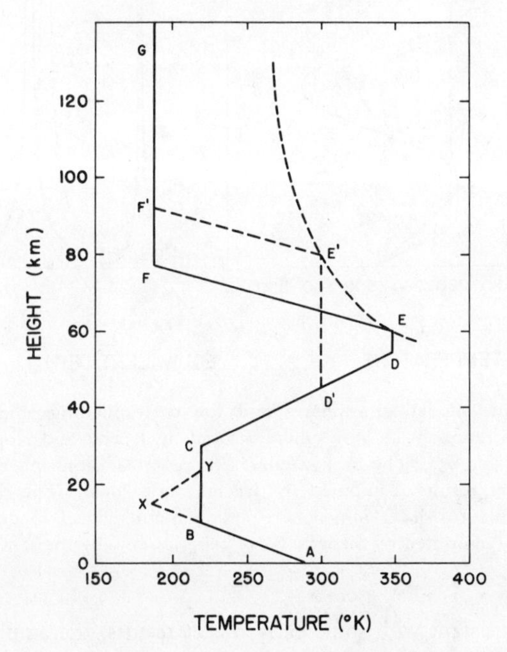

FIG. 8.5. Possible temperature profiles for an atmosphere with an eigenvalue near 7.85 km. Such an atmosphere might have the temperature profile ABCDEFG or ABCD′E′F′G, where E′ lies on the curve EE′. Substitution of the temperature distribution BXYC for BC makes no significant difference. (After Wilkes, 1949.)

The trouble with this explanation, which seemed quite reasonable up until 1952, is that rocket observations simply have not revealed temperatures at the stratopause as high as those required by the theory (see discussion at the beginning of Chapter 3).

The resonance amplification of a wave type can be expressed quantitatively by giving the amplitude of the surface pressure oscillation in units of $(-\rho_0(0)\Omega)$, which would be the amplitude of the surface pressure oscillation for the equilibrium tide (Lamb, 1932). In this unit, a purely gravitational solar tide would have to be amplified about 80 times to produce the observed $S_{2,2}^2$ oscillation. Figure 8.6, based on some calculations of Jacchia and Kopal (1952), shows that for an

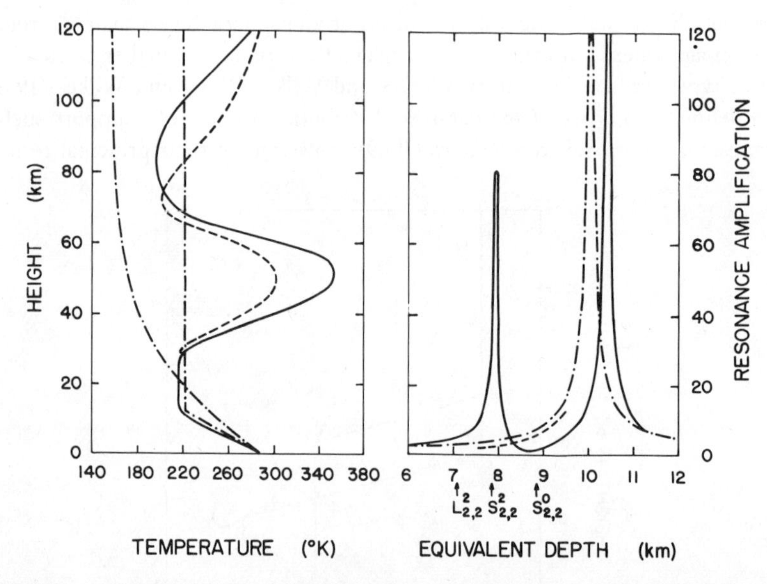

FIG. 8.6. Some model atmospheres and the corresponding amplification curves. The solid curves represent an atmosphere devised by Jacchia and Kopal (1952) to give large reasonance for $\Theta_{2,2}^2$. The dashed curves represent an atmosphere consistent with some early rocket results (computed by Jacchia and Kopal). The dashed-dot curves represent a simple two-layer atmosphere and a mathematical model of Siebert, the resonance curve (computed by Siebert, 1961) being essentially the same for both. (After Siebert, 1961.)

atmosphere consistent with some early rocket results, the amplification is only about 2 or 3, while a model atmosphere that would produce the required amplification involves considerably higher temperatures at the stratopause. Sen and White (1955) stressed this point. It should also be pointed out that the theory, despite its approximations, predicts roughly the observed amplification of L_2^2.

Chapman (1924) showed that the observed phase of $S_{2,2}^2$ at the ground requires a contribution from thermal tide-producing forces and, indeed, the importance of thermal effects had been suggested at the very beginning by Laplace, to account for the observed relative amplitudes of S_2 and L_2. However, when it was thought that a large degree of resonance could be invoked, the thermal and gravitational effects could be considered to be roughly comparable in importance. If the degree of amplification is as small as is now inferred, then the tide must be predominantly thermal in origin. In this circumstance, it is indeed surprising that the semidiurnal tide is so much more regular and even greater in magnitude than the diurnal tide, since periodic heating due to the sun has principally a 24-hour period.

Siebert (1961, p. 114) has concluded that the most likely explanation of the present situation is that the tides are thermally produced, but the diurnal tide, although more strongly excited than the semidiurnal, is somehow *suppressed* in the atmosphere. It is therefore necessary to consider the possible sources of thermal excitation and their effects, as we shall do in Subsection 8.1.3.

8.1.3 The Theory of Atmospheric Tides—Thermal Forces

The theory as presented above does not include the possibility of thermal forces and must be altered somewhat to allow their consideration.

Let J be a periodic function of time (and, in general, also a function of θ, λ, z) which represents the heat change per unit mass per unit time due to diabatic effects (for example, radiation). The temperature variations due to J are assumed to be of two types—a primary periodic temperature oscillation due directly to the periodic heating and cooling, which may drive an atmospheric tide, and a secondary oscillation connected adiabatically with the resulting tidal oscillations. The basic equations (8.6)–(8.9) are unchanged, but Eq. (8.11) has an additional term $[(\gamma - 1)\rho_0 J]$ on the right-hand side.

The resulting partial-differential equation corresponding to (8.13) contains terms involving J. If it is assumed that, for a given period and wave number, $J_l^m(\theta, z)$ can be adequately represented by a series of the form

$$J_l^m = \sum_n J_{l,n}^m(z)\Theta_{l,n}^m(\theta) \tag{8.18}$$

then the variables can still be separated and Laplace's tidal equation (8.15) is unchanged. The radial equation, however, is greatly complicated. Analysis of the radial equation is considerably facilitated by a change of variables (which could have been introduced in Subsection 8.1.2 but was not needed for the discussion there). Define new independent and dependent variables ζ and η by

$$\zeta = \int_0^z \frac{dz'}{H_0(z')} \tag{8.19}$$

$$\eta \exp(\zeta/2) = Z - \frac{\kappa J}{gH_0} \tag{8.20}$$

With these variables, the radial equation allowing for heating becomes

$$\frac{d^2\eta}{d\zeta^2} - \frac{1}{4}\left[1 - \frac{4}{h}\left(\kappa H_0 + \frac{dH_0}{d\zeta}\right)\right]\eta = \frac{\kappa J}{\gamma g h} e^{-\zeta/2} \tag{8.21}$$

In this equation, J is a function of ζ only and represents one of the $J_{l,n}^{m}$ in (8.18). Likewise, h is the equivalent depth for the corresponding wave type and Z (and therefore η) represents the vertical distribution of amplitude of the velocity divergence due to that wave type. However, for convenience of notation here and in what follows the identifying superscript and subscripts are sometimes omitted.

If (8.16) were written in terms of ζ and η [with $J = 0$ in the defining equation (8.20)], it would have exactly the same form as the left-hand side of (8.21). Therefore, the solution of (8.21) consists of a homogeneous part, which is the same (in terms of η) as the solution when heating is absent, and a particular solution that depends on the vertical dependence of J. Except for quite simple vertical distributions of H_0 and J, these solutions have to be obtained numerically. However, this poses no great problem nowadays. The real difficulty lies in specifying the heating function J with sufficient accuracy from our knowledge of the radiative processes (and temperature changes) in the atmosphere.

Three sources of periodic heating have been suggested and considered for this purpose. The first, and perhaps most obvious, is the daily temperature wave caused by radiative processes at the earth's surface and communicated to the atmosphere by thermal (eddy) conductivity. The second is the direct absorption of solar radiation by water vapor in the troposphere. And the third is the direct absorption of solar radiation by ozone in the stratosphere. We shall consider these in turn

Chapman (1924) was the first to study successfully the effect of sufarce heating. The model employed by him (and in subsequent studies of this effect) envisions that the periodic heat oscillation generated at the earth's surface spreads upward by eddy conduction, with a conduction coefficient that is constant with both time and altitude. Although the latter assumption constitutes a severe restriction, in view of the strong dependence of the eddy conductivity on hydrostatic stability, this is unlikely to change the main result of the calculation—namely, that the magnitude of the S_2^2 oscillation so induced is much too small to explain the observed one without a high degree of resonance magnification (Chapman, 1924; Wilkes, 1951; Sen and White, 1955; Siebert, 1961). The reason for this is that the heating, although quite large near the earth's surface, affects a relatively thin layer of the atmosphere and so is ineffectual as a tide-generating mechanism.

On the other hand, absorption of solar radiation by water vapor may be a significant factor, according to Siebert (1961). Air in the troposphere, of course, undergoes other radiative processes, such as absorption of long-wave radiation from the earth's surface as well as emission. However, to a first approximation, these are constant throughout the 24-hour day and therefore contribute nothing to a periodic variation. The amplitude of the heating due to absorption by water vapor is quite small; nevertheless, the heating affects the entire troposphere and is in phase at all levels. According to Siebert's calculations, when an idealized representation of this effect is used in the theory outlined above, the predicted amplitude of the $S_{2,2}^2$ oscillation (in terms of surface pressure) is about one-third of the observed amplitude. The computed phase angle gives maxima at 9 : 00 A.M. and P.M.

In this same paper Siebert pointed out the possibility of an important contribution due to the absorption of solar radiation by ozone in the stratosphere. The simplified model (of vertical temperature variation) used by Siebert is quite inaccurate at stratospheric levels and therefore is unable to elucidate this effect in detail. Small and Butler (1961) and Butler and Small (1963) have used a more realistic temperature model and, with the aid of numerical integration, studied the effect of ozone heating. They conclude that their model is capable of explaining the observed tidal oscillations quite satisfactorily.

In studies of this sort, it is first of all necessary to specify the heating function J that appears on the right side of (8.21). It is assumed that periodic variations of heating rate are due only to absorption of sunlight, net infrared emission being at approximately the same rate throughout the day. Heating rate and its distribution in time and space depend in the present case on ozone density and its distribution in time and space (and also, of course, on the solar zenith angle, whose variations are well known). As was discussed extensively in Chapter 5, the total amount of ozone and its latitudinal and seasonal variations are well known. However, very little is known about the vertical distribution of ozone above, say, 30 km. Theoretical calculations, *Umkehr* observations, and a very few rocket measurements (Sections 5.2, 5.3) have revealed the general character of this vertical distribution, but not much about the change of vertical distribution with latitude and season. It is therefore necessary to make rather arbitrary assumptions about this factor.

Butler and Small (1963), in keeping with the requirements of the theory, assumed that the ozone density ρ_3 could be represented by

$$\rho_3(z, \theta, t^+) = [\rho_3(z)]_{\text{equator}} \times f(\theta, t^+) \tag{8.22}$$

where t^+ is time of year measured from the northern spring equinox. They determined empirically a function $f(\theta, t^+)$ to fit the observed seasonal and latitudinal variations of *total amount* of ozone, thus assuming that ozone density at each level in the upper stratosphere is always proportional to the total amount of ozone. This assumption has been used previously for other purposes, but in fact has no real foundation in observation or theory. It is not clear how a more realistic evaluation of $f(\theta, t^+)$ would affect the final results of Butler and Small, but at least the details of their results must be considered suspect until this point is cleared up.

If the function $f(\theta, t^+)$ can be specified adequately, the further procedure is to compute J as a function of local time,* altitude, and latitude for a particular time of the year or for annual means (see, for example, Fig. 4.19); then to compute the Fourier coefficients of the daily variations for each altitude and latitude; then to analyze the coefficients of the oscillation under consideration (say $l = 2$, $m = 2$) in terms of Hough's functions at each altitude. This procedure gives a representation of $J_{l,n}^{m}(\zeta)$ for use in (8.21) for each wave type. For example, Fig. 8.7 shows results of Butler and Small for the latitudinal variations, at

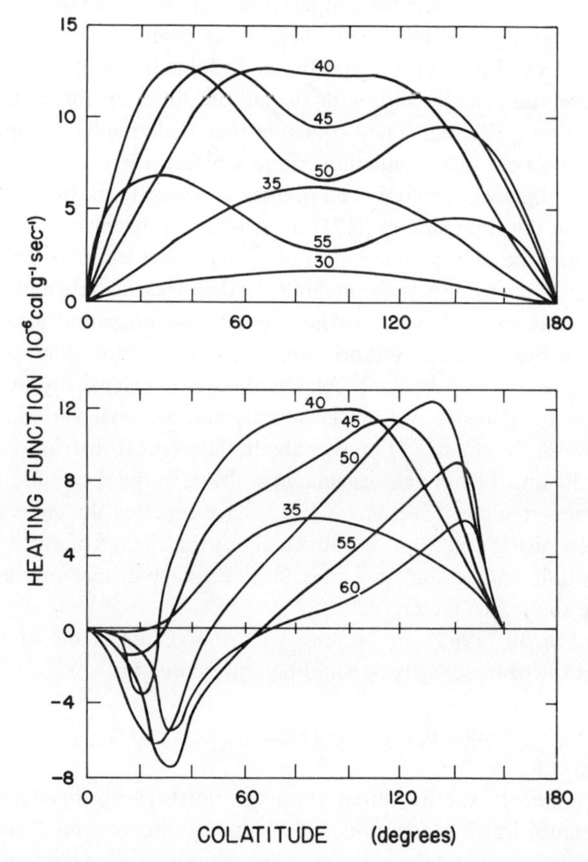

FIG. 8.7. Latitudinal variations of the semidiurnal Fourier coefficients of the daily variation of the absorption of solar radiation by ozone. Each curve refers to a different altitude, indicated above the curve in kilometers. The top figure is for the northern spring equinox and the bottom figure for the northern summer solstice. (After Butler and Small, 1963.)

* For the kind of direct heating under consideration, J is a function of local time only; therefore only oscillations with $l = m$ are excited and a separate calculation for longitudinal variations is unnecessary.

different altitudes and at two times of the year, of the semidiurnal Fourier coefficients. These are the variations that would have to be expressed in terms of Hough's functions.

Butler and Small's results seem to correspond in many important respects to the observational facts. They predict an S_2^2 oscillation at ground level that is similar to the observed one, a large difference† between S_2^2 at the ground and at high levels, the observed annual variation of S_2^2, and the observed behavior of the smaller S_3^3 oscillation. Perhaps most important and interesting, they predict a relatively small S_1^1 oscillation despite the relatively large amplitude of the J_1^1 heating term.

How the latter comes about is revealed clearly in Butler and Small's papers. [Siebert (1961) implied a similar explanation.] The complete solution of (8.21) is

$$\eta(\zeta) = C(\eta_1 + i\eta_2) + I(\zeta) \tag{8.23}$$

where C is a complex integration constant to be determined from boundary conditions, η_1 and η_2 are independent solutions of the homogeneous counterpart of (8.21), and $I(\zeta)$ is a particular solution of (8.21). The amplitude of a tidal oscillation depends critically on the magnitude of $I(\zeta)$. The particular solution $I(\zeta)$ is given by an expression that may be thought of as a special kind of "weighted average" of J in the vertical. For wave types with small equivalent depths, such as $\Theta_{1,1}^1$, the weighting factors are rapidly oscillatory in the vertical, that is, positive at some levels and negative at others. Thus, if J has the same sign over a moderate range of altitudes, contributions from different levels tend to "cancel out," owing to the variations with altitude of the weighting factors. This does not happen for $S_{2,2}^2$, where, indeed, the weighting factors have maximum size at just the altitudes where ozone heating is important. Furthermore, this same type of argument can be invoked to explain the relative regularity of the semidiurnal tide at the surface compared with the diurnal one.

Although many details remain to be filled in, it appears that a satisfactory theory of solar tidal oscillations is now emerging. This theory, contrary to the belief of a few years ago, does not require large resonance amplification of a small solar gravitational force, but rather involves a relatively large degree of thermal excitation. The preferential excitation of the semidiurnal tide over the diurnal tide, despite the greater magnitude of the force exciting the latter, is explained in terms of the atmosphere's response, and includes not only a moderate magnification of the 12-hour oscillation but also a suppression of the 24-hour oscillation.

---

† Both these results and the resonance theory as outlined, for example, by Wilkes (1949) predict an abrupt phase shift at 25 to 30 km. Unfortunately there is at present no adequate observational evidence to verify or disprove this prediction.

8.2 Measurements of Tidal Winds in the Lower Thermosphere

Separation of the tidal components of the wind from other components requires frequent measurements, well distributed diurnally. Only the radio-meteor method of wind measurement meets these requirements, and this method, unfortunately, has been used at only a few locations. The purpose of this section is to describe the method and the results that have been obtained with reference to tidal winds.

8.2.1 Measurement Techniques; Resolution in Time and Space

The basis of the radio-meteor method is that the ionized trail left in the wake of a meteor is capable of reflecting electromagnetic waves with frequencies in the vicinity of 30 Mc sec^{-1}. Owing to strong electrostatic attractive forces, the electron density in the ionized column remains high enough to accomplish the reflection for periods of up to a second (occasionally longer). This period decreases with altitude, as the rate of growth of the column by diffusion increases. During the period of reflection, the column drifts with the wind, and the rate of change of echo range (determined from the Doppler shift) is a measure of the wind component in the line of sight (radial component). Results pertain to the height interval 80–100 km, the lower limit being due to scarcity of suitable ionized trails and the upper limit to the rapid diffusive growth of the trails that are produced.

The two groups that have obtained most of the radio-meteor wind data, at Jodrell Bank, England, and at Adelaide, Australia, have used somewhat different techniques of measurement and analysis. These techniques are of interest to us chiefly with regard to the degree of resolution, in time and in altitude, exhibited by the wind data that are finally presented. Details of the radio equipment used can be found in the references to be cited.

The most that can be determined from a single echo is the range and direction of the reflection point, which give its position in space, and the radial velocity component. Three such components are, in general, necessary to define the velocity vector; from a single ground installation, these must be obtained from three different meteor trails, necessarily at different azimuths and times and, in practice, at somewhat different altitudes. Consequently, there is a certain minimum amount of time and space averaging inherent in the method.

The Jodrell Bank technique (Greenhow, 1954) employs a coherent pulse system at 36.3 Mc sec^{-1} and a beamed aerial, which is directed, for alternate periods of 10 minutes each, toward two azimuths at right angles to one another. For either of these directions, say the x direction (see Fig. 8.8), it is possible to determine for each echo the range R and the velocity component v_{xR}. On the assumption that the wind is horizontal, the velocity component in the

x direction is $v_x = v_{xR}/\cos\theta$, where θ is the elevation angle.* This angle is given by $\sin\theta = z/R$, in which the range R is measured and the height z is taken equal to 92.5 km, the mean height from which returns are received. In the same manner, values of v_y are determined for each echo during the next 10-minute period when the beamed aerial is directed in the y direction. All values of v_x and v_y obtained over a one-hour period are averaged to give a mean wind for the hour, for the height interval 85–100 km, and for the area from which returns are received. The number of echoes available for this purpose is usually a few hundred per hour with the Jodrell Bank equipment.

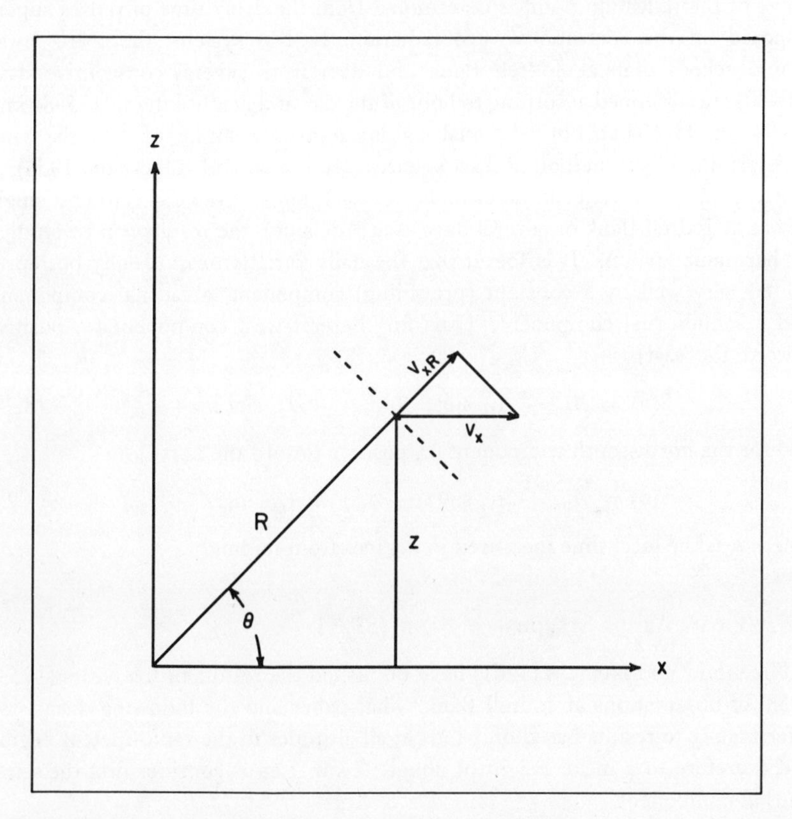

FIG. 8.8. Schematic diagram illustrating the determination of a horizontal wind component by the radio-meteor method at Jodrell Bank. The range R and the velocity component v_{xR} are determined directly. If the wind is assumed to be horizontal, then $v_x = v_{xR}/\cos\theta$.

---

* Verification that the wind is nearly horizontal at these levels is obtained by varying the elevation angle and noting that $v_{xR}/\cos\theta$ remains nearly constant (Greenhow, 1954). Results at Adelaide (Elford and Robertson, 1953) confirm this.

With the Jodrell Bank equipment, it is possible to get better resolution in both altitude and time and the results of special studies along this line will be given later. However, this degree of averaging turns out to be well suited to most studies of tidal components and prevailing winds.

At Adelaide (Robertson, Liddy, and Elford, 1953; Elford and Robertson, 1953), continuous wave radiation (at 27 Mc sec^{-1}) is emitted vertically within a cone of 40° half-angle. An echo from within this cone is detected at an aerial array on the ground, about 20 km from the transmitter, by means of which the direction of the echo and the Doppler shift of the wave can be determined. The range of the reflection point is determined from the delay time of pulses superimposed on the continuous-wave radiation. In this system, there are fewer usable echoes than at Jodrell Bank and data from several consecutive days (10–30) are combined according to hour of the day and height interval (75–84 km, 85–94 km, 95–104 km) prior to analysis. For a given grouping of data, the wind is determined by a method of least squares (see Elford and Robertson, 1953).

After the wind has been determined for each hour of the day (either a single day as at Jodrell Bank or several days as at Adelaide), the results can be studied by harmonic analysis. It is found that the daily variation can usually be represented very well by a constant (prevailing) component, a diurnal component, and a semidiurnal component. Thus for the east-west component (u, positive toward the east),

$$S(u) = A_{0u} + A_{1u} \sin(t' + a_{1u}) + A_{2u} \sin(2t' + a_{2u}) \tag{8.24}$$

and for the north-south component (v, positive toward the north),

$$S(v) = A_{0v} + A_{1v} \sin(t' + a_{1v}) + A_{2v} \sin(2t' + a_{2v}) \tag{8.25}$$

where t' is the local time measured in degrees from midnight.

8.2.2 Tidal Winds at Jodrell Bank (53°N)

Greenhow and Neufeld (1961) have published the results of five years (1953–1958) of observations at Jodrell Bank. That paper and the following discussion refer mainly to results based on winds at all altitudes in the radio-meteor region and therefore to a mean height of about 92 km. Let us consider first the semidiurnal component.

Table 8.5 gives the values, for different times of the year, of the amplitudes and phase angles, as defined in Eqs. (8.24) and (8.25), describing the semidiurnal variations of u and v. Data are subdivided into months, except for the autumn period, in which case shorter intervals are used because of what appears to be a rapid characteristic phase variation during that period. These shorter intervals do not correspond to precisely the same days each year. The data are based on results for all five years.

TABLE 8.5

AMPLITUDES AND PHASE ANGLES OF THE SEMIDIURNAL WIND VARIATION AT 92 KM OVER JODRELL BANK[a]

Time of year	A_{2u} (m sec^{-1})	A_{2v} (m sec^{-1})	a_{2u} (deg)	a_{2v} (deg)
January	19.0	20.0	162 (9.6)	252 (6.6)
February	12.5	16.5	129 (10.7)	228 (7.4)
March	7.0	5.5	180 (9.0)	270 (6.0)
April	6.5	10.5	228 (7.4)	312 (4.6)
May	5.0	8.0	162 (9.6)	294 (5.2)
June	6.0	10.5	237 (7.1)	306 (4.8)
July	6.5	10.0	282 (5.6)	300 (5.0)
August	14.0	15.5	237 (7.1)	318 (4.4)
September (1)	20.0	22.0	249 (6.7)	342 (3.6)
September (2)	16.5	17.5	309 (4.7)	36 (1.8)
October (1)	7.0	13.0	348 (3.4)	96 (11.8)
October (2)	5.5	5.0	219 (7.7)	294 (5.2)
October (3)	11.0	10.0	114 (11.2)	201 (8.3)
November (1)	13.5	11.0	96 (11.8)	186 (8.8)
November (2)	22.5	19.5	141 (10.3)	228 (7.4)
November (3)	27.0	28.0	186 (8.8)	273 (5.9)
December	21.0	25.0	192 (8.6)	270 (6.0)

[a] See Eqs. (8.24) and (8.25). The parenthesized values after the phase angles give the earlier local time (in hours) of maximum wind component toward east or north. Adapted from Greenhow and Neufeld (1961).

First of all, it is evident that the semidiurnal wind oscillation is large, its amplitude being comparable with the magnitude of the prevailing winds. The variation of the amplitudes throughout the year is also large, with minima in the spring and early summer and also for a short period in the autumn.

The phase is also variable through the year, but usually is such that maximum wind toward the east occurs between 7 and 11 A.M. (or P.M.) local time. According to the equations of tidal theory, maximum wind toward the east corresponds to the time of minimum pressure. At the ground, as we have seen, maximum pressure occurs between 9 and 10 A.M. (or P.M.). Therefore, the semidiurnal oscillation at 92 km is out of phase with the semidiurnal oscillation at the ground during most of the year, as required by the dynamo theory. According to the tidal theory of Butler and Small (1963), at 91.3 km minimum pressure and maximum wind toward the east should occur between 8 and 9 A.M. (or P.M.) for the $S_{2,2}^{2}$ oscillation.

A comparison of the phases for the u and v components reveals that maximum wind toward the north usually occurs about three hours earlier than maximum wind toward the east (to within $\pm\frac{1}{2}$ hour in all periods except May, June, July, and October (1)). A three-hour phase difference in this sense (in the Northern Hemisphere) is to be expected if the oscillation is to be described by a worldwide pattern migrating westward with the sun.

The annual phase variation is striking, the changes during the autum being most marked and puzzling. During September and early October, the phase angles increase rapidly to a maximum; during late October and early November they decrease rapidly to a minimum; during the rest of November they return to their "normal" values. There is no obvious factor in the tidal theory to explain this variation.

The data discussed above apply to average results over the entire five years of observations. They are rather difficult to evaluate because they are accompanied by no indication of the statistical reliability* of the determinations. Greenhow and Neufeld (1961) show also some results for the five individual years in the form of plotted points on harmonic dials.* These data are grouped to show for each year the semidiurnal oscillation in winter (December, January, February), in spring (March, April, May), in summer (June, July, August), in September, in October, and in November. It is possible to get a crude estimate of the radius of the probable-error circles for these seasonal determinations by regarding each year as a separate determination. This procedure is crude because some error is involved in reading values of the plotted data, because the determinations

TABLE 8.6

Amplitudes and Phase Angles of the Semidiurnal Wind Variation at 92 km over Jodrell Bank[a]

Time of year	A_{2u} (m sec^{-1})	A_{2v} (m sec^{-1})	a_{2u} (deg)	a_{2v} (deg)
Winter	16.4 (1.7–5)	19.7 (2.6–5)	160 (9.7)	248 (6.7)
Spring	5.2 (1.5–5)	7.4 (1.6–5)	168 (9.4)	289 (5.4)
Summer	9.2 (1.1–5)	13.7 (1.1–5)	242 (6.9)	309 (4.7)
September	16.3 (2.4–5)	18.0 (3.3–5)	269 (6.0)	2 (2.9)
October	3.9 (2.0–4)	3.5 (3.1–5)	10 (2.7)	69 (0.7)
November	16.4 (5.3–4)	19.6 (5.8–4)	144 (10.2)	238 (7.1)

[a] The parenthesized values after the phase angles give the earlier local time (in hours) of maximum wind component toward east or north. The parenthesized values after the amplitudes give the radius of the probable-error circle in meter (second)$^{-1}$, followed by the number of individual determinations on which it is based. Computed from data given by Greenhow and Neufeld (1960).

* Some of the terms used in this discussion are explained in Appendix I for the reader not familiar with them.

for some years may be based on more data than for other years, and especially because the number of cases is so small (five and, in a few instances, four). Nevertheless, the results are shown in Table 8.6. The seasonal determinations appear to be reliable, with the exception of October, when the radii of the probable-error circles are quite large relative to the amplitudes. The month of November, viewed as a whole rather than in separate parts, shows amplitudes and phases very similar to those of the other winter months.

The results discussed above apply to all wind observations in the 85- to 100-km region and therefore to a mean height of about 92 km. In a special study (of data for the year September 1954–August 1955) Greenhow and Neufeld (1956) examined the height varitation of the semidiurnal component. They found the amplitude increasing with height in the 85–100-km region at the rate of about 2 m sec^{-1} km^{-1} in winter, and nearly constant with height in the summer. The phase change is such that the oscillation at 100 km leads that at 85 km by about 3–4 hours in winter and 1–2 hours in summer.

Turning now to the diurnal variations, we find that at Jodrell Bank at 92 km, this component is highly variable and irregular, being on the average considerably smaller than the semidiurnal component. Table 8.7 gives the amplitude

TABLE 8.7

AMPLITUDES AND PHASE ANGLES OF THE DIURNAL WIND VARIATION AT 92 KM OVER JODRELL BANK[a]

Time of year	A_{1u} (m sec^{-1})	A_{1v} (m sec^{-1})	a_{1u} (deg)	a_{1v} (deg)
Spring	2.3	7.8	201 (16.6)	262.5 (12.5)
Summer	3.7	9.2	121.5 (21.9)	273.0 (11.8)
Autumn	3.4	4.2	214.5 (15.7)	259.5 (12.7)
Winter	3.9	3.6	12 (5.2)	55.5 (2.3)

[a] See Eqs. (8.24) and (8.25). The parenthesized values after the phase angles give the local time (in hours) of maximum wind component toward east or north. Adapted from Greenhow and Neufeld (1961).

and phase for each season determined from the five years of observations. Maximum wind toward the north occurs near local noon during most of the year, but about 10 hours earlier in winter. The east-west component is quite small in amplitude and variable in phase; its maximum value follows the maximum value of v but not, in any individual season, by the six hours that would be expected for an S_1^1 oscillation.

Greenhow and Neufeld also computed the amplitude and phase of the north-south component separately for each of four years, combining all seasons for each year. Comparison of the results for the separate years shows a reasonable degree of agreement. With regard to the east-west component, the seasonal variation of amplitude and phase is such that the resultant amplitude for the entire year is nearly zero.

8.2.3 Tidal Winds at Adelaide (35°S)

Elford (1959) has given the results of tidal wind determinations at Adelaide for certain months during the period October 1952 to February 1955. As at Jodrell Bank, these data show the importance of the periodic components relative to the prevailing wind. However, in contrast to the results at Jodrell Bank, they show a relatively large diurnal component and a semidiurnal component whose behavior seems to be quite irregular and difficult to interpret.

Figure 8.9 shows the seasonal variations of amplitude of the diurnal and semidiurnal wind components for each of three height intervals. It is to be noted that the amplitude of the 24-hour component is much larger than at Jodrell Bank. The amplitude of the 12-hour component is also larger than at Jodrell Bank and, during most of the year, increases with altitude in the 80–100-km region. Data from the different years seem to be reasonably consistent in terms of this parameter.

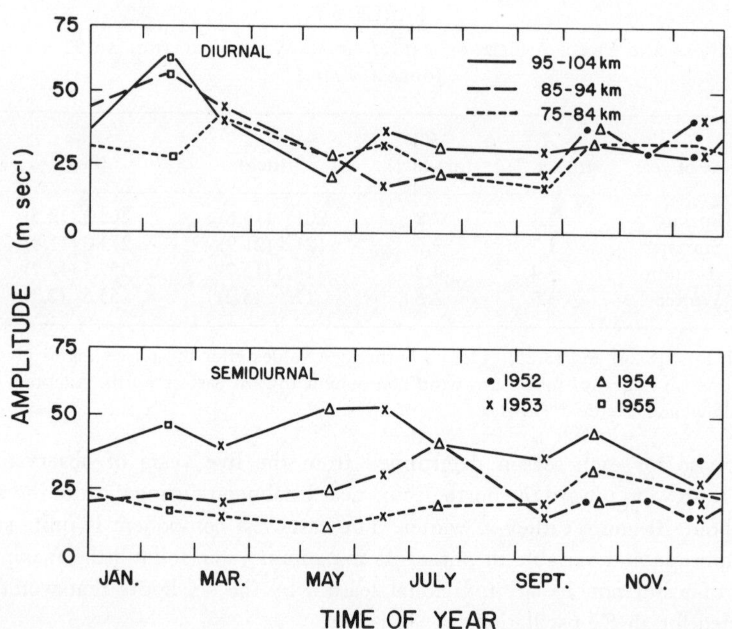

Fig. 8.9. Seasonal variations of the amplitudes of the diurnal and semidiurnal wind oscillations at Adelaide. Different curves refer to different altitudes as indicated. (After Elford, 1959.)

However, when the behavior of the periodic wind components determined at Adelaide is examined in detail, the results are irregular and inconsistent. Figure 8.10 shows these results for December 1952–December 1953 in polar form.

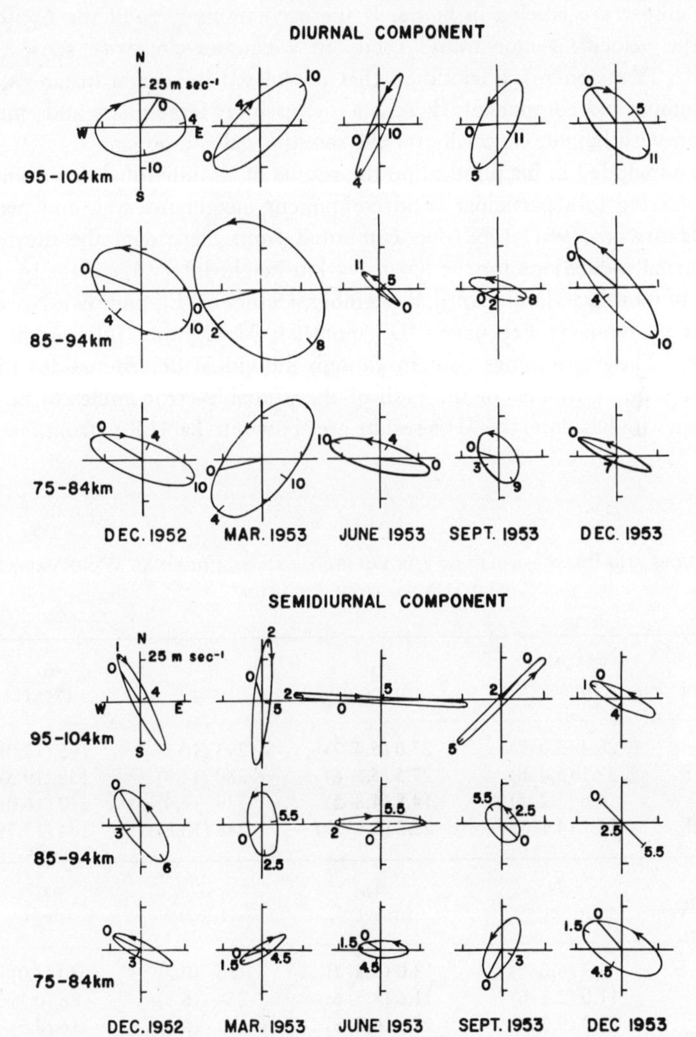

FIG. 8.10. The diurnal and semidiurnal components of the daily wind variations observed at Adelaide by the radio-meteor method. Each diagram refers to a particular component, altitude interval, and month, as indicated. The end of the wind vector, drawn from the origin, traces out an ellipse during a 24-hour (or 12-hour) period. The sense of rotation is given by an arrow on the ellipse, the phase by the times marked (in hours) outside the ellipse, and the amplitude of the wind vector by the scale marked on the N–S and E–W axes. (After Elford, 1959.)

For each component, month, and altitude range, the ellipse shows the locus of the end point of the wind vector during a 24-hour (or 12-hour) period. The arrow on the ellipse shows the sense of rotation of the vector, and certain points on the ellipse are labeled in hours. If the oscillations were of the S_1^1 (or S_2^2) type, the velocity vector would rotate in a counter-clockwise sense (in the Southern Hemisphere), a condition that is violated in several instances. For a given month and component, there is a distressingly large phase and amplitude variation with height, especially for the semidiurnal component.

Elford included in his publication the results of all individual determinations, each referring to a particular wind component, height interval, and period of time. Haurwitz (1961, 1962) has computed from these data the diurnal and semidiurnal oscillations for the 85- to 94-km height interval averaged for three groups of months: March, April, September, October ("E" months); November, December, January, February ("D" months); May, June, July, August ("J" months). These groupings contain enough individual determinations (5 to 7) to allow crude estimates of the radii of the probable-error circles to be made, and Haurwitz has done this. His results are shown in Table 8.8 (from Haurwitz, 1962).

TABLE 8.8

AMPLITUDES AND PHASE ANGLES OF THE DIURNAL AND SEMIDIURNAL WIND VARIATION AT 85 TO 94 KM OVER ADELAIDE[a]

Time of year	A_{1u} (m sec^{-1})	A_{1v} (m sec^{-1})	a_{1u} (deg)	a_{1v} (deg)
E	28.4 (5.0–7)	27.0 (5.4–7)	293 (10.5)	195 (17.0)
D	37.1 (5.4–6)	27.5 (5.5–6)	280 (11.3)	158 (19.5)
J	4.6 (5.2–5)	14.8 (4.8–5)	339 (7.4)	210 (16.0)
All	24.1 (4.1–18)	22.2 (3.5–18)	288 (10.8)	184 (17.7)

Time of year	A_{2u} (m sec^{-1})	A_{2v} (m sec^{-1})	a_{2u} (deg)	a_{2v} (deg)
E	3.7 (5.6–7)	3.0 (3.9–7)	85 (0.2)	30 (2.0)
D	11.0 (2.8–6)	11.6 (4.5–6)	254 (6.5)	86 (0.1)
J	13.9 (9.2–5)	17.1 (6.2–5)	84 (0.2)	316 (4.5)
All	1.9 (4.0–18)	4.8 (3.4–18)	106 (11.5)	16 (2.6)

[a] The parenthesized values after the phase angles give the local time in hours of maximum wind component toward east or north (the earlier time in the case of the semidiurnal oscillation). The parenthesized values after the amplitudes give the radius of the probable-error circle in meter (second)$^{-1}$, followed by the number of individual determinations on which it is based. After Haurwitz (1962).

A study of these data emphasizes the large magnitude of the diurnal component relative to the semidiurnal at Adelaide. It also indicates that, insofar as one can judge from probable-error circles based on such a small number of individual determinations, the semidiurnal component at Adelaide is not very reliably determined. The best determination appears to be for the D months (summer, in the Southern Hemisphere). Maximum wind toward the north should follow maximum wind toward the east (in the Southern Hemisphere) by six hours for the diurnal component and three hours for the semidiurnal component. The data fit this criterion reasonably well.

Determinations of the tidal wind components in the radio-meteor region have also been made at Mawson, Antarctica (68°S) (Elford and Murray, 1960). However, data for only three months (December 1957, December 1958, and January 1959) are available and these are not discussed here.

8.3 Irregular, Small-Scale Wind Variations in the Lower Thermosphere

The past 10 years have seen an increasing recognition of the importance of "small-scale" (to be defined below) irregularities* in the wind field at thermospheric levels. These show up primarily as large vertical shears in wind direction and speed. They have been noted in connection with photographic observations of meteor trains and of chemiluminescent vapor trails, as well as in the radio-meteor studies. An example was given in Chapter 3 (see Table 3.2). Here in this section, further observational material is cited (Subsection 8.3.1) and an explanation, due to Hines, of these irregularities as manifestations of internal gravity waves is presented (Subsection 8.3.2).

8.3.1 Observational Evidence

Striking examples of wind shears in the lower thermosphere were provided by Liller and Whipple (1954) from photographic observations of the distortion of long-enduring meteor trains. One of these is illustrated in Fig. 8.11, according to a representation by Hines (1960). The (horizontal) wind component perpendicular to the direction of view was determined by Liller and Whipple as a function of altitude from a series of photographs taken over a period of 17 seconds. The result is plotted in Fig. 8.11 with vertical and horizontal scales such that the graph would represent a direct photograph of the trail had it formed vertically at zero and been subject to distortion by the observed winds for a period of 200 seconds.

* It is interesting that these are sometimes referred to as "large-scale" irregularities in comparison with smaller turbulent motions. In the present context where we are considering them along with tidal and prevailing winds, they are relatively small in scale.

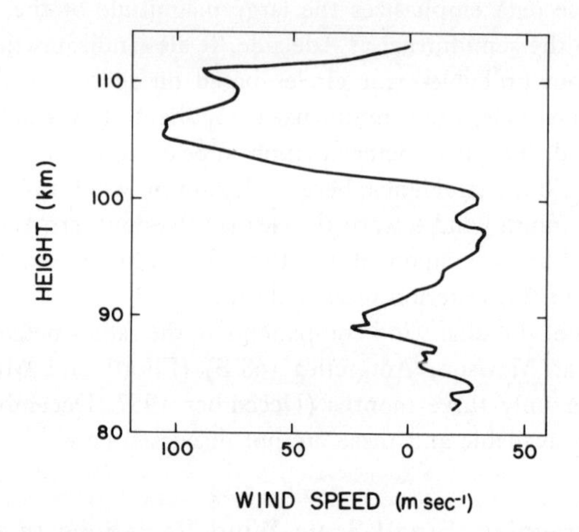

FIG. 8.11. Wind speeds from a meteor train photographed by Liller and Whipple (1954). (After Hines, 1960.)

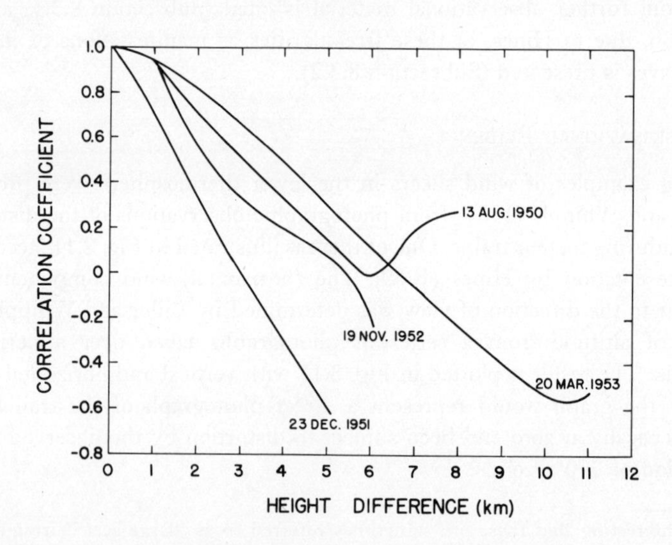

FIG. 8.12. Autocorrelation of wind speed as a function of height difference for winds from four meteor trains. (After Liller and Whipple, 1954.)

Apart from the over-all change from a direction toward the right near 80 km to a direction toward the left near 110 km, the wind component exhibits smaller-scale shears. The vertical distances associated with these shears are measured in kilometers and the magnitudes of the shears are very large by comparison with those encountered in the lower atmosphere. The trains photographed by Liller and Whipple showed maximum shears of 25–90 m sec^{-1} per kilometer of height.

Distortions of chemiluminescent trails of vapor released by rockets have revealed similar wind variations. Some photographs and data have been given, for example, by Blamont and de Jager (1961) and Blamont and Baguette (1961).

It is of great interest to determine the scale of the irregularities that are involved in these wind changes. One way to specify the scale is to compute correlation coefficients between winds separated by definite intervals and, by varying the interval, determine the interval for which the correlation coefficient falls to zero. Figure 8.12 shows such autocorrelation functions as computed by Liller and Whipple for four meteor trains. The vertical scale by this measure averages about 5 km for the four cases.

Extensive investigations of irregularities in the winds of the lower thermosphere have been conducted by Greenhow and Neufeld at Jodrell Bank by the radio-meteor method. The usual techniques of measurement and analysis employed at Jodrell Bank, as described in Section 8.2, tend to smooth out the irregularities, but special studies have been carried out to observe them.

Greenhow and Neufeld (1959) described one study that made use of a spaced-station technique to measure the wind velocity simultaneously at two points along a meteor trail. Spatial separations of the two points were variable up to 8 or 10 km. After substracting the measured wind components from mean values over an hour, to remove the prevailing and tidal components, Greenhow and Neufeld computed the autocorrelation coefficient as a function of both spatial separation and height separation. In the latter case, the vertical scale was found to be about 6 to 7 km, a result quite comparable with that of Liller and Whipple. In the case of spatial separation the autocorrelation function was observed to fall to a minimum of about 0.2 at separations of 5 to 6 km and then level off. Since the spatial separation consists of both a vertical and a horizontal component, and since the autocorrelation function for the vertical component falls to zero at 5 to 6 km, this behavior is to be interpreted as indicating a much larger horizontal than vertical scale. The largest spatial separations were observed for meteors traveling nearly horizontally. Greenhow and Neufeld estimated the horizontal scale to be more than 100 km.

Although in the usual single-station studies at Jodrell Bank winds are determined as averages from all meteor echoes occurring during an hour, better time resolution is possible when echoes are frequent. Figure 8.13 shows one such example for a 48-hour period (Greenhow and Neufeld, 1960). This figure shows the results for one component (positive toward the northeast), each plotted point representing a 10-minute average and the points being separated by 20 minutes in time. The dashed curve shows the prevailing and tidal components, determined by harmonic analysis of the data over the 48-hour period, and deviations of the actual wind from this curve are plotted separately on an expanded scale at the bottom.

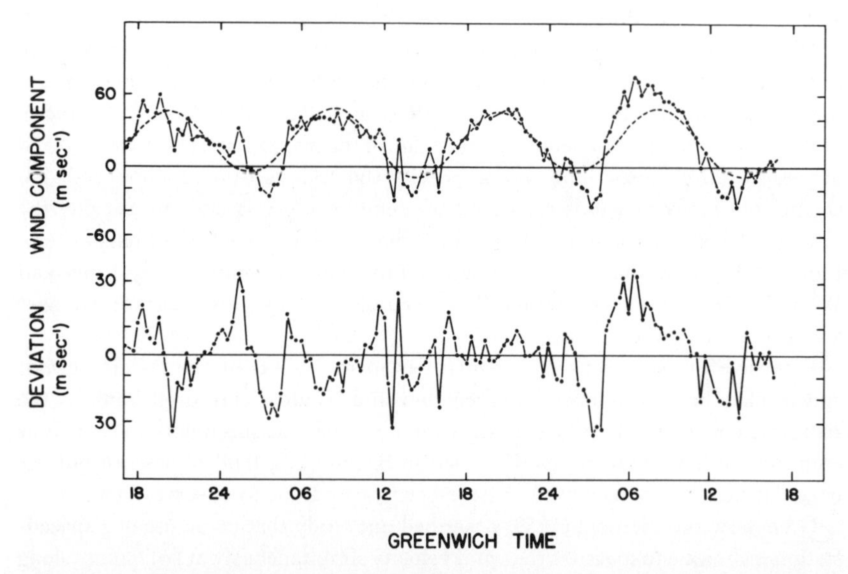

FIG. 8.13. Short-period wind variability determined from radio-meteor measurements at Jodrell Bank. In the upper graph, the solid curve shows wind speed (SW to NE) averaged over 10-minute intervals; the dashed curve shows the sum of the prevailing, diurnal, and semidiurnal components for the period. The lower curve shows, on an expanded scale, the deviations of observed winds from the sum of these components. (After Greenhow and Neufeld, 1960.)

From data like these, the behavior of the autocorrelation coefficient as a function of time can be investigated. For the particular data shown in Fig. 8.13, Greenhow and Neufeld (1960) found that the autocorrelation coefficient falls to zero in about 100 minutes. From similar investigations of data during other time intervals they found that this time scale varies between 100 and 200 minutes.

8.3.2 An Explanation of the Irregular Motions as Manifestations of Atmospheric Wave Motion

Hines (1959, 1960) has argued quite convincingly that much of the observational evidence on irregular motions in the lower thermosphere can be explained if the motions result from the superposition of various modes of a certain type of organized wave motion. This theory also accounts plausibly for many observed characteristics of ionospheric behavior, which will not be discussed here. The mathematical development of the theory is outlined here, and its applications to the thermospheric wind field briefly discussed. See Hines (1960, 1963) for a much more complete discussion.

In quantitative mathematical form, Hines' development is based on a rather simple atmospheric model. However, Hines has discussed extensively, from a qualitative point of view, the consequences to be expected from nonfulfillment in the actual atmosphere of some of the simplifying assumptions. In the model employed, the atmosphere is taken to be stationary in the absence of waves and to be of uniform temperature and composition. The wave motions to be studied are assumed to be of small enough horizontal extent for the employment of Cartesian coordinates and short enough in period for the neglect of Coriolis accelerations. They are furthermore assumed to be small enough to allow a perturbation treatment and are assumed to occur adiabatically. Viscous forces are neglected. With these assumptions, the equations governing the wave motion are (see Appendix K)

$$\rho_0 \frac{\partial u}{\partial t} = -\frac{\partial}{\partial x}(p') \tag{8.26}$$

$$\rho_0 \frac{\partial w}{\partial t} = -\frac{\partial}{\partial z}(p') - g(\rho') \tag{8.27}$$

$$\frac{\partial}{\partial t}(p') + w\frac{\partial p_0}{\partial z} = \gamma \frac{p_0}{\rho_0}\left[\frac{\partial}{\partial t}(\rho') + w\frac{\partial \rho_0}{\partial z}\right] \tag{8.28}$$

$$\frac{\partial}{\partial t}(\rho') + w\frac{\partial \rho_0}{\partial z} + \rho_0\left(\frac{\partial u}{\partial x} + \frac{\partial w}{\partial z}\right) = 0 \tag{8.29}$$

in which p' and ρ' are the deviations of pressure and density from the undisturbed values p_0 and ρ_0, u and w are perturbation velocity components in the x (horizontal) and z (vertical) directions, and γ is the ratio of the specific heats. The undisturbed atmosphere is described by the hydrostatic equation $(\partial p_0/\partial z) = -\rho_0 g$, and has p_0, $\rho_0 \propto \exp(-z/H_0)$ where H_0 is the scale height $p_0/\rho_0 g$.

Solutions are sought in the form

$$u = U \exp i(\omega t - K_x x - K_z z) \tag{8.30a}$$

$$w = W \exp i(\omega t - K_x x - K_z z) \tag{8.30b}$$

$$p' = p_0 P \exp i(\omega t - K_x x - K_z z) \tag{8.30c}$$

$$\rho' = \rho_0 R \exp i(\omega t - K_x x - K_z z) \tag{8.30d}$$

in which U, W, P, R, ω, K_x, and K_z are all constants. These are solutions of (8.26)–(8.29) only if the following relation (frequency equation) holds:

$$\omega^4 - \omega^2 \gamma g H_0(K_x^2 + K_z^2) + (\gamma - 1)g^2 K_x^2 + i\gamma g \omega^2 K_z = 0 \tag{8.31}$$

Of the various types of waves described by these equations, Hines has selected one type, which he calls internal atmospheric gravity waves, for further study. The chain of reasoning leading to this selection is as follows:

(1) It is first of all assumed that K_x is real ($= k_x$); that is, that no boundary conditions would arise to require the attenuation of the wave motion in the horizontal.

(2) In this case a study of Eq. (8.31) reveals either that K_z is purely imaginary or that its imaginary part is given by $(1/2H_0)$, that is,

$$K_z = k_z + i/2H_0 \tag{8.32}$$

where k_z is real. The first alternative allows no phase variation with height and is discarded in favor of the complex vertical wave number given by Eq. (8.32). With this choice the frequency equation (8.31) becomes

$$\omega^4 - \omega^2 \gamma g H_0 (k_x^2 + k_z^2) + (\gamma - 1) g^2 k_x^2 - \gamma g \omega^2 / 4H_0 = 0 \tag{8.33}$$

(3) The modified frequency equation (8.33) may be regarded as quadratic in ω^2 and thus has two positive roots. The larger one, corresponding to the choice of the plus sign in the solution for ω^2, is necessarily larger than

$$\omega_a = (\gamma g / 4H_0)^{\frac{1}{2}} \tag{8.34}$$

and the second is necessarily smaller than

$$\omega_g = [(\gamma - 1) g / \gamma H_0]^{\frac{1}{2}} \tag{8.35}$$

The latter choice defines the internal atmospheric gravity waves.

To compare the properties of these waves with the observed characteristics of irregular thermospheric motions, it is necessary to consider the latter in the light of a wave interpretation and to select characteristic wavelengths and periods. The dominant vertical wavelength, as reported by Millman (1959) from study of a large number of meteor trains and as is consistent, for example, with an inspection of Fig. 8.11, can be taken to be about 12 km in the radio-meteor region. It is to be noted that this is about half the value that would be implied (by considering that the autocorrelation function falls to zero in one-quarter of a wavelength) by the vertical autocorrelation function computed by Liller and Whipple (1954) and by Greenhow and Neufeld (1959); but those do not apply directly to the assumed wave motion, the correlation coefficients being increased by unresolved effects of the prevailing winds (see below). In the same manner the time scale of 100 to 200 minutes deduced from the autocorrelation analysis of Greenhow and Neufeld (1960) implies a dominant period of, say, 200 minutes. The horizontal wavelength has not been well resolved but must be measured in the hundreds of kilometers. These choices give, in terms of the parameters of the theory, $k_z \simeq 5 \times 10^{-6}$ cm^{-1}, $\omega \simeq 5 \times 10^{-4}$ sec^{-1}, and $k_x < 10k_z$.

These observed properties have, to a certain extent, already influenced the development of the theory. Thus the choice of a complex rather than purely imaginary wave number was indicated qualitatively by the observed vertical phase variation, and the choice of lower-frequency waves was indicated by the observed periods. In the latter case, for $H_0 = 6$ km, $\gamma = 1.4$, and $g = 10^3$ cm sec^{-2}, we have $\omega_a = 2.5 \times 10^{-2}$ sec^{-1} and $\omega_g = 2.2 \times 10^{-2}$ sec^{-1}, both considerably larger than the observed ω. On the other hand, other observed properties of the motions can be shown to be a consequence of the theory once the dominant vertical wavelength and period are specified. These include the relatively large horizontal scale of the motion, the relatively large horizontal component of motion, and the general increase of the amplitude with height.

An order-of-magnitude comparison of the terms in Eq. (8.33) for the values given in the preceding paragraphs shows that

$$k_z/k_x \simeq \omega_g/\omega$$

so that the horizontal wavelength is about 40 times the vertical wavelength, a result that is certainly in accord with the observations. The expressions for u and w, which follow from the theory but are not given explicitly here (see Hines, 1960), reveal a ratio of horizontal to vertical speed of the same order, a result also in accord with observations.

The large amplitude of the oscillations and the general growth of this amplitude with height in the radio-meteor region (as evident in Fig. 8.11 and other meteor trains shown by Liller and Whipple) are a more general property of these waves, independent of the observed frequency. It follows in fact from the positive value of the imaginary part of K_z [Eq. (8.32)]. This expression and Eqs. (8.30) show that the amplitude of the velocity components should increase upward as $\exp(z/2H_0)$ or as $(\rho_0)^{-\frac{1}{2}}$. This corresponds to a condition that the kinetic energy of the wave motions remains constant with height. Figure 8.14 shows the data of Fig. 8.11 reduced by Hines, first by the removal of the general "large-scale" vertical shear and second by multiplication of the residual by a factor proportional to $(\rho_0)^{\frac{1}{2}}$. The result indicates that the deduced amplification factor accounts not only qualitatively but even semiquantitatively for the observed vertical amplification. The vertical autocorrelation function for this reduced curve falls to zero at 4 km.

Various important physical factors not taken into account quantitatively in the theory have been discussed semiquantitatively by Hines. One of these is viscous dissipation of energy which must begin to become important in the radio-meteor region, at lower levels for the smaller-scale sizes. This indicates that both the smallest observed scale and the dominant scale should increase with height in the radio-meteor region, and both of these effects are apparent in the meteor train represented in Fig. 8.11 (and others). Using a semiquantitative criterion, Hines has shown that the smallest observed scales and the dominant scales are larger than those that would be expected to be damped rapidly by viscous forces.

The energy of the wave motions is considered by Hines to be derived from the tidal motions at the levels involved or from the upward propagation of amplifying waves generated in the lower atmosphere (or from some combination of the two). In the latter case, one might expect that not all modes would propagate freely to the thermosphere and that some would be reflected partially or wholly in the mesospheric region of temperature decline. Indeed, Hines has shown that this may limit the observed wave motions in the domain of larger vertical wavelengths and smaller horizontal wavelengths.

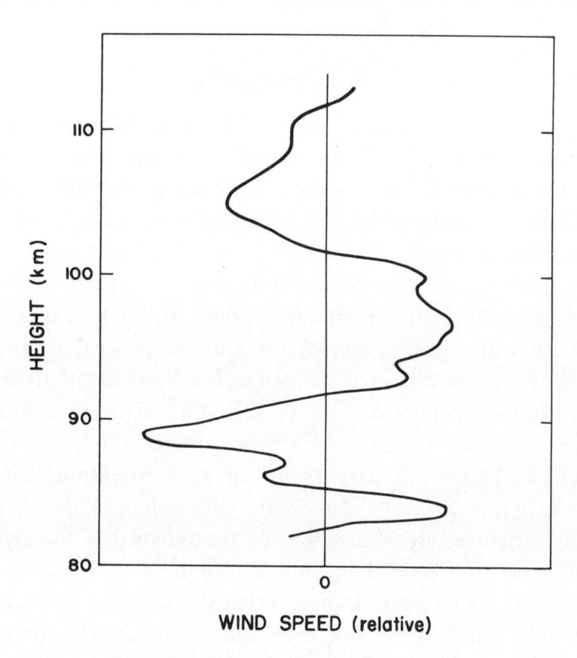

FIG. 8.14. The wind profile of Fig. 8.11, after removal of large-scale shear and weighting by the factor $(\rho_0)^{\frac{1}{2}}$. (After Hines, 1960.)

Nonlinear processes must become important for wave motion of such large amplitude. For this reason, as well as for a fuller treatment of the viscous effects and the effects of variable temperature in the undisturbed atmosphere, the theory is capable of important extensions. Nevertheless, it seems unlikely that these extensions will invalidate the general conclusion that semicoherent wave motions are capable of explaining the observed characteristics of the "irregular" motions of the radio-meteor region.

REFERENCES

APPLETON, E. V., and WEEKES, K. (1939). On lunar tides in the upper atmosphere. *Proc. Roy. Soc.* **A171**, 171–187.

BLAMONT, J. E., and BAGUETTE, J. M. (1961). Mesures déduites des déformations de six nuages de métaux alcalins formés par fusées dans la haute atmosphère. *Ann. Geoph.* **17**, 319–337.

BLAMONT, J. E., and DE JAGER, C. (1961). Upper atmospheric turbulence near the 100 km level. *Ann. Geoph.* **17**, 134–144.

BUTLER, S. T., and SMALL, K. A. (1963). The excitation of atmospheric oscillations. *Proc. Roy. Soc.* **A274**, 91–121.

CHAPMAN, S. (1918). The lunar atmospheric tide at Greenwich, 1854–1917. *Quart. J. Roy. Meteor. Soc.* **44**, 271–280.

CHAPMAN, S. (1924). The semidiurnal oscillation of the atmosphere. *Quart. J. Roy. Meteor. Soc.* **50**, 165–195.

CHAPMAN, S. (1951). Atmospheric tides and oscillations. *In* "Compendium of Meteorology" (T. F. Malone, ed.), pp. 510–530. American Meteorological Society, Boston, Massachusetts.

CHAPMAN, S. (1961). Regular motions in the ionosphere : electric and magnetic relationships. *Bull. Amer. Meteor. Soc.* **42**, 85–100.

CHAPMAN, S., and BARTELS, J. (1940). "Geomagnetism." Oxford Univ. Press (Clarendon), London and New York. (Reprinted with corrections, 1951 and 1962.)

CHAPMAN, S., and WESTFOLD, K. C. (1956). A comparison of the annual mean solar and lunar atmospheric tides in barometric pressure, as regards their worldwide distribution of amplitude and phase. *J. Atmos. Terr. Phys.* **8**, 1–23.

ELFORD, W. G. (1959). A study of winds between 80 and 100 km in medium latitudes. *Plan. Space Sci.* **1**, 94–101.

ELFORD, W. G., and MURRAY, E. L. (1960). Upper atmosphere wind measurements in the Antarctic. *In* "Space Research I" (H. Kallmann Bijl, ed.), pp. 158–163. North-Holland Publ. Co., Amsterdam.

ELFORD, W. G., and ROBERTSON, D. S. (1953). Measurements of winds in the upper atmosphere by means of drifting meteor trails, II. *J. Atmos. Terr. Phys.* **4**, 271–284.

GREENHOW, J. S. (1954). Systematic wind measurements at altitudes of 80–100 km using radio echoes from meteor trails. *Phil. Magaz.* [7] **45**, 471–490.

GREENHOW, J. S., and NEUFELD, E. L. (1956). The height variation of upper atmospheric winds. *Phil. Magaz.* [8] **1**, 1157–1171.

GREENHOW, J. S., and NEUFELD, E. L. (1959). Measurements of turbulence in the upper atmosphere. *Proc. Phys. Soc.* **74**, 1–10.

GREENHOW, J. S., and NEUFELD, E. L. (1960). Large scale irregularities in high altitude winds. *Proc. Phys. Soc.* **75**, 228–234.

GREENHOW, J. S., and NEUFELD, E. L. (1961). Winds in the upper atmosphere. *Quart. J. Roy. Meteor. Soc.* **87**, 472–489.

HARRIS, M. F., FINGER, F. G., and TEWELES, S. (1962). Diurnal variation of wind, pressure, and temperature in the troposphere and stratosphere over the Azores. *J. Atmos. Sci.* **19**, 136–149.

HAURWITZ, B. (1937). The oscillations of the atmosphere. *Beitr. Geoph.* **51**, 195–233.

HAURWITZ, B. (1956). The geographical distribution of the solar semidiurnal pressure oscillation. *Meteor. Papers* **2**, No. 5, 1–36. New York University.

HAURWITZ, B. (1961). Comments on tidal winds in the high atmosphere. *Plan. Space Sci.* **5**, 196–201.

Haurwitz, B. (1962). Wind and pressure oscillations in the upper atmosphere. *Archiv Meteor., Geoph., Biokl.* **A13**, 144–166.

Haurwitz, B. (1963). Personal communication.

Haurwitz, B., and Craig, R. A. (1952). Atmospheric flow patterns and their representation by spherical-surface harmonics. *Geoph. Res. Papers* No. 14, pp. 1–78.

Haurwitz, B., and Sepúlveda, G. M. (1957). Geographical distribution of the semidiurnal pressure oscillation at different seasons. *J. Meteor. Soc. Japan* 75th Anniv. Vol., pp. 149–155.

Hines, C. O. (1959). An interpretation of certain ionospheric motions in terms of atmospheric waves. *J. Geoph. Res.* **64**, 2210–2211.

Hines, C. O. (1960). Internal atmospheric gravity waves at ionospheric heights. *Canad. J. Phys.* **38**, 1441–1481.

Hines, C. O. (1963). The upper atmosphere in motion. *Quart. J. Roy. Meteor. Soc.* **89**, 1–42.

Hough, S. S. (1897). On the application of harmonic analysis to the dynamical theory of the tides. *Phil. Trans. Roy. Soc.* **A189**, 201–257.

Hough, S. S. (1898). On the application of harmonic analysis to the dynamical theory of the tides, II. *Phil. Trans. Roy. Soc.* **A191**, 139–185.

Jacchia, L. G., and Kopal, Z. (1952). Atmospheric oscillations and the temperature profile of the upper atmosphere. *J. Meteor.* **9**, 13–23.

Johnson, D. H. (1955). Tidal oscillations of the lower stratosphere. *Quart. J. Roy. Meteor. Soc.* **81**, 1–8.

Kelvin, Lord (W. Thomson) (1882). On the thermodynamic acceleration of the earth's rotation. *Proc. Roy. Soc. Edinb.* **11**, 396–405.

Kertz, W. (1957). Atmosphärische Gezeiten. *In* "Handbuch der Physik" (S. Flügge, ed.), Vol. 48, pp. 928–981. Springer, Berlin.

Lamb, H. (1910). On atmospheric oscillations. *Proc. Roy. Soc.* **A84**, 551–572.

Lamb, H. (1932). "Hydrodynamics," 6th ed. Cambridge Univ. Press, London and New York. (Reprinted, 1945, Dover, New York.)

Lenhard, R. W., Jr. (1963). Variation of hourly winds at 35 to 65 kilometers during one day at Eglin Air Force Base, Florida. *J. Geoph. Res.* **68**, 227–234.

Liller, W., and Whipple, F. L. (1954). High-altitude winds by meteor-train photography. *In* "Rocket Exploration of the Upper Atmosphere" (R. L. F. Boyd and M. J. Seaton, eds.), pp. 112–130. Pergamon Press, New York.

Martyn, D. F. (1955). Theory of height and ionization density changes at the maximum of a Chapman-like region, taking account of ion production, decay, diffusion and tidal drift. *In* "The Physics of the Ionosphere," pp. 254–259. Physical Society, London.

Millman, P. M. (1959). Visual and photographic observations of meteors and noctilucent clouds. *J. Geoph. Res.* **64**, 2122–2128.

Pekeris, C. L. (1937). Atmospheric oscillations. *Proc. Roy. Soc.* **A158**, 650–671.

Robertson, D. S., Liddy, D. T., and Elford, W. G. (1953). Measurements of winds in the upper atmosphere by means of drifting meteor trails, I. *J. Atmos. Terr. Phys.* **4**, 255–270.

Sen, H. K., and White, M. L. (1955). Thermal and gravitational excitation of atmospheric oscillations. *J. Geoph. Res.* **60**, 483–495.

Siebert, M. (1961). Atmospheric tides. *Advances in Geoph.* **7**, 105–187.

Simpson, G. C. (1918). The twelve-hourly barometer oscillation. *Quart. J. Roy. Meteor. Soc.* **44**, 1–18.

Small, K. A., and Butler, S. T. (1961). The solar semidiurnal atmospheric oscillation. *J. Geoph. Res.* **66**, 3723–3725.

Stewart, B. (1882). Hypothetical views regarding the connection between the state of the sun and terrestrial magnetism. *Encycl. Britannica* (9th ed.) **16**, 181–184.

Taylor, G. I. (1929). Waves and tides in the atmosphere. *Proc. Roy. Soc.* **A126**, 169–183.

Taylor, G. I. (1936). The oscillations of the atmosphere. *Proc. Roy. Soc.* **A156**, 318–326.

Weekes, K., and Wilkes, M. V. (1947). Atmospheric oscillations and the resonance theory. *Proc. Roy. Soc.* **A192**, 80–99.

Wilkes, M. V. (1949). "Oscillations of the Earth's Atmosphere." Cambridge Univ. Press, London and New York.

Wilkes, M. V. (1951). The thermal excitation of atmospheric oscillations. *Proc. Roy. Soc.* **A207**, 358–370.

CHAPTER 9

An Introduction to Some Other Aeronomic Problems

The topics that form the subject of this chapter play important roles in the physics of the upper atmosphere and have received considerable attention from aeronomers. Although it is beyond the scope of this book to attempt a detailed treatment of them, at least a brief introduction to some aspects of these subjects seems likely to be useful and convenient for the reader. The emphasis is on fundamentals and on those aspects that seem to the author to be the most pertinent to the subject matter of the previous chapters. The references emphasize books, monographs, review articles, and articles that are important from a historical point of view. Nevertheless some references that fit none of these categories are included when they are useful.

We consider in turn the earth's magnetic field and its variations (Section 9.1), some problems related to the ionosphere (Section 9.2), and the morphology and spectral characteristics of the airglow and aurora (Section 9.3).

9.1 An Introduction to the Earth's Magnetic Field and Its Variations

The main magnetic field of the earth, although produced primarily in regions interior to the earth's surface, is nevertheless of considerable interest in connection with the upper atmosphere because of its effect on the motion of charged particles. It is therefore useful to have a description of this field as in Subsection 9.1.1. Small transient variations of the field, on the other hand, are almost certainly indications of current systems in the upper atmosphere and thus possess an importance of another kind. The regular variations related to the solar or lunar day are described in Subsection 9.1.2 and the irregular variations related to magnetic disturbance are considered in Subsections 9.1.3 and 9.1.4. In Subsection 9.1.5 some remarks are made on the possible causes of these variations, with special reference to the dynamo theory.

The most comprehensive treatment of the subject of geomagnetism is that of Chapman and Bartels (1940). Although not current in certain respects, this treatise is still the starting point for studies of the subject. More recent discussions are by Mitra (1952), Massey and Boyd (1959), Vestine (1960), and Chapman (1963).

9.1.1 The Earth's Magnetic Field and Its Observed Variations

Observations of the earth's magnetic field are made by measuring the force acting on a small magnet near the earth's surface (or, more recently, at high levels from rockets and satellites). The magnetic field at a point is described by its intensity **H**, which is the force acting on a north-seeking magnetic pole of unit strength. The unit of intensity that is used in geomagnetism is the gauss* (Γ) or, frequently, a smaller unit called the gamma (γ): $1\ \Gamma = 10^5\ \gamma$. Regular measurements of **H**, for more than a century at some locations, have led to a wealth of knowledge about its spatial and temporal variations.

The earth's magnetic field has been observed to undergo slow and gradual variations which are called *secular changes*. These are presumably due to causes inside the earth and are not considered further here. The field consists mainly of a part that, except for the secular changes, is constant in time but has a characteristic spatial variation; this is spoken of as the earth's *main field*. Superimposed on the main field are relatively feeble (but very important) time variations, the characteristics of which also vary with location. These *transient variations* are considered later in Subsections 9.1.2, 9.1.3, and 9.1.4.

The vector intensity **H** at a point may be described by various combinations of a number of scalar quantities. Those in common use are as follows:

X = the horizontal northward component
Y = the horizontal eastward component
Z = the vertical downward component
H = the magnitude of the horizontal component
D = the direction of the horizontal component, measured in degrees from geographic north, positive toward the east
I = the angle between **H** and the horizontal, measured in degrees and taken to be positive if **H** is directed downward

The quantity Z is often called the *vertical intensity*, H the *horizontal intensity*, D the *magnetic declination* or *magnetic variation*, and I the *magnetic dip* or *magnetic inclination*.

The Equivalent Dipole and Geomagnetic Coordinates. Results of observations near the earth's surface show that, to an approximation sufficient for many purposes and certainly adequate for a gross description, the earth's main field resembles that of a magnetic dipole located at the center of the earth (or, equiva-

* Although the gauss is strictly a unit of magnetic induction (**B**), **B** and **H** are essentially equivalent for the conditions under which the measurements are made.

lently, that of a uniformly magnetized sphere). The equivalent dipole, does not, however, have its axis aligned along the axis of rotation of the earth. The dipole axis intersects the earth's surface at the antipodal points* 78.3°N, 69.0°W and 78.3°S, 11.0°E. These may be spoken of as the *geomagnetic poles* or the *axis poles.*

It is convenient for many purposes to define a geomagnetic coordinate system based on the geomagnetic poles. In this system θ_m is the geomagnetic colatitude, the angular distance measured at the center of the earth from the geomagnetic north pole; ϕ_m is the geomagnetic latitude, $\phi_m = 90° - \theta_m$; and λ_m is the geomagnetic longitude, measured eastward from the particular geomagnetic meridian which contains the geographic south pole. The geomagnetic equator is the locus of points where $\theta_m = 90°$. The geomagnetic colatitude and longitude can be determined from the geographic colatitude and longitude (θ, λ) by

$$\cos\theta_m = \cos\theta_p \cos\theta + \sin\theta_p \sin\theta \cos(\lambda - \lambda_p) \tag{9.1}$$

$$\sin\lambda_m = \frac{\sin\theta}{\sin\theta_m} \sin(\lambda - \lambda_p) \tag{9.2}$$

where $\theta_p = 11.7°$ and $\lambda_p = -69°$.

Description of the Earth's Main Field. The discussion in this and the following paragraph refers to the idealized dipole field (and only approximately to the observed field) and to geomagnetic coordinates. The horizontal component of intensity is everywhere directed northward; it has its greatest amplitude at the equator and zero amplitude at the poles. Its magnitude is given by $H = H_0 \sin\theta_m$, where H_0 is the equatorial value of H. The vertical intensity has maximum magnitude at the poles, in a direction downward at the north pole and upward at the south pole, and is zero at the equator. It is given by $Z = 2H_0 \cos\theta_m$. The factor H_0 decreases as the cube of the distance from the center of the earth; at the earth's surface a value $H_0 = 0.312\ \Gamma$ is appropriate to represent the observed field. The resulting directions of **H** near the earth's surface are shown graphically in Fig. 9.1.

The dipole field may also be represented by *lines of force* such that **H** at any point in space is tangent to the line of force through that point. Figure 9.2 gives an example. A line of force that intersects the earth's surface at geomagnetic latitude $\pm\phi_m$ is parallel to the geomagnetic axis at a distance from the center of the earth given (in units of earth radius) by $(1/\cos^2\phi_m)$. Its inclination i from the horizontal, at the point of intersection with the earth's surface, is given by $\tan i = Z/H = 2 \tan|\phi_m|$. The tangent to the line of force at a given point

* Because of the secular variations, the orientation of the dipole axis that gives the best fit to the observed field varies slowly. The values quoted here were given by Finch and Leaton (1957) and refer to the year 1955.

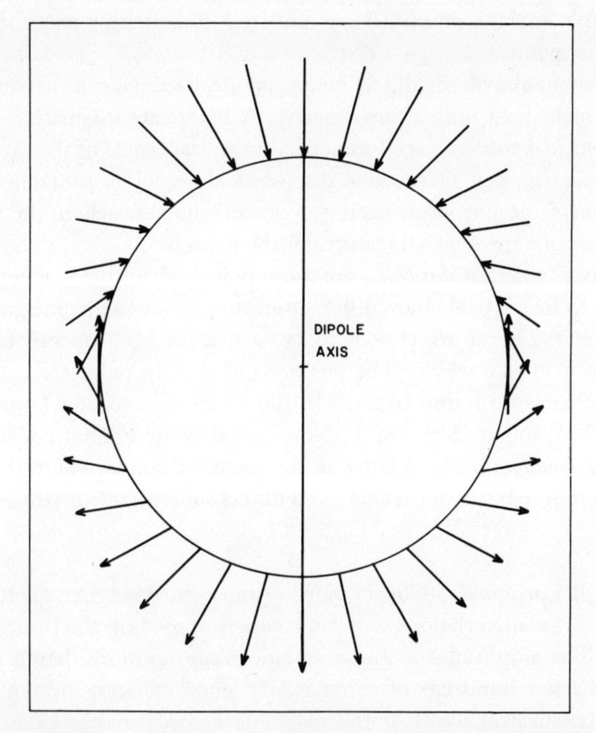

FIG. 9.1. Directions of magnetic intensity for a dipole field. Each arrow shows the direction at the geomagnetic latitude where it touches the circle. The upper half of the circle corresponds to the geomagnetic Northern Hemisphere.

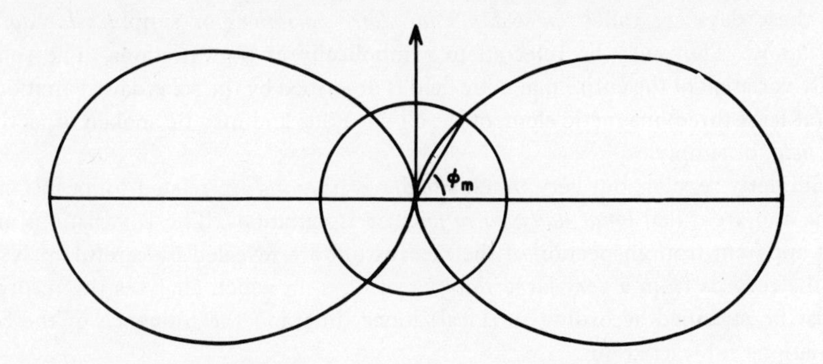

FIG. 9.2. Lines of force (for a dipole field) intersecting the earth's surface at geomagnetlc latitude ϕ_m. The particular ones shown are drawn for $\phi_m = 60°$ and are parallel to the dipole axis at a distance of four earth radii from the center.

on the earth's surface points (in an outward direction) toward the *magnetic zenith* for that point.

As pointed out above, the dipole field is an idealized concept which represents the earth's main field only approximately. A better approximation is obtained with an *eccentric dipole* displaced somewhat from the center of the earth but with its axis parallel to that of the one discussed above. The difference, which is significant for some purposes, need not concern us here where the intention is simply to describe the over-all features of the main field.

The *magnetic poles*, or *dip poles*, are those points where the magnetic intensity is observed to be vertical; they differ in location from the geomagnetic or axis poles because the latter are chosen so as to give the best representation of the earth's main field as a whole. The positions of the dip poles are not accurately known and change from time to time. In 1945, they were located (approximately) at 76°N, 102°W and at 68°S, 145°E. Note that they are not antipodal. Similarly, the *magnetic equator* or *dip equator* is the locus of points where the magnetic intensity is observed to be horizontal, and differs somewhat from the geomagnetic equator.

The Transient Variations. Observations of the earth's magnetic field at a given location reveal small variations with time superimposed on the (nearly) constant main field. The amplitudes of these variations are no more than a few tens or occasionally a few hundreds of gammas. By laborious and, in many instances, ingenious statistical analyses of the magnetic records, it has been possible to identify several physically meaningful types of time variations which, when combined, give the highly complex observed pattern.

On some days at a given location, the magnetic elements undergo regular daily variations that are characteristic of the elements and the location. These days are called magnetically *quiet* days and the variations of the magnetic elements on these days are called *quiet-day solar daily variations* or simply *solar daily variations*. They may be referred to symbolically as Sq variations. The solar daily variation of the entire magnetic field is described by the solar daily variations of (at least three) magnetic elements at all locations and may be spoken of as the Sq *field*, or simply Sq.

Similarly regular, but very much smaller, variations are related to local lunar time and are called *lunar daily variations*, or L variations. The L variations are not apparent from inspection of the records but are revealed by careful analysis of the records from a very large number of days, in which analyses the records must be stratified according to (local) lunar time and the influence of the Sq variations subtracted out.

The residual variation of the magnetic field, after allowance is made for the Sq field and the L field, is called the *disturbance* field and may be designated as the D field or simply D. The disturbance field is highly complex and variable.

On some days (the quiet days) it is largely absent, especially in middle and low latitudes. On some days it consists of a daily variation, in addition to Sq, which is designated SD but is considered to be part of the D field. On the other hand, there are days of large disturbance when, for example, the daily variations may be mostly obscured. These days mark the periods of *magnetic storms* and the disturbance field during these days has a large component that is related not to local time but to *storm time*, the time that has elapsed since the beginning of the storm.

9.1.2 Fields of the Regular Transient Variations, Sq and L

The Magnetic Character Figure. For the systematic study of Sq, it is necessary first of all to identify those days that are considered to be quiet. This is done by an international system of classification which results in the designation of five *international quiet days* (and also five *international disturbed days*) for each month. At each magnetic observatory, the record for a given day is classified as 0, 1, or 2 according to whether it is judged to be quiet, moderately disturbed, or unusually disturbed. The average of the numbers assigned by all observatories for a given day, computed to one decimal, is called the *international magnetic character figure*, C_i ; $C_i = 0.0, 0.1, \ldots, 2.0$. The classification at each station is somewhat subjective, and the criteria used are not necessarily exactly the same from station to station. There is, however, a reasonable degree of consistency in the classifications from station to station. On the basis of the international magnetic character figure and other (more objective) indices of magnetic disturbance (see Subsection 9.1.4), a Committee on Characterization of Magnetic Disturbances selects the five international quiet days and the five international disturbed days for each month.

The Sq *Field.* Consider (for a particular location) a particular magnetic element whose value is known for each hour of each day during a large number of days, say a month or more. For each hour of the day, form an average of the values occurring on the international quiet days. From each of these hourly averages, subtract the over-all average value of the element for all hours on the quiet days. This gives a sequence of 24 hourly values, whose sum is zero; such a sequence is sometimes called a mean daily inequality. It may be referred to as S_q. Since there may be some residual disturbance even on quiet days, S_q gives only an approximation to the quiet-day solar variations Sq. The approximation is, however, quite good in middle and low latitudes, especially in years of sunspot minimum. At high latitudes Sq should be determined only from exceptionally quiet days. The Sq field is obtained by synthesizing the results for all stations.

It is found that the Sq variations determined in this way are approximately independent of longitude; since they are derived as a function of local time, this means that the Sq field as viewed from the sun is approximately constant during the period for which it is determined. On the other hand, the Sq variation of a given element is in general different at different latitudes. Furthermore, the Sq field as a whole differs according to the season of the year and the phase of the sunspot cycle during which it is determined.

The nature of the Sq field is illustrated in a somewhat idealized diagram, based on analyses by Chapman (1919) and given by Chapman and Bartels (1940). This diagram, shown in Fig. 9.3, refers to the equinoctial months during a

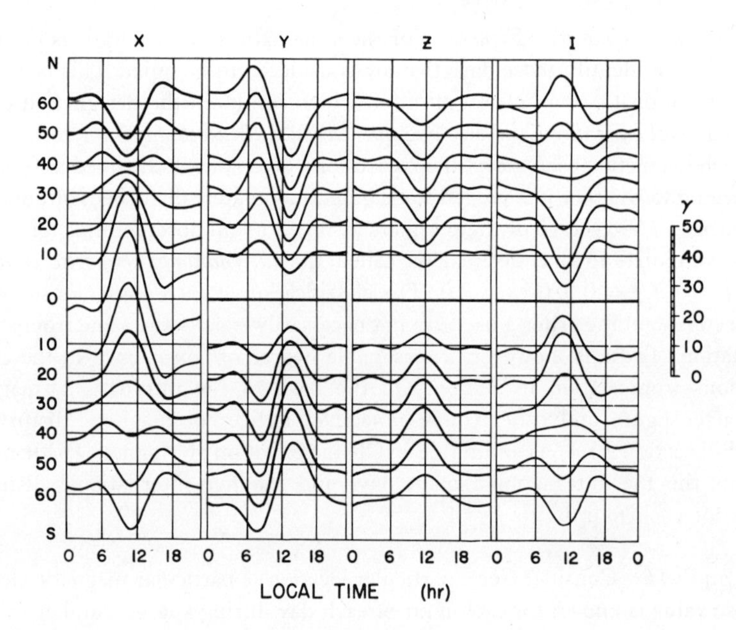

FIG. 9.3. The solar daily variations of northward component X, eastward component Y, downward component Z, and inclination I. The curves are based on analyses by Chapman (1919) and refer to the equinoxes in the sunspot-minimum year 1902. They are drawn for each 10° of geographic latitude. (After Chapman and Bartels, 1940.)

sunspot-minimum year. Variations of the northward component X are the same at corresponding latitudes in the two hemispheres, but change phase at latitudes $\pm$ 30 or so. The Y and Z variations are of opposite sign in the two hemispheres and change phase near the equator. The daily variations show signs of a semi-diurnal component, variations during the daylight hours being much more marked than during the dark hours.

A feature of the Sq field that is not shown in this presentation is the very large diurnal variation of X (or of horizontal intensity H) which has been observed at Huancayo, Peru, near the magnetic equator, since Chapman's analyses. The noon value of H is much higher than the midnight value. Vestine (1960) shows an average range at Huancayo of about 125 γ in his analyses of the Sq field based on 12 years of data.

The principal seasonal variation of the Sq field is such that the amplitude or range of the daily variations is augmented in the summer hemisphere and decreased in the winter hemisphere relative to equinoctial conditions. The principal sunspot-cycle variation of the Sq field is also in the amplitude of daily variations, which is increased by 50 to 100 per cent from sunspot minimum to sunspot maximum. This effect is discussed in considerably more detail by Chapman and Bartels (1940).

The L *Field.* The lunar daily variation, being rather small, is relatively more difficult to isolate and study. Methods and some of the results of studying L are discussed in considerable detail by Chapman and Bartels (1940, Chapter 8). Here, only the principal results are mentioned.

When the lunar daily variation is determined from magnetic data distributed evenly over a full lunation (that is, occurring at all phases of the moon) the result is a very simple one. The lunar daily variation in all elements at all locations consists of a regular semidiurnal oscillation of simple character. On the other hand, when the days from which the lunar daily variation is to be determined are selectively chosen so as to correspond to a particular phase of the moon (for example, new moon), then the oscillation still has a semidiurnal character, but with maxima and minima of unequal amplitudes. Study of this effect shows that the largest variations occur during that part of the lunar day that corresponds to hours of sunshine; for example, lunar noon near new moon. This is illustrated in Fig. 9.4, according to an analysis by Chapman (1925). The amplitude of the typical L variation is much smaller, by a factor of 10 or so, than that of the Sq variation.

Just as in the case of the solar daily variation, the lunar daily variation of horizontal intensity at Huancayo near the magnetic equator is much larger than at other locations. In fact, percentually, the abnormality of the range is even greater than in the case of Sq.

There have been studies of the dependence of L on various factors such as lunar distance, magnetic activity, season, and phase of the sunspot cycle. These studies are in general difficult because of the small amplitude of the oscillation. There is no doubt though that the seasonal variation is very pronounced, more so than in the case of Sq. On the other hand, L shows little dependence on the phase of the sunspot cycle, which is a rather surprising result in view of the behavior of Sq.

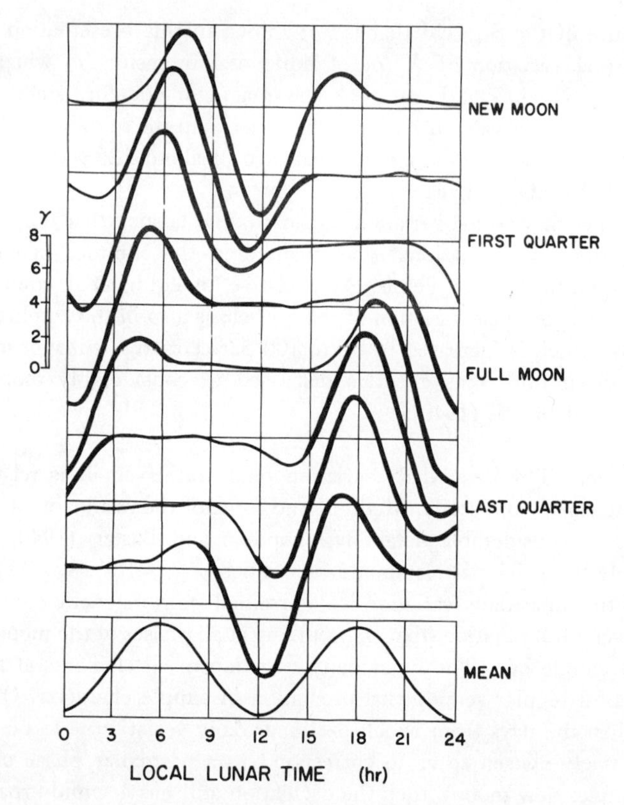

FIG. 9.4. The lunar daily variation of the magnetic west declination at Batavia. The top eight curves represent the variations at different phases of the moon; the thicker portion of each curve corresponds to daylight hours. The bottom curve represents a mean for all lunar phases. The scale in gammas is a measure of the westward component; 4γ corresponds to about 0.37′ in direction. [After Chapman and Bartels (1940), and based on an analysis by Chapman (1925).]

9.1.3 THE DISTURBANCE FIELD, D

There is a large variation with time in the degree of magnetic disturbance, as measured for example by the range of the magnetic variation associated with it. Even on the international quiet days, some residual disturbance is usually present, especially at high latitudes. On average days there is what may be designated as *slight* magnetic disturbance, and at times there is *intense* disturbance corresponding to the periods of magnetic storms. In general, slight disturbance is associated with the ordinary days of a month (neither international quiet days nor international disturbed days) and intense disturbance with the international disturbed days. However, it may happen that the degree of dis-

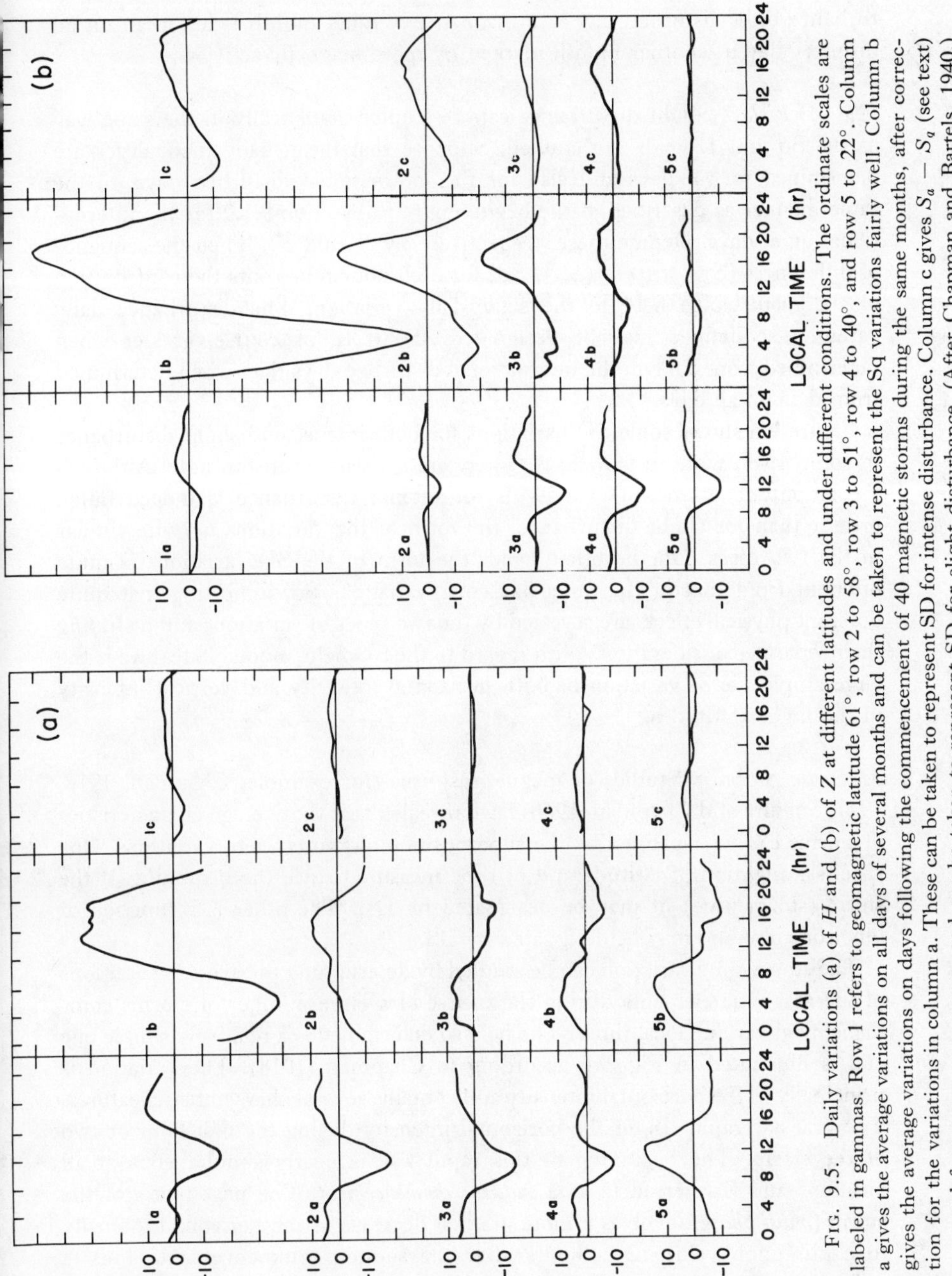

FIG. 9.5. Daily variations (a) of H and (b) of Z at different latitudes and under different conditions. The ordinate scales are labeled in gammas. Row 1 refers to geomagnetic latitude 61°, row 2 to 58°, row 3 to 51°, row 4 to 40°, and row 5 to 22°. Column a gives the average variations on all days of several months and can be taken to represent the Sq variations fairly well. Column b gives the average variations on days following the commencement of 40 magnetic storms during the same months, after correction for the variations in column a. These can be taken to represent SD for intense disturbance. Column c gives $S_a - S_q$ (see text) averaged for one or more years, and can be taken to represent SD for slight disturbance. (After Chapman and Bartels, 1940.)

turbance on a disturbed day in a magnetically quiet month is less than on an ordinary day in another month marked by more magnetic activity.

The SD *Field.* Slight disturbance can be studied statistically in the same way as the Sq and L fields are studied. Suppose that mean daily inequalities are determined for all days and also for the international disturbed days, in the same manner as described at the beginning of Subsection 9.1.2 for the international quiet days. Denote these, respectively, by S_a and S_d. Then the sequence of differences (S_a – Sq) or (S_d – Sq) for each hour represents the contribution of the disturbance field to the solar daily variation. The disturbance daily variation so defined may be designated SD. It is, of course, greater when determined from data on the international disturbed days than when determined from data on all days.

Figure 9.5 shows some SD variations for both intense and slight disturbance in comparison with Sq variations. There are several features to note. Although the amplitude of the SD variations for intense disturbance is indeed much greater than for slight disturbance, the form of the variations is quite similar in the two cases. On the other hand, the form of the SD variations is quite different from that of the quiet-day solar variations Sq, indicating that quite different physical effects are revealed by the two types of variations and justifying their separate consideration. With regard to the SD field, a notable feature is the large amplitude of variation of both horizontal intensity and vertical intensity in the higher latitudes.

Magnetic Storms. Studies of magnetic storms (for example, Chapman, 1918, 1927; Sugiura and Chapman, 1960) have revealed that the average characteristics of storms can be explained by the superposition of two types of variations. One type is a function of latitude and of time measured since the beginning of the storm (storm time); it may be designated by Dst. The other is a function of longitude also and may be designated by DS.

The storm-time variation can be studied by determining the average variations of the magnetic elements during the course of a large number of storms commencing at varying local times. The field so determined is a relatively simple one and is illustrated in Fig. 9.6 according to Chapman (1918). The variation is principally in the horizontal intensity and contains several characteristic features. The first is a rapid rise of the horizontal intensity during the first hour or two of the storm. The beginning of this rapid rise is nearly simultaneous at all locations and is referred to as a *sudden commencement*. The next phase of the storm (*main phase*) involves a more gradual decrease of the horizontal intensity to a minimum value 12–24 hours after the sudden commencement. This is followed by an even more gradual recovery during the course of the next day or two. These characteristics are apparent at all latitudes shown in Fig. 9.6, but with somewhat greater range of variation at the lower latitudes. The storm-time variation of the vertical intensity is very much smaller than that of the horizontal intensity and is in the opposite direction. The declination is essentially unaffected. There is a wide diversity in the range of variation among magnetic storms. Nevertheless, the storm-time variations are qualitatively the same for groups of widely varying range.

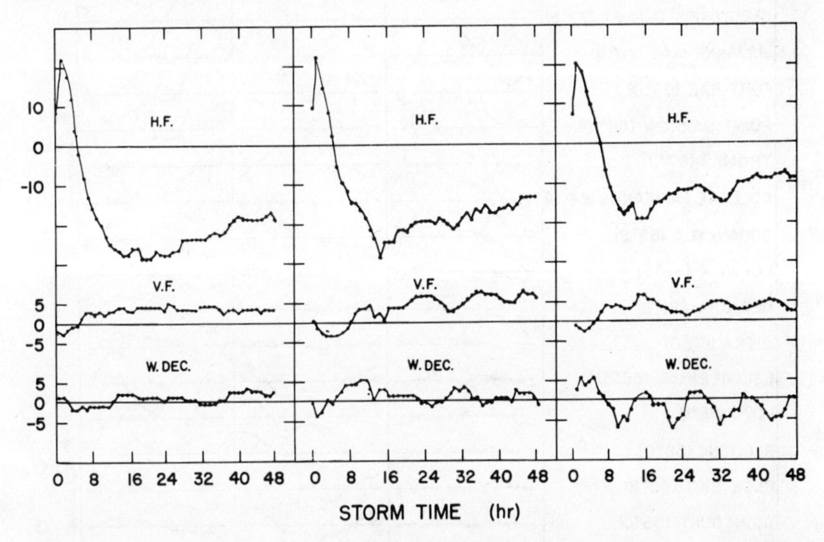

FIG. 9.6. Storm-time variations of horizontal force H, vertical force Z, and west declination at three different latitudes. The ordinate scales are labeled in gammas. The curves on the left are based on data from stations near $\phi_m = 22°$, those in the center from stations near $\phi_m = 40°$, and those on the right from stations near $\phi_m = 51°$. [After Chapman (1918) and Chapman and Bartels (1940).]

The part DS of the disturbance field associated with magnetic storms has at any given time a mean value of zero around any given latitude circle. Its main part represents a variation with longitude, measured from the midnight meridian, which is similar in form to the SD variation with local time, measured from local midnight. In fact, it is to be distinguished from SD simply because its amplitude changes with storm time, reaching a maximum in the first few hours of the storm and then decaying gradually to a low value during the next two days. Thus the representation of SD in Fig. 9.5 shows (in terms of local time at individual locations) the form and *average* amplitude of DS during the first day of a group of magnetic storms.

The disturbance–daily variation SD, the average manifestation of DS, is a very important feature of magnetic variations at high latitudes, often being quite apparent even on "quiet" days and overriding the storm-time variation during magnetic storms. The SD field is best studied in terms of geomagnetic coordinates, because distance from the geomagnetic rather than the geographic pole is an important parameter of the field. Figure 9.7 gives the SD variation at

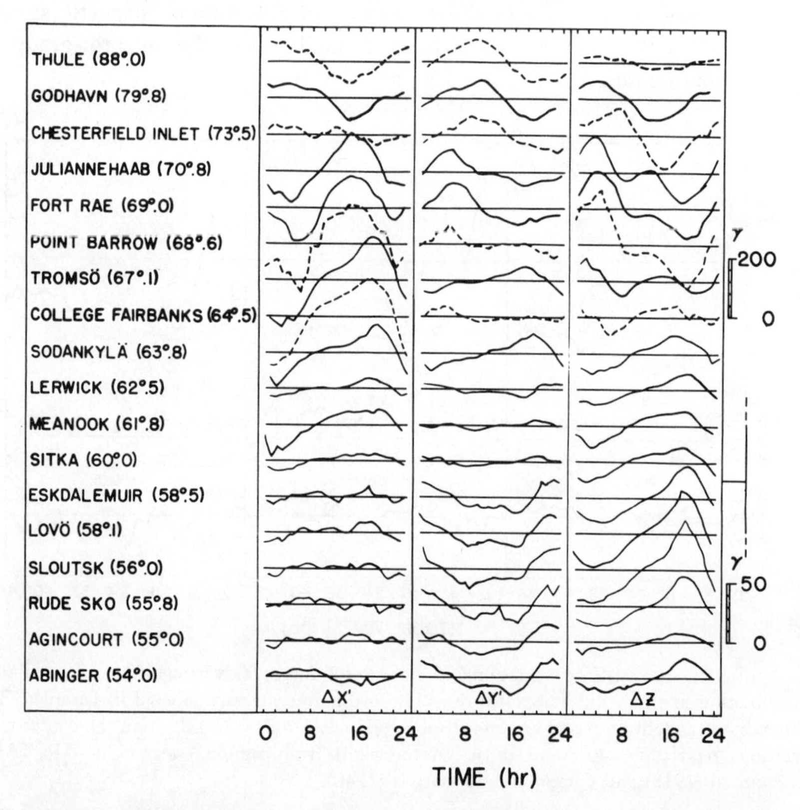

FIG. 9.7. Disturbance-daily variations ($S_d - S_q$) of the component toward geomagnetic north (X'), the component toward geomagnetic east (Y'), and the downward component. For each station, the geomagnetic latitude is given in parentheses. The abscissa is local geomagnetic time. Curves are based on 12 months of data in 1932–1933, with some minor variations of the time interval involved in the case of the dashed curves. (After Vestine *et al.*, 1947.)

several high-latitude stations as derived by Vestine *et al.* (1947). The form and amplitude of the variation change rapidly with increasing geomagnetic latitude and the phenomenon is evidently quite complex. Although largest in and near the auroral zones (near 67° geomagnetic latitude; see Section 9.3), it is still quite appreciable near the geomagnetic poles.

Other Magnetic Disturbance. This description in terms of averaged fields is an idealized one and obscures certain other features of magnetic disturbance.

Some storms are much slower in developing and are called gradual-commencement rather than sudden-commencement storms. Some run their course in a shorter or longer period of time than the average. And the variation of the magnetic elements at a given location during a given storm is much more irregular and complex than the variation implied by the averaged curves. In addition, there are certain characteristic features of magnetic disturbance that are not necessarily associated with magnetic storms and deserve special mention. These include magnetic *bays*, *pulsations*, and *crochets*.

Magnetic bays are observed on otherwise quiet days as well as during storms. The values of the elements, especially H, rapidly increase or decrease, attain maximum departure, and then return to "normal" values within an hour or two. According to whether H increases or decreases, a bay is spoken of as positive or negative. Positive bays are more frequent than negative ones. Bays are most pronounced at geomagnetic latitudes near 67°, the auroral zones, but large bays are worldwide phenomena. Several may occur during a magnetic storm and are sometimes referred to as *polar substorms*, or symbolically as DP.

Pulsations are rapid fluctuations of the magnetic elements, with periods of from a few seconds up to a few minutes. Those with periods of a few minutes occur most frequently at night. Oscillations of the different magnetic elements are not necessarily in phase.

A magnetic crochet is a small characteristic variation of the magnetic elements observed to occur in the sunlit hemisphere in connection with solar flares. The type of variation is such as to represent an increase of the variation normally associated with the Sq variation for the time, location, and element. For this reason this disturbance field is sometimes referred to as an Sq augmentation, Sqa.

9.1.4 Measures of Magnetic Disturbance; Certain Variations and Interrelationships

Measures of Magnetic Disturbance. In order to study magnetic disturbance from the point of view of its variation and its relationship to other geophysical parameters, it is desirable to have more objective measures of disturbance than that represented by the international character figure.

One of these, the u measure, is best suited to the numerical characterization of intense disturbance and, further, to its average occurrence over periods of, say, a month or a year, rather than on an individual day. It is based on the interdiurnal variability of H; at a given station and on a given day, the value of u is proportional to the difference (without regard to sign) between the average value of H for that day and the average value of H for the preceding day. As we saw in Subsection 9.1.3, one of the principal systematic effects of intense disturbance is a variation of H lasting for a few days; therefore a large average value of u for a month or longer signifies the frequent occurrence of magnetic storms. Actually a modified parameter u_1 is often used; u_1 increases with u, but not linearly and such that the effect of particularly intense disturbance manifests itself less in terms of u_1 than u.

Another parameter, more suited to the characterization of magnetic activity during short time periods, is the K index. For each three-hour period during the (Greenwich) day, each observatory characterizes its record by an integer running from 0 to 9 inclusive and based on the observed range of variation of the magnetic elements during the three-hour period. The K values are related to the observed range of variation in such a way that each value of K occurs with approximately the same frequency over a long period of time. A planetary three-hour index K_p is determined for each three-hour period by averaging the K values from certain stations, suitably weighted according to the average amount of disturbance at each station. The sum of the eight planetary three-hour indices is used to characterize an entire day in terms of the planetary daily indices A_p or C_p.

The annual variation of magnetic disturbance (as measured for example by the u index) shows two maxima, one in March and one in October. These maxima may be related to the geometry of the sun-earth system. As will be mentioned again in Subsection 9.1.5, magnetic disturbance is believed to be related to the interaction between the earth's upper atmosphere and streams of charged particles, especially electrons and protons, emitted from active regions on the sun. Near the times of the equinoxes, the relative orientation of the sun and earth is favorable for the earth to intercept such streams. However, recent measurements from space probes have cast doubt on this explanation, and the cause may be in the upper atmosphere itself.

Magnetic Disturbance and the Sun. It has long been known that there is a marked variation of the frequency of great magnetic storms over the sunspot cycle. Figure 9.8 shows the variation of the u_1 index with sunspot number and

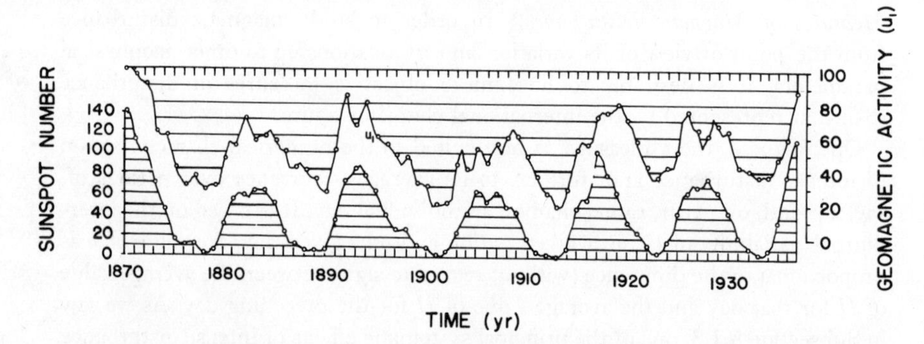

FIG. 9.8. Variations of annual means of sunspot number and of the geomagnetic index u_1. (After Chapman and Bartels, 1940.)

reveals a striking correlation. However, this correlation, impressive as it appears, must be interpreted carefully. In the first place, it does not imply a close connection between sunspot number and magnetic disturbance on particular days; evidently the causative solar feature is statistically connected with the occurrence of sunspots, but the presence of a large spot group or groups on the sun is neither a necessary nor a sufficient condition for magnetic disturbance. In the second place, it must not be thought that magnetic disturbance is nearly absent during sunspot-minimum years. Rather there is a change in the character of disturbance, such that intense storms of the sudden-commencement type (to which the u_1 measure is particularly sensitive; see above) are less frequent during sunspot minimum.

The general connection between solar activity and magnetic disturbance is revealed in another way by the 27-day recurrence phenomenon. This is a statistical tendency for disturbed or quiet magnetic days to recur after a 27-day period, presumably because this is the period of rotation of the sun (in low solar latitudes as seen from the earth). This tendency is particularly well marked in connection with the weaker magnetic storms that predominate during sunspot-minimum years.

Some, but by no means all, magnetic disturbances, and especially the most intense ones, occur in connection with a characteristic sequence of upper-atmosphere events accompanying particularly pronounced solar flares. These events include the magnetic crochets and radio fadeouts (see Section 9.2) which are believed to signal the arrival in the upper atmosphere of enhanced solar ultraviolet or X-ray radiation. Magnetic storms of the violent sudden-commencement type often follow by about a day, accompanied by especially strong auroral displays. The time lag is believed to reflect a period during which streams of charged particles are traveling from sun to earth. Interrelationships among magnetic disturbance, ionospheric disturbance, aurora, and solar activity are discussed further in Sections 9.2 and 9.3.

On the other hand, some magnetic disturbances, and especially the weaker disturbances associated with sunspot minimum and with 27-day recurrence, have no apparent correlation with any particular feature of solar disturbance. These are still usually ascribed to emission of charged particles from the sun, but from unidentified solar regions which are referred to as *M regions*, according to a suggestion of Bartels.

9.1.5 Geomagnetic Variations and Electric Currents in the Upper Atmosphere

Geomagnetic Variations and the Upper Atmosphere. Thus far, our discussion of the transient geomagnetic variations has been conducted from a purely descriptive point of view. However, these rather small perturbations on the earth's main field would have no interest here were it not for the fact that their causes lie in the earth's upper atmosphere. The processes that give rise to the magnetic variations are not understood in complete detail, but they have been identified in a general way and studied extensively. The wealth of geomagnetic data gathered in the past, present, and future from measurements at the earth's surface must be looked upon as geophysical information of great importance for studies of the upper atmosphere.

The first suggestion that geomagnetic variations might have their cause in the upper atmosphere was due to Stewart (1882), as mentioned in the introduction to Chapter 8. He was motivated primarily by the observation that the solar daily variation is much larger near sunspot maximum than near sunspot minimum (Subsection 9.1.2) and argued that such a direct linkage with solar activity must have its origin in the high atmosphere. He suggested that the magnetic variations are caused by electric currents flowing in the upper atmosphere. He further suggested that these currents are caused by periodic air motions which, in the presence of the earth's magnetic field, produce electric fields and cause currents to flow, as in a dynamo. This inference of a conducting layer in the upper atmosphere preceded by about 20 years the suggestion of an ionosphere to explain radio transmission, and by about 45 years a clear-cut experimental verification of its presence by radio-sounding methods.

Stewart's suggestion that periodic air motions cause the electric currents responsible for magnetic variations has come to be known as the *dynamo theory*. It is now accepted as the explanation of the solar and lunar daily variations. However, before discussing the dynamo theory itself, let us consider the more general question of an association between magnetic variations at the earth's surface and current systems in the upper atmosphere, whatever the cause of the current systems.

Internal and External Causes of the Earth's Magnetic Field. The magnetic intensity **H** near the earth's surface, in the essential absence of electric currents passing between ground and atmosphere, can be described by a scalar potential Φ, such that

$$\mathbf{H} = -\nabla\Phi \tag{9.3}$$

Thus, for example, the earth's main field, to the extent that it resembles the field of a dipole at the earth's center (Subsection 9.1.1), can be derived from the potential

$$\Phi = -M \cos \theta_m / r^2 \tag{9.4}$$

where M is the magnitude of the dipole moment (8.1×10^{25} gauss cm^3), θ_m is geomagnetic colatitude, and r is distance from the center of the earth.

More generally, if the matter which gives rise to the magnetic field near the earth's surface is contained entirely beneath the earth's surface, then the potential near the surface can be expressed as a sum of terms of the form

$$(r)^{-(n+1)}P_n{}^m(\cos\theta)(c_n{}^m \cos m\lambda + s_n{}^m \sin m\lambda) \tag{9.5}$$

where $P_n{}^m$ is an associated Legendre function (Appendix J). The coefficients of the individual terms in the series can be determined by spherical harmonic analysis (Appendix J) of the observed field. The first (and largest) term in such a series ($n = 1$, $m = 0$) corresponds to the idealized dipole field.

On the other hand, if the matter which gives rise to the magnetic field is exterior to the earth's surface, for example, in the upper atmosphere, then the appropriate expression for Φ is a sum of terms of the form

$$r^n P_n{}^m(\cos\theta)(c_n{}^m \cos m\lambda + s_n{}^m \sin m\lambda) \tag{9.6}$$

If causes both interior and exterior to the surface are involved, then a linear combination of (9.5) and (9.6) is required. By spherical harmonic analysis of the observed field it is possible to determine what fraction of the field is of internal origin, represented by terms of the type (9.5), and what fraction is of external origin, represented by terms of the type (9.6).

In 1839 Gauss used this method to determine that the origin of the main field is wholly (or almost so) within the earth. In 1889, Schuster, investigating Stewart's suggestion, applied Gauss's method to scanty data on the field of the solar daily variation S and concluded that the causes of this field lie principally above the earth's surface. As a result of additional investigations of this type (see especially, Chapman, 1919) it became clear that both the solar and lunar variations are mainly of external origin. The small parts of these fields that are of internal origin can be explained as a result of currents induced in the solid earth by the primary currents flowing in the outer atmosphere. It is now known that the disturbance variations also have an external origin.

Transient Variations and Current Systems in the Upper Atmosphere. From spherical harmonic analysis of any part of the earth's magnetic field (whether it be Sq, L, or any part of the D field), it is possible to deduce the currents which, flowing in a thin spherical shell at a certain distance from the center of the earth, could give rise to that field [see Chapman and Bartels (1940), especially pp. 630–631]. This procedure in no way reveals the causes of such currents, the altitudes at which they flow, or indeed whether the real currents do flow in a spherical shell at one altitude. Nevertheless, such idealized current systems synthesize the geomagnetic data in a convenient form for testing various theories and also furnish a very useful conceptual aid for visualizing the form of the field involved. As an example, current systems corresponding to the Sq field under some circumstances are shown in Fig. 9.9. The direction of current flow is shown by the arrows; the amount of current is shown by the spacing.

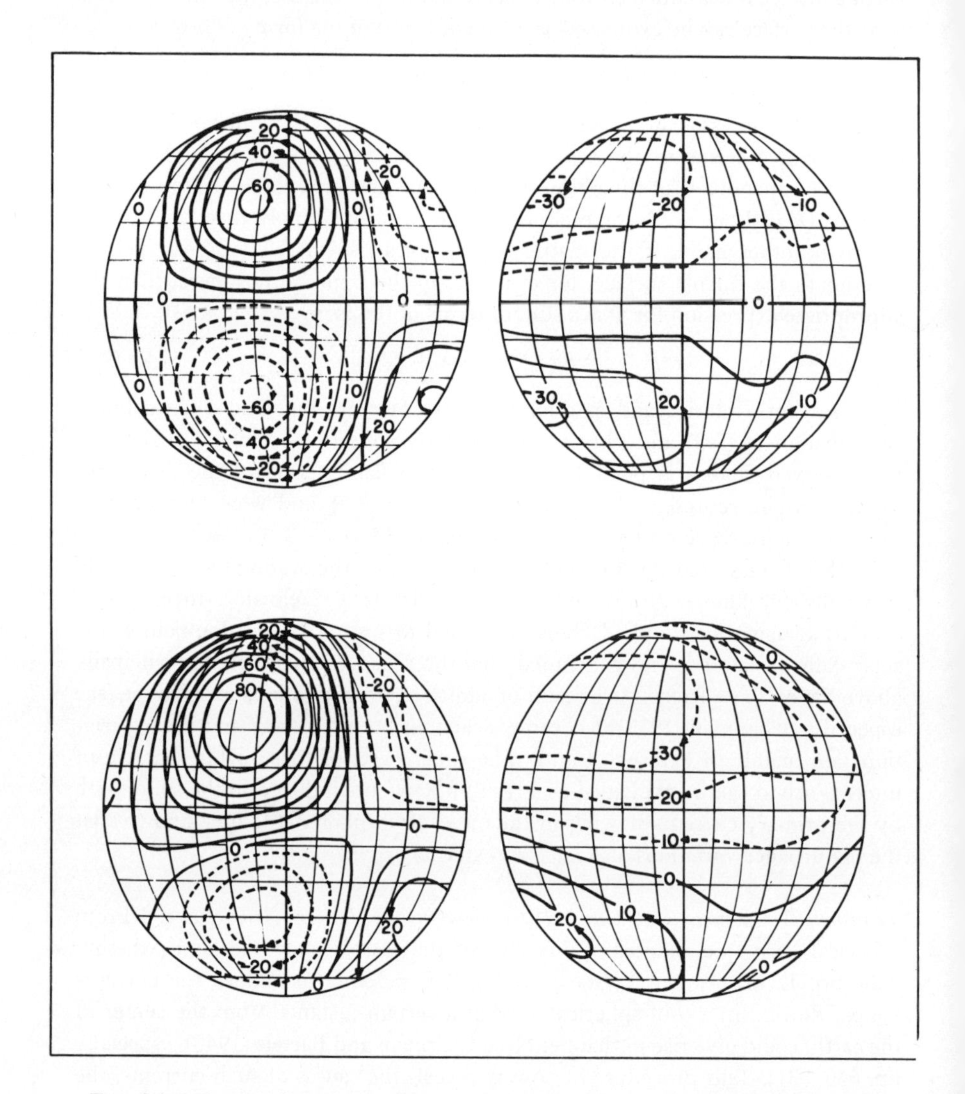

FIG. 9.9. Overhead current systems representing Sq: over the day hemisphere (left) and the night hemisphere (right), for the equinoxes (top) and for the northern summer solstice (bottom) at sunspot minimum. A current of 10,000 amperes flows between consecutive current lines in the direction indicated by the arrows. (After Chapman and Bartels, 1940.)

To a good approximation, the direction of the horizontal component of the field caused by these currents at a point below the current shell is perpendicular to the lines of current flow and toward higher values of the current function as labeled. Thus, for example, the northward component of the Sq field during the hours around local noon is positive between about 40°N and 40°S and negative poleward of those latitudes (compare with Fig. 9.3). Current systems corresponding to other variation fields have been computed but are not shown here (see general references cited at the beginning of this section).

The Dynamo Theory. Turning now to the dynamo theory of the causes of the Sq and L currents, we find that the situation is rather complicated and we shall not try to cover it in any detail. Useful references, in addition to the general ones cited at the beginning of this section, are Baker (1953), Baker and Martyn (1953), Fejer (1953), and Chapman (1956). The basic idea is that the motion of air in the upper atmosphere, in the presence of the earth's main field, gives rise to an electric field **E** in a direction perpendicular to both the magnetic field **H** and the direction of motion. In the case of Sq and L, these motions are of tidal origin. Free electrons and ions move in response to the electric field **E**, subject to collisions with the neutral particles, and their motion constitutes an electric current. The situation is complicated by the fact that the current in the direction of **E**, since it moves transverse to **H**, sets up an electromotive force in a direction perpendicular to both **E** and **H**, and a current flows in that direction also. The latter is known as the *Hall current.* The Hall current in turn modifies the current in the **E** direction. The electrical conductivity in the direction of **E**, taking account of both collisions and the Hall current, is known as the *Pedersen conductivity*, and the conductivity in the direction perpendicular to **E** and **H** is known as the *Hall conductivity*. Both depend on collision frequency and the number densities of free electrons, positive ions, and negative ions, and are functions of altitude, time, and position.

The situation is further complicated by polarization effects. Current can flow in the vertical direction only temporarily, because the conducting ionosphere rests on an effectively insulating base. Furthermore, the horizontal currents, viewed on a worldwide basis, are modified by polarization effects in such a way as to provide closed current circuits.

The Sq and L anomalies near the magnetic equator (see references to Huancayo in Subsection 9.1.2) can be explained in terms of a particularly intense current, the *equatorial electrojet*. Its presence can be explained, in a rather complicated way, from a consideration of the conductivities described above in a region where the main magnetic field is horizontal.

The equivalent spherical-sheet currents, which can be deduced from the magnetic variations, represent the integrated effects of dynamo currents wherever they may flow in the atmosphere. Rocket measurements seem to have established that the Sq currents flow between about 95 and 120 km. If the air motions that give rise to the dynamo currents were known accurately (and also the electron and ion densities and the collision frequencies) as a function of altitude, time, and position, then the net currents and their magnetic effects could be calculated. On the other hand, unfortunately, it appears to be impossible to infer the tidal winds from the magnetic variations, owing mainly to the fact that the magnetic variations reveal only the net integrated effect of the winds.

Magnetic Disturbance and Current Systems. Transient variations associated with magnetic disturbance can also be represented in terms of equivalent currents flowing in the upper atmosphere (see, for example, Chapman, 1963). In this case, however, there is much more uncertainty about the vertical distributions and causes of the currents that actually do flow.

The form of the Dst currents is quite simple, corresponding to the rather simple character of the Dst field, which consists mainly of changes in the magnitude of the northward intensity. During the initial phase of a storm, when the northward intensity increases, the currents flow in an eastward direction and encircle the earth. They are generally believed to flow at very great altitudes, perhaps 10 earth's radii, as a result of interaction between the earth's magnetic field and a stream of neutral but ionized gas from the sun. Further interaction leads to a westward current, the *ring current*, which probably flows at lower levels, and which produces the observed decrease in northward intensity during the main phase of the storm. The exact details of the interaction between the earth's outermost atmosphere and the solar plasma are not well understood and are the subject of very intensive investigation at the present time. Although much can be said about the subject, it lies well outside the scope of this book. (See, however, the brief discussion at the end of this chapter.)

The DS currents flow mainly in the auroral zones, partly eastward and partly westward, completing their circuits partly across the polar cap and partly in lower latitudes. They are much stronger than the Sq and L currents, at least during well-developed magnetic storms, and, contrary to the latter, are not noticeably stronger over the day hemisphere than over the night hemisphere. Their origin too is connected with the solar corpuscular flux and its effects on the upper atmosphere.

9.2 An Introduction to the Ionosphere

The expression *ionosphere* refers to the part of the earth's atmosphere that contains ions and free electrons in sufficient quantities to affect the propagation of radio waves. In practice, this means that it has its lower limit at about 50 to 70 km and no distinct upper limit. However, in keeping with the limitations of this book, we shall confine our discussion primarily to levels below the altitude of peak electron density, roughly 300 km.

It used to be thought that the ionosphere had a layered structure, with distinct maxima of electron density occurring in layers called (in order of increasing altitude) the *D* layer, the *E* layer, the *F*1 layer (in the daytime only), and the *F*2 layer. Rocket observations have revealed that this is not so; electron density increases monotonically (or nearly so) from the lowest levels to the *F*2 peak near 300 km. The terms *D region*, *E region*, and *F region* are now used to refer to altitude regions, respectively, below 90 km, between 90 km and 160 km, and above 160 km (although the altitudes dividing the regions are rather arbitrary and are not always taken to have exactly these values).

The existence of a conducting region in the earth's upper atmosphere was a subject of early speculation by Lord Kelvin (see Chalmers, 1962) and was invoked by Stewart (see Section 9.1) in connection with the daily magnetic variations. However, it was Marconi's demonstration of long-distance radio communication that stimulated widespread studies of the phenomenon. Kennelly (1902) and Heaviside (1902) independently postulated an ionized layer to explain the radio transmission, at a time when others were investigating diffraction effects as the explanation. In 1912, Eccles supplied the rudiments of the complicated theory of the transmission of a radio wave through an ionized medium. The problem was taken up again by Larmor (1924), whose work still gives an elementary, first-approximation explanation of the reflection of radio waves from the ionized regions.

It remained for Appleton and Barnett (1925a, b) and Breit and Tuve (1926) to demonstrate the existence of a reflected wave by detecting it experimentally. Since then theoretical and experimental studies of the ionosphere have been pursued at an accelerating pace until now a large volume would be necessary to do justice to the work. The subject not only is of great interest to students of the atmosphere but also has immense practical significance in the field of long-range communication. Here, we shall naturally enough emphasize the first topic in this short introduction. Various aspects of ionospheric studies have been reviewed by Mitra (1952), Rawer (1957), Ratcliffe and Weekes (1960), F. S. Johnson (1962), Bourdeau (1963), and MacDonald (1963), among others.

The present discussion begins with a consideration in Subsection 9.2.1 of the physical processes that govern the verical distribution of ions and electrons in the upper atmosphere: ionization, recombination by one means or another, and movement effects. Subsection 9.2.2 reviews the observational methods that are in use to study the ionosphere, both from the earth's surface and from instrumented vehicles. A discussion of some of the most important results of these studies—including some still unsolved problems—follows. Although this discussion could be organized in various ways, the most convenient for this survey is to consider the *D*, *E*, and *F* regions separately in Subsections 9.2.3, 9.2.4, and 9.2.5.

9.2.1 Processes Governing the Vertical Distribution of Ionization

Ionization of upper-atmospheric gases is accomplished largely by electromagnetic radiation from the sun, although we shall have occasion to refer to the effects of charged particles in the auroral zones and of cosmic rays. The number of electrons and ions actually present at any given time and place, however, depends also on the processes and rates by which the electrons and ions recombine into neutral particles. The net effect of ionization and recombination may at times be understood simply in terms of an equilibrium, within a given volume, between the two processes. For more detailed considerations one must take into account departures from equilibrium and the effects of electron transport due to diffusion and motions of various kinds. The present subsection discusses briefly all these topics.

The Production of Ionization. The classical paper on the ionizing effect of solar radiation is that of Chapman (1931a). This theory was discussed in Section 4.4 in the more general context of absorption of solar radiation. If, however, one considers the amount of energy absorbed per unit time and volume to be proportional to the number of photoionizations per unit time and volume, then the results derived there are applicable to the production of ionization. In particular, (4.44) gives the level of maximum production and (4.46) or (4.48) gives the vertical distribution of electron production for the assumptions of the Chapman theory. Equations (4.49) and (4.50) refer to an atmosphere with a (constant) scale-height gradient.

The ionosphere does not show distinct layers of electrons such as would follow if this simple theory accurately described the production of electrons and if electron concentration were actually directly related to the rate of production (see Fig. 4.11). This is due to several factors, including one that the ionizing radiation is by no means monochromatic and another that the recombination processes are complicated and vary with altitude. Nevertheless the theory of a *Chapman layer* has played a key role in studies of the ionosphere and anybody approaching the problems of the ionosphere for the first time would do well to be thoroughly familiar with it.

In discussions of the individual regions *D*, *E*, and *F* (in Subsections 9.2.3, 9.2.4, and 9.2.5) we shall refer to what is known and conjectured about the particular wavelengths of ionizing radiations and the particular atmospheric constituents that are ionized in each of the regions.

Loss Processes. The problem of recombination is a complex one. The simplest assumption, used by Chapman, is that recombination takes place in the same volume as the ionization and between the types of particles (electron and positive ion) previously dissociated.

Thus if $n(\mathrm{e})$ is the number of electrons per unit volume,

$$\frac{\partial n(\mathrm{e})}{\partial t} = q - \alpha n^2(\mathrm{e}) \tag{9.7}$$

where α is called the *recombination coefficient.* Chapman proceeded to integrate this equation by numerical methods to find $n(\mathrm{e})$ as a function of altitude and of the factors that govern the time change of zenith angle and therefore of q(time of day, latitude, and season).

The simplest process that would give rise to an equation like (9.7) is radiative combination,

$$M^+ + e \rightarrow M + h\nu \tag{9.8}$$

where M^+ is any atomic or molecular ion. If this were the only process operating, inasmuch as the ionosphere is electrically neutral, $n(\mathrm{e}) = n(\mathrm{M}^+)$ and recombination would proceed at the rate $\alpha n^2(\mathrm{e})$, with α the rate coefficient for radiative combination. An alternative is three-body combination [Eq. (4.9c)], in which case α would be proportional to the total number density of neutral particles and decrease with altitude.

Actually, neither of these processes is rapid enough to explain observed rates of recombination in the ionosphere. The question of recombination was explored in detail by Bates and Massey (1946, 1947) and their conclusions, although at the time based on somewhat uncertain knowledge of some of the rate coefficients and of atmospheric composition, are in the main still acceptable.

At high enough levels where attachment processes can be neglected (see below), the predominant process is dissociative recombination, Eq. (4.16). The most important examples are

$$N_2^+ + e \rightarrow N + N \tag{9.9a}$$

$$O_2^+ + e \rightarrow O + O \tag{9.9b}$$

$$NO^+ + e \rightarrow N + O \tag{9.9c}$$

When the ionized constituent is atomic, the most important example being atomic oxygen, this process is not, of course, directly applicable. A molecular ion must first be produced by charge transfer or ion-atom interchange, the most important examples of which are

$$O^+ + O_2 \rightarrow O + O_2^+ \tag{9.10a}$$

$$O^+ + N_2 \rightarrow NO^+ + N \tag{9.10b}$$

These and additional processes have been discussed in considerable detail by Nicolet (1963, section 6) and by Nicolet and Swider (1963), and the complete picture is more complicated than presented here. However, at least in the upper part of the E region and in the F region up to the $F2$ peak, ionization of molecular nitrogen (or molecular oxygen) is followed rather rapidly by dissociative recombination, which is described by the recombination equation (9.7) (but with α at least three orders of magnitude larger than would be appropriate for radiative recombination). In the case of the ionization of atomic oxygen, which is followed by a sequence of ion-atom interchange (or charge transfer) and dissociative recombination, the situation is more complicated. At low enough levels, ion-atom interchange takes place very rapidly and the rate of recombination depends on the rate of the dissociative recombination process. In this case, (9.7) still has the proper form. However, at high enough levels, where molecules are scarce, the formation of molecular ions is a slow process and the rate at which it proceeds is the limiting one in the over-all recombination process. In this case the loss of electrons is proportional to the product $n(O^+)n(M)$, or to $n(c)n(M)$, where M in this case represents a molecule. Thus the loss term in the equation for $\partial n(e)/\partial t$ is proportional not to $n^2(e)$ but to $n(e)$ and with a coefficient that is proportional to $n(M)$ and thus decreases with altitude. This rapid decrease with altitude of the effective recombination coefficient can be expected to lead to a maximum electron density well above the level of maximum electron production, and is usually invoked to explain the rather high altitude of the *F*2 peak, as originally suggested by Bradbury (1938).

Another mechanism for the disappearance of free electrons, which is certainly important in the *D* region and perhaps in the lower part of the *E* region, is radiative or three-body *attachment* to neutral particles to form negative ions,

$$e + M \rightarrow M^- + h\nu \tag{9.11a}$$

$$e + M + L \rightarrow M^- + L \tag{9.11b}$$

In the atmosphere, M may be O or O_2 (but not N or N_2). These negative ions, once formed, can be destroyed by a number of processes, including photodetachment ($M^- + h\nu \rightarrow M + e$) and neutralization ($M^- + L^+ \rightarrow M + L$). Nicolet (1963) and Nicolet and Swider (1963) have also discussed these processes in considerable detail. The net result of the formative and destructive processes is the presence of a certain number of negative ions. Although carrying a negative charge, the negative ion has so much greater inertia than a free electron, that its effect on radio propagation is unimportant.

A Layer in Equilibrium. Having discussed electron production and some processes that lead to recombination, we now turn to the question of the resulting electron concentration. Fixing attention on a volume at some fixed position, we can write the equation of continuity for electrons as

$$\partial N/\partial t = q - L - \operatorname{div}(N\mathbf{v}) \tag{9.12}$$

Here N refers to the number of electrons per unit volume, previously referred to as $n(e)$; q is the rate of production of electrons within the volume; L is the rate of loss of electrons within the volume; and the divergence term allows for the gain or loss of electrons by the volume as the result of various transport phenomena, **v** being the mean drift velocity of the electrons.

The simplest case would occur if the divergence term and time-change term could be neglected, in which case there would exist an equilibrium between electron production and electron loss so that $q = L$. If the loss process is dissociative recombination, then according to the discussion above, $L = \alpha N^2$ so that

$$N = (q/\alpha)^{\frac{1}{2}} \tag{9.13}$$

On the other hand, if the loss results from attachment [Eq. (9.11)] or from the charge transfer–dissociative recombination sequence in the case where the charge transfer process is slow, then $L = \beta N$ and

$$N = q/\beta \tag{9.14}$$

Since q varies as cos Z, according to Eqs. (4.44) and (4.46), an important question in ionospheric studies has been whether N, referring to a particular volume, varies as $(\cos Z)^{\frac{1}{2}}$ according to (9.13) or as $(\cos Z)$ according to (9.14).

In earlier studies of the ionosphere, it was not possible to measure N for a particular volume and study its variation with time. However, as discussed further in Subsection 9.2.2, there are large numbers of experimental data referring to quantities N_m, which were previously thought of as representing distinct local maxima of electron density in a vertical column, corresponding to well-defined "layers" of electrons. Except at the $F2$ peak, these correspond instead to levels of small vertical gradient of N ("ledges") or perhaps to poorly defined maxima of N. Nevertheless, because N_m is so easily measured, it is useful to consider the application of (9.12) in terms of this quantity, which we may speak of as representing a local maximum of electron density (in the vertical).

Small Departures from Equilibrium. Let us continue to neglect the divergence term and consider the effects of small departures from equilibrium. One effect of this is that the altitudes of maximum electron production and of maximum electron concentration do not coincide, but are separated by a small distance a. If a is expressed in terms of the height variable z_2 [defined by Eq. (4.47)], then Appleton and Lyon (1955) showed that, approximately,

$$dN_m/dt = q_m(0) \cos Z(1 - a^2/2) - \alpha N_m^2 \tag{9.15}$$

In practice, since a is small, the quantity $a^2/2$ may often be neglected, especially near local noon, so that Eq. (9.15) becomes identical in form with (9.12) (if we neglect the divergence term) but is applicable now to N_m. Appleton and Lyon (1961) have used this equation in studies of the E region, replacing N_m by the average value of N_m for two times, both having the same zenith angle, one before and one after local noon.

Another effect of a departure from equilibrium is a slight asymmetry in the diurnal curve of N_m, maximum N_m occurring a short time Δt after the noontime maximum of electron production. Appleton (1953) showed that, approximately,

$$\Delta t = 1/2\alpha N_m \tag{9.16}$$

and termed this quantity the "sluggishness." This relationship gives, in principle, a method of estimating α from ionospheric observations.

The Divergence Term. We have thus far neglected the divergence term in Eq. (9.12). It is customary in the consideration of this term to assume that vertical gradients are much larger than horizontal. It is important to remember that this term does not necessarily represent the divergence of all components of the air, but only of electrons. Differential drift of electrons might arise, for example, from diffusion or in response to an electric field. We shall return to the consideration of these possibilities later. Here, we shall simply note a simplified expression for the effect of vertical drift on the value of N_m in a Chapman layer. According to Appleton and Lyon (1955, 1957), the fractional change in N_m resulting from a vertical drift of electrons w is

$$\frac{\Delta N_m}{N_m} = -\frac{1}{2\alpha N_m}\frac{\partial w}{\partial z} - \frac{1}{4H^2}\left(\frac{w}{2\alpha N_m}\right)^2 \tag{9.17}$$

The change of the altitude of maximum electron concentration under the same conditions is

$$\Delta z = w/2\alpha N_m \tag{9.18}$$

9.2.2 Methods of Observing the Ionosphere

Most of our knowledge of the ionosphere has been obtained from observations of radio waves propagated through it. A large amount of this information stems from ground-based measurements of radio waves that have been reflected and returned to the surface. In recent years, the availability of rockets and satellites has made it possible to study propagation over one-way paths between ground and vehicle. In either case, interpretation of the measured quantities in terms of ionospheric structure involves the complicated theory of electromagnetic-wave propagation through an ionized gaseous medium in the presence of the earth's magnetic field. We shall first discuss this theory and its application to radio-propagation measurements, and then turn to some direct observations made with the aid of rockets and satellites.

Elements of the Magneto-Ionic Theory. Many of the phenomena of radio-wave propagation in the ionosphere can be described in terms of the *magneto-ionic theory*, defined by Ratcliffe (1959) to be the theory of "electro-magnetic waves passing through a gas of neutral molecules in which is embedded a statistically homogeneous mixture of free electrons and neutralizing heavy positive ions, in the presence of an imposed uniform magnetic field." The ionosphere is not "homogenous." The parameters important for radio propagation (electron density and collision frequency) are variable, particularly in the vertical direction. Nevertheless, in many ionospheric applications, the variations can be considered to be slow enough so that the wave behaves at each level as if it were in a homogeneous medium having the properties appropriate to that level. In such a slowly varying medium, under some circumstances, the reflection of a radio wave can be understood in terms of a simple ray treatment. The ultimate justification for this procedure rests, of course, on the consistency of the results it gives with the results based on a more complete theory, often spoken of as *full-wave theory*. However, here the magneto-ionic theory and the ray treatment are emphasized. The magneto-ionic theory is discussed in detail in Ratcliffe's excellent book (1959). Budden (1961) has given a discussion of the more complete theory.

When an electromagnetic wave passes through the ionosphere, the free electrons in the medium oscillate in response to the electric field of the wave. The motions of the positive ions, which are much less mobile than the electrons, are less important and usually may be neglected. The magnetic field of the wave has no important effect on the electronic motions, but the earth's magnetic field does. The situation is further complicated by the effect of collisions between the free electrons and the neutral particles of the medium; the collisions effectively reduce the energy of the electromagnetic wave in a process called *absorption*.

The resulting effect on the behavior of a wave can be calculated by the magneto-ionic theory. These calculations refer to a simple-harmonic, plane, progressive wave of angular frequency ω. In the simplest case, when collisions and the effects of the earth's magnetic field can be neglected, the refractive index describing the phase velocity of such a wave is given by

$$n^2 = 1 - X \tag{9.19}$$

where $X = 4\pi Ne^2/m\omega^2$, N is the number density of electrons, e is the electronic charge in electrostatic units, and m is the mass of the electron in grams.

In the case where collisions are important, the refractive index is complex, corresponding to the damping effect mentioned above, and is given by

$$n^2 = 1 - \frac{X}{1 - iZ} \tag{9.20}$$

where $Z = \nu/\omega$ and ν is the collision frequency.

The presence of the earth's magnetic field imposes a considerable additional complication. The behavior of the wave can usually be described in terms of two "characteristic" waves, each of which can be thought of as propagating independently through the medium with its own wave polarization and its own complex refractive index. These refractive indices are given by the Appleton–Hartree formula (Hartree, 1931; Appleton, 1932), which is derived and discussed in detail by Ratcliffe (1959). Nothing much would be gained by reproducing it here. It turns out that one of the characteristic waves, the *ordinary wave,* behaves very much as if the magnetic field were not present and can usually be described by (9.19) or (9.20). The other, the *extraordinary wave,* can give additional important information, but is usually ignored here in this elementary discussion.

Reflection of Vertically Incident Pulses. In the application of the magneto-ionic theory to the ionosphere, one supposes that the ionosphere is stratified in horizontal planes, the variation with altitude of electron density and collision frequency being so slow that the reflection of radio waves can be interpreted in terms of a simple ray theory. Although experiments can be conducted at oblique incidence, the usual experimental technique of sounding the ionosphere involves a pulse of energy transmitted nearly vertically and detected at the ground after reflection by the ionosphere.

Such a pulse of energy can be thought of as being made up of an infinite number of plane, harmonic waves of varying frequency centered at a dominant frequency ω. The refractive index μ [which is equal to n when n is real as in Eq. (9.19) or to the real part of n when n is complex as in Eq. (9.20)] describes the phase velocity of each wave and varies with frequency, the ionosphere being a dispersive medium. It is well known from the theory of wave motion that such a pulse travels with the group velocity V_g and group refractive index $\mu'(= c/V_g)$ described by

$$\mu' = \mu + \omega(d\mu/d\omega) \tag{9.21}$$

In the case of no magnetic field and no collisions, (9.21) applied to (9.19) gives

$$\mu' = (1/\mu) = (1 - \omega_N^2/\omega^2)^{-\frac{1}{2}} \tag{9.22}$$

where $\omega_N^2 = 4\pi Ne^2/m$ and ω_N is called the angular *plasma frequency.* At the level where $\omega = \omega_N$, the group velocity reaches zero and reflection occurs. In the case of an applied magnetic field (but without collisions), the group velocity still goes to zero for the ordinary wave when $\omega = \omega_N$. When the effect of collisions is considered, this is still closely correct if Z is small near the level where $\omega = \omega_N$. For frequencies greater than 1 or 2 Mc sec^{-1}, reflection occurs at altitudes above 100 km. For these frequencies and for values of collision frequency appropriate to these altitudes, Z is indeed small and (9.22) can still be applied for the ordinary wave.

Thus, the idea that a pulse of energy is reflected at the level where $4\pi Ne^2/m = \omega^2$ has rather wide applicability. This gives a relationship between the frequency of the reflected wave and the electron density N at the level of reflection. This relationship is usually expressed in terms of the frequency f rather than the angular frequency ω, where $f = \omega/2\pi$. In terms of f,

$$N = (\pi m/e^2)f^2 = 1.24 \times 10^{-8} f^2 \tag{9.23}$$

where the numerical value applies if N is in (centimeter) $^{-3}$ and f is in cycles (second)$^{-1}$.

Soundings of the Ionosphere. Suppose that in a region of the ionosphere N increases monotonically with z to a maximum value N_m at altitude z_m, above which N decreases (as in a Chapman layer). Then as the frequency of the wave is increased the altitude of reflection increases until there is reached a value of f (say f_0) for which N_m is just sufficient for reflection. For higher frequencies, no reflection occurs. The quantity f_0 is called the *critical frequency* or *penetration frequency* of the layer and is usually related to N_m by (9.23).

Next, suppose that the time t, required for the pulse to rise to its level of reflection and return to the ground, is measured. If h_1 is the altitude where reflection occurs, then clearly

$$t = 2 \int_0^{h_1} \frac{dz}{V_g} \tag{9.24}$$

It is customary to express t in terms of an *equivalent height* h', which is the height from which the pulse would have been reflected if it had traveled at all levels with the free-space velocity c. Then $t = 2h'/c$ and the equivalent height is related to the group refractive index (which depends on frequency f) by

$$h' = \int_0^{h_1} \mu' \, dz \tag{9.25}$$

If t and hence h' are measured as a function of frequency, the graph of h' vs. f is called an $h'(f)$ curve or an *ionogram*. The instrumentation which measures h' as a function of frequency is called an *ionosonde*. It must be emphasized that h' is not the geometric height of reflection, but exceeds it.

Figure 9.10 shows some examples of $h'(f)$ curves, after Ratcliffe and Weekes (1960). The critical frequencies show up as discontinuities in the curve and are typically observed at two or more frequencies. We shall discuss the significance of these critical frequencies in more detail in Subsection 9.2.3.

An ionogram contains much more information than simply a display of the critical frequencies. In principle, it contains enough information to calculate the variation of electron density with actual height z, although this is by no means a trivial problem. In the case of no magnetic field and no collisions, the variation of μ' with frequency takes the simple form of (9.22) which, when substituted into (9.25), leads to an integral equation with an analytical solution. For the real atmosphere, information about the extraordinary wave is used and numerical methods of solution are required [see, for example, Budden (1961), pp. 160–166].

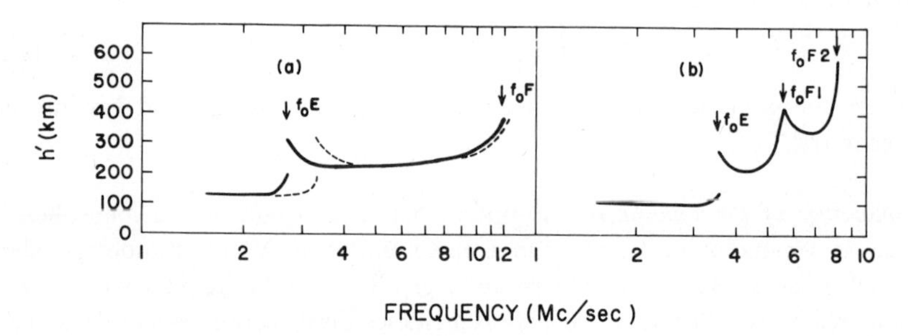

Fig. 9.10. Examples of ionograms. In (a) the dashed curve refers to the extraordinary wave. In (b) the critical frequency of the $F1$ layer is observed. (After Ratcliffe and Weekes, 1960.)

Development of such methods and routine publication by some observatories of $N(z)$ curves derived from $h'(f)$ curves represent a very important advance of recent years in ionospheric studies.

One of the most important facts revealed by the computation of $N(z)$ curves has already been mentioned in this discussion. This is that the value of N_m corresponding to a critical frequency does not necessarily represent a distinct maximum in the $N(z)$ curve. For example, Fig. 9.11 (after Ratcliffe and Weekes, 1960) shows the $N(z)$ curves corresponding to the $h'(f)$ curves of Fig. 9.10. The critical frequencies marked f_0E in Fig. 9.10a and f_0E, f_0F1 in Fig. 9.10b

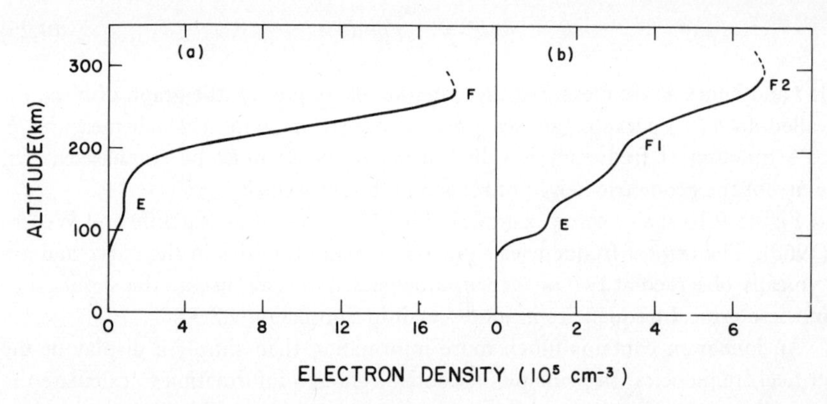

Fig. 9.11. Variations of electron density with height corresponding to the ionograms shown in Fig. 9.10. (After Ratcliffe and Weekes, 1960.)

correspond to rather indistinct maxima or simply ledges in the $N(z)$ curves. On the other hand, the F (or $F2$) critical frequency does correspond to a true maximum of electron density.

Observations of radio waves reflected from the ionosphere, besides giving delay times and hence equivalent height, are also sometimes analyzed for absorption effects. The general expression for the equivalent absorption coefficient, with consideration of the effects of the magnetic field, is quite complicated, and it is customary to employ certain approximations. This subject is discussed by Ratcliffe (1959, Chapters 12 and 13).

Other Ground-Based Measurements. Certain other methods of ground-based measurement, although not used as routinely as the vertical-incidence (or oblique-incidence) studies, can give important information and should be mentioned here. These include measurements of cross modulation, of incoherent back-scatter, and of atmospheric "whistlers."

In the cross-modulation experiment, two pulses of different frequency are sent through the same ionospheric path in quick succession. Absorption of the first supplies energy to the free electrons and raises their collision frequency. This results in additional absorption of the second and a resulting modulation of its amplitude. With powerful enough transmitters, it is possible to measure this effect and learn something about electron densities at different levels. This experiment has principal applicability to the D region.

A rather new and potentially very powerful method for sounding the ionosphere is based on measurements of the very small amount of energy back-scattered by free electrons at frequencies well above the plasma frequency. The amount of energy scattered from any level is related to the electron density at that level, and the length of time between transmission of a pulse and reception of the back-scattered energy is a measure of the altitude of the scattering level. Very powerful radars are required and the method has been used at only a few locations. It is capable of giving $N(z)$ curves from lower parts of the ionosphere to very great altitudes, perhaps ultimately to several earth radii, and the soundings can be repeated at frequent intervals to give time variations. For a description of one such installation and some sample results, see Bowles and Staff (1963).

Another technique, which is applicable at very great altitudes and therefore is not of primary concern here, is the observation of atmospheric "whistlers." Electromagnetic waves at low frequencies, say, 1 to 10 kc $\sec^{-1}$, such as are generated by lightning, are propagated along the magnetic field lines (see Fig. 9.2) and return to earth at the conjugate point in the other hemisphere. The higher frequencies from a lightning discharge (or from a man-made source) are propagated somewhat more rapidly than the lower; and the detected wave, when amplified at an audio frequency, has the characteristic sound of a whistle of descending pitch. Studies of the dispersion effect can give information about the electron density at the farthest point from the earth along the path of propagation, as much as a few earth radii. See, for example, Storey (1953) and Helliwell and Morgan (1959).

Measurements of Electron Density from Rockets and Satellites. Although ground-based observations are and will continue to be very useful, measurements from rockets and satellites are playing an increasingly important role in studies of the ionosphere. These include measurements of both electron and ion density, by a variety of methods, and of ionic composition by mass spectrometry. The former are discussed first.

The pioneer experiment to measure electron-density profiles and, according to Bourdeau (1963), still the most accurate one, was the Seddon continuous-wave (cw) experiment based on the Doppler shift of a radio wave transmitted from a vertically ascending rocket (Seddon and Jackson, 1958). This experiment was the first to reveal the essentially monotonic increase of electron density with height up to the $F2$ peak and also served to stimulate and verify the development of methods of deducing $N(z)$ curves from ground-based ionosondes. In this experiment two harmonically related waves are radiated continuously from the rocket, one at high enough frequency to be unaffected by the ionosphere and the other at low enough frequency so that its Doppler shift is affected by the local electron density. For example, in some of the early experiments, the first was at 46.5 Mc sec^{-1} and the second at a frequency exactly one-sixth this, or 7.75 Mc sec^{-1}. The first serves as a reference signal.

For a source moving radially away from an observer with speed V and radiating a signal of frequency f, the Doppler shift of frequency, Δf, is given by

$$\Delta f = -f\mu V/c \tag{9.26}$$

where μ is the refractive index at the source of the signal. For a rocket moving exactly vertically at known speed, it is possible to deduce from a comparison of the frequencies received at the ground under the rocket how the refractive index μ for the lower frequency varies with altitude. This gives, with the aid of the Appleton–Hartree formula, the electron density as a function of altitude.

When the vehicle emitting the radio waves has a component of motion that is not vertical, or, more precisely, is not perpendicular to the surfaces of constant refractive index, then the analysis is considerably complicated by effects of the integrated electron density along the path between source and observer. Owing to refraction, the ray travels a curved path and its angle of arrival at the ground depends on the ionospheric structure along the entire path. The observed Doppler shift is similarly affected because it depends on the angle between the motion of the source and the initial direction of the ray that is finally received at the ground. In the case of a rocket traveling nearly vertically, this effect can be taken into account and the variation of refractive index (or electron density) with height can still be deduced, but by a more complicated analysis.

In the case of a satellite that is moving nearly horizontally and therefore nearly parallel to surfaces of constant electron density and refractive index, the effect of the variation of refractive index along the ray path becomes the important one. The refractive index at the source cannot be uniquely determined. Measurements of radio transmission from satellites, either of angle of arrival or of Doppler shift, give information about the integrated electron density between the ground and the satellite. The theory of these measurements, which is quite involved and not an appropriate subject for elaboration here, is discussed, for example, by Garriott and Bracewell (1961).

A different type of measurement of radio waves transmitted from artificial satellites also gives the integrated electron density between ground and vehicle. This measurement is related to the effect of the magnetic field on radio-wave propagation, which was mentioned at the beginning of this subsection in the context of the ordinary and extraordinary waves. Another way of describing this phenomenon is in terms of a rotating plane of polarization of the wave. For example, a linearly polarized wave can be thought of as splitting into two circularly polarized components, each of which travels with a different phase velocity, and which can be recombined at any time into a wave linearly polarized in a new direction. The rotation of the plane of polarization during an ionospheric traverse is proportional, for high enough frequencies, to the integrated electron density along the path. Garriott and Bracewell (1961) also describe this phenomenon in detail.

Still another kind of measurement that can be made with radio waves emitted from a satellite is easy to describe at this stage because it is nothing more than an adaptation of the ground-based ionosonde. A pulse directed downward from a satellite from above the $F2$ peak is reflected back up to the satellite if its carrier frequency is equal to the plasma frequency at any level above the $F2$ peak. If frequency is varied continuously over the range of plasma frequencies and delay times are measured, then an $h'(f)$ curve is obtained that describes the electron-density profile above the $F2$ peak.

In addition to all of these techniques based on radio propagation through the ionosphere, methods have been developed and are being perfected to allow direct measurements of electron (or positive-ion) density at the position of the rocket or satellite (see, for example, Bourdeau *et al.*, 1960; Smith, 1961; Aono *et al.*, 1963). In general, these are considerably complicated by interactions between the vehicle and the environment. Bourdeau (1963) reviews some of these techniques and problems.

Measurements of Ion Composition from Rockets. Quite apart from the measurement of electron (or ion) density, some very useful and revealing measurements of ion composition can be made with mass spectrometers. The results of American, British, and Soviet experiments are generally consistent in broad outline. They show that the three predominant positive ions are those of atomic mass 16 (atomic oxygen), 30 (nitric oxide), and 32 (molecular oxygen). Nitric oxide is the principal ionized constituent in the *E* and lower *F* regions, while atomic oxygen becomes the principal one above about 200 km. Ionized molecular oxygen has its maximum concentration in the *E* region, where, however, it is usually found in lower concentration than is nitric oxide. To illustrate these points, some results of Taylor and Brinton (1961) are shown in Fig. 9.12. See

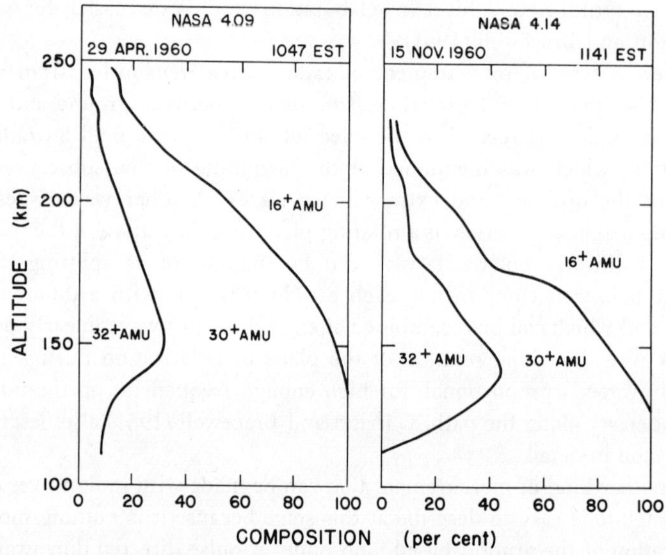

FIG. 9.12. Ionic composition measured by mass spectrometers on rockets at the indicated times. At any altitude, the horizontal distance from the left border to the left curve gives the proportion of O_2^+, the distance from the left curve to the right curve gives the proportion of NO^+, and the distance from the right curve to the right border gives the proportion of O^+. The proportions refer to the total number of positive ions. (After Taylor and Brinton, 1961.)

also C. Y. Johnson *et al.* (1958) and Newell (1960) for results of measurements made by the Naval Research Laboratory.

These results, of course, have no 1-to-1 relationship with the composition of the neutral atmosphere. For example, nitric oxide is a trace constituent, but has a low ionization potential and can also be formed by reactions like (9.10b). Neither can the essential absence of N_2^+ be taken to mean that this ion is not formed in important quantities by photoionization. The ions thus formed disappear rapidly by dissociative recombination so that the number present at any time is small.

9.2.3 Some Characteristics of the *D* Region

Electron Densities and Their Regular Variation. There are not many quantitative measurements relating to the *D* region. The electron densities are low and are not easily measured by direct probes, by the cw rocket experiment, or even by vertical-incidence soundings. The collision frequencies are high, which complicates the theory and interpretation of wave propagation through the region. Furthermore, it is clear that there are wide variations of electron density from day to night and with changing solar activity, so that such measurements as are available must be considered to be suggestive only of particular conditions which are subject to wide variations. For example, Nicolet and Aikin (1960) have suggested a variation of electron density by one to two orders of magnitude from a time of overhead, very quiet sun to a time of a strong solar flare (see Fig. 9.14 later in this subsection).

Figure 9.13 shows two electron-density profiles, one inferred by Fejer (1955)

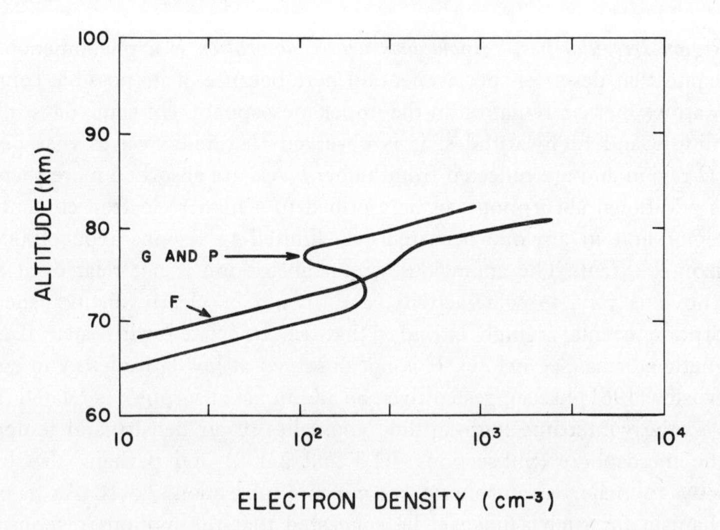

Fig. 9.13. Electron-density profiles in the *D* region measured by Gardner and Pawsey (1953) and by Fejer (1955).

from a cross-modulation experiment and the other by Gardner and Pawsey (1953) from partial reflections at a frequency higher than the plasma frequencies for the region. Both curves are representative of daytime conditions with a rather quiet sun.

Nighttime electron densities have not been measured but they must be very much lower than during the day. This effect shows itself qualitatively in the behavior of commercial long-wave radio transmission. During the day these waves are strongly absorbed in the *D* region and ordinarily reception is limited to broadcasts from relatively close transmitters. However, at night, the absorption effect is greatly reduced and the waves can traverse rather long paths through the lower ionosphere at oblique incidence before being returned to the earth by refraction and being detected at rather distant points.

There are no satisfactory direct measurements to define the variation of electron density with the sunspot cycle. However, measurements of absorption show that there is such a variation (Appleton and Piggott, 1954). The results are obtained from observation of the absorption of a radio wave that penetrates the *D* region and is reflected from higher levels, under conditions such that changes of absorption represent changing effects of the *D* region. The results give only $\int N\nu \, dz$ through the *D* region and indicate that (with respect to noontime conditions at Slough, England) this quantity varies by a factor of about 3 from sunspot minimum to sunspot maximum; the variation from minimum to maximum is somewhat greater for summer measurements than for winter measurements.

D-Region Irregularities. *Anomalous winter absorption* is a phenomenon of the *D* region that deserves special mention here because of its possible connection with atmospheric circulation in the upper mesosphere. On some days in winter in middle and high latitudes, it is observed that radio waves that penetrate the *D* region and are reflected from higher levels are absorbed more than usual. This additional absorption can be ascribed to a higher electron content of the *D* region and in any one occurrence is limited to regions about 1000 km in horizontal extent. The anomalous winter absorption is not related, at least in any obvious way, to solar activity, and should be clearly distinguished from absorption events at high latitudes that are associated with solar flares and magnetic storms (see below). It is not observed at low latitudes or in summer. Mawdsley (1961) has suggested that the additional absorption is related in some way to the wintertime high-latitude anomalies of air density and temperature in the mesosphere (Subsections 3.3.3 and 3.3.4); and perhaps also but less directly to major stratospheric warmings (Subsection 2.5.2). As a possible mechanism for such a linkage, he suggested that the motions responsible for high densities and temperatures in the mesosphere might also be responsible for temporary enhancements of the nitric oxide content of the mesosphere, leading to greater electron production by Lyman α.

There have been a large number of studies of echoes received from within the *D* region at vertical incidence in the general frequency range 0.5 to 2.0 Mc sec^{-1}. These have different characteristics than have the usual echoes from the *E* and *F* regions. They are very weak, they may be received at any given time from several different heights, and reflections at different frequencies can be received from the same height. They cannot generally be explained on the basis of a total reflection described by the simple magneto-ionic theory; instead they are attributed to irregularities of ionization in the D (and lower E) region. These same irregularities are believed to be responsible for the phenomenon of forward scattering in which a wave of frequency high enough to penetrate the entire ionosphere (on the ray theory), say at 50 Mc sec^{-1}, is nevertheless returned to earth at oblique incidence.

An interesting feature of the irregularities responsible for forward scattering and for weak reflection at vertical incidence is their tendency to occur at height intervals separated by 5 to 10 km. The observational evidence is not clear as to whether they occur at definite preferred heights which are relatively independent of time, or whether the heights vary while retaining their characteristic vertical separation. Titheridge (1962) has argued for the former point of view and Ellyett and Watts (1959) for the latter. In connection with these irregularities, one is naturally reminded of the vertical wind shears in the upper mesosphere and lower thermosphere, with a vertical scale of the same order (Section 8.3). If the phenomena are connected, it is not obvious that there should be preferred heights for the irregularities.

The D Region and Solar Activity. A large amount of observational evidence shows that the D region is particularly responsive to changes of solar activity. We have already mentioned the variation of electron density with the sunspot cycle. In addition there are several important effects associated with short-period changes of solar activity. The oldest and best known of these is the sudden ionospheric disturbance (SID) following an important solar flare. Another phenomenon associated with flares, studied extensively only in recent years, is polar-cap absorption (PCA). A third is the ionospheric storm, which accompanies magnetic storms and auroral activity; although this has been studied more extensively in connection with the F region (see Subsection 9.2.5), it also affects the D region.

The SID is observed to begin simultaneously with the observation in Hα of an intense solar flare (Subsection 4.5.2). Its effects are confined to the sunlit hemisphere, especially near the subsolar point, and it lasts only an hour or so. Various phenomena observed during an SID point to the occurrence of abnormal electron production in the D (and lower E) region. These include but are not limited to a sudden decrease of strength of radio waves that traverse the D region before and after reflection at higher levels (short-wave radio fadeout, SWF), and of radio waves arriving from extraterrestrial sources; both of these effects are attributed to intense absorption in the D region. The near simultaneity of the SID with the flare shows that the ionizing radiation must be electromagnetic radiation. As a result of solar observations (see Subsection 4.5.2) from rockets and satellites, this radiation is believed to be in the hard X-ray region (< 10 A), energy which can penetrate to D-region levels and which can ionize all atmospheric constituents there.

Polar-cap absorption, on the other hand, is observed only in the auroral zones, follows solar flares by periods of up to an hour, and occurs during the night as well as the day. As the name implies, this phenomenon, like the SID, shows up most clearly in terms of greatly enhanced absorption in the *D* region. In addition there are distinct radio reflections from levels as low as 60 km during PCA. However, unlike the SID, the PCA event is attributed not to electromagnetic energy but to ionization by very energetic (10^8 ev) protons emitted by the sun at the time of a flare. In the presence of the earth's magnetic field, these particles can penetrate deeply into the atmosphere only along magnetic lines of force that reach the *D* region in the auroral zones.

Ionospheric storms usually accompany magnetic storms of the sudden-commencement type, although they appear to reach their full development only during the main phase (decrease of H) of the magnetic storm. With regard to the *D* region, their most obvious manifestation is increased absorption in high latitudes. Their effects in middle latitudes are not so easily described here, but show that the *D* region is disturbed, sometimes for periods of days during the recovery phase of the magnetic storm.

Theory of the D Region. The theory of *D*-region ionization, as of ionization in other regions, involves first of all the dual question of the production and loss of ionization. In the case of production, the relevant questions have to do with the identification of the ionizing radiations and the ionized constituents. In the case of loss, the problem is to distinguish the most important of the many possible loss processes. Nicolet and Aikin (1960) have discussed the *D* region from this point of view, and their conclusions are in the main still accepted.

The possible spectrum of ionizing radiation is, of couse, limited by absorption at higher levels. The *D* region is effectively shielded from extreme ultraviolet radiation at $\lambda < 1026$ A, which corresponds to the first ionization potential of O_2. From about 1225 A to about 1800 A, it is similarly shielded by O_2 absorption in the Schumann–Runge continuum and bands. Energy at wavelengths longer than 1800 A can ionize no atmospheric constituent except sodium, whose ionization potential corresponds to a wavelength of about 2410 A. Owing to uncertainties about loss processes, the contribution of this ionization is not known, but is generally believed to be small. Between 1026 A and about 1225 A, solar radiation can reach the upper *D* region only in certain atmospheric windows, notably the one near 1216 A corresponding to Lyman α (Subsection 4.5.3, Fig. 4.17). Lyman α can ionize no major atmospheric constituent, but it can ionize nitric oxide. The importance of NO ionization by Lyman α in the *D* region was first suggested by Nicolet (1945) and is still generally accepted.

Rocket and satellite observations of solar X-ray radiation and its variations with solar activity have led in recent years to the recognition of the importance of these radiations in the formation of the ionosphere. With regard to the

D region, only radiation at $\lambda < 6$ A can penetrate below 90 km in significant quantities. Photons with these energies can each produce many electrons. Rocket and satellite measurements have shown (Subsection 4.5.2) that this part of the X-ray spectrum is highly variable. Under quiet conditions near solar minimum, it is essentially absent. Near sunspot maximum and especially when the sun is disturbed, these hard X-rays must play an important role. For example, the additional *D*-region ionization during SID's is attributable to them.

It appears that under ordinary conditions even the small numbers of electrons found below 65 to 70 km cannot be caused by solar radiation. Nicolet and Aikin suggest that cosmic rays can produce the observed ionization.

The processes controlling the recombination of electrons and positive ions are difficult to elucidate at any level of the ionosphere, owing to the large number of possible reactions, for some of which the rate coefficients are imperfectly known, and also to uncertainties about composition. In the *D* region the possib-

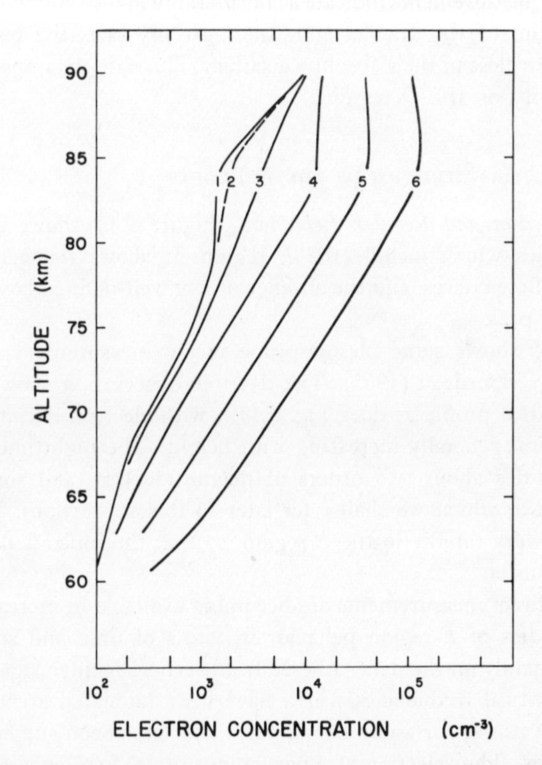

FIG. 9.14. Electron-density profiles in the *D* region computed by Nicolet and Aikin (1960) for overhead sun and different conditions of solar activity. The curves correspond to solar conditions as follows: 1, very quiet sun; 2, quiet sun; 3, lightly disturbed sun; 4, disturbed sun; 5, special events; 6, strong flares. (After Nicolet and Aikin, 1960.)

ility of formation of negative ions by attachment, principally to O_2, provides a particular complication that is absent at higher levels. Nicolet and Aikin's analysis (1960) was based on equilibrium conditions during daytime and took into account the following processes affecting free electrons: three-body attachment to O_2 and photodetachment, resulting in a ratio of negative ions to electrons as high as 7.5 at 60 km, but only 0.03 at 80 km; dissociative recombination of N_2, O_2, and NO; and ionic recombination by collisions between positive ions and negative ions. Production of electrons was computed, as discussed above, by a consideration of ionizations of N_2, O_2, and NO by cosmic rays and X-rays at $\lambda < 6$ A, and of ionization of NO by Lyman α.

Their results for an overhead sun are shown in Fig. 9.14. The inferred electron densities even for a very quiet sun are higher than the measured ones (Fig. 9.13), but a difference is not surprising in view of the uncertainties involved in both observation and theory. Bourdeau (1963) believes, for example, that recent rocket measurements indicate a considerably higher abundance of negative ions than estimated by Nicolet and Aikin. In any case, the results shown in Fig. 9.14, regardless of their absolute accuracy, illustrate the important influence of solar activity on the *D* region.

9.2.4 Some Characteristics of the *E* Region

Electron Densities and Regular Behavior. Figure 9.11 shows some electron-density profiles which include the *E* region. It shows the generally smooth character of these curves, and the absence of any well-defined layers of electrons below the *F*2 peak.

Figure 9.15 shows some plasma-probe rocket measurements due to Smith and shown by Bourdeau (1963). The daytime observation shows pretty much the same kind of profile as does Fig. 9.11b, with electron density of the order of 10^5 cm^{-3} and generally increasing with height. The nighttime profile shows electron densities about two orders of magnitude less (and some interesting fine structure to which we shall refer later in this subsection). Loss processes evidently act very rapidly in the *E* region, so that the ionization decays to low values after sunset.

Although direct measurements are becoming available in increasing numbers, statistical studies of *E*-region behavior in terms of time and space variations still depend mainly on the data gathered from vertical-incidence radio soundings. These give critical frequencies which have been tabulated at numerous ionospheric observatories for a number of years and have been subjected to a great deal of analysis. The selection of a particular critical frequency to be tabulated as f_0E is not without its ambiguities, when sometimes several appear within the *E* region on the same ionogram. Nor is its interpretation unambiguous in the absence of a pronounced *E* layer. Nevertheless, the relatively regular behavior of the tabulated values of f_0E indicates that they are meaningful and that they reveal electron densities at some physically significant level, even if not at a pronounced maximum. It is still convenient to speak of this level as the "E layer."

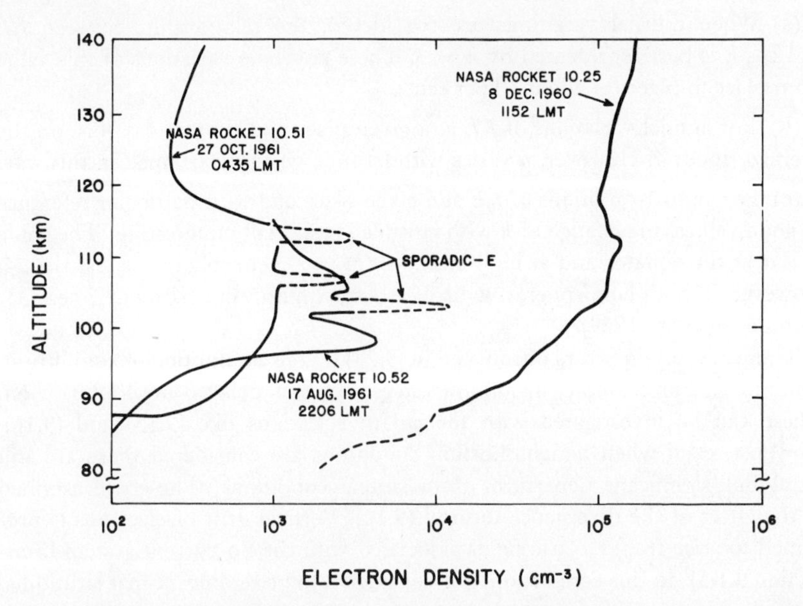

FIG. 9.15. Electron-density profiles measured at the indicated times, two at night and one in the daytime. (After Bourdeau, 1963.)

If the loss process is proportional to N^2, as in (9.7), with an effective recombination coefficient α, and if the production function q is given by Chapman's theory, then one speaks of an α Chapman layer or simply Chapman layer. For a Chapman layer in equilibrium, according to (9.13) and to (4.46) and (4.44),

$$N^2 \propto \cos Z/\alpha \tag{9.27}$$

If further, the E-layer critical frequency is presumed to refer to a particular level, or to a local maximum of N [see Eq. (9.15) and the discussion thereof], then according to (9.23)

$$(f_0E)^4 \propto \cos Z/\alpha \tag{9.28}$$

There has been a great deal of work devoted to tests of this simple relationship, much of which has been summarized by Appleton (1959a, b).

These studies have shown that, to a first approximation, (9.28) is verified. However, the exponent of (f_0E) varies somewhat from 4, and its value depends upon whether the variations of $\cos Z$ refer to diurnal, latitudinal, or seasonal variations of solar zenith angle. It is convenient to call this coefficient n and to describe the studies in terms of the value of n revealed for different kinds of variations.

(a) When diurnal variations are considered, the relationship between f_0E and $\cos Z$ is best represented by $n \simeq 3$. There are some variations of this value from place to place (of about 10 per cent).

(b) For annual variations of f_0E at a given place and time of day n is, on the average, about 4. However, n varies with latitude when determined in this way.

(c) For annual variations of f_0E at a given place and with particular reference to noon values, the change of n with latitude is especially interesting. The value of n near the equator and at high latitudes (in both hemispheres) is less than 4. However, in both hemispheres, n displays a maximum value of about 5 near 35° latitude (Beynon, 1959).

There are certain errors introduced in (9.28) by the assumption of equilibrium conditions. For example, maximum value of f_0E is delayed until after noon. These can be investigated with the aid of equations like (9.15) and (9.16). However, even when nonequilibrium conditions are considered there are still small but significant departures from simple conditions. These are ascribed to the effect of the divergence term in (9.12). Vertical drift of electrons is presumed to arise from electric fields associated with the Sq current system (Subsection 9.1.5). In this connection, it is considered to be significant that latitudinal anomalies in n are greatest at latitudes near 35° where the Sq current system has its foci (see Fig. 9.9).

Apart from variations with zenith angle, the critical frequency f_0E varies over the sunspot cycle, such that $(f_0E)^4/\cos Z$ is approximately linearly related to sunspot number. In terms of simple theory, this means that the intensity of the ionizing radiation is approximately linearly related to sunspot number.

The critical frequency of the E layer at night is low, corresponding to the low electron density (Fig. 9.15). Such studies of critical frequencies as are available show considerable variation from time to time and no regular behavior that can be studied in the same way as that of critical frequencies applicable to daylight hours.

Sporadic E and Other Irregularities. Some echoes received from the E region, even in daytime, show a highly irregular behavior. They are caused by reflections from relatively transient layers of abnormal ionization. This phenomenon is called *sporadic E* and sometimes referred to as *Es*. Sporadic-E ionization occurs simultaneously over areas a few hundred kilometers in extent and usually lasts an hour or more.

The occurrence of *Es* ionization can be studied statistically by considering the percentage of time for which f_0Es exceeds some particular frequency. The character of this statistic varies with latitude. Near the equator, *Es* occurs predominantly in the daytime, with little seasonal variation. In middle latitudes it is less frequent, has a maximum in summer, and is somewhat more likely during the day than at night. In high latitudes it occurs most frequently during the nighttime hours and has little seasonal variation.

Ionized layers responsible for occurrences of sporadic *E* have been detected from rockets and some examples are included in Fig. 9.15, after Bourdeau (1963); see also Seddon and Jackson (1958). According to Bourdeau, two of the peaks shown in Fig. 9.15 (at 102 km on the 17 August flight and at 112 km on the 27 October flight) were detected at the same altitude on both the upward and downward portions of the rocket trajectory, at positions separated horizontally by 72 km. On the other hand, the peak at 108 km on the 27 October flight and others not shown were observed only once and presumably represent patches of small horizontal dimensions.

There is no single accepted explanation for *Es* ionization and in view of its variable characteristics perhaps several different causes are involved. Among suggestions that have been advanced are ionization by meteors, the effect of thunderstorms, and the stratification of ionization in connection with gravity waves (Section 8.3).

Traveling irregularities of ionization in the *E* region have been identified by the observation at different places and times of some characteristic change in the ionosonde records. By observing such times at three stations separated by known distances, it is possible to deduce the velocity of travel of the irregularity (see Briggs and Spencer, 1954). If this velocity is presumed to represent the net motion of the neutral gas at the level in question, then such observations give measures of wind in the *E* region. There is some uncertainty about the latter point, because the exact nature and causes of the irregularities are not known. For example, apparent drift of ionization might correspond to the phase speed of a wavelike disturbance which distorts the electron distribution, and not to motion of the plasma itself. Furthermore, in the *E* region, with decreasing ion-molecule collision frequency, the motion of the plasma becomes less and less likely with increasing altitude to represent motion of the neutral air molecules. Neither the exact dependence on altitude of this effect nor the exact altitude of any individual irregularity is known.

In any case, studies of the regular horizontal drifts of *E*-region irregularities have revealed an important semidiurnal component of motion, which appears to correspond to an upward extrapolation of the tidal components observed at meteor heights at Jodrell Bank (Section 8.2).

The E Region and Solar Activity. Variations of the *E* region in response to short-period solar effects appear to be less pronounced than are those of the *D* region. It was mentioned in Subsection 9.2.3 that the enhanced ionization during an SID appears to extend upward at least into the lower part of the E region, and there is also some increase of f_0E during an SID; but the greatest percentual change is probably at lower levels. Ionospheric storms affect the E region less than either D or F. There is apparently some decrease of f_0E during such storms, but it does not represent a very marked change. On the other hand, there is a rather well-marked correlation between $(f_0E)^4/\cos Z$, which is sometimes called the E layer character figure, and 10-cm solar radiation.

Theory of the E Region. There are several possibilities for the electromagnetic radiations producing ionization in the E region. The relative importance of these at different heights is not known with certainty, partly because of incomplete knowledge about atmospheric composition in the thermosphere.

Important contributions must be made by X-rays which can reach the lowest parts of the E region in the spectral regions near 10 A and also near 30 to 40 A. The solar flux is greater in the latter spectral interval and is probably responsible for ionization at the base of the E region, say 85 to 100 km. Soft X-rays at longer wavelengths up to 100 A are absorbed at higher levels in the E region and contribute to the ionization of those levels.

The first ionization potentials of O_2, O, and N_2 correspond, respectively, to 1026 A, 911 A, and 796 A. Extreme ultraviolet radiation short of 796 A is absorbed mainly in the F region by N_2 and O and is not important for the E region, except perhaps in its uppermost parts. Radiation between 911 A and 1026 A is likely to penetrate into the E region and ionize O_2, especially in certain of the solar emission lines such as Lyman β at 1025 A. The relative importance of this radiation at different levels depends on the vertical distribution of O_2, but there is some evidence (Watanabe and Hinteregger, 1962) that it makes an important contribution at levels as low as 100 km. Ionization of O by radiation at $796 \text{ A} < \lambda < 911 \text{ A}$ must also be considered. Some of this radiation is absorbed at higher levels by N_2 bands but some can reach the E region in absorption windows. This spectral region contains the important solar emission in the Lyman continuum short of 912 A.

Although other reactions must play some part (see Subsection 9.2.1), the molecular ions produced must disappear mainly by processes of dissociative recombination [Eqs. (9.9)]. Ionized oxygen atoms produced by X-ray or ultraviolet radiation must participate in an interchange process [Eqs. (9.10)] followed by dissociative recombination. The relative abundance of NO^+ ions in the E region (Fig. 9.12) points to the importance of ionization of atomic oxygen, since most of these NO^+ ions are probably produced by such ionization followed by reaction (9.10b).

9.2.5 Some Characteristics of the *F* Region

The F1 Layer. From within the *F* region, critical frequencies are sometimes observed from a level which is spoken of as the *F*1 layer or ledge, and which is located near 180 km. The *F*1 layer is absent at night and is weakest in the winter and near the maximum of the sunspot cycle. Statistical studies of *F*1 critical frequencies have shown that they behave pretty much as would be expected if the *F*1 layer were an equilibrium α Chapman layer. The value of n in the relationship $(f_0F1)^n \propto \cos Z/\alpha$ varies somewhat, however, from the theoretical value of 4. For seasonal variations of Z, referred to noon, n varies from station to station but is on the average greater than 4 near sunspot minimum and less than 4 near sunspot maximum.

The value of N corresponding to f_0F1 is about 2.5×10^5 cm^{-3} at noon near the equator, at the equinoxes, and at sunspot minimum. This value increases linearly with sunspot number, but less (percentually) than does the value appropriate to the *E* layer.

The Anomalous Behavior of F2 Critical Frequency. It is clear from ionosonde data and also from rocket measurements that the *F*2 layer near 300 km represents a definite maximum of electron density, greater than at any other level in the vertical column. Consequently, the critical frequency f_0F2 has an unambiguous interpretation as the penetration frequency of a definite layer, and is related to the maximum electron density of the layer by (9.23). It might be thought that the interpretation of the f_0F2 variations would be correspondingly simple, but such is not the case. These variations show many "anomalous" features, that is, differences from what would be expected for a simple Chapman layer. These anomalies have been reviewed, for example, by Martyn (1959) and by Ratcliffe and Weekes (1960). Only some of the more suggestive features are mentioned here.

Figure 9.16, after Martyn (1955), shows local time-latitude cross sections of f_0F2 for equinoctial conditions at sunspot minimum and maximum. For comparison, Fig. 9.17 shows a theoretical cross section for a Chapman layer. Note first of all that Fig. 9.16 has magnetic latitude as ordinate; f_0F2 shows more symmetry with respect to the magnetic equator than it does with respect to the geographic equator. This indicates a certain dependence on the earth's magnetic field, in addition to dependence on the solar zenith angle. There is also a peculiar minimum very near the magnetic equator, especially during the daytime. Especially in low latitudes, the diurnal maximum is delayed until well after noon and the over-all decay of electron density during the night hours is relatively small. This indicates a rather slow rate of recombination and conditions that are far from equilibrium. There are other anomalies that show up near the solstices. For example, near sunspot minimum at moderate latitudes in the

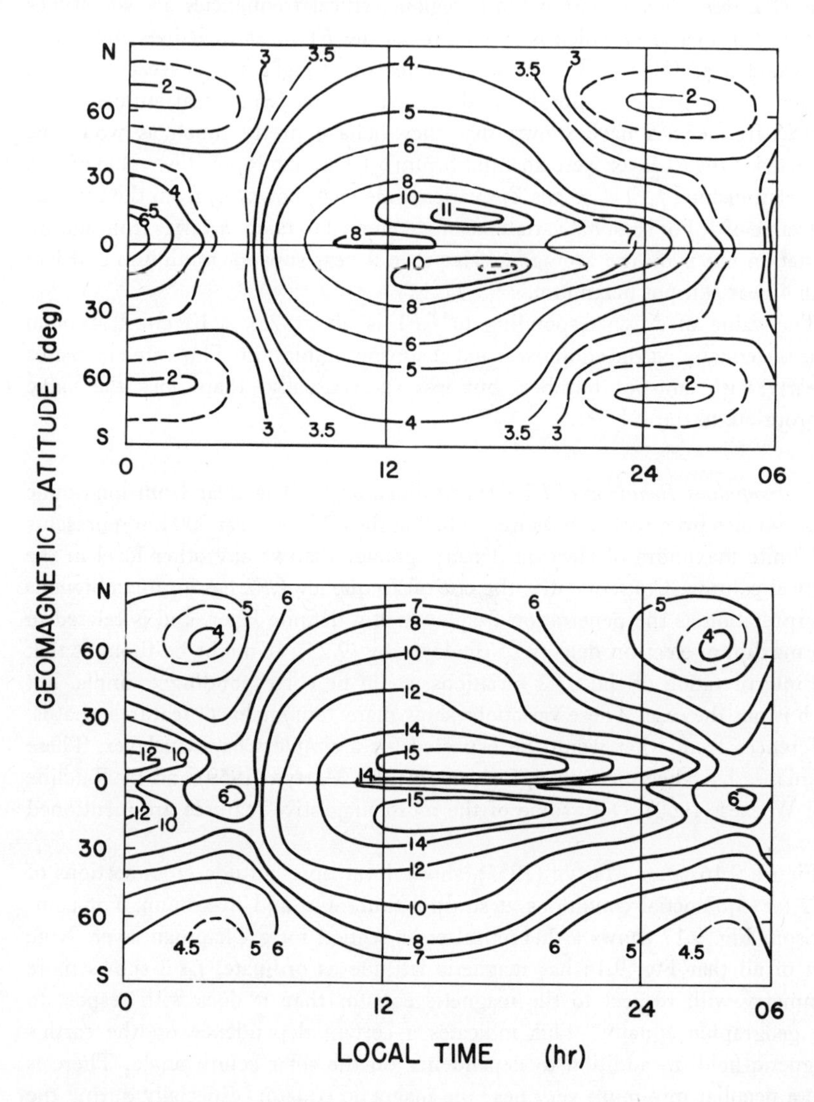

FIG. 9.16. Critical frequency of the $F2$ layer (in megacycles [second]$^{-1}$) as a function of geomagnetic latitude and local time. The values apply to the equinoxes and to sunspot minimum (*upper*) and sunspot maximum (*lower*). (After Martyn, 1955.)

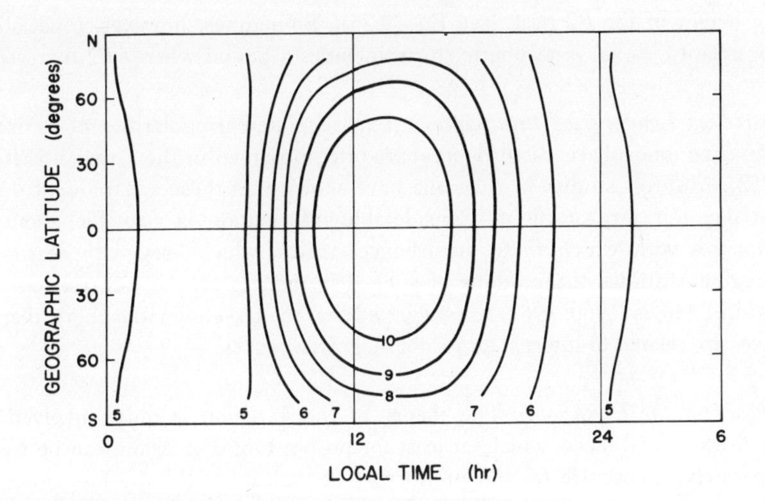

FIG. 9.17. Theoretical critical frequency of a Chapman layer as a function of geographic latitude and local time. (After Martyn, 1955.)

winter hemisphere, f_0F2 actually increases during a few hours of the night. These examples will suffice to illustrate that the behavior of the $F2$ peak is not at all simple and must be related to factors other than the local rate of production of ionization.

The height of the $F2$ peak, although not studied as extensively as the critical frequency, also shows peculiarities. For example, it is higher in summer than in winter in middle latitudes, contrary to the effect to be expected from the corresponding variation of solar zenith angle. It is also considerably higher near the magnetic equator.

The F2 Peak and Solar Activity. The variation of f_0F2 with the sunspot cycle is illustrated by Fig. 9.16. Contrary to the situation for f_0E and f_0F1, it appears in this case that $(f_0F2)^2$ varies linearly with sunspot number. If it is considered that the production of ionization q varies linearly with sunspot number, then f_0F2 varies as $q^{\frac{1}{2}}$ and N_m varies as q, which indicates an attachment-like loss process [see Eq. (9.14)].

During an SID, there are no well-marked effects on the critical frequency of the $F2$ layer. However, ionospheric storms are most noticeable in terms of this parameter. The usual change is an important decrease of f_0F2, most pronounced during the main phase of the accompanying magnetic storm. This is accompanied by a characteristic change in the $h'(f)$ curve, including an increase in the equivalent height of reflection, which is due to some combination of smaller group velocity below the $F2$ peak and an actual geometric increase in the height of the $F2$ peak [see Eq. (9.25)]. Sometimes, however, especially in lower latitudes, an ionospheric storm includes a period when f_0F2 increases.

Traveling Ionospheric Disturbances. Large isolated irregularities of ionization (traveling ionospheric disturbances) are often observed in the F region (Munro, 1958). Munro's studies in Australia have shown that these are frequent during the day but rare at night. Considerable information is available about the velocities with which these disturbances travel, which vary with season. At F-region altitudes the motions of such disturbances are not representative of "wind." Hines (1960, 1963) has suggested that the traveling ionospheric disturbances are related to internal atmospheric gravity waves.

Theory of the F Region. The theory of the F region is quite involved and includes several facets which, at least for purposes of discussion, can be treated separately. These are (a) the production of ionization, (b) the loss processes, (c) the over-all electron density profile that leads to an $F1$ layer and an $F2$ peak, and (d) the anomalous behavior of the $F2$ peak. These are discussed briefly in turn.

The production of ionization throughout the F region is due primarily to extreme ultraviolet radiation in the spectral interval 200–800 A. This is reasonable in view of the ionization cross sections of atmospheric gases in this part of the spectrum and has been confirmed by the flux measurements of Hinteregger (Watanabe and Hinteregger, 1962). There are, however, uncertainties about some of the absorption cross sections and especially about the spectral distribution of the solar flux and the variability thereof. It is not possible to specify the vertical distribution of electron production in great detail, but it does appear to have a maximum value well below the level of the $F2$ peak, probably more nearly in the vicinity of the $F1$ layer.

The principal loss processes are almost certainly the same as in the E region—that is, dissociative recombination in the case of molecular ions (principally N_2^+ in the region of interest), and ion-atom interchange followed by dissociative recombination in the case of atomic ions (principally O^+ in the region of interest). Although N_2^+ ions are undoubtedly formed in considerable quantity (at least in the lower part of the F region), they are not observed to be present in significant numbers (Fig. 9.12) and must disappear rapidly, once formed. The principal ionic constituent is O^+. As pointed out in Subsection 9.1.2, at high levels where there are few molecules, the limiting process in the eventual recombination of O^+ and an electron is the ion-atom interchange process, so that the loss process is proportional to N, not N^2, with a height-dependent coefficient. If we denote this coefficient as $\beta(z)$, we can write (9.12) as

$$\partial N/\partial t = q - \beta(z)N - \text{div}(N\mathbf{v}) \tag{9.29}$$

Ratcliffe *et al.* (1956), from a study of the time variations of electron densities at different heights in the F region, determined that this type of loss process is the most likely one. They further deduced that between 250 and 350 km, $\beta(z)$ can be represented reasonably well by

$$\beta(z) = 10^{-4} \exp[(300 - z)/50] \tag{9.30}$$

where z is in kilometers and β in (second)$^{-1}$.

Such a height variation of the loss coefficient can lead to a peak electron density at an altitude much higher than the altitude of peak electron production. The generally accepted, first-order theory of the F region is that the $F1$ layer is at or near the level of peak production, but that, because of the upward decrease of β, the electron density continues to increase upward to the level of the $F2$ peak.

If the upward decrease of β were rapid enough relative to the upward decrease of q, then the production term and this loss term alone would never produce a peak in the electron-density profile. Electron density would continue to increase to the outer edge of the atmosphere. This would presumably happen in the F region if other factors did not come into play. It is clear, however, that at some upper level radiative recombination (of the form αN^2, with $\alpha \simeq 10^{-12}$ sec^{-1}) becomes important relative to the $[\beta(z)N]$ type of loss, and sets a limit to how small the loss term can be. Such an explanation leads to the prediction of an $F2$ peak, but at higher levels than is generally observed.

The most favored explanation for the existence of an $F2$ peak at the observed altitude involves a consideration of diffusive effects, which enter (9.29) through the divergence term. The important influence of diffusion on the vertical distribution of neutral constituents was emphasized in Chapter 6. In the case of the diffusion of electrons, the theory is somewhat different. The heavier positive ions move under the influence of the gravitational field and through electrostatic attraction carry the electrons with them. The diffusive-equilibrium vertical distribution of electrons turns out to be described by a scale height that is twice that of the neutral constituent corresponding to the predominant positive ion. An additional complication is that the ions do not drift transverse to the magnetic field, but only along the field. Thus near the magnetic equator, where the field lines are nearly horizontal, vertical diffusion is inhibited. However, for latitudes not too close to the magnetic equator, calculations seem to show (see, for example, Chandra, 1963) that the effects of diffusion become comparable with those of the loss process at altitudes near 300 km and are therefore important in determining the altitude of the $F2$ peak. However, it should be pointed out that some calculations (Sagalyn *et al.*, 1963) minimize the importance of diffusion.

The complicated behavior of the $F2$ peak, with all its anomalies, is not understood in detail and must represent the net effect of many diverse processes. In addition to the effects of diffusion, the plasma will drift under the influence of an electromagnetic field. In particular, the importance of the electric polarization field associated with the dynamo currents at lower levels (Subsection 9.1.5) has been emphasized, especially by Martyn [for example, Martyn (1959), which contains references to earlier work].

The vertical distribution of electron density above the *F*2 peak is now being measured by backscatter experiments and from rockets and satellites. Although this part of the atmosphere lies above the limits set for this book, it is pertinent to note that to a first approximation the vertical distribution of electrons above, say, 350 to 400 km seems to correspond to a diffusive-equilibrium distribution in an isothermal atmosphere. The predominant ion is O^+ up to a certain height, above which He^+ and, at still higher altitudes, H^+ predominate.

9.3 An Introduction to the Airglow and Aurora

This section is an introduction to the study of electromagnetic energy emitted by the upper atmosphere in the ultraviolet, visible, and near-infrared parts of the spectrum. This energy includes components known as the airglow and the aurora. The airglow is present at all times and places, although its detection during the day in the presence of sunlight is in most cases not possible. The aurora occurs predominantly in high latitudes, less in middle latitudes, and only occasionally, during great auroral displays, in low latitudes. The aurora is typically much brighter than the airglow, and its spectrum shows features that attest to high levels of excitation of the emitting atoms and molecules.

Both have been the subject of a great deal of research. The most comprehensive treatment of the airglow and aurora (Chamberlain, 1961) lists over 1600 references to the literature, and those were compiled only through 1959. Such research has a great deal of interest in itself, for it presents certain unique problems and challenges in areas of both observation and interpretation. The results of such research are of great importance in the general context of upper-atmospheric studies, because they bear on the composition and state of excitation of the upper atmosphere and give valuable information about the physical processes that are occurring there.

This necessarily selective introduction emphasizes such results and refers only briefly to the techniques and problems of observation and interpretation. Subsection 9.3.1 contains the latter references along with some other background considerations. Subsection 9.3.2 takes up the airglow spectum and the morphology and possible explanations of some of its features. Auroral morphology, being based largely on visual observations, can be discussed in some generality, as in Subsection 9.3.3. But the aurora is a complex phenomenon which must be studied also in connection with the details of its rich spectrum. Some of these are described in Subsection 9.3.4.

In addition to Chamberlain's treatise mentioned above, the reader is referred to the authoritative and relatively recent review articles by Bates (1960a, b, c) and Barbier (1963) and to the books of Harang (1951) and Störmer (1955).

9.3.1 Some General Considerations

Lights of the Night Sky. The airglow and aurora are observed at night (or at twilight). The night sky also contains light from astronomical sources, the most obvious being moonlight and starlight, including their components due to scattering by the earth's atmosphere. There are, in addition, *zodiacal light* and *galactic light.* The zodiacal light is concentrated along the ecliptic near the horizon after sunset and before sunrise. It results from the scattering of sunlight by interplanetary matter. Galactic light is concentrated along the galactic plane (Milky Way) and results from the scattering of light from distant stars by interstellar matter. In studies of airglow and aurora, the extent to which these other sources contribute to the results of observations must be taken into account. This can be done rather successfully for most purposes, but in some cases, such as the measurement of absolute or relative intensities of airglow features, the unwanted background light can introduce serious uncertainties.

Even if the contribution of astronomical sources can be eliminated, there is a certain ambiguity in the distinction between airglow and aurora, as discussed at some length by Chamberlain (1961, pp. 345–347). Historically, the distinction between the two arose because of the usual great difference in the intensity of emission; aurora is usually thought of as a spectrum of emissions, some of whose components in the visible are bright enough to be detectable by the human eye. This is a loose and unsatisfactory definition, especially when some emissions that fall below the visual threshold exhibit certain characteristics otherwise associated with aurora (*subvisual aurora*). However, it will suffice for the purposes of this section, where the discussions of auroral characteristics are confined mainly to those observed during active, visible auroral displays.

Observational Methods and Difficulties. Airglow and auroral observations, except for a few (but important) rocket measurements, are made at the ground. Thus the energy is observed after its passage through the atmosphere. Although our atmosphere is largely transparent in the visible, near ultraviolet, and near infrared, there are important examples of lines and bands that must be emitted but are absorbed by the atmosphere and thus not observed at the ground. In addition, all radiations, and especially those at the shorter wavelengths, are scattered by the atmosphere. This process includes not only scattering out of the column that lies along the line of sight, but also scattering into that column of energy originally emitted in other directions. The ground observations should be interpreted with full consideration of the effects of the intervening atmosphere.

Certain auroral observations, for example, of frequency of occurrence, color, form, and movement, can be made visually or with all-sky cameras. However, airglow and other auroral observations require techniques that are more complicated and difficult. Such observations are usually intended to provide information about the spectrum, the emission altitude, and the intensity, and the variations thereof.

Spectrographic studies of the airglow are difficult because of the low intensity of the source. Even with long exposures, it is necessary in some cases to resort to spectrographs with relatively low dispersion. For this reason, positive identification of some lines and bands was delayed until quite recently and there is still uncertainty about a few features. On the other hand, the aurora is usually much brighter, and spectrographic studies are correspondingly easier from this point of view. Nevertheless, the auroral spectrum is so complex, with overlapping of lines and bands, that the identification of some of the auroral emissions has also been delayed. Presumably, the important features of both have now been identified and are discussed later in Subsections 9.3.2 and 9.3.4.

Altitudes of auroral features have been determined many times by triangulation after simultaneous photography from two or more locations, as will be discussed in Subsection 9.3.3. Determination of the altitudes of airglow emissions is a much more difficult matter and several methods have been used.

The oldest, called the van Rhijn method, compares the intensities of emission at different zenith angles. Let the observer's line of sight make the angle Z with the zenith direction at the earth's surface and the angle ψ with the zenith direction at the height h. If the emitting layer is thin, it can be considered to have one effective angle ψ such that $\sin \psi/\sin Z = r_e/(r_e + h)$, where r_e is the radius of the earth and h is the mean height of the emitting layer. If the emitting layer is spherically symmetric and homogeneous, and if the effects of the lower atmosphere are neglected, then the intensity of emission at different directions from the zenith varies as the distance through the layer along the line of sight, and therefore as $\sec \psi$. A comparison of intensities for different values of Z gives then, in principle, a method of determining h. In practice, the variation with Z due to this effect is not very large except when Z is large and then the effects of the intervening atmosphere become important. The assumption of horizontal homogeneity is also often not fulfilled. The van Rhijn method has given widely conflicting results for some spectral features.

A second technique involves triangulation, as in the case of the aurora, from two different stations on some particular feature (for example, a bright region) of the airglow. This in principle is a very powerful method but in practice is difficult to apply to the airglow because of the difficulty of finding distinguishable features for triangulation. However, some useful results of St. Amand *et al.* (1955), who compared regions with similar time variations of intensity, will be quoted in Subsection 9.3.2.

A third technique that has had wide application is the determination of the temperature of the emitting layer by observing the Doppler broadening of lines or the relative intensities of the rotational lines in bands, both of which are temperature dependent. Of course, such methods can yield altitudes only when some independent information about the variation of temperature with altitude is available (as, for example, from a standard atmosphere).

Finally, the most significant is the rocket observations that have become available through the efforts of the Naval Research Laboratory (Tousey, 1958; Packer, 1961). The photometrically determined intensity remains constant as the rocket approaches the base of the emitting layer (in the absence of atmospheric absorption and scattering), decreases as the rocket passes through it, and becomes zero above it for a photometer pointing up. Rocket measurements have determined the usual range of altitudes for some types of emission, but, of course, are not applicable for economic reasons to studies of short-term variations.

Measurements of Intensity; the Rayleigh. Ground-based measurements of the intensitites of airglow and auroral emissions present several difficulties. Apart from those inherently involved in the absolute photometry of a weak energy source, there are problems of interpretation which affect relative as well as absolute measurements. One of them is caused by the contribution of background energy (starlight, zodiacal light, etc.), which is observed along with the emitted by the upper atmosphere. Another involves the effect of the atmosphere between the altitude of emission and the ground. Still another has to do with the contribution of neighboring lines and bands which may not be completely eliminated by the filtering process. In the case of the aurora, rapid changes of pattern and brightness often make photometry difficult.

A ground-based photometer measures the specific intensity $I(\Delta\nu)$ (integrated over a line or band) emerging from the atmosphere in its field of view. This represents (see Appendix E and the discussion of a similar problem in Subsection 5.2.2) the sum of the unidirectional emission rates from all volume elements in a unit-area column along the line of sight, if atmospheric attenuation is neglected. If such radiations are isotropic, then $4\pi I(\Delta\nu)$ gives the sum of the omnidirectional emission rates for the column. It is customary to express the quantity $4\pi I(\Delta\nu)$ in a unit called the *rayleigh* (R): if the measured quantity $I(\Delta\nu)$ is expressed in units of 10^6 photon cm^{-2} sec^{-1} sterad^{-1}, then $4\pi I(\Delta\nu)$ is in rayleighs. It represents the *apparent emission rate* of a (centimeter)^2 column along the line of sight. It represents the true emission rate only if the radiation is isotropic and if radiative-transfer effects can be neglected. Even in this case, it represents the integrated emission of an entire column and does not give the volume-emission rate at any particular altitude without additional information. The apparent emission rates of airglow features are usually referred to the zenith by multiplying $4\pi I$ by the factor $\cos Z$.

The rayleigh is named after the fourth Lord Rayleigh, who pioneered the measurement of absolute intensity of airglow features (Rayleigh, 1930). Its use and name were suggested by Hunten, Roach, and Chamberlain (1956).

Radar Observations of Aurora. Observations related to the aurora are sometimes made by observing radar echoes (in the frequency range 20–800 Mc sec^{-1}) from ionization associated with the occurrence of visual aurora. These observations and their results are not covered in this book. See, however, Booker (1960) and Chamberlain (1961, Chapter 6).

9.3.2 The Airglow Spectrum

Discussions of the airglow are sometimes conveniently divided into discussions of *nightglow*, *twilightglow*, and *dayglow*. Most observations are made at night and refer to the nightglow, but those observations made at twilight reveal some characteristic differences from nightglow. The dayglow spectrum has been considered theoretically, but hardly measured. The discussion here is in terms of individual features of the airglow spectrum and refers usually to nightglow. A short discussion of the twilightglow and the dayglow is included at the end of this subsection.

The Green Line of Atomic Oxygen. The green line of atomic oxygen at 5577 A is one of the most prominent and extensively studied features of the airglow. Its presence in the spectrum of the night sky, even in the absence of aurora, was noted nearly 100 years ago, but its identification was not definitely established until about 1930 (Frerichs, 1930). The green line arises from the forbidden transition $^1D \leftarrow {}^1S$ in the ground configuration of the oxygen atom (see Fig. 4.5). It is difficult to produce this line in the laboratory, because the excited ^1S atom generally loses its excitation energy by collisions with other particles or with the walls of its container before it undergoes this improbable radiative transition. However, in the near-vacuum of the upper atmosphere collisions are less frequent and the atom has "time" to radiate. This feature of the airglow tends to be somewhat brighter at twilight, but its twilight enhancement is not very marked and not always observed.

The apparent emission rate of λ_{5577} is on the average about 250 R, but shows wide fluctuations between about 100 R and about 500 R. The search for systematic components of this fluctuation has been extensive, but not very successful. Certain systematic variations with local time, latitude, season, and phase of the sunspot cycle have been suggested from time to time, but none is very well marked. A tendency for a maximum around local midnight was reported early, but this seems to be noticeable only in middle latitudes during the winter months. The annual variations reported by individual stations are rather complex and do not fit any simple over-all pattern. Some observations suggest that there may be a latitudinal maximum in middle latitudes which shifts with the season. There is better evidence for a greater intensity near sunspot maximum than near sunspot minimum.

One factor that hampers the search for such systematic variations is of considerable interest in itself. This is the large variation of apparent emission rate that is often observed at a station, either with direction at any one time or with time in any one direction. These variations have been interpreted as a tendency for bright regions to occur in patches or cells, which are of diameter about 2000 km and are in motion. Such a cell is about twice as large as the field of view (from a single station) of the sky at 100 km (about where λ_{5577} is emitted; see below). It may cover the sky at a particular time and then only part of the sky as it drifts over the station. Motions on the order of 100 m sec^{-1} have been inferred for such cells, but since the origin and cause of the phenomenon are not known it is not safe to interpret these motions as being due to "winds." However, the cells do suggest a structure of some kind near the 100-km level with a scale not unlike that of disturbances in the lower atmosphere.

The altitude of the layer emitting λ_{5577} does appear now to be rather well established. Although early applications of the van Rhijn method gave highly variable results, Roach and Meinel (1955a, b) with newer data and re-analysis of older data suggested an altitude between 62 and 104 km for the emitting layer. St. Amand *et al.* (1955) with their triangulation technique estimated an altitude of 80 to 100 km. Various spectroscopic determinations based on Doppler broadening of the line [for example, Armstrong (1959) and Wark (1960)] have suggested temperatures of around 190°K and therefore altitudes in the vicinity of the mesopause. Rocket determinations (Tousey, 1958; Packer, 1961) give consistent results of a layer centered at about 97 km and emitting most of the λ_{5577} radiation between 85 and 105 km.

The problem of the excitation mechanism responsible for the production of ^1S oxygen atoms for emission of the green line is in a state of some confusion. A mechanism which has been quite generally accepted until recently was suggested by Chapman (1931b) and is

$$O + O + O \rightarrow O_2 + O(^1S) \tag{9.31}$$

In this case an oxygen atom serves as the third body during the three-body association of oxygen and is raised to the ^1S term by some of the energy released in the association. However, recent laboratory measurements (for example, Barth and Hildebrandt, 1961) have indicated that the rate coefficient for the production of ^1S oxygen atoms by this mechanism is too small to allow (9.31) to explain the observed emission rate of the green line.

The Green-Line Covariance Group. Certain other features of the airglow spectrum are observed to have apparent emission rates that are correlated with those of the green line. The correlations are quite good for fluctuations during a single night or from night to night during a period of a few weeks, but the relationships are apt to be different during different time periods.

These other emissions are mostly from molecular oxygen and include

(a) Herzberg bands ($^3\Sigma_u^+ \rightarrow {}^3\Sigma_g^-$) in the near-ultraviolet and blue parts of the spectrum;

(b) Chamberlain bands ($^3\Delta_u \rightarrow {}^1\Delta_g$) in the blue (Chamberlain, 1958);

(c) a continuum in the green whose origin is discussed below;

(d) the 0–1 band of the Atmospheric system ($^1\Sigma_g^+ \rightarrow {}^3\Sigma_g^-$) at 8645 A, sometimes called the Kaplan–Meinel band.

The emission rate of the Herzberg bands in the spectral region where they can be detected (they are, of course, absorbed by ozone at short enough wavelengths) is about 500 R. The Chamberlain bands emit only about 100 R and the continuum about 1000 R(1 kR). The apparent emission rate of the 0–1 band is about 1.5 kR. However, this by no means represents the actual emission of O_2 in the near infrared. The 0–0 band of the Atmospheric system must emit much more than the 0–1 band, but is not observed at the ground because of absorption by the atmosphere. In addition the 0–0 band of the Infrared Atmospheric system ($^1\Delta_g \rightarrow {}^3\Sigma_g^-$) at 1.27 μ is not observed, presumably for the same reason. The 0–1 band of the latter system might be observed if it could be separated from a stronger overlying OH band (see below). Altogether the total emission (not apparent emission) due to molecular oxygen according to Chamberlain (1961) is in the range 10–80 kR. Latitudinal and seasonal variations of all these emissions are, like those of the green line, complex and poorly defined.

There is strong evidence that the emissions of the green-line covariance group, as noted above for the green line, originate mostly near 100 km. Rocket measurements have been made of the wavelength region 2500–2950 A, which includes members of the Herzberg system not observed at the ground owing to ozone absorption, and on the 0–0 band of the Atmospheric system also not observed at the ground owing to oxygen absorption. Apart from the intensity increase with altitude below the emitting layer, resulting from passage through the lower absorbing layers, the resulting intensity curves for both systems are remarkably similar to that for the green line (Packer, 1961). The emitting layer lies mainly between 85 and 100 km with maximum luminosity near 97 km. Rotational temperatures for the Herzberg bands between 3000 and 4000 A (Barbier, 1947; Chamberlain, 1955) and for the Kaplan–Meinel band (Meinel, 1950c; Dufay and Dufay, 1951; Mironov *et al.*, 1958) are low and indicate an origin near the mesopause. Rocket measurements (Heppner and Meredith, 1958; Tousey, 1958; Packer, 1961) have also indicated an origin near this level for the continuum radiation.

The excited O_2 states that give rise to the O_2 emissions can be produced during the three-body association

$$O + O + M \rightarrow O_2' + M \tag{9.32}$$

where the prime indicates one of the excited states of O_2 (see for example, Bates, 1957). Probably more energy is available from this process than is observed or inferred for the O_2 nightglow emissions and the difference is to be explained by collisional deactivation.

There are various processes that could give rise to continuum radiation. One which has been proposed (for example, Bates, 1954) is

$$NO + O \rightarrow NO_2 + h\nu \tag{9.33}$$

Others, such as radiative association of oxygen, have also been suggested.

The Red Line of Atomic Oxygen. The red "line" is actually a doublet, with lines at 6300 A ($^3P_2 \leftarrow {}^1D_2$) and at 6364 A ($^3P_1 \leftarrow {}^1D_2$). This feature was first recorded by Slipher (1929) and identified a few years later. Since ^1D atoms are certainly present after the emission of the green line, one might expect the red lines to arise at the same elevation and exhibit the same behavior as the green line. Such, however, is not at all true.

The red lines are much brighter in the twilightglow than in the nightglow, with an apparent emission rate of 500 to 1000 R. The average in the nightglow is only about one-tenth of this. The decrease after twilight to the typical nightglow value is rather slow, a feature referred to as *post-twilight enhancement*. The red lines are also somewhat brighter again just before dawn (*predawn enhancement*). However, this type of variation is not universally present [for example, Barbier (1956) found that the predawn enhancement disappears during the summer]. Seasonal and latitudinal variations are not pronounced and there is less night-to-night flunctuation than in the case of the green-line covariance group.

The altitude of emission of the red line is rather uncertain, but it is evidently considerably higher than that of the green line. The van Rhijn determinations of Roach and Meinel (1955a) place it between 116 and 143 km, but this appears to be lower than the results of other determinations. Various spectroscopic studies (for example, Wark, 1960) have placed it between 200 and 300 km. Barbier (1959), using a method based on twilight observations, found an altitude of 275 km. Rocket measurements have been to some extent inconclusive, partly because the filters used have admitted some light due to nearby OH bands (see below) as well as the red oxygen lines, partly because the rockets evidently did not reach the level of maximum red-line emission. The results (Heppner and Meredith, 1958; Tousey, 1958) have been interpreted to indicate that the red line originates somewhere above 163 km.

There are two aspects to the explanation of the red line. In the first place, one must explain why it is not emitted from the same height as the green line, since, as mentioned above, OI ^1D atoms must be present at that height. The explanation for this must be that the metastable ^1D term (with a radiative lifetime much longer than that of the ^1S term) is deactivated before radiating near the 100-km level. One such possibility, suggested originally by Bates and Dalgarno (1954), is collision with an oxygen molecule in the ground state, the energy of the excited atom going to raise the molecule to the $^1\Sigma_g^+$ state at the second vibrational level [see Eq. (4.15)]; and there are other possibilities. Seaton (1958) has discussed this general problem.

In the second place, mechanisms for the production of ^1D atoms must be operative at the high altitudes where the energy in the red line originates. Two possibilities are dissociation in the Schumann–Runge continuum, which may be particularly important during twilight,

$$O_2 + h\nu \rightarrow O(^3P) + O(^1D) \tag{9.34}$$

and dissociative recombination of either O_2^+ or NO^+.

Emission by Hydroxyl. The red and infrared portions of the nightglow spectrum are rich in vibrational-rotational bands of the $^2\Pi$ ground state of hydroxyl, OH. Discovery and identification of the infrared bands of this system were due largely to Meinel (1950a, b). Many features were formerly attributed to other molecules, for example, to N_2. The various bands that have been detected and identified correspond to transitions from vibrational levels with $v' < 10$.

The total emission rate for the Meinel hydroxyl system has been estimated by Chamberlain to be about 4500 kR. The variation during the night and from night to night is irregular. Variations during the year are relatively small. There is some evidence for cellular structure of the OH airglow similar to that observed for the green line.

Temperature determinations by Meinel (1950b) and by Dufay and Dufay (1951) indicate an emitting layer not far above the stratopause (or, alternatively, near 110 km where the temperature is nearly the same). On the other hand, Packer (1961), from a review of several rocket determinations that sampled parts of different bands, concluded that the luminosity originates largely in the layer 70–95 km with a maximum near 85–90 km.

The vibrationally excited OH may arise from a collision between ozone and hydrogen (Bates and Nicolet, 1950; Herzberg, 1951), or, if the glow originates at higher levels, from the collision of a vibrationally excited oxygen molecule and hydrogen as proposed by Krassovsky (see his discussion, 1958). Packer (1961) points out that if his assessment of the rocket observations is correct, both mechanisms may be operative.

Emission by Sodium. An important feature of the airglow spectrum is the D doublet of sodium, with lines at 5890 and 5896 A. The D doublet arises from the transition from the lowest excited term of sodium $3p\ ^2P^o_{\frac{3}{2},\frac{1}{2}}$ to the ground level $3s\ ^2S_{\frac{1}{2}}$. Noted first by Slipher in 1929, the D doublet was identified in 1938 (Cabannes *et al.*, 1938; Bernard, 1939).

This feature of the airglow spectrum was discussed in Subsection 5.5.4, where it was pointed out that the maximum rate of emission appears from rocket measurements to be in the vicinity of 90 km, with contributions from levels between about 70 km and about 110 km. The van Rhijn height determinations, according to Roach and Meinel (1955a), indicate a level between 108 and 129 km.

The apparent emission rate is rather difficult to determine because of a nearby OH band. It appears to be around 100 to 300 R in winter and less in summer. There are no well-established daily variations, except the very pronounced enhancement (by a factor of 15 or so) in the twilightglow as a result of scattering.

Production of the ^2P atoms that radiate the D lines may occur (Dalgarno, 1958) through one or more of the processes

$$NaO + O \rightarrow Na(^2P) + O_2 \tag{9.35}$$

$$NaH + O \rightarrow Na(^2P) + OH \tag{9.36}$$

$$NaH + H \rightarrow Na(^2P) + H_2 \tag{9.37}$$

suggested, respectively, by Chapman (1939), Bates and Nicolet (1950), and Bates (1954). However, reaction rates for these processes, and for others that might form NaO and NaH, are not very well known.

The Twilightglow and the Dayglow. Twilightglow refers to the airglow at a time when the surface is in shadow and the upper atmosphere is still illuminated (from below) by the sun's rays. It has already been noted in the above discussion that the oxygen red lines and the sodium D lines are greatly enhanced at twilight, although probably not for the same reason.

An additional feature of the twilightglow, which is not observed in the nightglow (but is observed in the auroral spectrum) is the First Negative band system of $N_2^+(B^2\Sigma_u^+ \rightarrow X\ ^2\Sigma_g^+$; see Fig. 4.7). The 0–0 band of this system is at 3914 A and the 0–1 band at 4278 A. The apparent emission rate is not very well known but is of the order of 1 kR. This system most likely arises from scattering of sunlight.

Other (weaker) emissions of the twilightglow are the forbidden transition at 5199 A in the ground configuration of atomic nitrogen ($2p\ ^4S \leftarrow 2p\ ^2D$), the 0–1 band of the Infrared Atmospheric system of O_2 at 1.58 μ, and some lines of ionized calcium.

The dayglow spectrum must contain the features of the twilightglow and perhaps others. It has been discussed theoretically but, for obvious reasons, observed hardly at all. The sodium *D* lines have, however, been detected (see Subsection 5.5.4). More recently Noxon and Goody (1962) have reported very interesting observations of the red lines, indicating an apparent emission rate of about 35 kR, much greater than at twilight.

9.3.3 Appearance and Morphology of the Aurora

In contrast to the case of the airglow, auroral emissions are ordinarily observed only from middle and high latitudes. Also in contrast to the airglow, an auroral display, when it does occur, may be a vivid phenomenon, interesting and striking even to the casual observer. Easily visible at night and often displaying rapid changes in form, color, and appearance, the aurora is undoubtedly the oldest known phenomenon of the upper atmosphere. To the layman, it is still probably the most interesting. To the scientist, the aurora is also extremely interesting, because it represents the visible effects of a very complex interaction, still imperfectly understood, between extraterrestrial charged particles and the particles of the earth's upper atmosphere in the presence of the earth's magnetic field.

It is interesting to read accounts of the visual impressions made by the aurora on early observers. Chapman and Bartels (1940, Chapter 14) and Bates (1960a) have quoted some of these. An account that was recorded by Rev. Ezra Stiles, then president of Yale College, in his "meteorological journal" on 14 November 1789 is included in Appendix L.

Visual Appearance of Aurora. The appearance and form of an aurora can take on any of several generally distinguishable types. More than one type may occur simultaneously in the same display or at different times during a display. These types are classified and named according to a system* devised by Störmer (1930), although a given auroral form may not fall sharply and distinctly into one or another of the classifications.

Any attempted verbal description of the appearance of auroral types is bound to be somewhat unsatisfactory without illustration, preferably in color. However, the description of Chamberlain (1961) is graphic and brief and is quoted below. The interested reader should supplement this description with an inspection of photographs, for example, in Störmer (1921, 1926, 1930, 1953), in Chamberlain's book or in Harang's book. Massey and Boyd (1959) show a few excellent color photographs due to Montalbetti, and Gartlein (1947) has repro-

* Some changes are incorporated in a new classification system published in "International Auroral Atlas" for the International Union of Geodesy and Geophysics by Edinburgh University Press during 1963. See also a review by D. M. Hunten in *Science*, **144**, 706 (May 8, 1964).

duced in color some beautiful paintings of Crowder and included also some photographs.

I. *Forms without ray structure*

Homogeneous arc.* A luminous arch usually streching from magnetic east to west (approximately: ...) with its highest point near the magnetic meridian. The lower edge of an arc appears rather sharp; the top portion fades more gradually with height.

Homogeneous band. Similar to an arc, its shape is less uniform and it generally shows active apparent motion along its length. The band may have one or more "horseshoe" turns.

Pulsating arc. Part or all of an arc may pulsate in brightness with a period of a few seconds to a minute or longer.

Diffuse surfaces are amorphous glows without distinct boundaries or isolated luminous patches resembling clouds.

Pulsating surfaces remain nearly constant in position and shape, but pulsate in brightness irregularly with periods of several seconds.

Feeble glow is a term applied to auroral light seen near the horizon. It is not a true auroral form, and may arise, for example, from an arc or band whose lower border is below the horizon.

II. *Forms with ray structure*

Rayed arc. Similar to the homogeneous arc except that the luminosity is broken by numerous vertical striations.

Rayed band. A band composed of numerous vertical rays.

Drapery is a band composed of very long rays. Often it has a horseshoe fold, giving the appearance of a hanging curtain.

Rays sometimes appear singly or (more often) in isolated bundles or more extended groups.

Corona is a rayed aurora seen near the magnetic zenith. Since rays are aligned more or less along the dipole lines of force, they appear to converge toward the magnetic zenith. This railroad-track effect produces the illusion of a dome or, if it is developed on one side only, a fan.

III. *Flaming aurora*

Waves of light move rapidly (less than a second) upward, one after the other, from the base of the aurora toward the magnetic zenith.

---

* The word homogeneous is used in this connection to signify the absence of perceptible ray structure; it must not be taken too literally.

Note that the orientations of auroral features are related to the magnetic field. Arcs and bands tend to be aligned along circles of magnetic latitude and rays along lines of force of the field. The latter appear to converge toward the magnetic zenith and in intense displays form the corona in the magnetic zenith. Arcs and bands in a quiet phase may extend 1000 km or so in an east-west direction and cover 100 km or so of height; but they are only a few kilometers thick in the other (geomagnetic north-south) direction. At times during the active phases of an auroral display this thickness may be reduced by a factor of 10.

Geographical Distribution of Aurora. In 1881, Fritz compiled from records that had accrued since 1700 the first systematic account of the frequency with which aurora is visible from various locations in the Northern Hemisphere. He presented his results in the form of a map with lines of equal frequency (isochasms), expressed in percentage of total nights during the year. These isochasms turned out to be nearly circular and concentric about a point that coincided with the north geomagnetic pole. Minimum values of occurrence were found in low latitudes, maximum values at a distance of about 23° from the geomagnetic pole, and a secondary minimum near that pole. The details of these variations in the polar regions were not very clear at the time but have been clarified by observations since then.

Figure 9.18 shows a revised version of such a map based on more data and corrected for the occurrence of cloudiness, prepared by Vestine (1944). These results show that aurora is rarely seen in low latitudes but that near geomagnetic latitude 67° it is visible on every clear night. Note that the isochasms, by definition, include occurrence of visible aurora in any part of the sky and not necessarily overhead. Chapman (1953) has pointed out that a similar diagram showing the frequency of occurrence of *overhead* aurora would be very useful and has suggested the name "isoauroral diagram" for such a compilation. Detailed observations during the IGY in Alaska (Davis, 1962) have confirmed that the frequency of overhead aurora has a maximum near $66\frac{1}{2}$°, at least for the longitude and time period studied. According to these observations, only 5° separates the geomagnetic latitudes at which the frequency of overhead aurora is half that at the latitude of peak occurrence.

Southern Hemisphere data are relatively scarce. The scattered observations available in 1945 were summarized by Vestine and Snyder (1945) in a tentative isochasm chart that showed again the close relationship between isochasms and geomagnetic-latitude circles.

According to terminology suggested by Chapman and now frequently used, one can speak of *auroral regions*, *subauroral belts*, and a *minauroral belt*. With reference to geomagnetic latitude, the auroral regions extend from 60°N or S to the poles, the subauroral belts lie between 45° and 60°N or S, and the minauroral belt is between 45°N and 45°S. The *auroral zones* are regions of maximum

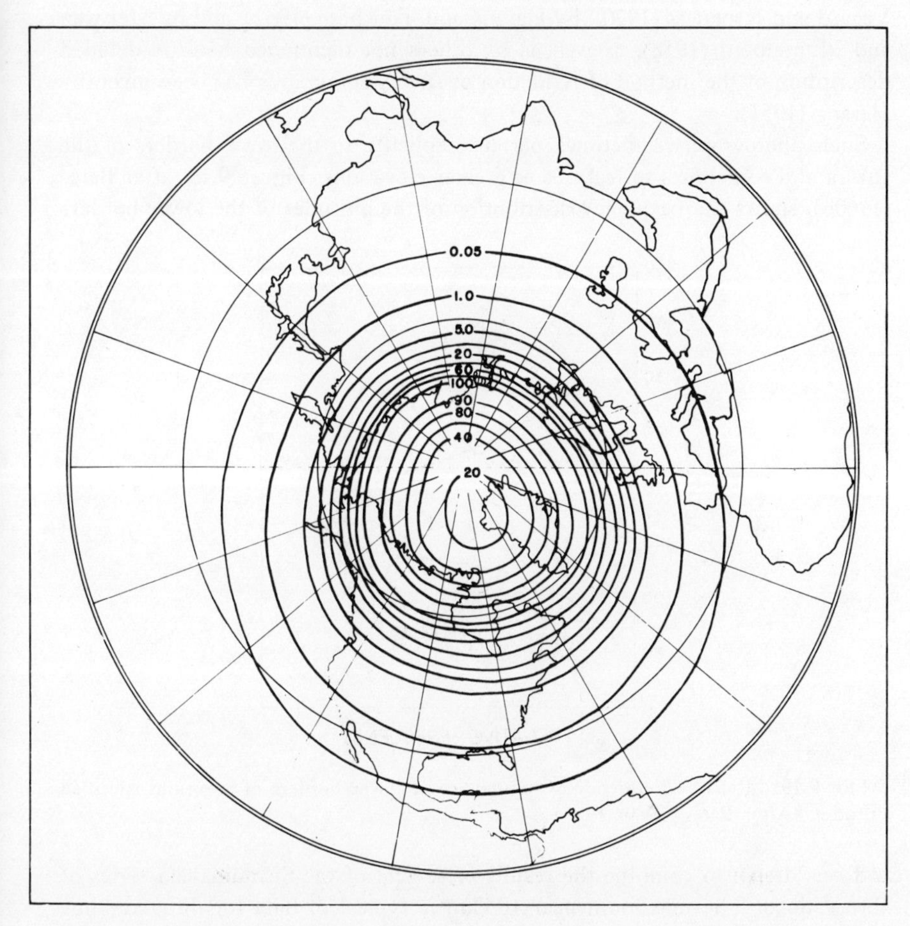

FIG. 9.18. Frequency of occurrence of aurora in the Northern Hemisphere. The curves (isochasms) are labeled in percentage of nights when aurora would be visible in the absence of cloudiness. (After Vestine, 1944.)

occurrence within the auroral regions, and the *auroral caps* lie within the auroral zones.

Altitude of Aurora. Altitudes of auroral emissions have been determined many times by the method of simultaneous photography from two or more locations. The pioneer worker in this field was Störmer, who made his first measurements in 1910 and more extensive and reliable determinations in 1913. The results of the latter observations were described completely by Störmer (1921). Other extensive measurements have been reported by Störmer (1926, 1953), by

Vegard and Krogness (1920), by Harang and Tönsberg (1932), and by McEwen and Montalbetti (1958), as well as by others not mentioned here. A detailed description of the method of reduction of such photographs has been given by Harang (1951).

Such photographs sometimes pertain explicitly to the lower borders of the aurora and sometimes to features at greater elevations. Figure 9.19, after Bates (1960b), shows the frequency distribution of the altitudes of the lower borders

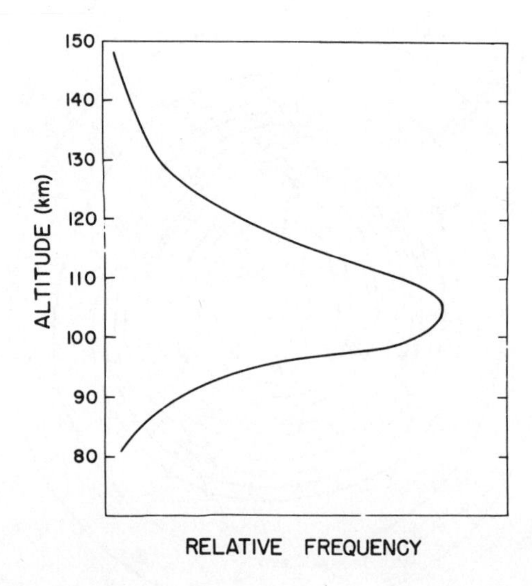

FIG. 9.19. Relative frequency of occurrence of lower borders of aurora at specified altitudes. (After Bates, 1960b.)

and was drawn to combine the results of several of the Scandinavian series of observations. The maximum near 105 km is typical of data for different time periods and different locations. Aurora is seldom observed with a lower border below 80 to 90 km.

There have been a number of studies to determine whether different auroral forms, on the average, have typically different lower limits. No very marked trend has emerged from these studies. Comparison of the results of different series of measurements shows variations reported by different observers for the same forms to be nearly as great as variations among different forms (see, for example, Bates, 1960b, table 1). The rays, however, do appear to occur at consistently higher altitudes than other forms.

However, for the same form, the lower limit appears to be lower for an aurora of stronger intensity. For example, Fig. 9.20 gives (after Harang, 1951) some results of Störmer with respect to draperies. Harang (1945) and McEwen and Montalbetti (1958) have found the same effect.

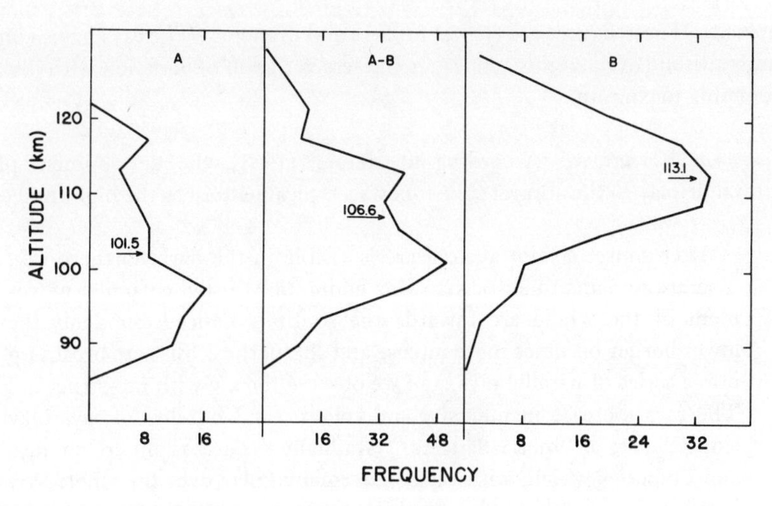

FIG. 9.20. Frequency with which the lower limits of draperies occur at various altitudes: A, very intense, many-colored draperies; B, faint, quiet draperies; A–B, draperies with intensities between A and B. (After Harang, 1951.)

The intensity of emission from aurora falls off rather slowly above the level of maximum emission and the "top" of an aurora is usually ill defined. Table 9.1 gives, after Harang (1946), the distance between the level of maximum emission and the upper and lower levels at which the emission has one-half the maximum

TABLE 9.1

THE DISTANCE FROM LEVEL OF MAXIMUM AURORAL EMISSION TO THAT LEVEL ABOVE AT WHICH THE EMISSION HAS ONE-HALF ITS MAXIMUM VALUE (l_u) AND TO THE CORRESPONDING LEVEL BELOW (l_e)[a]

Altitude of max. emission (km)	l_u (km)	l_e (km)
100	13.5	6.8
110	15.2	7.2
120	18.4	8.8
130	22.0	11.0
140	25.5	13.4
150	30.8	18.0
160	—	27.6

[a] After Harang (1946).

intensity. These figures are typical for arcs and draperies. The rays have a higher mean altitude of maximum emission and less variation of emission with distance from this maximum.

Temporal Variations. According to Harang (1951), the development of an auroral display in the auroral zone follows a typical pattern as the night proceeds:

> After sunset a faint auroral arc is visible in the dark northern sky. The arc remains quiet perhaps for hours, only a slight parallel movement of the whole arc towards the south is noticed. Suddenly the lower border becomes more intense and sharp, the diffuse arc breaks up into a series of parallel rays, and we observe the arc with ray-structure. The rays increase in intensity and colours, and bundles of rays play along the arc as "merry dancers." Gradually the arc is split up into rays and draperies which, with rapid movements, play over the whole sky. During the climax of the display the draperies and rays cover the whole sky, and the ray-structure converges towards the radiation point of a corona. This climax of the displays usually lasts only a short time, perhaps some minutes. The forms fade away and the sky is covered with a faint diffuse luminosity, against which fainter bands, draperies and rays now and then appear. As a last phase in the display we often again observe a faint arc in the north. Such a display may be repeated one or more times during the night.
>
> This general development of an auroral display is less characteristic for the great aurorae occasionally occurring at lower latitudes, where the aurorae often appear as series of draperies and coronae. The faint diffuse luminosity which usually covers the northern sky after a strong auroral display, has, at lower latitudes, often a reddish colour which has given the aurora borealis its name, "the northern dawn."

A number of studies (for example, Vegard, 1912; Malville, 1959; and Davis 1962) have revealed that the maximum frequency of auroral forms occurs around local midnight. However, according to Davis (1962), if the total intensity of auroral emission is used as an index, the nocturnal maximum is less pronounced and activity continues at a high level for several hours after midnight. Malville (1959), from Antarctic data, has pointed out a trend for the time of maximum occurrence of visible aurora to be earlier as one approaches the pole within the auroral zone and to be as much as nine hours before geomagnetic midnight at the pole.

The annual variation of the frequency of auroral occurrence in middle latitudes shows two well-marked maxima near the equinoxes and two well-marked minima near the solstices. This is illustrated in Fig. 9.21 from the results of Meinel, Negaard, and Chamberlain (1954) at geomagnetic latitude 53°N. Nearer the auroral zone, however, where aurora is such a frequent phenomenon, and where, furthermore, the duration of the dark hours varies so drastically from winter to summer, the variation is more obscure. There a winter maximum and summer minimum are shown by most analyses, but this may be due in some measure to the effect of the longer winter nights.

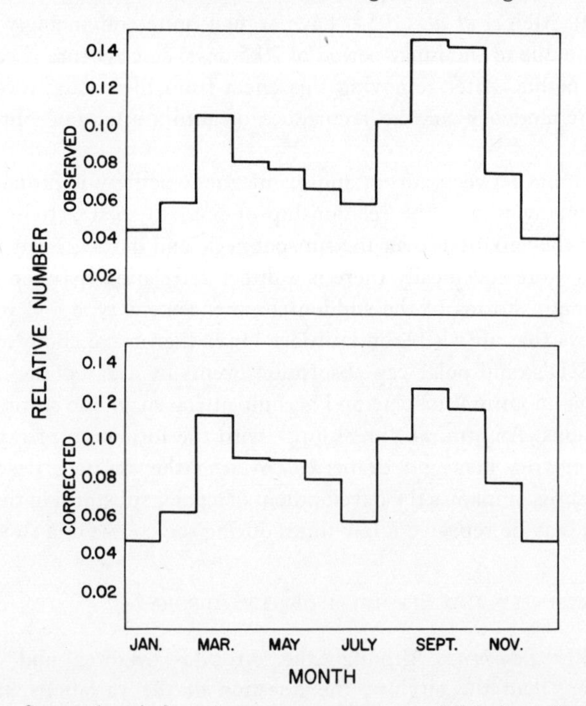

FIG. 9.21. Seasonal variation of occurrence of aurora visible from Yerkes Observatory. The top graph shows the observed proportion for each calendar month; the bottom graph shows the distribution after correction for cloudiness and number of dark hours in each month. (After Meinel *et al.*, 1954.)

Association with Solar and Geomagnetic Activity. Variation from year to year of the frequency of aurora observed from middle latitudes bears a pronounced relationship to the variation of sunspot number. Maximum frequency occurs a year or two after sunspot maximum and minimum frequency occurs at sunspot minimum (for example, Clayton, 1940; Meinel *et al.*, 1954). However, the frequency may be considerably greater at some sunspot maxima than at others (or at some sunspot minima than at other minima). Furthermore, the variation from one sunspot maximum to another is not well correlated with the corresponding variation of maximum sunspot number (Meinel *et al.*, 1954).

Occurrences of unusually bright aurora, such as would be visible in middle latitudes, are usually associated with the passage of large sunspot groups across the sun's central meridian (as seen from earth). These occurrences usually follow such passages by 1–2 days (for example, see Clayton, 1940). However, by no means do all sunspot groups produce such aurora.

There is, furthermore, some tendency for auroral displays to recur after a lapse of about 27 days, the synodic rotation period of the sun in its low latitudes (Dixon, 1939). Meinel *et al.* (1954) have argued quite convincingly that some of this must be due to the lunar period of 29.5 days, since aurora is easier to see on moonless nights. After removing this effect from their data, these workers found a slight tendency for the recurrence of aurora after one but not more solar rotation.

The association between aurora and geomagnetic activity is pronounced and shows up in many ways. The relationship of both to solar activity shows up statistically in the variations over the sunspot cycle and in the 27-day recurrence phenomenon. More specifically there is a direct correlation between the occurrence of magnetic storms of the sudden-commencement type and pronounced auroral displays that are visible in latitudes lower than usual. Both follow large solar flares, SID's, and polar-cap absorption events by a day or so.

Even during an auroral display and accompanying magnetic storm, there are close correlations. An auroral "break-up," with the formation of ray structure and rapidly moving rays and draperies covering the sky (see description of Harang above), accompanies the development of a polar substorm in the magnetic records. Both may be repeated a few times during the course of a storm.

9.3.4 The Intensity and Spectrum of the Aurora

Brightness of the Aurora. Although the aurora is brighter, and often very much brighter, than the airglow, the question of the variations of apparent emission rates from feature to feature, from time to time, and from place to place is a difficult one. There are enormous variations in the over-all level of auroral brightness from night to night, and large short-period fluctuations, some lasting only seconds, during a given display. Even relative intensities of some of the spectral features undergo large variations. Consequently there are not many well-defined quantitative relationships.

The over-all brightness of a visual aurora is classified as I, II, III, or IV (International Brightness Classification, IBC). These classes correspond, respectively, to the brightness of the Milky Way, the brightness of thin moonlit cirrus clouds, the brightness of thin moonlit cumulus clouds, and the brightness (at the ground) due to an unobscured full moon. Hunten (1955), following a discussion by Seaton (1954), has proposed a more quantitative definition in terms of the apparent emission rate of the green line, which is one of the brightest of the auroral features. According to this proposal, apparent emission rates of 1, 10, 100, and 1000 kR for the green line correspond approximately to the four classifications. Chamberlain (1961, p. 197) has estimated the average intensities of other auroral features relative to the green line for an aurora of brightness III.

The question of subvisual aurora was raised in Subsection 9.3.1. One important example which has received considerable study in recent years is the midlatitude stable arc. This feature is observed only in the red lines of atomic oxygen between the altitudes of about 300 and 700 km and in geomagnetic latitude 40°–60°. Although its brightness usually falls below the visual threshold, its orientation with respect to the magnetic field and its variation with magnetic activity are features characteristic of aurora and not nightglow.

Atomic Emissions in the Aurora. Among the more prominent features of the auroral spectrum, as of the airglow spectrum, are the green line (5577 A) and the red lines (6300–6364 A) of neutral atomic oxygen. The $^3P \leftarrow {}^1S$ transition in the ground configuration would give rise to emission near 2972 A which is not observed because of absorption by ozone between the emitting region and the surface. In any case, this transition has a probability considerably less than that of the transition which gives the green line and would appear only weakly.

In contrast to the airglow spectrum, the auroral spectrum also contains lines arising from transitions involving low-lying metastable terms of N I and even, weakly, of N II and O II. For N I (see Fig. 4.4), the $^4S \leftarrow {}^2D$ doublet at 5198 to 5201 A and the $^4S \leftarrow {}^2P$ line at 3466 A have been clearly identified. The $^2D \leftarrow {}^2P$ doublet at 10,395 to 10,404 A should give rather strong emission. However, it appears in the same portion of the spectrum as a strong band due to N_2 and is difficult to identify. Some of the corresponding lines from N II and O II are observed weakly, and some not at all. Among those that have been observed are O II, $^4S \leftarrow {}^2D$ (3728 A); O II, $^2D \leftarrow {}^2P$ (7319–7330 A); and N II, $^1D \leftarrow {}^1S$ (5755 A).

In addition to these forbidden transitions, a large number of permitted lines from N I, N II, O I, and O II have been observed and identified. None of these is particularly intense, but their presence attests to high levels of excitation since all the responsible transitions originate from orbitals with $n \geqslant 3$. Bates (1960c) and Chamberlain (1961) have given lists of allowed lines that are certainly or probably present in the auroral spectrum.

Perhaps the most significant of the observed lines, in terms of an interpretation of the exciting mechanisms, are the Hα (6563 A), Hβ (4861 A), and Hγ (4340 A) lines of the Balmer series of hydrogen. Vegard (1939) in a short letter to *Nature* announced the detection of Hα and Hβ. These identifications were not universally accepted for over a decade. The features had not shown up on earlier spectra (Vegard, 1940) and the available dispersion was such that their wavelengths could not be specified very exactly. In 1950, Gartlein discussed observations of these and of Hγ on several auroral spectra. He pointed out that "these three lines rise and fall in intensity together and not in step with other lines," which strongly suggested that they arose from the same element, hydrogen.

Both Vegard and Gartlein observed that the lines were quite broad, indicating rapid motion of the emitting atoms in the line of sight with resultant Doppler broadening. These observations were made in the general direction of the magnetic horizon. Meinel (1951a) observed Hα in the direction of the magnetic zenith (along the lines of force of the earth's magnetic field) and made the very significant observation that in this direction the line was not only Doppler broadened but also noticeably displaced toward the violet, indicating a rapid average motion of the emitting atoms toward the earth. This evidence, verified by subsequent work, shows that the atoms responsible for the emission of the auroral hydrogen lines are entering that part of the earth's atmosphere where the aurora originates from much higher levels or from outer space at high velocities, and suggests that the high energy of these impinging particles may contribute to the auroral emissions.

Molecular Emissions in the Aurora. Emissions of molecular nitrogen, mostly absent from the airglow spectrum, are very prominent in the auroral spectrum. Among the most intense bands in the visible part of the spectrum are those from the transition $B\,{}^3\Pi_g \rightarrow A\,{}^3\Sigma_u{}^+$ (see Fig. 4.7), which is called the First Positive system. The bands in the infrared are particularly strong. The Second Positive system $C\,{}^3\Pi_u \rightarrow B\,{}^3\Pi_g$ is prominent in the ultraviolet part of the spectrum, with the 0–0 band at 3371 A. Even the forbidden transition $A\,{}^3\Sigma_u{}^+ \rightarrow X\,{}^1\Sigma_g{}^+$ is observed, giving rise to the Vegard–Kaplan bands in the ultraviolet.

Emissions from singly ionized molecular nitrogen are also among the recognized band systems in the auroral spectrum. The First Negative system $N_2{}^+\,{}^2\Sigma_u{}^+ \rightarrow {}^2\Sigma_g{}^+$, with its 0–0 band at 3914 A has long been recognized and is a very prominent part of the ultraviolet emission. On the other hand, bands due to the forbidden transition ${}^2\Pi \rightarrow {}^2\Sigma_g{}^+$ were identified relatively recently by Meinel (1951b) and are called after him the Meinel Negative system.

Emission from molecular oxygen, although present, is relatively weak. The most prominent system to be expected is the Atmospheric system, ${}^1\Sigma_g{}^+ \rightarrow {}^3\Sigma_g{}^-$ (see Fig. 4.8). The 0–0 band of this system is missing, because of absorption between the emitting layer and the surface. However, the 0–1 band at 8645 A was detected by Meinel (1951b) and the 1–1 band at 7708 A by Chamberlain, Fan, and Meinel (1954). In addition the 0–1 band of the infrared system of $O_2{}^+$, ${}^4\Sigma_g{}^- \rightarrow {}^4\Pi_u$, has been recognized.

Theory of Auroral Excitation. It is quite apparent from a consideration of the auroral spectrum that a source of appreciable energy is required to produce the high levels of excitation involved. This source is associated ultimately with streams of neutral but ionized gas, mostly protons and electrons, emitted by the sun. The importance of the particle emissions in producing aurora (and magnetic activity) has been appreciated for many years; but the interaction of these particles with the earth's magnetic field and the ionized constituents of the earth's upper atmosphere is vastly more complex than was imagined in some of the earlier theories. In fact, it is still imperfectly understood and may be for some years to come.

Measurements from interplanetary probes have shown the presence of a steady flux of plasma from the sun (the *solar wind*). Times of magnetic disturbance appear from the most recent measurements to correspond to times when this plasma has a particularly high velocity. The earth's magnetic field prevents the ionized particles from entering the atmosphere directly and carves out a hollow in the plasma at a distance from the earth of about 10 earth radii, perhaps more on the night side. At the boundary of the plasma the character of the magnetic field (as observed from space probes) changes suddenly. Within the boundary, the strength of the field decreases steadily with distance from the earth, although somewhat less rapidly than predicted by the inverse-cube law of a dipole field. At and beyond the boundary there are turbulent, rapidly changing magnetic fields, which merge gradually into a quiet interplanetary field.

At the same time it is known that there are energetic charged particles trapped in the earth's magnetic field at distances of more than 1 earth radius and to some extent concentrated in the Van Allen belts. The origin of these particles has not been established. Some of them may be of solar origin, but it seems more likely that they come from the terrestrial atmosphere. By some mechanism not well understood, these particles acquire higher energy at the time of magnetic activity and aurora, presumably as a result of local accelerations induced in some way by the incident solar plasma.

There is evidence that aurora are accompanied by and caused at least in part by the penetration to low levels of high-energy electrons (10 kev or more). These have been detected directly by rockets during auroral displays and inferred from measurements at balloon altitudes of *Bremsstrahlung* (X-ray radiation produced by the acceleration of energetic electrons passing through the Coulomb field of an atomic nucleus). They seem to be present in the auroral zones at all times, but with greater intensity during auroral displays. Exactly where such electrons originate and how they acquire their energies is not really clear, but it seems likely that they are somehow accelerated at levels above several hundred kilometers and precipitated into the auroral zones along the magnetic lines of force. There is direct evidence also, from rockets and from the observation of hydrogen lines in the auroral spectrum, of a similar bombardment by protons during an aurora, but the electrons seem to carry more energy.

Regardless of their origin, energetic electrons or protons can produce a wealth of reactions as they penetrate to 100 km or so. Bates (1960c) and Chamberlain (1961, Chapter 7) have discussed some of these. With the different types of particles that might be involved—protons, neutral hydrogen atoms, primary electrons, and secondary electrons (released at lower energies after collisions involving the higher-energy particles)—and with the number of atmospheric constituents that are affected and with the number of emissions that result, the discussion becomes quite complicated and encompasses many uncertainties.

In addition to the direct influx of energy associated with charged particles, it has been suggested that energy can reach the auroral zones in the form of hydromagnetic waves, originating at high levels and propagated along the magnetic lines of force.

REFERENCES

AONO, Y., HIRAO, K., and MIYAZAKI, S. (1963). Profile of charged particle density in the ionosphere observed with rockets. *In* "Space Research III" (W. Priester, ed.), pp. 221–227. North-Holland Publ. Co., Amsterdam.

APPLETON, E. V. (1932). Wireless studies of the ionosphere. *J. Instn. Elect. Eng.* **71**, 642–650.

APPLETON, E. V. (1953). A note on the "sluggishness" of the ionosphere. *J. Atmos. Terr. Phys.* **3**, 282–284.

APPLETON, E. V. (1959a). Global morphology of the *E*- and *F*1-layers of the ionosphere. *J. Atmos. Terr. Phys.* **15**, 9–12.

APPLETON, E. V. (1959b). The normal *E* region of the ionosphere. *Proc. Inst. Rad. Eng.* **47**, 155–159.

APPLETON, E. V., and BARNETT, M. A. F. (1925a). Local reflection of wireless waves from the upper atmosphere. *Nature* **115**, 333–334.

APPLETON, E. V., and BARNETT, M. A. F. (1925b). On some direct evidence for downward atmospheric reflection of electric rays. *Proc. Roy. Soc.* **A109**, 621–641.

APPLETON, E. V., and LYON, A. J. (1955). Ionospheric layer formation under quasi-stationary conditions. *In* "The Physics of the Ionosphere," pp. 20–39. Physical Society, London.

APPLETON, E. V., and LYON, A. J. (1957). Studies of the *E* layer of the ionosphere. I, Some relevant theoretical relationships. *J. Atmos. Terr. Phys.* **10**, 1–11.

APPLETON, E. V., and LYON, A. J. (1961). Studies of the *E*-layer of the ionosphere. II, Electromagnetic perturbations and other anomalies. *J. Atmos. Terr. Phys.* **21**, 73–99.

APPLETON, E. V., and PIGGOTT, W. R. (1954). Ionosphere absorption measurements during a sunspot cycle. *J. Atmos. Terr. Phys.* **5**, 141–172.

ARMSTRONG, E. B. (1959). The temperature in the atmospheric region emitting the night-glow O I 5577 line and in regions above faint auroral arcs. *J. Atmos. Terr. Phys.* **13**, 205–216.

BAKER, W. G. (1953). Electric currents in the ionosphere. II, The atmospheric dynamo. *Phil. Trans. Roy. Soc.* **A246**, 295–305.

BAKER, W. G., and MARTYN, D. F. (1953). Electric currents in the ionosphere. I, The conductivity. *Phil. Trans. Roy. Soc.* **A246**, 281–294.

BARBIER, D. (1947). Mesures spectrophotométriques sur le spectre du ciel nocturne (λλ4600–3100). *Comptes Rendus Acad. Sci. Paris* **224**, 635–636.

BARBIER, D. (1956). Résultats préliminaires d'une photométrie en huit couleurs de la lumière du ciel nocturne. *In* "The Airglow and the Aurorae" (E. B. Armstrong and A. Dalgarno, eds.), pp. 38–62. Pergamon Press, New York.

BARBIER, D. (1959). Recherches sur la raie 6300 de la luminescence atmosphérique nocturne. *Ann. Geoph.* **15**, 179–217.

BARBIER, D. (1963). Introduction à l'étude de la luminescence atmosphérique et de l'aurore polaire. *In* "Geophysics, the Earth's Environment" (C. DeWitt, J. Hieblot, and A. Lebeau, eds.), pp. 301–368. Gordon & Breach, New York.

BARTH, C. A., and HILDEBRANDT, A. F. (1961). The 5577 A airglow emission mechanism. *J. Geoph. Res.* **66**, 985–986.

BATES, D. R. (1954). The physics of the upper atmosphere. *In* "The Earth as a Planet" (G. P. Kuiper, ed.), pp. 576–643. Univ. of Chicago Press, Chicago, Illinois.

BATES, D. R. (1957). Theory of the night airglow. *In* "Threshold of Space" (M. Zelikoff, ed.), pp. 14–21. Pergamon Press, New York.

BATES, D. R. (1960a). The airglow. *In* "Physics of the Upper Atmosphere" (J. A. Ratcliffe, ed.), pp. 219–267. Academic Press, New York.

BATES, D. R. (1960b). General character of auroras. *In* "Physics of the Upper Atmosphere" (J. A. Ratcliffe, ed.), pp. 269–296. Academic Press, New York.

BATES, D. R. (1960c). The auroral spectrum and its interpretation. *In* "Physics of the Upper Atmosphere" (J. A. Ratcliffe, ed.), pp. 297–353. Academic Press, New York.

BATES, D. R., and DALGARNO, A. (1954). Theoretical considerations regarding the dayglow. *J. Atmos. Terr. Phys.* **5**, 329–344.

BATES, D. R., and MASSEY, H. S. W. (1946). The basic reactions in the upper atmosphere, I. *Proc. Roy. Soc.* **A187**, 261–296.

BATES, D. R., and MASSEY, H. S. W. (1947). The basic reactions in the upper atmosphere. II, The theory of recombination in the ionized layers. *Proc. Roy. Soc.* **A192**, 1–16.

BATES, D. R., and NICOLET, M. (1950). The photochemistry of atmospheric water vapor. *J. Geoph. Res.* **55**, 301–327.

BERNARD, R. (1939). The identification and the origin of atmospheric sodium. *Ap. J.* **89**, 133–135.

BEYNON, W. J. G. (1959). Vertical drift effects in region-*E*. *J. Atmos. Terr. Phys.* **15**, 13–20.

BOOKER, H. G. (1960). Radar studies of the aurora. *In* "Physics of the Upper Atmosphere" (J. A. Ratcliffe, ed.), pp. 355–375. Academic Press, New York.

BOURDEAU, R. E. (1963). Ionospheric research from space vehicles. *Space Sci. Rev.* **1**, 683–728.

BOURDEAU, R. E., JACKSON, J. E., KANE, J. A., and SERBU, G. P. (1960). Ionospheric measurements using environmental sampling techniques. *In* "Space Research I" (H. Kallmann Bijl, ed.), pp. 328–339. North-Holland Publ. Co., Amsterdam.

BOWLES, K. L., and Staff (1963). Equatorial electron density profiles to 5000 km, using the incoherent scatter technique. *In* "Space Research III" (W. Priester, ed.), pp. 253–264. North-Holland Publ. Co., Amsterdam.

BRADBURY, N. E. (1938). Ionization, negative-ion formation, and recombination in the ionosphere. *Terr. Magn. Atmos. Elect.* **43**, 55–66.

BREIT, G., and TUVE, M. A. (1926). A test of the existence of the conducting layer. *Phys. Rev.* **28**, 554–575.

BRIGGS, B. H., and SPENCER, M. (1954). Horizontal movements in the ionosphere. *Rep. Prog. Phys.* **17**, 245–280.

BUDDEN, K. G. (1961). "Radio Waves in the Ionosphere." Cambridge Univ. Press, London and New York.

CABANNES, J., DUFAY, J., and GAUZIT, J. (1938). Sodium in the upper atmosphere. *Ap. J.* **88**, 164–172.

CHALMERS, J. A. (1962). The first suggestion of an ionosphere. *J. Atmos. Terr. Phys.* **24**, p. 219.

CHAMBERLAIN, J. W. (1955). The ultraviolet airglow spectrum. *Ap. J.* **121**, 277–286.

CHAMBERLAIN, J. W. (1958). The blue airglow spectrum. *Ap. J.* **128**, 713–717.

CHAMBERLAIN, J. W. (1961). "Physics of the Aurora and Airglow." Academic Press, New York.

CHAMBERLAIN, J. W., FAN, C. Y., and MEINEL, A. B. (1954). A new O_2 band in the infrared auroral spectrum. *Ap. J.* **120**, 560–562.

CHANDRA, S. (1963). Electron density distribution in the upper *F* region. *J. Geoph. Res.* **68**, 1937–1942.

CHAPMAN, S. (1918). An outline of a theory of magnetic storms. *Proc. Roy. Soc.* **A95**, 61–83.

CHAPMAN, S. (1919). The solar and lunar diurnal variations of terrestrial magnetism. *Phil. Trans. Roy. Soc.* **A218**, 1–118.

CHAPMAN, S. (1925). The lunar diurnal magnetic variation at Greenwich and other observatories. *Phil. Trans. Roy. Soc.* **A225**, 49–91.

CHAPMAN, S. (1927). On certain average characteristics of world wide magnetic disturbance. *Proc. Roy. Soc.* **A115**, 242–267.

CHAPMAN, S. (1931a). The absorption and dissociative or ionizing effect of monochromatic radiation in an atmosphere on a rotating earth. *Proc. Phys. Soc.* **43**, 26–45.

CHAPMAN, S. (1931b). Some phenomena of the upper atmosphere. *Proc. Roy. Soc.* **A132**, 353–374.

CHAPMAN, S. (1939). Notes on atmospheric sodium. *Ap. J.* **90**, 309–316.

CHAPMAN, S. (1953). Polar and tropical aurorae : and the isoauroral diagram. *Proc. Ind. Acad. Sci.* **A37**, 175–188.

CHAPMAN, S. (1956). The electrical conductivity of the ionosphere : a review. *Nuovo Cimento* [10] **4**, Suppl., 1385–1412.

CHAPMAN, S. (1963). Solar plasma, geomagnetism and aurora. *In* "Geophysics, the Earth's Environment" (C. DeWitt, J. Hieblot, and A. Lebeau, eds.), pp. 371–502. Gordon & Breach, New York.

CHAPMAN, S., and BARTELS, J. (1940). "Geomagnetism." Oxford Univ. Press (Clarendon), London and New York. (Reprinted, with corrections, 1951 and 1962.)

CLAYTON, H. H. (1940). Auroras and sunspots. *Terr. Magn. Atmos. Elect.* **45**, 13–17.

DALGARNO, A. (1958). The altitudes and excitation mechanisms of the night airglow. *Ann. Geoph.* **14**, 241–252.

DAVIS, T. N. (1962). The morphology of the auroral displays of 1957–1958. 1. Statistical analyses of Alaska data. *J. Geoph. Res.* **67**, 59–74.

DIXON, F. E. (1939). A 27.3-day period in the aurora borealis. *Terr. Magn. Atmos. Elect.* **44**, 335–338.

DUFAY, J., and DUFAY, M. (1951). Étude du spectre d'émission du ciel nocturne, de 6800 à 9000 A. *Comptes Rendus Acad. Sci. Paris* **232**, 426–428.

ECCLES, W. H. (1912). On the diurnal variations of the electric waves occurring in nature, and on the propagation of electric waves round the bend of the earth. *Proc. Roy. Soc.* **A87**, 79–99.

ELLYETT, C., and WATTS, J. M. (1959). Stratification in the lower ionosphere. *J. Res. Nat. Bur. Stand.* **63D**, 117–134.

FEJER, J. A. (1953). Semidiurnal currents and electron drifts in the ionosphere. *J. Atmos. Terr. Phys.* **4**, 184–203.

Fejer, J. A. (1955). The interaction of pulsed radio waves in the ionosphere. *J. Atmos. Terr. Phys.* **7**, 322–332.

Finch, H. F., and Leaton, B. R. (1957). The earth's main magnetic field-Epoch 1955.0. *Mon. Not. Roy. Ast. Soc.*, Geoph. Suppl. 7, 314–317.

Frerichs, R. (1930). The singlet system of the oxygen arc spectrum and the origin of the green auroral line. *Phys. Rev.* **36**, 398–409.

Fritz, H. (1881). "Das Polarlicht." Brockhaus, Leipzig.

Gardner, F. F., and Pawsey, J. L. (1953). Study of the ionospheric *D*-region using partial reflections. *J. Atmos. Terr. Phys.* **3**, 321–344.

Garriott, O. K., and Bracewell, R. N. (1961). Satellite studies of the ionization in space by radio. *Advances in Geoph.* **8**, 85–135.

Gartlein, C. W. (1947). Unlocking secrets of the northern lights. *Nat. Geogr. Magaz.* **92**, 673–704.

Gartlein, C. W. (1950). Aurora spectra showing broad hydrogen lines. *Trans. Amer. Geoph. Union* **31**, 18–20.

Harang, L. (1945). A study of auroral arcs and draperies. *Terr. Magn. Atmos. Elect.* **50**, 297–306.

Harang, L. (1946). The auroral luminosity-curve. *Terr. Magn. Atmos. Elect.* **51**, 381–400.

Harang, L. (1951). "The Aurorae." Wiley, New York.

Harang, L., and Tönsberg, E. (1932). Investigations of the aurora borealis at Nordlys Observatoriet, Tromsö, 1929–1930. *Geofys. Publ.* **9**, No. 5, 1–50.

Hartree, D. R. (1931). The propagation of electromagnetic waves in a refracting medium in a magnetic field. *Proc. Camb. Phil. Soc.* **27**, 143–162.

Heaviside, O. (1902). Telegraphy. I, Theory. *Encycl. Brittanica* (10th ed.) **33**, 213–218.

Helliwell, R. A., and Morgan, M. G. (1959). Atmospheric whistlers. *Proc. Inst. Rad. Eng.* **47**, 200–208.

Heppner, J. P., and Meredith, L. H. (1958). Nightglow emission altitudes from rocket measurements. *J. Geoph. Res.* **63**, 51–65.

Herzberg, G. (1951). The atmospheres of the planets. *J. Roy. Ast. Soc. Canada* **45**, 100–123.

Hines, C. O. (1960). Internal atmospheric gravity waves at ionospheric heights. *Canad. J. Phys.* **38**, 1441–1481.

Hines, C. O. (1963). The upper atmosphere in motion. *Quart. J. Roy. Meteor. Soc.* **89**, 1–42.

Hunten, D. M. (1955). Some photometric observations of auroral spectra. *J. Atmos. Terr. Phys.* **7**, 141–151.

Hunten, D. M., Roach, F. E., and Chamberlain, J. W. (1956). A photometric unit for the airglow and aurora. *J. Atmos. Terr. Phys.* **8**, 345–346.

Johnson, C. Y., Meadows, E. B., and Holmes, J. C. (1958). Ion composition of the arctic ionosphere. *IGY Rocket Rep. Ser.* No. 1, pp. 120–130.

Johnson, F. S. (1962). The physical properties of the earth's ionosphere. *In* "Progress in Astronautical Sciences" (S. F. Singer, ed.), pp. 51–91. North-Holland Publ. Co., Amsterdam.

Kennelly, A. E. (1902). On the elevation of the electrically conducting strata of the Earth's atmosphere. *Elec. World and Eng.* **39**, 473.

Krassovsky, V. I. (1958). The nature of emissions of the upper atmosphere. *Ann. Geoph.* **14**, 395–413.

Larmor, J. (1924). Why wireless electric rays can bend round the earth. *Phil. Magaz.* [6] **48**, 1025–1036.

MacDonald, G. J. F. (1963). The structure of the ionosphere. *In* "Geophysics, the Earth's Environment" (C. DeWitt, J. Hieblot, and A. Lebeau, eds.), pp. 279–299. Gordon & Breach, New York.

McEwen, D. J., and Montalbetti, R. (1958). Parallactic measurements on aurorae over Churchill, Canada. *Canad. J. Phys.* **36**, 1593–1600.

Malville, J. M. (1959). Antarctic auroral observations, Ellsworth Station, 1957. *J. Geoph. Res.* **64**, 1389–1393.

Martyn, D. F. (1955). Geomagnetic anomalies of the $F2$ region and their interpretation. *In* "The Physics of the Ionosphere," pp. 260–264. Physical Society, London.

Martyn, D. F. (1959). The normal F region of the ionosphere. *Proc. Inst. Rad. Eng.* **47**, 147–155.

Massey, H. S. W., and Boyd, R. L. F. (1959). "The Upper Atmosphere." Philosophical Library, New York.

Mawdsley, J. (1961). Air density variations in the mesosphere, and the winter anomaly in ionospheric absorption. *J. Geoph. Res.* **66**, 1298–1299.

Meinel, A. B. (1950a). OH emission bands in the spectrum of the night sky, I. *Ap. J.* **111**, 555–564.

Meinel, A. B. (1950b). OH emission bands in the spectrum of the night sky, II. *Ap. J.* **112**, 120–130.

Meinel, A. B. (1950c). O_2 emission bands in the infrared spectrum of the night sky. *Ap. J.* **112**, 464–468.

Meinel, A. B. (1951a). Doppler-shifted auroral hydrogen emission. *Ap. J.* **113**, 50–54.

Meinel, A. B. (1951b). The auroral spectrum from 6200 to 8900 A. *Ap. J.* **113**, 583–588.

Meinel, A. B., Negaard, B. J., and Chamberlain, J. W. (1954). A statistical analysis of low-latitude aurorae. *J. Geoph. Res.* **59**, 407–413.

Mironov, A. V., Prokudina, V. S., and Shefov, N. N. (1958). Some results of investigations of night airglow and aurorae. *Ann. Geoph.* **14**, 364–365.

Mitra, S. K. (1952). "The Upper Atmosphere," 2nd ed. Asiatic Society, Calcutta.

Munro, G. H. (1958). Travelling ionospheric disturbances in the F region. *Austr. J. Phys.* **11**, 91–112.

Newell, H. E., Jr. (1960). The upper atmosphere studied by rockets and satellites. *In* "Physics of the Upper Atmosphere" (J. A. Ratcliffe, ed.), pp. 73–132. Academic Press, New York.

Nicolet, M. (1945). Contribution à l'étude de la structure de l'ionosphère. *Mem. Inst. Meteor. Belg.* No. 19, p. 83.

Nicolet, M. (1963). La constitution et la composition de l'atmosphère supérieure. *In* "Geophysics, the Earth's Environment" (C. DeWitt, J. Hieblot, and A. Lebeau, eds.), pp. 199–277. Gordon & Breach, New York.

Nicolet, M., and Aikin, A. C. (1960). The formation of the D region of the ionosphere. *J. Geoph. Res.* **65**, 1469–1483.

Nicolet, M., and Swider, W., Jr. (1963). Ionospheric conditions. *Plan. Space Sci.* **11**, 1459–1482.

Noxon, J. F., and Goody, R. M. (1962). Observation of day airglow emission. *J. Atmos. Sci.* **19**, 342–343.

Packer, D. M. (1961). Altitudes of the night airglow radiations. *Ann. Geoph.* **17**, 67–75.

Ratcliffe, J. A. (1959). "The Magneto-ionic Theory and Its Applications to the Ionosphere." Cambridge Univ. Press, London and New York. (Reprinted, 1962.)

Ratcliffe, J. A., and Weekes, K. (1960). The ionosphere. *In* "Physics of the Upper Atmosphere" (J. A. Ratcliffe, ed.), pp. 377–470. Academic Press, New York.

RATCLIFFE, J. A., SCHMERLING, E. R., SETTY, C. S. G. K., and THOMAS, J. O. (1956). The rates of production and loss of electrons in the *F* region of the ionosphere. *Phil. Trans. Roy. Soc.* **A248**, 621–642.

RAWER, K. (1957). "The Ionosphere, Its Significance for Geophysics and Radio Communications." Ungar, New York. [Translation of K. Rawer (1952) "Die Ionosphäre."]

RAYLEIGH, Lord (Strutt, R. J.) (1930). Absolute intensity of the aurora line in the night sky, and the number of atomic transitions required to maintain it. *Proc. Roy. Soc.* **A129**, 458–467.

ROACH, F. E., and MEINEL, A. B. (1955a). The height of the nightglow by the van Rhijn method. *Ap. J.* **122**, 530–553.

ROACH, F. E., and MEINEL, A. B. (1955b). Nightglow heights: a reinterpretation of old data. *Ap. J.* **122**, 554–558.

SAGALYN, R. C., SMIDDY, M., and WISNIA, J. (1963). Measurement and interpretation of ion density distributions in the daytime *F* region. *J. Geoph. Res.* **68**, 199–211.

ST. AMAND, P., PETTIT, H. B., ROACH, F. E., and WILLIAMS, D. R. (1955). On a new method of determining the height of the nightglow. *J. Atmos. Terr. Phys.* **6**, 189–197.

SCHUSTER, A. (1889). The diurnal variation of terrestrial magnetism. *Phil. Trans. Roy. Soc.* **A180**, 467-518.

SEATON, M. J. (1954). Excitation processes in the aurora and airglow. I, Absolute intensities, relative ultra-violet intensities and electron densities in high latitude aurorae. *J. Atmos. Terr. Phys.* **4**, 285-294.

SEATON, M. J. (1958). Oxygen red lines in the airglow. II, Collisional deactivation effects. *Ap. J.* **127**, 67–74.

SEDDON, J. C., and JACKSON, J. E. (1958). Ionospheric electron densities and differential absorption. *Ann. Geoph.* **14**, 456–463.

SLIPHER, V. M. (1929). Emissions in the spectrum of the light of the night sky. *Publ. Ast. Soc. Pacific* **14**, 262–263.

SMITH, L. G. (1961). Electron density measurements by the asymmetrical bipolar probe. *J. Geoph. Res.* **66**, 2562.

STEWART, B. (1882). Hypothetical views regarding the connection between the state of the sun and terrestrial magnetism. *Encycl. Britannica* (9th ed.) **16**, 181–184.

STOREY, L. R. O. (1953). An investigation of whistling atmospherics. *Phil. Trans. Roy. Soc.* **A246**, 113–141.

STÖRMER, C. (1921). Rapport sur une expédition d'aurores boréales à Bossekop et Store Korsnes pendant le printemps de l'année 1913. *Geofys. Publ.* **1**, No. 5, 1–269.

STÖRMER, C. (1926). Résultats des mesures photogrammétriques des Aurores Boréales observées dans le Norvège méridionale de 1911 à 1922. *Geofys. Publ.* **4**, No. 7, 1–107.

STÖRMER, C. (1930). "Photographic Atlas of Auroral Forms and Scheme for Visual Observations of Aurorae," 1st ed. A. W. Bröggers, Oslo.

STÖRMER, C. (1953). Results of the observations and photographic measurements of aurora in southern Norway and from ships in the Atlantic during the Polar Year 1932–1933. *Geofys. Publ.* **18**, No. 7, 1–117.

STÖRMER, C. (1955). "The Polar Aurora." Oxford Univ. Press (Clarendon), London and New York.

SUGIURA, M., and CHAPMAN, S. (1960). The average morphology of geomagnetic storms with sudden commencement. *Abh. Akad. Wiss. Göttingen, Math.-Phys. Kl.* **1**, No. 4, 1–53.

TAYLOR, H. A., Jr., and BRINTON, H. C. (1961). Atmospheric ion composition measured above Wallops Island, Virginia. *J. Geoph. Res.* **66**, 2587–2588.

TITHERIDGE, J. E. (1962). The stratification of the lower ionosphere. *J. Atmos. Terr. Phys.* **24**, 283–296.

TOUSEY, R. (1958). Rocket measurements of the night airglow. *Ann. Geoph.* **14**, 186–195.
VEGARD, L. (1912). On the properties of the rays producing aurora borealis. *Phil. Magaz.* [6] **23**, 211–237.
VEGARD, L. (1939). Hydrogen showers in the auroral region. *Nature* **144**, 1089–1090.
VEGARD, L. (1940). On some recently detected important variations within the auroral spectrum. *Terr. Magn. Atmos. Elect.* **45**, 5–12.
VEGARD, L., and KROGNESS, O. (1920). The position in space of the Aurora Polaris from observations made at the Haldde-Observatory, 1913–14. *Geofys. Publ.* **1**, No. 1, 1–172.
VESTINE, E. H. (1944). The geographic incidence of aurora and magnetic disturbance, northern hemisphere. *Terr. Magn. Atmos. Elect.* **49**, 77–102.
VESTINE, E. H. (1960). The upper atmosphere and geomagnetism. *In* "Physics of the Upper Atmosphere" (J. A. Ratcliffe, ed.), pp. 471–512. Academic Press, New York.
VESTINE, E. H., and SNYDER, E. J. (1945). The geographic incidence of aurora and magnetic disturbance, southern hemisphere. *Terr. Magn. Atmos. Elect.* **50**, 105-124.
VESTINE, E. H., LANGE, I., LAPORTE, L., and SCOTT, W. E. (1947). The geomagnetic field, its description and analysis. *Carnegie Instn. Wash. Publ.* **580**, 1–390.
WARK, D. Q. (1960). Doppler widths of the atomic oxygen lines in the airglow. *Ap. J.* **131**, 491–501.
WATANABE, K., and HINTEREGGER, H. E. (1962). Photoionization rates in the *E* and *F* regions. *J. Geoph. Res.* **67**, 999–1006.

CHAPTER 10

The Transport of Properties in the Upper Atmosphere

Discussions in various portions of this book have emphasized the important role that transport processes must play in the explanation of several observed phenomena of the upper atmosphere. Thus, water vapor (Section 2.3), ozone (Section 5.4), and the heat budget (Sections 7.1 and 7.2) imply the importance of transport processes and, at the same time, indicate some of their effects. However, these discussions are carried out in the contexts of specific problems and are spread through the book in a rather disconnected fashion. Furthermore, they do not cover all of the relevant information; especially, no mention has been made of some highly significant studies of the transport of radioactive material.

For these reasons, this short final chapter is devoted to the general question of transport within the upper atmosphere. Its brevity is by no means an indication of the relative importance of the subject. Rather, the chapter is short partly because brief references can be made to earlier extended discussions, and partly because there are woefully few answers to some of the questions that are raised.

Appendix F contains a discussion classifying the processes responsible for transport in the meridional and vertical directions—mean circulations, standing eddies, and transient eddies. The formulation is in principle appropriate for any scale of motion. However, the latter two are usually thought of with reference to large eddies of synoptic scale, because of the essential role that these large disturbances play in the troposphere. They are spoken of here with this meaning. However, it is not necessarily true that concepts verified for lower levels will also apply to the higher ones. It is perfectly clear (Section 7.3) that molecular diffusion plays a vital role at high enough levels in the thermosphere, and it may be that turbulence of an intermediate scale is important in parts of the upper atmosphere at times. For example, relatively small-scale motions in the lower thermosphere and perhaps also the mesosphere may turn out to play an active, dynamic role in the general circulation.

Useful recent surveys of some of the topics of this chapter have been prepared by Newell (1963a) and by Sheppard (1963).

10.1 Transport through the Tropopause

The problem of transport through the tropopause is not a new one to meteorologists; it is raised, for example, in connection with the presence of ozone in the troposphere. But it has been brought into sharp focus and has acquired a more practical aspect as a result of the development of thermonuclear weapons.

A high-yield bomb burst at the earth's surface or in the troposphere results in the deposition of large amounts of radioactive material in the atmosphere. Some of this radioactive material is attached to large particles (those with significant terminal velocities) and falls out rather quickly as a result of gravity. Some is attached to smaller particles in the troposphere and reaches the earth's surface in a matter of weeks as a result of turbulent mixing or, more important, precipitation. However, a significant amount of radioactive material is deposited in the stratosphere and remains there, meanwhile spreading quasi-horizontally from its source, until it is transported downward through the tropopause. Once in the troposphere it can reach the earth's surface relatively quickly, just like the material originally deposited in the troposphere. How long the material can be expected to remain in the stratosphere is a matter of considerable practical importance as well as scientific interest. Equally important and interesting is the question of whether its transport into the troposphere, after injection at some point in the stratosphere, shows any preferential variation with latitude and season. Large numbers of measurements have been made, both at the earth's surface and in the free atmosphere, to ascertain the amount of radioactive material present at different times and places.

Measurements at the earth's surface have typically showed large deposition rates in late spring in middle latitudes (of both hemispheres, but especially the northern). (See, for example, Martell, 1959; Machta, 1960.) However, the interpretation of this has its ambiguities. In the first place, there is sometimes difficulty in identifying the source of the radioactive material that is finally observed. For example, in some instances it may have come either from Soviet tests at high latitudes or from American or British tests at low latitudes. In the second place, material finally reaching the earth's surface must have entered the troposphere some time earlier and at a different location, perhaps systematically different with respect to latitude because of latitudinal variations of the scavenging precipitation process. Nevertheless, the late-spring maximum is similar to the late-spring maximum of tropospheric ozone in middle latitudes, and both are consistent with the assumption of maximum stratosphere-troposphere interchange during the winter and early spring.

It is now quite clear that important transfer between stratosphere and troposphere takes place impulsively at preferred times and places, and not, as has sometimes been supposed, as a result of quasi-steady small-scale diffusion through a continuous boundary surface (Subsection 2.2.1). There is much evidence for this and, in particular, evidence that points to the importance of the tropopause breaks as preferred regions of such transfer. There have been several revealing case studies of the complex atmospheric circulation in the vicinity of tropopause breaks and jet streams. Among the early ones were those of Reed and Sanders (1953) and Reed (1955), who studied two pronounced individual cases in middle latitudes. They found (among other things) that air of stratospheric origin entered the troposphere in important quantities in the cases studied. Staley (1960) carried out a careful study of the dynamics of air motions near the tropopause. He concluded on the basis of reconstructed isentropic trajectories that air may descend rather rapidly from the lower stratosphere above the low tropopause of high latitudes into the middle-latitude troposphere.

More recently, observational studies of the vertical distribution of radioactive debris and of ozone have lent added weight to these suggestions of flow through the "tropopause." Staley (1962) found that locally high concentrations of radioactive debris in the troposphere tend to be associated with stable layers. In some cases, he was able to trace trajectories of air with these high concentrations back into the stratosphere of higher latitudes. Anderson (1962) found a similar association, but traced no trajectories. Kroening and Ney (1962) have pointed out the same association in connection with a few measurements of the vertical distribution of ozone. From aircraft measurements, Briggs and Roach (1963) have on some occasions found a layer of ozone-rich air extending from stratosphere into troposphere in the region of the break between polar tropopause and middle tropopause.

It is not entirely clear to what extent such interchange is confined to clear-cut tropopause breaks and to what extent the entire middle tropopause may be penetrated by the stable laminae described by Danielsen (Subsection 2.2.1). Many more studies of these complex phenomena are needed. It seems safe to predict that these studies will explain such things as the maxima of tropospheric ozone and radioactive fallout in rather late spring in middle latitudes, the immediate source region for both being the higher-latitude lower stratosphere.

10.2 Meridional and Vertical Transport in the Stratosphere

The classical problem defining the importance and some consequences of transport processes in the stratosphere is the ozone problem. This was discussed at length in Subsection 5.4.2 and there is no point in repeating that discussion here. The principal conclusions are as follows:

(a) Ozone is transported from the middle-latitude or low-latitude upper stratosphere to the high-latitude lower stratosphere during the winter months. Such transport is missing or much weaker during the rest of the year.

(b) This transport could be accomplished by either a meridional circulation or eddy processes. Rough calculations of the latter by Newell (1961, 1963a) indicate that they are in the right direction and strong enough to accomplish the required transport. But this is by no means conclusive because only total amounts of ozone can presently be used in such calculations.

(c) Large amounts of ozone seem to appear in the high-latitude lower stratosphere about the time of stratospheric warmings, indicating a connection between large-scale disturbances and ozone transport.

Some of the most exciting results pertaining to transport in the stratosphere have come from studies of radioactive tracers. A particularly useful study of this type was reported by Feely and Spar (1960). It refers to the distribution of a radionuclide, tungsten-185, injected into the lower stratosphere at 12°N between May and July of 1958 during the "Hardtack" series of U. S. bomb tests. So far as is known, this particular radionuclide entered the stratosphere only at this one place and during this one period of time. Subsequent measurements of its distribution, from U-2 aircraft, therefore could be compared with a known source. Figure 10.1 shows the distribution of tungsten-185 averaged over longitude and time during three later time periods—September-October 1958, November-December 1958, and November-December 1959. For all these time periods the maximum concentration of tungsten-185 remained in low latitudes, but the material spread northward (and presumably also southward) and downward into the higher-latitude stratosphere. The results are consistent with a mechanism of eddy transfer rather than with a large-scale meridional circulation.

Another interesting experiment with a radioactive tracer (Kalkstein, 1962, 1963) was initiated during the Hardtack series. Rhodium-102 was placed in the atmosphere above Johnston Island (16°N) on 11 August 1958 from an air burst at 43 km. The material, according to Kalkstein, reached an altitude greater than 100 km shortly after burst. Sampling measurements in the lower stratosphere above 19.4 km showed the appearance of rhodium-102 in significant quantities only after more than a year. In the Northern Hemisphere the increase began in the fall of 1959 and the rhodium-102 concentration at 20 km had reached a high steady value by late winter (perhaps somewhat earlier at 60° to 70°N). After the spring of 1960, it remained at a relatively steady value through 1961, but with significantly lower concentrations in low latitudes (0°–15°N). In the Southern Hemisphere, large concentrations above 19.4 km appeared about

---

FIG. 10.1. Average distributions of tungsten-185 during the indicated time periods. Units are 10^3 disintegrations per minute per 1000 ft^3 of air at standard temperature and pressure. Data for all time periods are corrected for radioactive decay to a standard time of 15 August 1958. [From data of Feely and Spar, presented by Feely and Spar (1960), and by Newell (1961).]

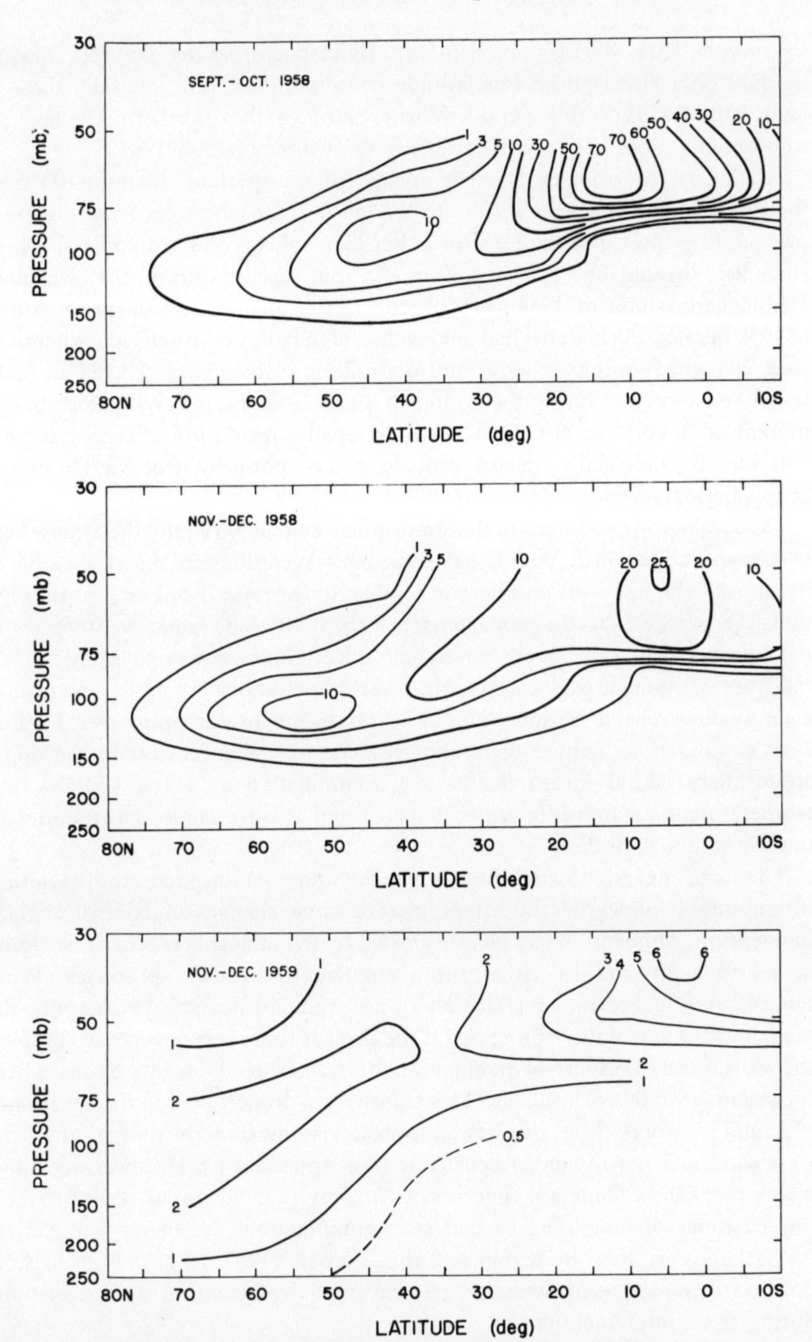
SEPT.-OCT. 1958
PRESSURE (mb)
LATITUDE (deg)
NOV.-DEC. 1958
PRESSURE (mb)
LATITUDE (deg)
NOV.-DEC. 1959
PRESSURE (mb)
LATITUDE (deg)
80N
70
60
50
40
30
20
10
0
10S
30
50
75
100
150
200
250

six months later, during the (Southern Hemisphere) winter of 1960. Also in the Southern Hemisphere, low-latitude concentrations were less than those at high latitudes after this period. With regard to the Northern Hemisphere troposphere, large rises occurred during the winter of 1960–1961.

It appears from the results that downward transport of rhodium-102 from the thermosphere and mesosphere to the lower stratosphere occurred preferentially during the winter months (of either hemisphere) and at middle and high latitudes. Presumably such transfer did not occur during the Northern Hemisphere winter of 1958–1959 or during the Southern Hemisphere winter of 1959 because the material had not reached high latitudes in sufficient quantities soon enough. Presumably also, the tropospheric increase was delayed an additional year because during the winter of 1959–1960 the material (even though present at 20 km) had not reached the tropopause region in sufficient quantity.

Evidently, radioactive tracers provide a very powerful tool for studies of atmospheric transport.

Meteorological conditions in the stratosphere evidently require the atmosphere to transport heat in a meridional direction. According to the discussion in Section 7.1 the upper stratosphere in high latitudes appears to lose heat through radiative processes in the winter and to gain it through radiative processes in the summer. The calculations have some uncertainties but there is little doubt that they are qualitatively correct. However, observations are up to now insufficient to allow reliable computations of heat transport by eddy processes, because such computations require wind and temperature measurements in the upper stratosphere. Wind measurements are accumulating at a few stations, but temperature measurements are still scarce and involve some uncertainties as to their accuracy (Section 3.4).

The large changes of wind direction in the upper stratosphere from easterlies in summer to westerlies in winter involve large changes of relative angular momentum. Momentum can be transported by the atmosphere, either vertically or meridionally and by either mean circulations or eddy processes. Wind observations in the upper stratosphere are quite inadequate to compute the magnitude of any of these processes at the present time. It is possible to compute the meridional transport of momentum by transient eddies at stations where rocketsonde winds are available. This transport is proportional to the covariance of u' and v', where these are instantaneous deviations from the time mean of the west wind and north wind, respectively (see Appendix F). However, locations where this can be done are concentrated mainly near 30° to 40°N and cover a limited range of longitude, so that such computations are merely suggestive. Nevertheless, it may be significant that Newell (1963b) has found sizeable northward transports of (westerly) relative angular momentum at these stations during the winter months.

Most of these studies seem to point to the importance of large-scale disturbances and eddy mechanisms in the transport of atmospheric properties in the stratosphere. They do not rule out a possible role for mean meridional circulations, although any such circulation involving the low-latitude lower stratosphere is hard to reconcile with the tungsten-185 data.

It should be pointed out that the distribution of water vapor in the stratosphere is still mystifying. There are no observations for the upper stratosphere and those for the lower stratosphere (Subsection 2.3.1) do not yet give an unambiguous picture of exactly what does have to be explained by atmospheric transport processes.

10.3 Transport in the Mesosphere

The outstanding problem of transport with respect to the mesosphere is that posed by the heat budget. Doubtless, other problems will come to light when more and varied observations become available.

The subject of the warm mesopause of winter and the cold mesopause of summer in high latitudes has received frequent mention throughout this book (and elsewhere). In particular it is discussed at some length in Subsection 7.2.3, to which the reader is referred. Nothing much can be added here except references to two attempts to compute patterns of meridional circulation presumed to be responsible for the situation.

Murgatroyd and Singleton (1961) used the first law of thermodynamics, with a heating term deduced from the radiative calculations of Murgatroyd and Goody (1958) (see Fig. 7.9), and also the equation of continuity. From these they were able to deduce a mean meridional circulation for the upper stratosphere and mesosphere, on the assumption that all eddy transport processes could be neglected. They neglected also any effects of oxygen recombination, such as have been suggested by Kellogg. Their results showed (for the solstices and the mesosphere) a flow from summer pole to winter pole, with upward motion in the summer hemisphere and downward motion in the winter hemisphere. The meridional component of this circulation was measured in meters (second)$^{-1}$ and the vertical component in centimeters (second)$^{-1}$. However, the details of the calculations should not be taken very seriously because, apart from the question of assumptions, these details evidently depend on the details of the assumed temperature and radiative-heating distributions, neither of which is known very accurately.

Haurwitz (1961) computed a meridional circulation for the Northern Hemisphere upper stratosphere and mesosphere during winter and summer. His scheme was based on an assumed balance (in the horizontal component of the equation of motion) among the Coriolis term, the pressure-gradient term, and a frictional term. The latter was assumed, in accordance with classical mixing-length theory, to be proportional to the curvature of the mean vertical wind profile. This balance is the same one that gives rise to the Ekman spiral in boundary-layer studies. He found, for the winter, poleward motion between about 50 and about 80 km, with equatorward motion above and below.

There is no direct evidence about the importance of eddy-transfer processes in the mesosphere and there will not be until observations become available, because such evidence is necessarily of an empirical type. There is not even much evidence for or against the existence of synoptic-scale disturbances, of the kind so important at lower levels, though in this connection one must recall the wind changes with periods of a few days at 90 km (determined from radio-meteor winds; Subsection 3.2.4), anomalous winter absorption (Subsection 9.2.3), and airglow cells (Subsection 9.3.2). But it would be foolish to discount the possibility that eddy processes play an important role, in view of the past evolution of ideas about transport processes in the troposphere and, more recently, in the lower stratosphere.

REFERENCES

ANDERSON, K. A. (1962). Thin atmospheric layers of radioactive debris during September 1961. Report UCB-62-2. Department of Physics, Univ. of California, Berkeley, California.

BRIGGS, J., and ROACH, W. T. (1963). Aircraft observations near jet streams. *Quart. J. Roy. Meteor. Soc.* **89**, 225–247.

FEELY, H. W., and SPAR, J. (1960). Tungsten-185 from nuclear bomb tests as a tracer for stratospheric meteorology. *Nature* **188**, 1062–1064.

HAURWITZ, B. (1961). Frictional effects and the meridional circulation in the mesosphere. *J. Geoph. Res.* **66**, 2381–2391.

KALKSTEIN, M. I. (1962). Rhodium-102 high-altitude tracer experiment. *Science* **137**, 645–652.

KALKSTEIN, M. I. (1963). Movement of material from high altitude deduced from tracer observations. *J. Geoph. Res.* **68**, 3835–3839.

KROENING, J. L., and NEY, E. P. (1962). Atmospheric ozone. *J. Geoph. Res.* **67**, 1867–1875.

MACHTA, L. (1960). Meteorology and radioactive fallout. *WMO Bull.* (April, 1960) pp. 64–70.

MARTELL, E. (1959). Atmospheric aspects of strontium-90 fallout. *Science* **129**, 1197–1206.

MURGATROYD, R. J., and GOODY, R. M. (1958). Sources and sinks of radiative energy from 30 to 90 km. *Quart. J. Roy. Meteor. Soc.* **84**, 225–234.

MURGATROYD, R. J., and SINGLETON, F. (1961). Possible meridional circulations in the stratosphere and mesosphere. *Quart. J. Roy. Meteor. Soc.* **87**, 125–135.

NEWELL, R. E. (1961). The transport of trace substances in the atmosphere and their implications for the general circulation of the stratosphere. *Geof. Pura et Appl.* **49**, 137–158.

NEWELL, R. E. (1963a). Transfer through the tropopause and within the stratosphere. *Quart. J. Roy. Meteor. Soc.* **89**, 167–204.

NEWELL, R. E. (1963b). Preliminary study of quasi-horizontal eddy fluxes from meteorological rocket network data. *J. Atmos. Sci.* **20**, 213–225.

REED, R. J. (1955). A study of a characteristic type of upper-level frontogenesis. *J. Meteor.* **12**, 226–237.

REED, R. J., and SANDERS, F. (1953). An investigation of the development of a mid-tropospheric frontal zone and its associated vorticity field. *J. Meteor.* **10**, 338–349.

SHEPPARD, P. A. (1963). Atmospheric tracers and the study of the general circulation of the atmosphere. *Rep. Prog. Phys.* **26**, 213–267.

STALEY, D. O. (1960). Evaluation of potential-vorticity changes near the tropopause and the related vertical motions, vertical advection of vorticity, and transfer of radio-active debris from stratosphere to troposphere. *J. Meteor.* **17**, 591–620.

STALEY, D. O. (1962). On the mechanism of mass and radioactivity transport from stratosphere to troposphere. *J. Atmos. Sci.* **19**, 450–467.

APPENDIX A

Some Useful Constants

Electronic charge	$e = 4.803 \times 10^{-10}$ esu
	$= 1.602 \times 10^{-20}$ emu
Electronic mass	$M_e = 9.108 \times 10^{-28}$ g
Speed of light	$c = 2.998 \times 10^{10}$ cm sec^{-1}
Planck constant	$h = 6.625 \times 10^{-27}$ erg sec
Avogadro number	$N = 6.025 \times 10^{23}$ molecule mole^{-1}
Loschmidt number	$n_0 = 2.687 \times 10^{19}$ molecule cm^{-3} at NTP
Boltzmann constant	$k = 1.380 \times 10^{-16}$ erg molecule^{-1} deg^{-1}
Universal gas constant	$R = 8.317 \times 10^{7}$ erg mole^{-1} deg^{-1}
Gas constant for dry air at sea level	$R_0 = 2.871 \times 10^{6}$ erg g^{-1} deg^{-1}
Ratio of the specific heats of dry air	$\gamma = 1.400$
Gravitational constant	$G = 6.668 \times 10^{-8}$ dyne cm^{2} g^{-2}
Acceleration of gravity (at sea level and 45° latitude)	$g = 9.806 \times 10^{2}$ cm sec^{-2}
Radius of the earth (of a sphere with the same volume)	$r_e = 6.371 \times 10^{8}$ cm
Angular speed of rotation of the earth	$\omega = 7.292 \times 10^{5}$ radians sec^{-1}
First constant in the Planck radiation law (radiance as a function of frequency)	$C_1 = 2h/c^2 = 1.474 \times 10^{-7}$ g sec
Second constant in the Planck radiation law (frequency)	$C_2 = h/k = 4.801 \times 10^{-11}$ sec deg
Stefan–Boltzmann constant	$\sigma = 5.669 \times 10^{-5}$ erg cm^{-2} sec^{-1} deg^{-4}
Constant in the Wien displacement law	$C = .2898$ cm deg
Mean distance, earth to sun	$d = 1.496 \times 10^{13}$ cm
Mean radius of the sun (visible disk)	$r_s = 6.96 \times 10^{10}$ cm

APPENDIX B

Additional Comments on Spectroscopy and Quantum Mechanics

B.1 The Schrödinger Wave Equation

According to wave mechanics, the behavior of a corpuscle is described by the Schrödinger wave equation

$$\nabla^2\psi + \frac{8\pi^2 M}{h^2}(E - E_p)\psi = 0 \tag{B.1}$$

where M is the mass, E the total energy, and E_p the potential energy of the corpuscle. ψ is the amplitude of the wave function Ψ, where $\Psi = \psi \exp[-\pi i(E/h)t]$. A particular function (of the coordinates) ψ_i that satisfies (B.1) and is everywhere single valued, finite, continuous and vanishes at infinity is called an *eigenfunction* of the problem. Equation (B.1) with the specified boundary conditions is soluble only for certain specific values of E, the *eigenvalues* of the problem.* The eigenvalues and the eigenfunctions depend on the form of the potential energy function E_p of the corpuscle whose behavior is being investigated. For an arbitrary many-electron atom or especially for a molecule, E_p may be a very complicated function of the coordinates of all the electrons and nuclei involved and the wave equation takes a more complicated form.

It is not always possible to solve the wave equation for complicated cases. However, in the cases where an exact or approximate solution can be found corresponding to a sufficiently accurate expression of the potential energy function for a given situation, the eigenvalues E are the observed energy levels. To each eigenvalue there corresponds one or more eigenfunctions related to the probable position of the corpuscle. Transition probabilities between two energy levels can also be calculated from wave-mechanical considerations, but we shall not attempt to discuss that problem here.

One simple application of the Schrödinger wave equation is to the hydrogen atom, with one electron moving under the influence of the Coulomb attraction

* This statement is true when E is negative, that is, when the corpuscle moves in a potential field. Corresponding to the ionization continuum discussed in Subsection 4.1.1, the equation is soluble for all positive values of E.

of the nucleus. To a very good approximation, the potential energy of the electron is given by

$$E_p = -e^2/r \tag{B.2}$$

where e is the charge on the electron and r is the distance between electron and nucleus. Actually both electron and nucleus move about their center of mass, but this can be taken into account by replacing the M of Eq. (B.1) by the *reduced mass* $M' = M_e M_n/(M_e + M_n)$, where M_e is the mass of the electron and M_n is the mass of the nucleus. Since $M_n \cong 1836\, M_e$, $M' \cong M_e$.

The eigenvalues of this problem are

$$E_n = -\frac{R_H hc}{n^2} \tag{B.3}$$

which is exactly Eq. (4.2). The Rydberg constant is given by $2\pi^2 M' e^4/ch^3$. The eigenfunctions of the problem involve other quantum numbers than n, which do not (in this case) enter into the expression for the energy. For any other single-electron atom (such as singly ionized helium, or doubly ionized lithium) the problem is the same except that the reduced mass is slightly different and the right sides of (B.2) and (B.3) must be multiplied by Z^2, where Z is the number of elemental charges (of magnitude e) on the nucleus.

Another simple application is to the vibrational energy levels of a diatomic molecule [see Eq. (4.5)]. The force between the two atoms depends on the difference $(r - r_E)$ between the actual separation and the equilibrium separation. The wave-mechanical problem for the two atoms can be reduced to the problem of a single particle of mass M' whose displacement x from an equilibrium position is equal to $(r - r_E)$. M' is defined as $M_1 M_2/(M_1 + M_2)$, if M_1 and M_2 are the masses of the two atoms. The simplest assumption about the restoring force is that it is given by

$$F = -Cx \tag{B.4}$$

where C is a constant of proportionality. This formulation corresponds to the simple harmonic oscillator of classical mechanics. In this case

$$E_p = \tfrac{1}{2}Cx^2 \tag{B.5}$$

and the wave equation (B.1) becomes

$$\frac{d^2\psi}{dx^2} + \frac{8\pi^2 M'}{h^2}(E - \tfrac{1}{2}Cx^2)\psi = 0 \tag{B.6}$$

Appropriate solutions of this equation exist only for the eigenvalues given by (4.5), where $\nu_e = (C/M')^{\frac{1}{2}}/2\pi$. More accurate approximations for the potential energy lead to better representations of the vibrational energy levels.

A third simple application, and the last considered here, is to the rotational energy levels of a diatomic molecule. The simplest model of rotation is one in which the atoms are taken to be point masses a (constant) distance r apart, connected by a rigid weightless rod and rotating about an axis perpendicular to the center of mass. This problem too can be reduced to the problem of one mass M', defined as in the case of the vibrator, rotating in a circle of radius r. For such a simple rigid rotator the potential energy is zero. The eigenvalues of the wave equation are

$$E_r = hcBJ(J+1) \tag{B.7}$$

where $B = h/8\pi^2 M' r_e^2 c$. When the effect of simultaneous vibration is considered, the situation is more complicated. The leading term in the expression for E_r is analogous to (B.7), but with B replaced by B_v. B_v is somewhat smaller than the B defined above, the more so the larger the vibrational quantum number v.

B.2 Electron Quantum Numbers

The principal quantum number n is a measure of the distance of the electron from the nucleus. Although it is not possible to specify exact electronic orbits, as was supposed in the older Bohr theory of the hydrogen atom, a larger value of n corresponds to a greater likelihood that the electron is at a greater distance from the nucleus. The azimuthal quantum number, l, specifies the angular momentum of the electron, which is given by $\sqrt{l(l+1)}(h/2\pi)$. In a magnetic or electric field the component of angular momentum in the direction of the field can take on only the discrete values $m_l(h/2\pi)$. The angular momentum associated with electron spin is $\sqrt{s(s+1)}(h/2\pi)$ where s has the value $\frac{1}{2}$. In a magnetic field, the component of this momentum in the direction of the field is $m_s(h/2\pi)$ where m_s can have only the values $\pm\frac{1}{2}$.

B.3 Term Symbols for Atoms

The procedure described here is referred to as "Russell–Saunders coupling" and is valid for the atoms we shall consider. For atoms with very many electrons a different procedure is necessary.

Consider first nonequivalent electrons. The quantum number L, for two such electrons with azimuthal quantum numbers l_1 and l_2, can take on any of the integral values $(l_1 + l_2)$, $(l_1 + l_2 - 1)$, ..., $|l_1 - l_2|$. For example, for two nonequivalent p electrons, $L = 2, 1, 0$. For three nonequivalent electrons, L is obtained by combining any two in the manner just described and then combining the resulting possible values with the third. For example, if a d electron were combined with the above two p electrons, $L = 4, 3, 2, 1, 0; 3, 2, 1; 2$. The quantum number S for two nonequivalent electrons can be either 0 or 1;

in the former case the spin momentum vectors are in opposite directions; in the latter they are in the same direction. Thus, either singlet or triplet terms can result. If a third electron is added, then the resultant S is $\frac{1}{2}$ (obtained by adding $\frac{1}{2}$ to 0), $\frac{1}{2}$ (obtained by subtracting $\frac{1}{2}$ from 1), or $\frac{3}{2}$ (obtained by adding $\frac{1}{2}$ to 1), giving doublets and quartets. The combination of these L's and S's gives altogether six terms for the two p electrons and 27 terms (some of which have the same L and S but arise in different ways) for the ppd combination.

In the case of equivalent electrons, which by definition have the same values of n and the same values of l, the Pauli exclusion principle severely limits the number of possible terms. For example, for two *equivalent* p electrons only the terms ^1S, ^1D, ^3P are possible, although the manner of determining this will not be considered here. Equivalent electrons in a *closed shell* need not be considered at all in determining the term symbol for the atom.

The vector angular momenta of L and S can be oriented to each other only in such a way that their resultant has the value $\sqrt{J(J+1)}(h/2\pi)$ where J is an integer. J can be no larger than $L + S$ and no smaller than $|L - S|$ and can take on all integral values between. For $L \geqslant S$, there are $(2S + 1)$ possible values of J, to each of which there corresponds a slightly different amount of energy. For $L < S$, there are $(2L + 1)$ possible values. In all cases, however, the factor $(2S + 1)$ is called the multiplicity of the term.

B.4 The Helium Spectrum

With reference to Table 4.1 and Fig. 4.3, each combination of L and S gives rise to a series of terms with different values of n for the emission electron. According to Table 4.1, all terms with $S = 0$ have only one possible value of J and thus only one energy level for each term. These are called singlets. On the other hand, terms with $S = 1$ $(2S + 1 = 3)$ have three possible values of J (except when $L = 0$) and thus three energy levels associated with each term. These are called triplets. Note that the ^3S terms have only one value of J (because $S > L$) but nevertheless are called triplet terms and have different energies than the corresponding ^1S terms. In general the variation among these energies is small (for atoms with not many electrons) and the levels corresponding to different J values are not shown in Fig. 4.3.

If all transitions among this large number of energy levels were equally probable, many more lines would occur in the helium spectrum than are actually observed. However, the transition rules "forbid" or render unlikely a large number of them. According to (a) (see Subsection 4.1.2), transitions between singlets and triplets do not occur. Furthermore, within the triplet system or within the singlet system, L changes only by ± 1 [$\Delta L = 0$, in this case, would violate rule (d)]. There are, however, several permitted series of transitions that are similar in many respects to the Rydberg series of the H atom. For example, transitions from the ^1S term with $n = 1$ to the ^1P term with $n \geqslant 2$ give rise to intense lines in the far ultraviolet between 500 and 600 A; transitions from ^1P with $n = 2$ to ^1S or ^1D with $n \geqslant 3$ give lines in the visible; transitions from ^3P with $n = 2$ to ^3S or ^3D with $n \geqslant 3$ also give lines in the visible region.

Note that the ^1S and ^3S terms with $n = 2$, which lie well above the ground term, are metastable terms.

B.5 Electronic Quantum Numbers for Molecules

In the case of molecules, as opposed to atoms, there is a strong electric field in the direction of the internuclear axis and the energy of the electronic state depends markedly on the component of the orbital angular momentum in this direction. This component is quantized and its magnitude is given by $\Lambda(h/2\pi)$ where Λ is zero or a positive integer.

Again, as with the atom, the resultant spin of all the electrons is characterized by the quantum number S, which is half-integral if the number of electrons in the molecule is odd and integral if the total number of electrons is even. The component of the spin vector along the internuclear axis is quantized and has the value $\Sigma(h/2\pi)$ where $\Sigma = S, S - 1, S - 2, \ldots, -S$. Note that Σ can be positive or negative, which is another way of saying that the component of spin angular momentum along the internuclear axis can be in the same direction as or in the opposite direction from the component of orbital angular momentum. The quantum number of the resultant electronic angular momentum about the internuclear axis is $\Omega = |\Lambda + \Sigma|$, which is always positive and can take on $(2S + 1)$ values if $\Lambda > 0$ and $\Lambda \geqslant S$.

APPENDIX C

Comments on Radiation Nomenclature

Nomenclature and symbolism for radiation quantities are quite confusing. In many cases different authors use different words to refer to the same quantity, and in some cases they use the same word to refer to different quantities. The purpose of this brief discussion is to point out the more obvious pitfalls to the student or new worker in the field, and not to imply any judgment of the reasons or justifications for the present state of affairs.

Let us start with the recommendations of the Commission for Symbols, Units, and Nomenclature (SUN Commission) of the International Union of Pure and Applied Physics (Wolfe, 1962). This system distinguishes between quantities

TABLE C.1

SOME OF THE RECOMMENDATIONS OF SUN COMMISSION FOR QUANTITIES, NAMES, AND SYMBOLS IN PHOTOMETRY AND RADIATION

Quantity	Photometry		Radiation	
	Name	Symbol	Name	Symbol
1. Energy (emitted[a] by an object) per unit time	Luminous flux	Φ	Radiant flux	Φ_e or P
2. Energy (emitted)[a] per unit time, per unit solid angle	Luminous intensity	I	Radiant intensity	I_e
3. Energy (emitted)[a] per unit time, per unit area	Luminous emittance	M	Radiant emittance	M_e
4. Energy (emitted[a] in one direction) per unit time, per unit solid angle, per unit area normal to the direction considered	Luminance	L	Radiance	L_e
5. Energy (received on or passing through a real or imaginary surface) per unit time, per unit area	Illumination	E	Irradiance	E_e

[a] In general these may include reflected radiation, but in the present context we are interested in emission.

referring to visible light, when photometric units may be appropriate, and quantities referring to radiation in general, when the ordinary physical units are appropriate. Table C.1 lists the quantities, names, and symbols recommended in this system. The use of photometric units is seldom, if ever, necessary in atmospheric studies, and the rest of this discussion ignores this complication.

In the first place, some authors consider that these and similar quantities refer to energy contained within a finite spectral interval. When these authors refer to radiation at one frequency (in the infinitesimal frequency interval ν to $\nu + d\nu$) they specify this by using one of the prefixes "monochromatic" or "spectral." On the other hand, some authors consider that these and similar quantities refer to monochromatic radiation in the first place. When these authors refer to radiation in a finite spectral interval they specify this by using the prefix "integrated" or "average." This difference does not lead to very much confusion, because the meaning is usually stated or is clear in context.

Most authors in the atmospheric sciences nowadays use (with varying degrees of consistency) the system used by Chandrasekhar (1950). The basic quantity in this system is the *specific intensity* which is defined as the energy (transported in one direction across a real or imaginary surface) per unit time, per unit solid angle, per unit area normal to the direction considered, per unit frequency. It is usually given the symbol I_ν. Several points are of interest when this is compared with Table C.1.

(1) In the first place, this quantity is not specifically defined in Table C.1, although it is the geometrical counterpart of and may be used to mean "radiance."

(2) In the second place, it is not the same thing as "radiant intensity," although it has a similar name and symbol. Geometrically and dimensionally it is the "radiant intensity" divided by an area (although in physics the word "specific" usually denotes division by a mass; for example, "specific volume").

(3) In the third place, the prefix "specific" is often not included and the quantity is referred to as simply "intensity."

(4) In the fourth place, even if the definition is in terms of monochromatic radiation, the subscript ν is sometimes dropped from the symbol I.

A second basic quantity used by Chandrasekhar and, with variations, by others is the *net flux*. It is defined by Chandrasekhar as the energy (transported in all directions across a real or imaginary surface) per unit time, per unit area, per unit frequency interval. Chandrasekhar (and some others) use the symbol F_ν to denote the net flux divided by π. Some authors use F_ν to mean the net flux itself. Some authors restrict the direction included to a hemisphere rather than "all directions" in which case the quantity should be designated as simply "flux." Apart from these complications, several points are of interest when this is compared with Table C.1.

(1) In the first place, flux as defined here is the same thing as "irradiance" and may be used to mean "radiant emittance."

(2) In the second place, it is not the same thing as "radiant flux," although it has a similar name. Geometrically and dimensionally it is the "radiant flux" divided by an area.

(3) Sometimes it is referred to as "flux per unit area," but this is cumbersome. Sometimes it is referred to as "flux density," but this is unfortunate because "density" usually denotes division by a volume and not an area. Sometimes it is referred to as "intensity," which is particularly unfortunate, because this usage adds a third possible meaning to an overworked word.

(4) In the fourth place, even if the definition is in terms of monochromatic radiation, the subscript ν is sometimes dropped from the symbol F_ν.

There are other variations in both nomenclature and symbolism, which one might mention. This is enough, however, to emphasize the importance of proper definitions by authors and careful interpretation by readers.

REFERENCES

CHANDRASEKHAR, S. (1950). "Radiative Transfer." Oxford Univ. Press (Clarendon), London and New York. (Reprinted, 1960, Dover, New York.)

WOLFE, H. C. (1962). Symbols, units and nomenclature in physics. *Phys. Today* **15**, No. 6, 19–30.

APPENDIX D

Absorption by an Exponential Atmosphere; Grazing Incidence

Section 4.4 is concerned with the problem of absorption of monochromatic radiation by an exponential atmosphere when the earth is considered to be flat. The purpose of this appendix is to generalize the results obtained there to the case of a spherical earth, following Chapman (1931).

The geometry of the problem is shown in Fig. D.1. It is desired to compute

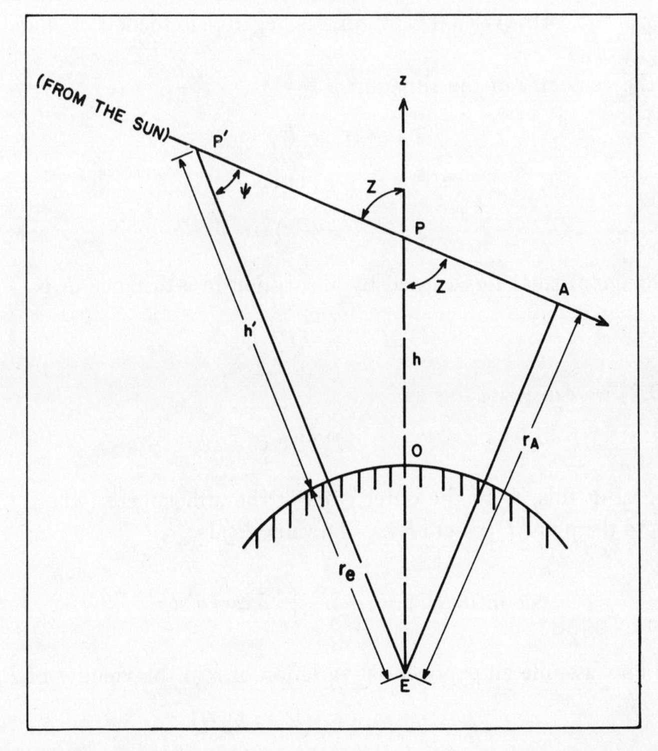

FIG. D.1. Geometry for computation of absorption at grazing incidence. The height of P above the earth's surface, relative to the earth's radius, is exaggerated for illustrative purposes.

the absorption (per unit time, per unit volume) at a point P, which is at a height h above a point O on the earth's surface. At P (or at any point in the vertical above O) the zenith angle is Z. However, owing to the sphericity of the earth, the angle which a ray from the sun makes with the local vertical varies during the ray's passage through the atmosphere. Let $\overline{P'PA}$ represent a ray passing through P and let P' be any point on this ray to the left of P. The point P' is at a height h' above the earth's surface and a distance $(h' + r_e)$ from the center of the earth, where r_e is the earth's radius. At P', the ray makes an angle ψ with the local vertical. As the position of the point P' is varied continuously from an infinite distance to the point P, then ψ varies continuously from 0 to Z. The angle ψ may therefore be thought of as a variable which specifies position on the ray (and also the height h' above the earth's surface).

Let s be a coordinate representing distance along the ray, positive in the direction from P' to P to A. It is convenient to take the origin at the point A, the point on the ray at which a line from the center of the earth (E) is perpendicular to the ray (so that at A, $\psi = 90°$). Let the point A be a distance r_A from the center of the earth. With this origin, s is negative to the left of A and positive to the right of A.

From the geometry of the situation,

$$r_A = (r_e + h') \sin \psi \tag{D.1}$$

$$s(P') = -r_A \operatorname{ctn} \psi \tag{D.2}$$

$$r_A = (r_e + h) \sin Z \tag{D.3}$$

The change of solar flux caused by absorption in a distance ds is

$$dF = -kF\rho \, ds \tag{D.4}$$

Using (D.2) we can write this as

$$dF = -kF\rho r_A \csc^2 \psi \, d\psi \tag{D.5}$$

The integral of this, from the outer edge of the atmosphere (where F has the value F_∞) to the point P (where F has the value F_h) is

$$\ln(F_h/F_\infty) = -kr_A \int_0^Z \rho \csc^2 \psi \, d\psi \tag{D.6}$$

Let us now assume an exponential variation of ρ in the vicinity of P, so that

$$\rho = \rho_h \exp[-(h' - h)/H] \tag{D.7}$$

where ρ_h is the density at height h. With the use of (D.1) this can be rewritten as

$$\rho = \rho_h \exp[(r_e - r_A \csc \psi + h)/H] \tag{D.8}$$

Equation (D.6) becomes, when this expression is used,

$$F_h = F_\infty \exp\left\{-kr_A\rho_h \int_0^Z \csc^2\psi \exp[(r_e - r_A \csc\psi + h)/H]\, d\psi\right\} \quad \text{(D.9)}$$

Finally, if we let $x \equiv (h + r_e)/H$ and replace r_A by $(xH \sin Z)$ according to (D.3), we get

$$F_h = F_\infty \exp\left[-kH\rho_h x \sin Z \int_0^Z \csc^2\psi \exp(x - x \sin Z \csc\psi)\, d\psi\right] \quad \text{(D.10)}$$

This may be written as

$$F_h = F_\infty \exp[-kH\rho_h \operatorname{Ch}(x, Z)] \quad \text{(D.11)}$$

where $\operatorname{Ch}(x, Z)$ is known as the *Chapman function*

$$\operatorname{Ch}(x, Z) \equiv x \sin Z \int_0^Z \csc^2\psi \exp(x - x \sin Z \csc\psi)\, d\psi \quad \text{(D.12)}$$

It has been tabulated by Wilkes (1954).

If we compare (D.11) with (4.41) and note that $[\rho_0 \exp(-z/H)]$ in (4.41) is the same thing as ρ_h in (D.11), we find that the sphericity of the earth can be accounted for simply by replacing sec Z by $\operatorname{Ch}(x, Z)$. This is a very useful result.

The absorbed energy per unit time and volume at height h is

$$q_h = (-dF/ds)_h = -kF_h\rho_h = -k\rho_h F_\infty \exp[-kH\rho_h \operatorname{Ch}(x, Z)] \quad \text{(D.13)}$$

which is the same as the expression given by (4.42) if sec Z there is replaced by $\operatorname{Ch}(x, Z)$.

REFERENCES

Chapman, S. (1931). The absorption and dissociative or ionizing effect of monochromatic radiation in an atmosphere on a rotating earth. II, Grazing incidence. *Proc. Phys. Soc.* **43**, 483–501.

Wilkes, M. V. (1954). A table of Chapman's grazing incidence integral $\operatorname{Ch}(x, \chi)$. *Proc. Phys. Soc.* **B67**, 304–308.

APPENDIX E

The Götz Umkehr Method; Basic Equation

With reference to Fig. E.1, consider an instrument at the ground and pointing toward the zenith, with effective light-gathering area A' and field of view ω' such that it "sees" an area A at height z. The light reaching A' is scattered from the solar beam at all levels in the vertical column above A', but first we confine our attention to the light scattered from a particular volume of air between z and $z + dz$. Furthermore, the discussion is confined to primary scattering by air molecules (Rayleigh scattering).

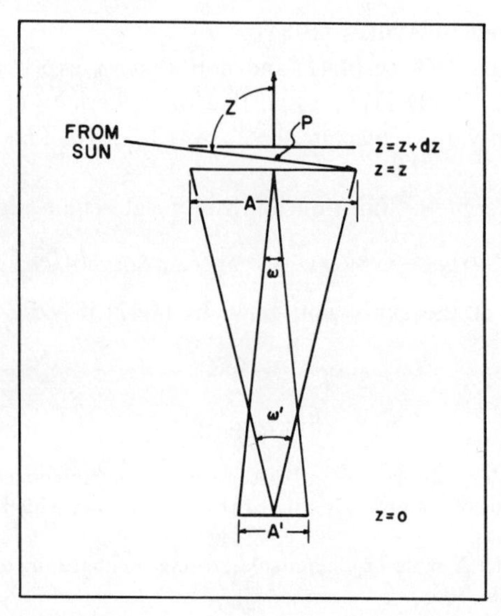

FIG. E.1. Diagram illustrating how light scattered from a height z contributes to the specific intensity observed at the ground.

At the point P, let the flux of solar radiation (through an area normal to the beam) be F_ν, let the air density be ρ, and let the zenith angle be Z. An elemental volume ($dA \sec Z\, dz$) at P scatters energy in the frequency interval $d\nu$ at the rate

$$F_\nu K_\nu \rho \; dA \; dz \; d\nu$$

in all directions, where K_ν is the mass scattering coefficient. If we consider a narrow spectral interval centered at wavelength λ, we can write this as

$$F(\lambda)K(\lambda)\rho \, dA \, dz$$

where $F(\lambda)$ is the integrated flux over the wavelength interval and $K(\lambda)$ is a mean value of the scattering coefficient.

For Rayleigh scattering, the rate at which energy is scattered into the (small) solid angle ω toward the instrument is

$$F(\lambda)K(\lambda)\rho \, dA \, dz \frac{\omega}{4\pi} [\tfrac{3}{4}(1 + \cos^2 Z)]$$

where the quantity in brackets is the phase function for Rayleigh scattering. Assume that the area A is small enough so that ω is essentially the same at any point on A and has the value A'/z^2. Then the rate at which energy is scattered toward the instrument by the entire volume ($A \, dz$) is

$$F(\lambda)K(\lambda)\rho A \, dz \frac{A'}{4\pi z^2} [\tfrac{3}{4}(1 + \cos^2 Z)]$$

Neglecting, for the time being, absorption and scattering by the air between the height z and the instrument, we note that the specific intensity detected at the instrument due to scattering at height z is simply the above expression divided by A' and by ω', and that the latter is equal to A/z^2. To take account of the absorption and scattering, multiply the resulting expression by $\tau_{0,z}(\lambda)$, the transmissivity at λ for a path between O and z. The result is

$$F(\lambda)K(\lambda)(\rho/4\pi)[\tfrac{3}{4}(1 + \cos^2 Z)]\tau_{0,z}(\lambda) \, dz$$

The specific intensity integrated over $\Delta\lambda$ at the instrument $I(\lambda)$ is obtained by integrating the above expression over all altitudes. If we write this integral for two wavelength intervals centered at λ_1 and λ_2 and take the ratio of the two, we obtain

$$\frac{I(\lambda_1)}{I(\lambda_2)} = \frac{K(\lambda_1) \int_0^\infty F(\lambda_1)\rho\tau_{0,z}(\lambda_1) \, dz}{K(\lambda_2) \int_0^\infty F(\lambda_2)\rho\tau_{0,z}(\lambda_2) \, dz} \tag{E.1}$$

We can write $F(\lambda_1)$ as $F_\infty(\lambda_1)\tau_{z,\infty}(\lambda_1)$ where $F_\infty(\lambda_1)$ is the solar flux at the outer edge of the atmosphere and $\tau_{z,\infty}(\lambda_1)$ is the transmissivity of the atmosphere (along the slant path) above z; and similarly for λ_2. With this notation

$$\frac{I(\lambda_1)}{I(\lambda_2)} = \frac{K(\lambda_1)F_\infty(\lambda_1) \int_0^\infty \rho\tau_{z,\infty}(\lambda_1)\tau_{0,z}(\lambda_1) \, dz}{K(\lambda_2)F_\infty(\lambda_2) \int_0^\infty \rho\tau_{z,\infty}(\lambda_2)\tau_{0,z}(\lambda_2) \, dz} \tag{E.2}$$

The transmissivity along a vertical path between the surface and z is given by

$$\tau_{0,z}(\lambda) = \exp\left[-k(\lambda)\int_0^z \rho_3\, dz - K(\lambda)\int_0^z \rho\, dz\right] \tag{E.3}$$

where the first term accounts for absorption by ozone, k being the absorption coefficient for ozone and ρ_3 the ozone density, and the second term accounts for Rayleigh scattering.

The transmissivity along a slant path between z and the outer edge of the atmosphere is given by

$$\tau_{z,\infty}(\lambda) = \exp\left[-k(\lambda)\int_z^\infty \rho_3 \sec\psi\, dz - K(\lambda)\int_z^\infty \rho \sec\psi\, dz\right] \tag{E.4}$$

where ψ is the angle the solar beam makes with the (local) vertical at the height z (see Appendix D). Because *Umkehr* measurements are made at very large zenith angles, it is not possible to assume ψ constant and equal to the zenith angle Z at the point of observation.

APPENDIX F

Large-Scale Transport of Atmospheric Properties

In many instances it is clear that air motions must effect a significant net transport, in either a north-south or a vertical direction, of certain quantities. These include, for example, ozone, momentum, and heat. A central problem connected with the understanding of the general circulation of the atmosphere is to determine how atmospheric circulations accomplish these transports. This problem has received extensive consideration with respect to the lower atmosphere, particularly by Starr and his collaborators. It is an equally valid and important problem with respect to the upper atmosphere, but has received less attention there, owing to lack of the necessary observations. (See, however, Chapter 10.)

It is customary to divide the possible mechanisms into two broad categories—the one connected with a slow net circulation in a vertical-meridional plane and the other connected with quasi-horizontal large-scale eddy processes. A mathematical representation of the two processes is achieved rather simply and is set forth below for reference in various portions of the text. The argument is given in terms of north-south transport, but is equally valid for vertical transport.

Let Q represent the value (per unit volume) at a given point and time of the quantity being transported, and let v represent the meridional component of wind at that point and time. Then the instantaneous northward flux is vQ. We can write $v = \bar{v} + v'$ and $Q = \bar{Q} + Q'$, where the bar represents an average value over a certain period of time and the prime represents the instantaneous deviation from the average. Therefore, we can write the instantaneous northward flux as

$$vQ = \bar{v}\bar{Q} + \bar{v}Q' + v'\bar{Q} + v'Q' \tag{F.1}$$

The time-mean value of vQ, represented by $\overline{vQ}$, is given by

$$\overline{vQ} = \bar{v}\bar{Q} + \overline{v'Q'} \tag{F.2}$$

Equation (F.2) is simply a statement of the fact that the mean value of the product of two quantities is given by the product of the two mean values plus the mean value of the product of the deviations.

The value of $\overline{vQ}$ averaged around a latitude circle can be written

$$[\overline{vQ}] = [\bar{v}][\bar{Q}] + [\bar{v}''\bar{Q}''] + [\overline{v'Q'}] \tag{F.3}$$

where the brackets represent averages around the latitude circle and the double primes represent deviations from that average.

The first term on the right of Eq. (F.3) represents the contribution of a mean meridional circulation to the time-averaged and longitudinally averaged northward transport. It is to be noted that for strictly geostrophic motion, and for transport along a constant-pressure surface, this term is exactly zero, because $[\bar{v}]$ (or $[v]$) is zero. This does not necessarily mean that the term is negligible, even if the wind appears to obey the geostrophic equation to a first approximation. It does mean that the term is very difficult to measure, because wind observations are usually not accurate enough to define the deviation of the actual wind from the geostrophic wind.

The second term $[\bar{v}''\bar{Q}'']$ represents the contribution of *standing eddies*. $\bar{Q}''$ (and similarly for $\bar{v}''$) is the deviation of $\bar{Q}$ from the longitudinally averaged value of $\bar{Q}$. For this term to contribute, there must be a spatial correlation between $\bar{v}''$ and $\bar{Q}''$. For example, a net northward transport of ozone occurs if, around a latitude circle, time-averaged south winds tend to coincide with time-averaged ozone ridges.

The third term $[\overline{v'Q'}]$ represents the contribution of *transient eddies*. For this term to contribute, there must be a time correlation between v' and Q' at any one point and the average value of this time correlation around a latitude circle must be different from zero.

Numerical evaluation of these terms in (F.3) requires data at stations as evenly distributed and as closely spaced as possible around a latitude circle. Computations of $\bar{v}\bar{Q}$ and $\overline{v'Q'}$ at only one station [Eq. (F.2)] can be considered to be only suggestive. In particular, in the case of $\bar{v}\bar{Q}$, a good deal of cancellation can be expected when values of $\bar{v}\bar{Q}$ are averaged around a latitude circle.

APPENDIX G

Some Early Studies of Radiative Effects on Thermal Structure

Gold (1908) recognized that "any explanation of the existence of the isothermal layer must take into consideration the effect of atmospheric radiation." Gold's own explanation (1909), although not internally consistent in all details, contained some elements of considerable interest.

Gold pointed out that, in an atmosphere heated from below, only the lower part of the atmosphere can have an equilibrium state characterized by a lapse of temperature at the dry-adiabatic rate. This is so because any layer in such an atmosphere, for equilibrium, must radiate more energy than it absorbs in order to balance the heat gained from below by convection. But, at some upper level, the temperature will have fallen to such a low value that this is impossible, the air will gain heat, and convection will cease. He showed that, if the atmosphere were of uniform composition, the convective layer could extend no higher than that height where the pressure is one-half the surface pressure. He then took a model atmosphere with two layers, a lower one (troposphere) heated by convection and having an adiabatic lapse rate, and an upper one (stratosphere) in radiative equilibrium and isothermal. For certain assumptions about the lapse of water-vapor concentration, he found that the lower layer would extend up to a height where the pressure is between one-half and one-fourth the surface pressure. This level he associated with the tropopause, and its temperature with the temperature of an overlying isothermal stratosphere.

The concept of a troposphere whose temperature distribution is controlled primarily by convection and a stratosphere whose temperature distribution is controlled primarily by radiation has endured. The radiative processes are, of course, enormously more complex than visualized by Gold. However, even apart from this, there are certain difficulties with Gold's model that were pointed out by Emden (1913) and Milne (1922) in their studies of a gray atmosphere irradiated from above and bounded at the bottom by a black surface. These studies were rather similar in concept, although different in detail, and the discussion below combines some of the features of both.

Consider an atmosphere irradiated by vertically incident solar radiation. Let the atmosphere have a certain absorption coefficient (independent of frequency) for its own low-temperature radiation and another (smaller) one for the solar radiation. Then under conditions of radiative equilibrium and with Kirchhoff's law, the temperature must increase with increasing optical depth, measured positive inward from zero at the outer edge of the atmosphere. In terms of altitude above the earth's surface, which is more meaningful in the present context, the temperature must decrease with height at all levels. For a relationship between optical depth and altitude which is appropriate for our atmosphere, the radiative-equilibrium temperature gradient in the lowest few kilometers turns out to be strongly superadiabatic. In any case, any atmosphere bounded at the bottom by a black surface must be convectively unstable, because the black surface under conditions of radiation equilibrium would assume a higher temperature than the air immediately above it and convection would immediately begin. In addition to this important result, the analysis shows that a model like Gold's with an isothermal top in radiative equilibrium is impossible for this type of atmosphere.

An additional important factor pointed out by Milne has to do with the possible location of a lower boundary of that part of the atmosphere in radiative equilibrium. If the lower atmosphere is assumed to have some specified temperature distribution controlled by convective as well as radiative transfer of heat, and is assumed to be overlain by a layer in radiative equilibrium, then the temperature at the interface of the two layers is not necessarily continuous. Goody later made this sort of argument more quantitative and specific in his well-known theory of the tropopause (see Subsection 2.2.2).

REFERENCES

Emden, R. (1913). Über Strahlungsgleichgewicht und atmosphärische Strahlung. *S. B. Akad. Wissenschaften, Munich* pp. 55–142.

Gold, E. (1908). Contribution to a discussion of "The isothermal layer of the atmosphere." *Nature* **78**, 550–552.

Gold, E. (1909). The isothermal layer of the atmosphere and atmospheric radiation. *Proc. Roy. Soc.* **A82**, 43–70.

Milne, E. A. (1922). Radiative equilibrium: the insolation of an atmosphere. *Phil. Magaz.* [6] **44**, 872–896.

APPENDIX H

Thermal Equilibrium and the Upper Atmosphere

For an atmosphere in strict thermal equilibrium, the temperature is everywhere the same and there are no motions or redistribution of mass. Clearly these conditions are not met in our atmosphere. Nevertheless, some of the consequences of thermal equilibrium are for particular purposes applicable to atmospheric problems up to great altitudes—how great is determined by the nature of the problem.

One important consequence is a Maxwellian distribution of velocities for the particles. If dn/n is the fraction of particles of molecular mass μ with velocity between V and $V + dV$, then

$$dn/n = 4\pi V^2(\mu/2\pi kT)^{\frac{3}{2}} \exp(-\mu V^2/2kT)\, dV \tag{H.1}$$

where T is the (local) kinetic temperature. In terms of translational kinetic energy E of the particles, the dependence on mass disappears and the fraction of all particles with kinetic energy between E and $E + dE$ is

$$dn/n = 2E^{\frac{1}{2}}\pi^{-\frac{1}{2}}(1/kT)^{\frac{3}{2}} \exp(-E/kT)\, dE \tag{H.2}$$

A Maxwellian velocity distribution is usually assumed to apply in the upper atmosphere up to the exosphere. Without it, the word "temperature" loses its usual meaning. However, there are some known or inferred exceptions. For instance, there is considerable evidence that electrons above the F peak do not have the same kinetic temperature as the neutral particles.

Another consequence of thermal equilibrium (really a more general statement of the above) is a Boltzmann distribution of particles in various energy levels. The fraction of particles (n_i/n) in a level with energy E_i above the ground level is proportional to $\exp[-E_i/kT]$. A Boltzmann distribution may apply, for example, to the population of vibrational levels within the ground electronic state of a molecule. Such a distribution is continually upset by interaction with the low-temperature radiation field and can be maintained only by frequent collisions. It is not, in general, applicable to as high altitudes as is the Maxwellian distribution of thermal motions. The question of its applicability under various circumstances constitutes a serious problem, one aspect of which is discussed in Section 7.2.

APPENDIX I

Harmonic Analysis and Significance Criteria

This discussion* refers specifically to a harmonic analysis designed to study tidal variations, which by definition have a fundamental period of 24 hours. However, harmonic analysis can be applied just as well to the variation of any scalar with a known or suspected fundamental period in terms of any independent variable; for example, to the variation of any meteorological quantity with longitude.

Let $Q_1, Q_2, ..., Q_N$ be the values of a meteorological quantity at the times t_1, $t_2, ..., t_N$, which times are equally spaced during the 24-hour day. For example, one might take $t_1 = 0100$, $t_2 = 0200$, etc., in which case $N = 24$; or one might take $t_1 = 0200$, $t_2 = 0400$, etc., in which case $N = 12$. The method of harmonic analysis represents $Q(t)$ by

$$Q(t) = \sum_{l=0}^{L} (c_l \cos lt + s_l \sin lt) \tag{I.1}$$

where t is expressed in angular measure at the rate of 15° per hour, so that there are 360° in the fundamental period of 24 hours. The coefficients c_l and s_l are determined by the method of least squares. The formulae are very simple when the times are equally spaced, which provides orthogonality for the sine and cosine functions. They are

$$c_l = \sum_{i=1}^{N} [Q_i \cos(lt_i)] \Big/ \sum_{i=1}^{N} [\cos^2(lt_i)] \tag{I.2}$$

$$s_l = \sum_{i=1}^{N} [Q_i \sin(lt_i)] \Big/ \sum_{i=1}^{N} [\sin^2(lt_i)] \tag{I.3}$$

In practice, tidal variations are usually expressed not in terms of the harmonic

---

* In the preparation of this appendix, I have consulted principally a useful summary discussion by Professor B. Haurwitz, "Tidal phenomena in the upper atmosphere," which constitutes a report to the World Meteorological Organization. Professor Haurwitz kindly made this paper available to me prior to its publication by that organization.

coefficients c_l and s_l as in (I.1), but in terms of an amplitude and phase angle, as in (8.1) and (8.2). Thus, one can write

$$(Qt) = \sum_{l=0}^{L} A_l \sin(lt + a_l) \tag{I.4}$$

The appropriate relationships between the harmonic coefficients and the amplitudes and phase angles are

$$A_l = (c_l^2 + s_l^2)^{\frac{1}{2}} \tag{I.5}$$

$$a_l = \tan^{-1}(c_l/s_l) \tag{I.6}$$

or

$$c_l = A_l \sin a_l \tag{I.7}$$

$$s_l = A_l \cos a_l \tag{I.8}$$

Of course, (I.4) could be written in terms of a cosine function or with a minus sign before the phase angle, in which cases (I.6)–(I.8) would be correspondingly changed.

A given harmonic component determined in this manner can be represented by a point P on a diagram which has s_l as abscissa and c_l as ordinate. According to (I.5) and (I.6), A_l is the distance from the origin to P and a_l is the angle measured counterclockwise from the s_l axis to the line between the origin and P. See Fig. I.1. Maximum value of Q_l occurs when $(lt + a_l) = 90°$, or when

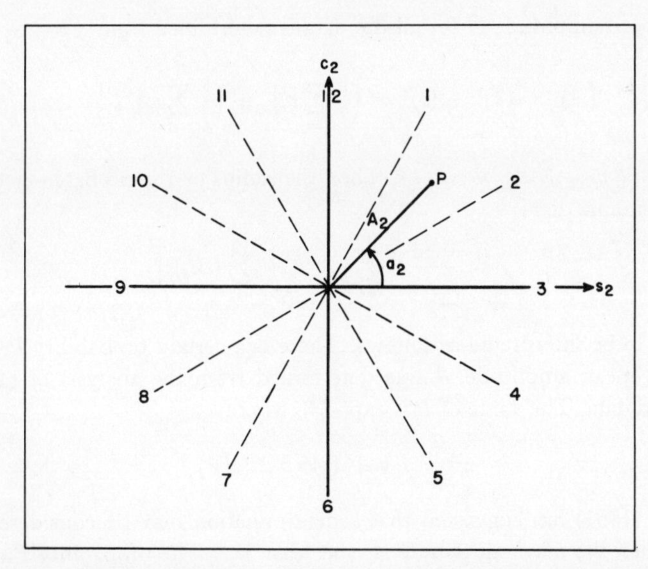

FIG. I.1. A harmonic dial for $l = 2$. The point at P describes an oscillation with amplitude A_2, phase angle a_2, and maximum value at 0130 (and 1330) local time.

$lt = (90° - a_l)$. Therefore the angle measured clockwise from the c_l axis to this line is equal to lt and values of t, usually expressed in hours, can be marked off and labeled on the diagram. When $l = 1$, there are 24 hours in the complete circle; when $l = 2$, there are 12 hours; etc. Such a diagram is called a *harmonic dial.*

These formal procedures can be applied to any set of data and will always yield formal answers. However, it is important to decide whether the computed amplitudes and phase angles have any physical significance. In general, the variation of Q reflects other causes than tides and other periods than those that are submultiples of a day. Therefore harmonic analysis for tidal variations should be applied to values of Q that have been averaged over enough days to eliminate or minimize the effect of other variations. The required number of days depends on the relative amplitudes of the tidal variations compared with those of other variations. A simple statistical criterion is used to determine the probability that the results are real, that is, the probability that the results did not arise from an analysis of randomly distributed data.

Divide the data into J groups of days, each group consisting of Q values averaged over a given number of days (which might be 1). We confine our attention to a particular value of l and dispense with the subscript l. Let the subscript j refer to values determined from one group. Then

$$A_j^2 = c_j^2 + s_j^2 \tag{I.9}$$

and the mean amplitude $\bar{A}$ for all the data is determined from

$$(\bar{A})^2 = (\bar{c})^2 + (\bar{s})^2 = \left(\frac{1}{J}\sum_{j=1}^{J} c_i\right)^2 + \left(\frac{1}{J}\sum_{j=1}^{J} s_j\right)^2 \tag{I.10}$$

Let $\Delta c_j = c_j - \bar{c}$, $\Delta s_j = s_j - \bar{s}$. Then the radius of the probable-error circle on the harmonic dial is

$$r_p = \frac{0.83}{J}\left[\sum_{j=1}^{J} (\Delta c_j)^2 + \sum_{j=1}^{J} (\Delta s_j)^2\right]^{\frac{1}{2}} \tag{I.11}$$

This is to be interpreted as follows. There is a certain probability P that the computed mean amplitude $\bar{A}$ was determined from the analysis of randomly distributed data. The value of P is $\exp(-\eta^2)$, where

$$\eta^2 = J[1 + 1.45 J r_p^2/(\bar{A})^2]^{-1} \tag{I.12}$$

Chapman (1951) has suggested that a determination may be considered satisfactory when the mean amplitude $\bar{A}$ is at least $3r_p$. The probability P depends also on the value of J; for $\bar{A} = 3r_p$ it has a value of only 1/15 when $J = 5$ and approaches 1/500 as J gets indefinitely large.

To illustrate these considerations, Fig. I.2 shows on a harmonic dial the results of some determinations of the diurnal and semidiurnal tide in surface pressure at Seawell Airport, Barbados (13°04′N, 59°30′W). The points to the right of the origin show 10 determinations of the diurnal tide from individual days (1–10 August 1963) and the cross with its probable-error circle shows the

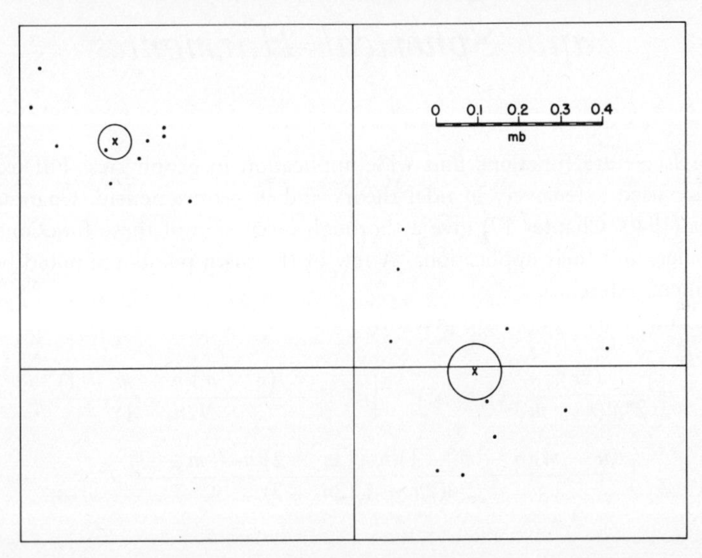

FIG. I.2. Some determinations of the diurnal and semidiurnal tide in surface pressure at Barbados. Each point refers to a determination for a single day and the crosses give the determinations for 10 days. The probable-error circles refer to the mean determinations.

mean for the 10 days. In this case $\bar{A}_1 = 0.29$ mb and $r_p = 0.064$ mb. As indicated by the position of the cross the pressure maximum for this component occurs at about 0600 local mean time. The points to the left of the origin show determinations of the semidiurnal tide for the same 10 days and the cross with its probable-error circle shows the mean for the 10 days. In this case $\bar{A}_2 = 0.78$ mb and $r_p = 0.040$ mb and maximum pressure for this component occurs at about 1030 (and 2230) local time.

REFERENCE

CHAPMAN, S. (1951). Atmospheric tides and oscillations. *In* "Compendium of Meteorology" (T. F. Malone, ed.), pp. 510–530. American Meteorological Society, Boston, Massachusetts.

APPENDIX J

Legendre Functions and Spherical Harmonics

The Legendre functions find wide application in geophysics. For example, they are used extensively in tidal theory and in geomagnetism. Chapman and Bartels (1940, Chapter 17) give a thorough discussion of these functions, their properties, and their applications. A few of the main points are noted here for convenient reference.

The functions $P_{n,m}(\mu)$ are given by

$$P_{n,m}(\mu) = \frac{(2n)!}{2^n n!(n-m)!}(1-\mu^2)^{m/2}\Big[\mu^{n-m} - \frac{(n-m)(n-m-1)}{2(2n-1)}\mu^{n-m-2} + \frac{(n-m)(n-m-1)(n-m-2)(n-m-3)}{2\cdot 4(2n-1)(2n-3)}\mu^{n-m-4} - \cdots\Big] \tag{J.1}$$

where n and m are zero or positive integers, such that $n \geqslant m$. The factor in brackets is a polynomial of degree $(n - m)$. When $m = 0$, the functions $P_{n,0}(\mu)$, or simply $P_n(\mu)$, are called *zonal functions* or *Legendre functions*. When $m > 0$, the functions $P_{n,m}(\mu)$ are called *spherical functions* or *associated Legendre functions*. The functions $P_{n,m}(\mu)$ are said to be of order m and degree n.

One reason for the usefulness of the Legendre and associated Legendre functions in geophysics is their orthogonality property. For the same order m, the functions of degree n are an orthogonal set in the range $\mu = -1$ to $\mu = 1$, that is,

$$\int_{-1}^{1} P_{n,m}(\mu)P_{n',m}(\mu)\, d\mu = 0 \tag{J.2}$$

when $n \neq n'$. If the argument μ is identified with $\cos\theta$, where θ is the colatitude,

$$\int_{0}^{\pi} P_{n,m}(\cos\theta)P_{n',m}(\cos\theta)\sin\theta\, d\theta = 0 \tag{J.3}$$

The functions $P_{n,m}(\cos\theta)\cos m\lambda$, $P_{n,m}(\cos\theta)\sin m\lambda$, where λ is the longitude, are called *spherical surface harmonics*. The Legendre functions $P_n(\cos\theta)$ are special cases of spherical surface harmonics for $m = 0$. Because of the orthogonality property (J.3) and also the orthogonality properties of the sine and cosine functions

$$\int_0^{\pi}\int_0^{2\pi} P_{n,m}(\cos\theta)\begin{Bmatrix}\cos\\ \sin\end{Bmatrix} m\lambda P_{n',m'}(\cos\theta)\begin{Bmatrix}\cos\\ \sin\end{Bmatrix} m'\lambda \sin\theta\, d\lambda\, d\theta = 0 \qquad \text{(J.4)}$$

unless $n = n'$ and $m = m'$. Thus spherical surface harmonics are convenient functions for representing the distribution of geophysical quantities over the surface of a sphere by the method of least squares. This process is called spherical harmonic analysis (see Chapman and Bartels, 1940, Chapter 17).

The mean value over the surface of a sphere of the square of a spherical surface harmonic is given (for $m \neq 0$), by

$$\frac{1}{4\pi}\int_0^{\pi}\int_0^{2\pi}\left[P_{n,m}(\cos\theta)\begin{Bmatrix}\cos\\ \sin\end{Bmatrix} m\lambda\right]^2 \sin\theta\, d\lambda\, d\theta = \frac{1}{2(2n+1)}\,\frac{(n+m)!}{(n-m)!} \qquad \text{(J.5)}$$

For the special case $m = 0$, the corresponding mean value is $(2n+1)^{-1}$. The mean values given by (J.5), for a given n, vary greatly with m. For example, the ratio of this quantity for $n = 3$, $m = 1$ to that for $n = 3$, $m = 3$ is 1 : 60. This is an inconvenience in spherical harmonic analysis, because the coefficients of the terms resulting from such an analysis differ greatly because of this factor alone; differences due to the inherent importance of the terms in the distribution are obscured. Therefore, the *seminormalized* associated Legendre functions of A. Schmidt are almost universally used in geophysics. To distinguish them from the functions defined above, it is customary to use the notation $P_n^m(\cos\theta)$ (which is, however, sometimes used outside geophysics to designate the ordinary associated Legendre functions). The functions $P_n^m(\cos\theta)$ are identical with $P_{n,m}(\cos\theta)$ when $m = 0$ and, for $m > 0$, are defined by

$$P_n^m(\cos\theta) = \left[2\,\frac{(n-m)!}{(n+m)!}\right]^{\frac{1}{2}} P_{n,m}(\cos\theta) \qquad \text{(J.6)}$$

Thus the mean value over the sphere of the square of $P_n^m(\cos\theta)\begin{Bmatrix}\cos\\ \sin\end{Bmatrix} m\lambda$ is $(2n+1)^{-1}$ for any n and m. Schmidt's functions are used exclusively in the text of this book and henceforth in this appendix.

The zonal functions $P_n(\cos\theta)$ always have the value 1 at the north pole ($\theta = 0$). When $n = 0$, $P_0(\cos\theta) = 1$ everywhere. When n is odd, $P_n(\cos\theta) = 0$ at the equator and the function is antisymmetric about the equator: $P_n(\cos\theta) = -P_n[\cos(\pi-\theta)]$. When n is even, $P_n(\cos\theta)$ is symmetric about the equator: $P_n(\cos\theta) = P_n[\cos(\pi-\theta)]$. In either case, the function has n zeroes: one at the equator and $(n-1)/2$ in each hemisphere, when n is odd; and $n/2$ in each hemisphere when n is even.

The spherical functions $P_n^m(\cos\theta)$ always have the value zero at both poles. When $(n-m)$ is odd, $P_n^m(\cos\theta) = 0$ at the equator and the function is antisymmetric about the equator. When $(n - m)$ is even, $P_n^m(\cos\theta)$ is symmetric about the equator. The function has $(n - m)$ zeroes. When $n = m$, $P_n^m(\cos\theta)\cos m\lambda$ or $P_n^m(\cos\theta)\sin m\lambda$ is sometimes called a *sectorial harmonic*. When $n > m$, it is sometimes called a *tesseral harmonic*.

A few useful relations connecting seminormalized Legendre functions of argument $(\cos\theta)$ are the following:

$$(2n+1)\cos\theta\, P_n^m = [(n+1)^2 - m^2]^{\frac{1}{2}} P_{n+1}^m + (n^2 - m^2)^{\frac{1}{2}} P_{n-1}^m \tag{J.7}$$

$$\begin{aligned} 2m\cos\theta\, P_n^m &= [(n-m)(n+m+1)]^{\frac{1}{2}} \sin\theta\, P_n^{m+1} \\ &\quad + [\delta(n+m)(n-m+1)]^{\frac{1}{2}} \sin\theta\, P_n^{m-1} \end{aligned} \tag{J.8}$$

(valid only for $m > 0$; $\delta = 2$ when $m = 1$, $\delta = 1$ when $m > 1$)

$$\frac{2n+1}{n(n+1)} \sin\theta \frac{dP_n^m}{d\theta} = \frac{[(n+1)^2 - m^2]^{\frac{1}{2}}}{n+1} P_{n+1}^m - \frac{(n^2 - m^2)^{\frac{1}{2}}}{n} P_{n-1}^m \tag{J.9}$$

Others are given by Chapman and Bartels (1940, Chapter 17) and by Jahnke and Emde (1945).

The value of the associated Legendre functions arises not only from their usefulness in the numerical representation of functions on a spherical surface, but more fundamentally from their mathematical properties as solutions of important differential equations that arise in geophysics (and physics). For example, Laplace's equation in spherical coordinates is

$$\nabla^2\Phi = \frac{1}{r^2}\frac{\partial}{\partial r}\left(r^2 \frac{\partial\Phi}{\partial r}\right) + \frac{1}{r^2\sin\theta}\frac{\partial}{\partial\theta}\left(\sin\theta \frac{\partial\Phi}{\partial\theta}\right) + \frac{1}{r^2}\sin^2\theta \frac{\partial^2\Phi}{\partial\lambda^2} = 0 \tag{J.10}$$

and a particular solution of this is

$$\Phi = r^n P_n^m(\cos\theta) \begin{Bmatrix} \cos \\ \sin \end{Bmatrix} m\lambda \tag{J.11}$$

REFERENCES

CHAPMAN, S., and BARTELS, J. (1940). "Geomagnetism." Oxford Univ. Press (Clarendon), London and New York. (Reprinted, with corrections, 1951 and 1962.)

JAHNKE, E., and EMDE, F. (1945). "Tables of Functions," 4th ed. Dover, New York.

APPENDIX K

The Perturbation Equations

The equations usually taken to describe the behavior of the atmosphere are the vector equation of motion in a rotating frame, the equation of continuity, the adiabatic form of the first law of thermodynamics, and the equation of state for an ideal gas. These are

$$\frac{\partial \mathbf{V}}{\partial t} + \mathbf{V} \cdot \nabla \mathbf{V} + 2\boldsymbol{\omega} \times \mathbf{V} + \frac{1}{\rho} \nabla p + \nabla \Phi = 0 \tag{K.1}$$

$$\frac{\partial \rho}{\partial t} + \mathbf{V} \cdot \nabla \rho + \rho \nabla \cdot \mathbf{V} = 0 \tag{K.2}$$

$$\frac{\partial p}{\partial t} + \mathbf{V} \cdot \nabla p = \gamma \frac{p}{\rho} \left(\frac{\partial \rho}{\partial t} + \mathbf{V} \cdot \nabla \rho \right) \tag{K.3}$$

$$p = \rho RT/m \tag{K.4}$$

where Φ is a scalar potential representing the combined effects of gravitation and centripetal acceleration due to the earth's rotation with a component g in the vertical (only) and the other symbols are as used throughout this book. In tidal theory, Φ may also include a scalar potential Ω describing the gravitational tide-producing force. Viscous forces have been neglected in (K.1).

A treatment of these equations, short of numerical integrations, usually requires their linearization on the basis of certain assumptions. Such a linearization, for example, is involved in the derivation of the basic equations of tidal theory (8.6)–(8.11) and of gravity waves (8.27)–(8.30). The purpose of this appendix is to outline briefly the method by which this is done, with special reference to these two applications. For a more complete and more general treatment, see Haurwitz (1951).

The method of linearization is usually referred to as the *perturbation method.* It is assumed that each dependent variable $\mathbf{V}$, ρ, p, T consists of two parts. The first part refers to an "undisturbed" state, usually a highly idealized atmosphere whose properties are specified for any particular problem. The undisturbed variables alone are assumed to satisfy (K.1)–(K.4). The second part refers to an additional component of each variable connected with a small disturbance or "perturbation." It is assumed that the total state, undisturbed plus disturbed, also satisfies (K.1)–(K.4). The perturbation quantities are assumed to be so small that second-degree (or higher degree) terms involving them and their derivatives can be neglected in comparison with first-degree terms.

In tidal theory, the ellipticity of the earth is neglected and spherical coordinates are the appropriate ones to use. In these coordinates (see, for example, Haurwitz, 1951), the three components of the equation of motion are

$$\frac{\partial V_\theta}{\partial t} + \mathbf{V} \cdot \nabla V_\theta + \frac{V_r V_\theta}{r} - V_\lambda^2 \left(\frac{\cot \theta}{r}\right) - 2\omega \cos \theta \, V_\lambda = -\frac{1}{\rho r} \frac{\partial p}{\partial \theta} - \frac{1}{r} \frac{\partial \Phi}{\partial \theta} \tag{K.5}$$

$$\frac{\partial V_\lambda}{\partial t} + \mathbf{V} \cdot \nabla V_\lambda + \frac{V_r V_\lambda}{r} + V_\theta V_\lambda \left(\frac{\cot \theta}{r}\right) + 2\omega \cos \theta \, V_\theta + 2\omega V_r \sin \theta = -\frac{1}{\rho r \sin \theta} \frac{\partial p}{\partial \lambda} - \frac{1}{r \sin \theta} \frac{\partial \Phi}{\partial \lambda} \tag{K.6}$$

$$\frac{\partial V_r}{\partial t} + \mathbf{V} \cdot \nabla V_r - \frac{V_\theta^2}{r} - \frac{V_\lambda^2}{r} - 2\omega \sin \theta \, V_\lambda = -\frac{1}{\rho} \frac{\partial p}{\partial r} - \frac{\partial \Phi}{\partial r} \tag{K.7}$$

where θ is colatitude measured positive southward from zero at the north pole, λ is longitude measured positive eastward, r is distance from the center of the earth, and V_θ, V_λ, V_r are the linear velocity components in these directions. The operator $(\mathbf{V} \cdot \nabla)$ is given by

$$\mathbf{V} \cdot \nabla = \frac{V_\theta}{r} \frac{\partial}{\partial \theta} + \frac{V_\lambda}{r \sin \theta} \frac{\partial}{\partial \lambda} + V_r \frac{\partial}{\partial r} \tag{K.8}$$

and the velocity divergence is given by

$$\nabla \cdot \mathbf{V} = \frac{1}{r \sin \theta} \frac{\partial}{\partial \theta} (V_\theta \sin \theta) + \frac{1}{r \sin \theta} \frac{\partial V_\lambda}{\partial \lambda} + \frac{1}{r^2} \frac{\partial}{\partial r} (r^2 V_r) \tag{K.9}$$

Let the variables in the undisturbed state be represented by subscripts zero, let the perturbation components of velocity be given by v_θ, v_λ, v_r and the perturbation values of p, ρ, and T by p', ρ', T'. For the undisturbed atmosphere used in tidal theory and described in Subsection 8.1.2, $V_{\theta_0} = V_{\lambda_0} = V_{r_0} = 0$; also p_0, ρ_0, T_0 are functions of r only and are connected by the hydrostatic equation and the equation of state for an ideal gas. Substituting $V_\theta = v_\theta$, $V_\lambda = v_\lambda$, $V_r = v_r$, $p = p_0 + p'$, $\rho = \rho_0 + \rho'$, $T = T_0 + T'$ in (K.5), (K.6),

(K.7), (K.2), (K.3) and (K.4), and neglecting all terms of second or higher degree in the perturbation quantities, we get,

$$\frac{\partial v_\theta}{\partial t} - 2\omega \cos\theta\, v_\lambda = -\frac{1}{r}\left(\frac{1}{\rho_0}\frac{\partial p'}{\partial \theta} - \frac{\partial \Omega}{\partial \theta}\right) \tag{K.10}$$

$$\frac{\partial v_\lambda}{\partial t} + 2\omega \cos\theta\, v_\theta + 2\omega \sin\theta\, v_r = -\frac{1}{r\sin\theta}\left(\frac{1}{\rho_0}\frac{\partial p'}{\partial \lambda} - \frac{\partial \Omega}{\partial \lambda}\right) \tag{K.11}$$

$$\frac{\partial v_r}{\partial t} - 2\omega \sin\theta\, v_\lambda = -\underline{\frac{1}{\rho_0}\frac{\partial p_0}{\partial r}} - \frac{1}{\rho_0}\frac{\partial p'}{\partial r} - \frac{\rho'}{{\rho_0}^2}\frac{\partial p_0}{\partial r} - \underline{g} - \frac{\partial \Omega}{\partial r} \tag{K.12}$$

$$\frac{\partial \rho'}{\partial t} + \mathbf{v}\cdot\nabla\rho_0 + \rho_0\nabla\cdot\mathbf{v} = 0 \tag{K.13}$$

$$\frac{\partial p'}{\partial t} + \mathbf{v}\cdot\nabla p_0 = \gamma\frac{p_0}{\rho_0}\left(\frac{\partial \rho'}{\partial t} + \mathbf{v}\cdot\nabla\rho_0\right) \tag{K.14}$$

$$(\underline{p_0} + p') = R(\rho_0 T' + T_0\rho' + \underline{\rho_0 T_0})/m \tag{K.15}$$

The scalar tidal potential $\Omega(\theta, \lambda, r)$ has been included in these equations.

Because the undisturbed atmosphere obeys the hydrostatic equation and the perfect gas law, the underlined terms in (K.12) and (K.15) cancel. Further simplifications in accordance with the assumptions stated in Subsection 8.1.2 are made as follows:

(a) The radius vector r is replaced everywhere by a constant value, the radius of the earth (derivatives of other quantities with respect to r, such as $\partial p_0/\partial r$, are, of course, retained).

(b) The vertical acceleration term in (K.12) is neglected.

(c) The Coriolis term ($2\omega \sin\theta\, v_\lambda$) in (K.12) and the Coriolis term involving v_r in (K.11) are neglected.

Equations (K.9)–(K.14) then reduce to Eqs. (8.6)–(8.11).

In the case of tangent-plane Cartesian coordinates, tangent to the earth's surface at latitude ϕ, the components of the vector equation of motion (K.1) are

$$\frac{\partial u}{\partial t} + \mathbf{V}\cdot\nabla u - 2\omega(v\sin\phi - w\cos\phi) = -\frac{1}{\rho}\frac{\partial p}{\partial x} \tag{K.16}$$

$$\frac{\partial v}{\partial t} + \mathbf{V}\cdot\nabla v + 2\omega u\sin\phi = -\frac{1}{\rho}\frac{\partial p}{\partial y} \tag{K.17}$$

$$\frac{\partial w}{\partial t} + \mathbf{V}\cdot\nabla w - 2\omega u\cos\phi = -\frac{1}{\rho}\frac{\partial p}{\partial z} - g \tag{K.18}$$

The operator $(\mathbf{V} \cdot \nabla)$ is given by

$$u \frac{\partial}{\partial x} + v \frac{\partial}{\partial y} + w \frac{\partial}{\partial z} \tag{K.19}$$

and the velocity divergence is

$$\frac{\partial u}{\partial x} + \frac{\partial v}{\partial y} + \frac{\partial w}{\partial z} \tag{K.20}$$

In the model atmosphere employed by Hines $u_0 = v_0 = w_0 = 0$, T_0 is constant, and ρ_0 and p_0 are functions of z only. The x axis is taken in the direction of the horizontal component of the perturbed motion and the components of the perturbed motion are indicated by u and w. All Coriolis terms are neglected. In the undisturbed atmosphere $\partial p_0/\partial z = -\rho_0 g$ and, since temperature is independent of height, p_0, $\rho_0 \propto \exp(-z/H_0)$ where H_0 is the scale height $p_0/\rho_0 g$. With these assumptions and the neglect of nonlinear terms in the perturbation quantities, it is a straightforward matter to derive (8.26)–(8.29) from (K.16), (K.18), (K.19), (K.20), (K.2), (K.3).

REFERENCE

Haurwitz, B. (1951). The perturbation equations in meteorology. *In* "Compendium of Meteorology" (T. F. Malone, ed.), pp. 401–420. American Meteorological Society, Boston, Massachusetts.

APPENDIX L

An Early Account of an Auroral Display

The following description of a vivid auroral display at New Haven, Connecticut, on 14 November 1789 was recorded in his "meteorological journal" by Rev. Ezra Stiles. This, along with many other of Rev. Stiles' observations, was published by Loomis (1866).

> A grand Aurora Borealis and Australis, the striae and coruscations shooting up and playing all over and round every part of the hemisphere. Luminous spots and beautiful red and flaming pyramidal columns ascending to the zenith. At 10^h was formed a grand auroral crown exactly on the meridian: the center of which was near Beta in Aries, and about 4° S.W. from Alpha Arietis, 20° south of the zenith of Yale College. From this center proceeded every way, N., S., E., and W., radial beams half way down to the horizon; besides a red beam or arch or broad stream which then crossed from east to west through the coronal center from horizon to horizon, or very nearly. A dim arch, or luminous circle or belt, round through the southern board, crossing the meridian about two-thirds the way from the zenith to the south horizon. Below which was a penumbral darkness, which in five or six minutes became tinged with red to the southern horizon. About 10^h10^m or sooner the vertical crown became faint and evanishing. It lighted up again with new playing, varying and alternately evanishing striae or radii. I examined the celestial globe, and found the stars near the center to be the two stars in the head of Aries, which were on the meridian at 10^h35^m nearly; at which time the crown was much obliterated, though the heavens are full of aurora. At 11^h the radii continue, proceeding to S.E., S., and S.W., but chiefly S.E. and nearly S., then a chasm, and then streams to S.W. and W., but the extremities of the radial beams are dissipated about the center. Yet there is still a redness in the southern board, down to the horizon, and perhaps an hour and a half high. In the northern board the whole semi-hemisphere is covered with a dim whitish sheet of light like the dawn of day. In the height of the phenomenon I read the New Haven Gazette by the auroral light. I never before saw so splendid and universal an aurora. At 12^h aurora north and northwest.

REFERENCE

Loomis, E. (1866). Notices of auroras extracted from the meteorological journal of Rev. Ezra Stiles, S. T. D., formerly President of Yale College. *Trans. Connecticut Acad. Arts Sci.* 1, Part 1, 155–172.

APPENDIX M

Glossary of Symbols

This list includes the principal symbols used in this book. Some symbols formed by adding subscripts or primes to principal symbols are not listed. Symbols that appear only once are also not listed, unless they are commonly used with the same meaning in important references. No attempt is made to include standard chemical symbols or standard nomenclature for atomic terms and molecular states (see Chapter 4).

a Acceleration
a_l Phase of the lth harmonic (with reference to the solar tide)
A Area
A Symbol designating a point in space
A Symbol designating any atom (in chemical reactions)
A Unit of wavelength, the Angstrom
A_l Amplitude of the lth harmonic (with reference to the solar tide)
A_p An index of magnetic activity

b_l Phase of the lth harmonic (with reference to the lunar tide)
B Symbol designating any atom (in chemical reactions)
B_l Amplitude of the lth harmonic (with reference to the lunar tide)
B_v Rotational constant

c Speed of light in a vacuum
c_D Drag coefficient
c_l Coefficient of the lth cosine term in a Fourier series
c_n^m Coefficient of the cosine term in a spherical harmonic
C Speed of sound
C Symbol designating any atom (in chemical reactions)
C Constant of proportionality
C_i International magnetic character figure
C_p An index of magnetic activity

d Distance between two points
D Diffusion coefficient
D Magnetic declination
D Symbol for reference to the magnetic disturbance variations or the magnetic disturbance field
Dst, DS, DP Symbols for reference to portions of D

e Charge on an electron
e Symbol designating an electron (in chemical reactions)
E Internal energy of an atom or molecule
E_p Potential energy of an atom or molecule
E_r Rotational energy of a molecule
E_v Vibrational energy of a molecule

f Coriolis parameter, $2\omega \sin \phi$
f Frequency (in cycles [second]$^{-1}$)
f_0 Critical frequency
f Ratio of number density of O to equilibrium number density of O
f A function of independent variables specified in parentheses

f A parameter in tidal theory [see Eq. (8.14)]
F Flux (per unit area)
F An operator (in tidal theory)

g Acceleration of gravity
g Ratio of number density of O_3 to equilibrium number density of O_3

h Distance, usually above the earth's surface
h Planck's constant
h Equivalent depth
$\bar{h}$ An atmospheric eigenvalue
H Scale height
H_0 Scale height of undisturbed atmosphere
H_ρ Density scale height
H Change of heat content per unit volume per unit time
H Magnetic intensity
H Magnitude of the horizontal component of **H**
H_0 Equatorial value of H (as defined immediately above)

i Angle of incidence in ray theory
i Angle between line of force and the horizontal
I Specific intensity
I Magnetic dip
$I(\zeta)$ A particular solution of the radial equation (in tidal theory)

j Emission coefficient
J Rotational quantum number
J Source function
J Change of heat content per unit mass per unit time (in tidal theory)
J_2 Number of quanta absorbed by O_2 per unit volume per unit time per molecule of O_2
J_3 Number of quanta absorbed by O_3 per unit volume per unit time per molecule of O_3

k Boltzmann's constant
k Absorption coefficient
k_x, k_z Real parts of complex angular frequency (in Section 8.3 only)
K Scattering coefficient
K Coefficient of conductivity
K, K_p Indices of magnetic activity
K_x, K_z Complex angular frequency (in Section 8.3 only)

l An integer designating a Fourier component
l Azimuthal quantum number
L Radiance
L_B Black-body radiance
L Loss term in the equation of continuity for electrons
L Symbol designating any atom or molecule (in chemical reactions)
L Symbol for reference to the magnetic lunar variations or field
L_l Symbol for reference to the lunar tide with period l^{-1} of a lunar day

m Gram-molecular weight
m_0 Sea-level value of the gram-molecular weight of dry air
m Air mass (as given, for example, by Bemporad's function)
m An integer representing the longitudinal wave number
M Mass
M Emittance
M Symbol designating any atom or molecule (in chemical reactions)

n Number density (usually with a subscript or parenthesized symbol to indicate the constituent in question)
n_0 Loschmidt number
n Principal quantum number
n An integer in various contexts
n Refractive index
$\tilde{n}_3$ Parameter related to the equilibrium distribution of ozone [see Eq. (5.28)]
N Avogadro's constant
N Number density of electrons

O Symbol designating a point in space

p Pressure
p_3 Partial pressure of ozone
p_e Equivalent pressure in the Curtis–Godson approximation
P Symbol designating a point in space
P Probability
$P_{n,m}$ Ordinary Legendre function
P_n^m Schmidt's seminormalized Legendre function

q Energy absorbed per unit volume per unit time
q Number of photoionizations per unit volume per unit time
Q A scalar quantity

r Distance, often distance from the center of the earth
r_e Radius of the earth
r Separation ratio (in Section 6.3 only)
r_3 Ozone mixing ratio
R Universal gas constant
R_0 Gas constant for dry air at sea level
R_H Rydberg constant for hydrogen
R Unit of emission, the rayleigh

s Distance, usually distance measured in the direction of energy propagation
s_l Coefficient of the lth sine term in a Fourier series
s_n^m Coefficient of the sine term in a spherical harmonic
S Intensity of an absorption line
S_l Symbol for reference to the solar tide with period l^{-1} of a solar day
Sq Symbol for reference to magnetic quiet-day solar variations or field
S_q, S_a, S_d Symbols for reference to magnetic solar daily variations determined from quiet days, all days, or disturbed days

t Time
t' Time, with the special meaning of local time
t Optical thickness
T Temperature
T_m Molecular-scale temperature
T Radiative half-life

u Speed in x direction, usually toward the east (west wind)
u Integrated mass (of a unit-area column)
u, u_1 Indices of magnetic activity

v Speed in y direction, usually toward north (south wind)
v_θ, v_λ Speed in θ and λ directions when spherical coordinates are used
v Vibrational quantum number
V or $\mathbf{V}$ Speed or velocity
V_g Group velocity

w Speed in vertical or radial direction, positive upward or outward
w Mean speed in vertical direction of molecules of a particular kind or of electrons

x Horizontal coordinate, usually positive toward the east
x Distance
x A dimensionless parameter in infrared theory [see Eq. (7.13)]
x A dimensionless measure of height (Appendix D)
X Northward component of **H**
X Vertical distribution factor [see Eq. (6.16)]
X A dimensionless parameter in magneto-ionic theory [see Eq. (9.19)]

y Horizontal coordinate, usually positive toward the north
Y Eastward component of **H**

z Vertical coordinate
z_1, z_2 Normalized vertical coordinates [see Eqs. (4.45) and (4.47)]

Z Number of elemental charges on a nucleus
Z Zenith angle, usually with reference to the earth's surface
$Z(z)$ A function of z in tidal theory
Z Downward component of **H**
Z A dimensionless parameter in magneto-ionic theory [see Eq. (9.20)]

α Half-width of a Lorentz line
α_D Half-width of a Doppler line
α Recombination coefficient

β Gradient of scale height
β A dimensionless parameter in infrared theory [see Eq. (7.13)]
β Rate coefficient for attachment-like recombination

γ Ratio of the specific heats
γ Unit of magnetic intensity, the gamma
$\varGamma$ Lapse rate
Γ Unit of magnetic intensity, the Gauss

δ Distance between lines in an absorption band

ϵ An angle
ϵ Emissivity
ϵ Thermal efficiency

ζ Height variable

η A dependent variable in tidal theory [see Eq. (8.20)]

θ Colatitude
θ_m Geomagnetic colatitude
θ An angle, usually measured from the direction normal to a plane
$\varTheta(\theta)$ A function of θ in tidal theory

κ Extinction coefficient
κ Rate coefficient
κ $(\gamma - 1)/\gamma$, where γ is the ratio of specific heats
λ Wavelength
λ Longitude
λ_m Geomagnetic longitude

μ Mass of a molecule
μ Unit of wavelength, the micron
μ $\cos\theta$, where θ is the colatitude
μ Real part of the refractive index describing phase velocity
μ' Refractive index describing group velocity

ν Frequency of electromagnetic radiation
ν Collision frequency
ν_e Vibrational frequency
$\tilde{\nu}$ Wave number

π 3.14159...

ρ Density
ρ_3 Ozone density
ρ_a Density of an absorbing constituent

σ Stefan–Boltzmann constant
σ Absorption cross section
σ Collision diameter of a molecule

τ Transmissivity

ϕ Latitude
ϕ_m Geomagnetic latitude
ϕ An angle, usually measured from a reference line in a plane
$\varPhi$ Geopotential
$\varPhi$ A scalar potential

χ Velocity divergence

ψ An angle, often the zenith angle at heights above the earth's surface
ψ Wave function

ω Solid angle
ω or $\boldsymbol{\omega}$ Angular speed or velocity of the earth's rotation
ω Angular frequency
ω_N Plasma frequency

$\varOmega$ Total amount of ozone (in a vertical unit-area column above the earth's surface)
$\varOmega$ A scalar potential describing gravitational tidal forces

Author Index

Numbers in italic show the page on which the complete reference is listed.

Subject Index